AN AGENDA FOR ACTION

Recommendations for School Mathematics of the 1980s

The National Council of Teachers of Mathematics recommends that

1. problem solving be the focus of school mathematics in the 1980s;
2. basic skills in mathematics be defined to encompass more than computational facility;
3. mathematics programs take full advantage of the power of calculators and computers at all grade levels;
4. stringent standards of both effectiveness and efficiency be applied to the teaching of mathematics;
5. the success of mathematics programs and student learning be evaluated by a wider range of measures than conventional testing;
6. more mathematics study be required for all students and a flexible curriculum with a greater range of options be designed to accommodate the diverse needs of the student population;
7. mathematics teachers demand of themselves and their colleagues a high level of professionalism;
8. public support for mathematics instruction be raised to a level commensurate with the importance of mathematical understanding to individuals and society.

Guiding Each Child's Learning of Mathematics

A Diagnostic Approach to Instruction

Robert B. Ashlock
Graduate School of Education
Reformed Theological Seminary

Martin L. Johnson
University of Maryland

John W. Wilson

Wilmer L. Jones
Baltimore City School District

Charles E. Merrill Publishing Company
A Bell & Howell Company
Columbus Toronto London Sydney

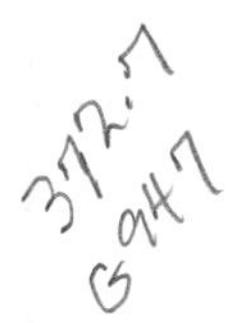

Published by
Charles E. Merrill Publishing Company
A Bell & Howell Company
Columbus, Ohio 43216

This book was set in Times Roman.
Cover Design by Tony Faiola.
Production Coordination by Martha Morss.

Photos: cover, Jean Greenwald; 3, photo by Richard Khanlian; 9, The Bettmann Archive, Inc.; 21, © Phillips Photo Illustrators; 41, Paul Conklin; 65, © 1980 Steven C. Tuttle; 83, © Phillips Photo Illustrators; 115, Strix Pix; 143, © 1983 Anne E. Schullstrom; 167, © 1983 Joanne Meldrum; 207, Strix Pix; 237, © Phillips Photo Illustrators; 265, © 1983 Anne E. Schullstrom; 291, © C. Quinlan; 317, © 1983 Anne E. Schullstrom; 347, © 1980 Greg Miller; 373, © Phillips Photo Illustrators; 393, Leo M. Johnson/Corn's Photo Service; 409, Strix Pix; 425, Barbara Lagomarsino; 445, © Phillips Photo Illustrators; 465, Tom McGuire.

Verbal problem types (back inside cover) reprinted with permission from Instructor.

Library of Congress Catalog Card Number: 82-62867
International Standard Book Number: 0-675-20023-7

Printed in the United States of America

4 5 6 7 8 9 10 – 87 86

Guiding Each Child's Learning of Mathematics

John W. Wilson (1929–1979)

. . . is remembered for his ideas and for his enthusiasm. His ideas served as guides for many classroom teachers over the years. Those ideas are very much a part of this book where other teachers will find them to be useful guides. His enthusiasm inspired many, including his coauthors.

Contents

PREFACE

How can I teach math so that *each* child understands? What do I do to enable children to *use* the math they learn? How can I teach a skill so that children will not forget it so quickly? What can I do to help each child become a real problem solver? How can I motivate a child to enjoy the study of mathematics? This text was written to help teachers and future teachers find answers to these and similar questions they often ask.

As teachers, we want the preservice and inservice teachers who read this introductory methods text to think through their own concerns and beliefs about teaching mathematics as they consider our views and those of others.

Careful planning for instruction, planning based on sound theory, is emphasized. In keeping with our own beliefs, we present guides for teaching that show with specific content and hundreds of illustrations how teachers can plan math lessons so that each child learns. Along with a cognitive emphasis, there is an emphasis on the varied pupil behaviors that enable us to make inferences about a child's learning. The text focuses on the needs of *each* child; the approach to instruction in elementary school mathematics is diagnostic.

Sample content objectives for elementary school mathematics are presented (they are also collected at the end of the text for easy reference), and selected mathematical content is summarized and explained for teachers who need a content review. Our aim, however, is to help you plan instruction. There is a direct connection between specific content ideas and the instructional activities we describe. We also offer many sample activity plans. Many chapters include specific guidance for planning assessment procedures, as well as sections entitled, "Tips on Managing the Classroom." The closing chapter, "In Your Future," recognizes the fact that teacher education is a continuing process.

A separate student handbook is available for students using this text. In this supplement, essential ideas are highlighted, many ideas are extended, and more classroom activities are provided.

The number of people who have been involved in one way or another in the preparation of this book is too great to acknowledge publicly, but we are deeply grateful to all of them. A special thank you goes to those who served as reviewers: Dr. Jack Beal, University of Washington; Dr. Jon M. Engelhardt, Arizona State University; Dr. E. Glenadine Gibb, University of Texas at Austin; Dr. Carol Larson, University of Arizona; Dr. Jane Ann McLaughlin, Trenton State College; Dr. Bernadette H. Perham, Ball State University; and Dr. Alice Robold, Ball State University. We do want to thank our wives for their encouragement and patience: Julia, JoAnn, Nancy, and Carol. We also appreciate the many hours that Barbara, Tammie, Vivianne, and Patricia spent working with the manuscript. The many teachers who used and helped refine the ideas contained in this book have been a special source of inspiration.

You, Children, and the Teaching and Learning of Mathematics

I

1

Your Concerns and Beliefs about Teaching Children Mathematics

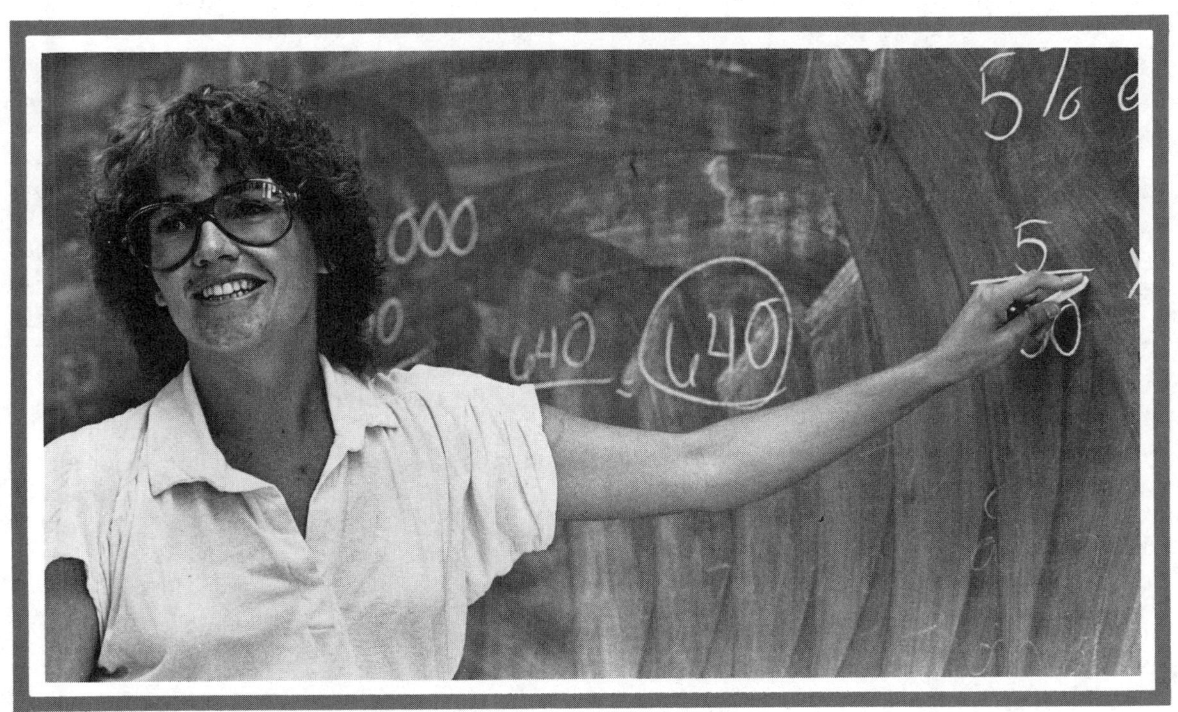

The first chapter of a book is a place for introductions and for getting acquainted. As we (the authors and you) begin to think together about teaching children mathematics, we must face many tasks and issues.

WHO ARE YOU? WHO ARE WE?

The authors have written this book with a number of beliefs about you, the reader. These beliefs are a result of our varied experiences with children, teachers, and schools.

You may be either a preservice or an inservice teacher; you may or may not have already taught mathematics. You want to teach and you enjoy working with children; in fact, you are enthusiastic about teaching and want to be very good at it. You often think about *what* and *how* you are going to teach. You even think about the many *whys* of teaching, learning, and schools: Why do I need to teach math to younger children? Why do so many children have difficulty remembering basic facts? Why are some children disruptive in the classroom? And so forth.

You already have your own beliefs and feelings about yourself as a learner. These concern your personal experiences with learning and your expectations about the future. You also have beliefs and feelings about yourself as a teacher. You may even be a bit overwhelmed as you think of future responsibilities.

You may be hesitant about teaching mathematics. You may even be rather apprehensive about mathematics yourself. Because you have many beliefs about teaching mathematics that help shape the planning you do, you probably have a number of questions and concerns.

These questions may include: What math am I supposed to cover in the grade I am teaching? How do I teach those topics? How is it possible to work with 25–30 children? What *are* the basics? Should children use calculators? Why should children learn math anyway? Why should *I* teach it—I don't like it very much; shouldn't a "specialist" teach it? It is important to keep questioning. Pose the questions to your colleagues, to your instructors, and to authors of books and articles that you read.

We all have such questions, and how we answer them will be shaped, in part, by our beliefs about learning, mathematics, children, and teaching. Our own beliefs as authors about the tasks and issues facing all of us have helped shape this book. These beliefs are presented as models to guide the planning we do as teachers, and as we look at different areas of content together, we will use the models to help us decide on specific procedures.

This book is more than a manual of rules and procedures. Although they are included, the authors regard them as suggestions designed to help you think through the problems you face in teaching and to assist in planning and implementing your own teaching procedures. The book is, therefore, designed as a tool to help you become a professional.

FOUR TYPES OF TASKS WE FACE AS TEACHERS

Your questions suggest the many specific tasks that we must deal with as teachers; they imply decisions to be made and actions to be taken. These tasks tend to fall into four major categories: assessment procedures and evaluation, instructional objectives, instructional procedures, and managerial procedures.

Tasks Concerning Assessment

As teachers we need to find out what our students already know, what they can do, and how they feel. We want to help them by planning our teaching so they will be able to relate what they are learning to what they already know.

As we teach, we need to find out if our students are learning what we have set as the objectives. If each child understands, we can plan different types of activities in which the learned concept is related to other ideas, reinforced, and applied in varied settings. If a child does not understand, our instructional procedures may need to be changed or entirely different objectives set. Many questions then come to mind: What topics are causing the most difficulty for the child? Which materials seem to help and which inhibit? Is there one instructional strategy that will be especially productive with this child? Should I modify my assessment procedures in order to be more precise?

We also need to find out if factors other than the chosen objectives and instructional procedures need attention. These factors may include the physical and psychological well-being of the child.

In order to get the necessary information about each child, we need to select or create procedures for assessing. We need to choose and design appropriate tests and plan problems for interviews that will provide specific information.

Tasks Concerning Instructional Objectives

The tasks concerned with instructional objectives include selection of the mathematical content. Further, selecting, relating, and sequencing objectives for instruction are planning tasks we face even when using a rather detailed curriculum guide or following a text. *Each* child should be planned for. Our assessment activities should provide us with ample evidence that each knows a somewhat different set of concepts and skills.

Time is fixed, yet children learn at different rates. Realistically, we cannot teach everything in the text to all children. Any teacher who skips a page in a text is making a decision about which content to teach, whether the decision is made consciously or not. But upon what do we base such decisions? What criteria should be considered as we select? (What do you the reader believe at present would affect your decisions?) The authors hope this book will help you develop your own beliefs. However, we cannot avoid selecting content for each of our students.

If the concepts and skills selected as objectives for instruction are taught as isolated bits and pieces of information, we create an impossible situation. Such teaching makes it difficult for a child to remember what is being learned from one day to the next. Rather, we must relate one idea to another, a new skill to skills already known, and so on; and we need to help children do the same. This way, learning is easier and more interesting, and children remember what they learn. Therefore, as we plan, we cluster objectives together in a logical way for instruction.

The objectives or clusters of objectives selected cannot be taught at random but must be sequenced in a way that will facilitate learning. Even so, there is no single, rigid hierarchy to use as a road map. The mathematics of the elementary school involves a rich mileau of concepts and skills, and when considered against the background of varied experiences and learning styles of individual children, it soon becomes obvious that there is not just one sequence or even a single best sequence for learning. Yet we cannot teach everything at once, and there are some things that are logically learned before others. In summary, we have to select, relate, and sequence curriculum content as we plan to teach.

Tasks Concerning Instructional Procedures

Many tasks concern instructional procedures. Teaching uses materials in *activities*—situations in which a teacher and one or more learners take part in a planned event in which some kind of learning is meant to occur. Materials must be selected or created, and organized and sequenced for most effective use.

The task of selecting materials is inevitably related to decisions about curriculum content for each child and the results of planned and administered assessment. Materials are chosen or created not only to provide an interesting activity but also to help a child learn a specific mathematical thing. Of course, our goals for instruction sometimes involve affective considerations related to mathematics, such as a child's attitude toward the study of some topic or how a child feels about his or her ability to learn.

The materials we select often exemplify a mathematical concept or principle. However, mathematical ideas are abstractions, and materials typically have attributes, such as color or shape, that are irrelevant to the idea being represented. Sometimes we must choose the order of presenting different materials for a single topic so that each child will be led most easily to an accurate understanding of the concept or process.

Also, instructional activities are of different types and are designed to serve varied purposes. In some, a child encounters only examples of the kind of thing to be learned. In others, a child will come to understand a concept or skill and learn to use symbols. Still others help a child relate what was just learned to previous learnings. In yet another type, a child reinforces what was previously learned. And in some activities a child uses what he or she knows while solving problems in varied or novel settings. For a given concept or skill, most of these different types of activities are possible. As we plan, a decision on how to teach a sequence of different types of activities for the concept or process must be made.

These different activities may also require the use of very different materials. Therefore, as we select and create materials, both the mathematical content and the purpose of the activity must be considered.

Tasks Concerning Management-Orchestration

Teaching usually occurs in a setting that involves many children, and much can happen to hinder rather than facilitate learning during a lesson. Even in a resource room where the teacher is working with only one or a few children at a time, hindrances are always possible. Therefore, certain decisions regarding management of the overall instructional setting must be made. We plan and teach much as a conductor would work with an orchestra: we make decisions about who will be involved and when, and the nature of their activity. We are concerned with keeping behavior appropriate and the classroom in balance. We identify with our students in their excitement of discovery and their frustrations. In doing this, we involve not only the students but a wider community of parents, principal, and other teachers.

In addition to those tasks cited in preceding sections, the tasks for planning and orchestrating instruction for a particular group of learners include:

- Establishing and maintaining order,
- Grouping students for different activities,
- Arranging the room,
- Storing and distributing materials,
- Keeping records,
- Relating to students and to parents, and
- Relating to professional colleagues.

These tasks require decision making and planning if the teaching and instruction are to be as pleasing and effective as a harmonious orchestra.

For example, materials must be correctly stored and adequately stocked; learning is hindered if during a lesson we suddenly discover that they are insufficient or unavailable. Even the distribution of materials needs to be planned in advance to maintain order and promptly involve children in a new activity.

YOUR THEORIES IN USE

The number of tasks confronting you may seem overwhelming, especially when you consider the need to plan assessment procedures, curriculum content, and instructional procedures and to make decisions regarding the management and orchestration of a classroom. A "cookbook" with "recipes" of clear-cut rules for each planning task and for solving problems might at first seem desirable. But the authors believe there is no *one* way to do planning tasks. You will, however, receive many suggestions on specific techniques and general procedures. They will come from your instructors, colleagues, and books such as this.

In addition, you already have many personal beliefs or theories that will affect how you approach the above-mentioned tasks. And you have had courses in mathematics, psychology, and possibly even philosophy that have also affected your thinking.

Your convictions may be either consciously or subconsciously formed; they may be clearly consistent or somewhat contradictory. But they affect the decisions you make. Therefore, the authors call them *theories in use.* They affect the way we phrase the questions that guide our thinking and planning and how we attack the tasks before us. They determine in part not only how we behave as planners but also what we do while conducting lessons and evaluating. Our beliefs actually have the greatest influence on how we plan and carry out our teaching tasks and will have a greater affect than even our knowledge of specific techniques.

The next chapter describes some alternative theories in use and the effects that these beliefs can have on instruction.

FIVE ISSUES TO BEGIN THINKING ABOUT

Before you begin Chapter 2, there are at least five major issues to start thinking about. As you read the following section, relate them to your own beliefs. Write down your thoughts concerning the questions posed. Then, in the succeeding chapters of this book, compare your beliefs with those of the authors. At some point, you might want to compare your future beliefs with those you hold at present.

Our beliefs concerning the five major issues greatly determine how we teach mathematics: the content, instructional activities, assessment procedures, and general management of the classroom. That is why the issues are especially important.

What Is Learning?

What does it mean to *learn* something? Can you describe learning? What are its characteristics? Does the process we call learning differ from person to person, or is it essentially the same for everyone? Does learning differ from age to age so that for you, for the children you work with, and for older adults, learning is dissimilar? Is human learning unique as contrasted with the learning of animals studied in laboratories and natural habitats?

Consider the learning of mathematics in particular. Some people have difficulty with mathematics although they learn history or English quite easily. Is the learning of mathematics different from other subject areas? Many topics are studied within mathematics, such as algebra and geometry. Is the process of learning mathematics essentially the same, regardless of the topic? Is, for example, the process of learning to recall basic multiplication facts the same as the process of learning to solve open number sentences like $\square + 23 = 41$?

What Can Be Learned?

What is learnable? Is it possible to learn simple, arbitrary associations? ideas or concepts? knowledge about procedures? Can the relatedness of ideas—a conceptual schema – be learned? If the above can be learned, is skill in using it also teachable? Are attitudes toward knowledge learned?

We often read about cognitive processes: attending, distinguishing likenesses and differences, classifying, hypothesizing, testing, and similar behaviors. Can these processes, or even the ability to simply reflect on an idea, be learned? Is it even possible to learn attitudes toward cognitive processes? Are values and simple motor behaviors learned?

Consider mathematics specifically. What does a person learn when he or she learns mathematics? That is, what *is* mathematics? Is it a set of procedures? a system of related ideas? a process? a way of thinking? What is learned when a child in elementary school or a student in college learns mathematics?

What Factors Affect Learning?

What conditions affect the learning process?[1] Regarding subject matter itself, is the process of learning affected by the type of content being learned? Is it affected by *what* is being studied? Does the level of abstraction of the subject affect the process?

As we plan instructional activities, we select ways to represent ideas. These representations include objects that can be moved about,

1. See Robert B. Ashlock, John W. Wilson, and Barton Hutchings, "Identifying and Describing the Remedial Mathematics Student," in *Remedial Mathematics: Diagnostic and Prescriptive Approaches,* ed. Jon L. Higgens and James W. Heddens (Columbus, Oh.; ERIC Center for Science, Mathematics, and Environmental Education, Ohio State University, 1976).

diagrams or pictures, and written symbols. Does our choice of representation affect the learning process? That is, will some facilitate while others inhibit? Do certain instructional strategies such as explaining an idea as opposed to leading a child to discover the idea affect the process of learning?

Are there cognitive factors such as capacity or cognitive style *within* each child that affect learning? Do personality and social factors or even specific attitudes toward mathematics affect the learning process? Is the process also affected by physical factors, including the ability to receive and process sensory data?

Finally, do environmental factors, such as the child's family situation, affect learning? What effect do socioeconomic factors have?

What Ought to be Learned?

Of all that can be learned, what is of most value? Is this decided according to societal or the individual child's needs? Who decides–the faculty? children? parents? What criteria can be used to determine the most important things to teach?

Consider the content of elementary school mathematics. If we use a school district curriculum guide or a textbook, some decisions have already been made about what should be taught. But even the curriculum guide and textbook provide more than each child can learn, for children differ in how much they can learn within the available time. When choices must be made because of limited time, which areas of curriculum content should have priority? Upon what criteria do we base such a decision?

Who Learns?

Do we as teachers somehow instill knowledge and skills in our students as if they were passive beings, or is each child learning in an active way? Is each child busy perceiving, contrasting, relating, testing, and concluding as he or she learns? In short, is it really the *child* who learns? If it is the child who does the learning, what is teaching?

How we deal with issues such as selecting and sequencing content, assessment, instruction, and classroom management reflects our philosophy towards learning. The authors believe that we can become master teachers if we carefully study these issues and the different approaches to their resolution, then attempt to apply them to our classrooms.

Models are available that can guide us as we plan for each child. For example, in Chapter 3, the authors introduce useful models for selecting and sequencing content, planning instruction, and planning assessment procedures.

Throughout this book, the emphasis is on *planning* for teaching and assessment and on making informed decisions based on data from each child. Numerous activity plans are included to help you prepare your lessons. In many of the chapters, the authors also include "Excursions", or extensions of the main ideas. Excursions provide additional information about the topic; they also suggest ways of varying the classroom routine.

2

Beliefs of Others about Teaching Mathematics

You may already have some answers to the questions posed in Chapter 1. Others in your class probably have answers that differ from yours.

There are many views, some of which are contradictory, about teaching and learning and how they interact and complement each other. Different opinions also abound concerning how mathematics content should be chosen, how it should be sequenced and taught in general, the content that is most valuable to a particular child, and the factors that must be considered in preparing a lesson. Many outstanding people, among them mathematics and general educators and psychologists, have expressed their diverse theories on the questions that we asked you to answer.

In this chapter, the authors present a brief introduction to the theoretical positions that currently influence decisions about teaching elementary school mathematics, including proponents for each position and points of agreement and disagreement.

As a teacher, it is important to be aware of these major theoretical positions and to know the implications of each. Such information will help you formulate your own philosophy of teaching.

THEORETICAL POSITIONS

Several theoretical positions will be discussed; however, it is beyond the scope of this book to describe each one in detail.

The positions to be discussed emanate from different theories of learning. Of special interest are those inherent in behaviorally oriented theories, cognitively oriented theories, and sociologically oriented theories. Information-processing theories and neurologically oriented theories are also relevant because of their increasing impact on research in mathematics education.

Behaviorally Oriented Theory

Central to all behaviorally oriented learning theories is the definition of learning as *a change in overt behavior.*

> The child is born empty of psychological content into a world of coherently organized content. Like a mirror, however, the child comes to reflect his environment; like an empty slate he is written upon by external stimuli; like a wax tablet he stores the impressions left by these stimuli; and like a machine he may be made to react in response to stimulating agents.[1]

Behaviorists rely exclusively on behavioral change as an indicator of learning. Overt behavioral acts, such as staring, pointing, writing, and demonstrating, are used as examples of knowledge that has been acquired and can be measured.

Perhaps the most descriptive term in behavioristic learning theory is *stimulus-response* (S-R). Many theorists have used this term in attempting to explain human behavior. Locke, in the seventeenth century, proposed that nothing could be in the mind except that which came from the senses. Hull argued that stimuli are always physical and that the objects in the environment provide a variety of potential stimuli.[2] Others, such as Kendler and Kendler, and Bijou and Baer argue that a child's own behavior

---

1. Jonas Langer, *Theories of Development* (New York: Holt, Rinehart, and Winston, 1969), p. 51.

2. Clark Hull, *Conditioning: Outline of a Systematic Theory of Learning,* Yearbook of the National Society for the Study of Education 41, no. 2 (1942): 61–67.

EXCURSION
Burrhus Frederic Skinner

B. F. Skinner was born March 20, 1904, in Pennsylvania. He received his A.B. degree (1926) from Hamilton College and his M.A. (1930) and Ph.D (1931) degrees from Harvard.

Skinner wanted to make a living as a writer, but he was unsuccessful and turned to the study of psychology. He became the chief exponent of *operationism* or *operant behaviorism.* Toward the end of World War II, he developed an air crib, testing it on his own daughter. He also developed the Skinner box, which was used in psychological research on the behavior of pigeons and rats. He also refined techniques for producing desired behavior from animals and introduced concepts such as shaping, chaining, and operant conditioning. In addition, he developed programmed instruction techniques and the teaching machine.

Skinner's most popular works include *Walden Two* (1948), *Science and Human Behavior* (1953), and *Beyond Freedom and Dignity* (1971).

responses may be a source of stimuli for himself.[3,4]

In the 1920s, E. L. Thorndike proposed a theory that attempted to tie together the positions of many earlier behaviorists. According to Thorndike, when an organism is subjected to a stimulus, a response naturally occurs.[5] Stimuli may be physical or abstract, such as a word. Particular stimuli may be selected in order to elicit a desired, specific response. Modern theorists, such as B. F. Skinner and S. Englemann, have perfected training procedures that reliably result in a desired response, or what behaviorists label as a *product.*[6,7] If the desired learning (the product) can be clearly specified, then a sequence of prerequisite learnings can be identified, and the individual who learns these prerequisites in sequence will likely acquire the desired learning.

3. Howard H. Kendler and T. S. Kendler, "Vertical and Horizontal Processes in Problem Solving," *Psychological Review* 69, (1962): 1–16.

4. Sidney W. Bijou and Donald M. Baer, *Child Development,* vol. 1, *A Systematic and Empirical Theory* (New York: Appleton-Century-Crofts, 1961).

5. Edward L. Thorndike, *The Psychology of Arithmetic* (New York: Macmillan, 1922).

6. B. F. Skinner, *Science and Human Behavior* (New York: Macmillan, 1953).

7. S. Englemann, *Conceptual Learning* (San Rafael, Ca.: Dimensions, 1969).

R. Gagné[8] has extensively applied the above approach to mathematics topics. His work on behavioral objectives and in *task analysis* has specified the products that teachers desire of learners and has identified sequences of behaviors to be learned if each product is to be acquired.

Cognitively Oriented Theory

The cognitivists define learning as *an internal act,* not as a simple change in overt behavior. Such a definition admits that, although overt behavior can indicate that learning has occurred, much has been learned that is not outwardly expressible. Cognitivists speak of the learner organizing information through the use of mental structures or *schemata.* Central to this theory is the influence of factors such as environment and experience upon the development of cognitive structures.

Some cognitivists study the actual development of cognitive structures in humans. Through deliberate and extensive research, J. Piaget identified major *stages of development.*[9]

8. Robert Gagné, *The Conditions of Learning,* 3rd ed. (New York: Holt, Rinehart, and Winston, 1977).

9. Jean Piaget, *The Psychology of Intelligence* (New York: Harcourt Brace Jovanovich, 1950).

EXCURSION
Jean Piaget

Jean Piaget was born in Switzerland on August 9, 1896, and died September 16, 1980, at the age of 84. At the age of ten, he published a scientific paper on a partly albino sparrow. After an early career in science, he became interested in the development of children and began to write in this area in the early 1920s, publishing *Language and Thought of the Child* in 1926.

Among Piaget's major contributions was his theory that children pass through distinct stages of mental and emotional development. These stages—sensorimotor, concrete operations, and formal operations—represent distinctive differences in the qualitative thinking abilities.

Piaget's research methodology is described as quasi-clinical, primarily one-to-one interviews and direct observation in classrooms. He also studied epistemology, the study of how knowledge is acquired, and regarded the child's incorrect responses to be as important as the correct ones.

Piaget's most famous works include *The Language and Thought of the Child* (1926), *Judgment and Reasoning in the Child* (1928), *The Construction of Reality in the Child* (1954), *The Child's Conception of Number* (1952), and *The Child's Conception of Space* (1956).

Along with the approximate ages associated with each stage, they are as follows:

1. Sensorimotor stage (0–1½ years)
2. Preoperational stage (1½–7 years)
   a. Preconceptual thought (1½–4 years)
   b. Intuitive thought (4–7 years)
3. Operational stage (7–16 years)
   a. Concrete operational thought (7–11 years)
   b. Formal operational thought (11–16 years and beyond)

Piaget's cognitive developmental theory attempts to explain how humans modify cognitive structures through the interaction of four major factors: physical experience, social experience, biological maturation, and equilibration. Other theorists with a similar philosophy include J. Bruner,[10] R. Skemp,[11] and Z. Dienes.[12]

Cognitivists are concerned with more than *what* is learned. They are also concerned with *how* it is learned. The environment is viewed as the *setting* for development, not as the agent of development, as behaviorists claim. The environment provides the catalyst for the child's developing emotional and cognitive self. Through interaction with his environment, the child slowly develops schemata to solve new problems. Learning experiences that allow development of new schemata and application of already acquired schemata, and objects from the environment that clearly exemplify the concepts must be carefully chosen.

10. Jerome Bruner, *The Process of Education* (New York: Random House, 1960).
11. Richard Skemp, *The Psychology of Learning Mathematics* (New York: Penguin Books, 1971).
12. Zolton Dienes, *The Power of Mathematics* (London: Hutchinson Educational, 1964).

Sociologically Oriented Theory

Instead of giving a definition of learning, advocates of a sociological approach discuss criteria by which content should be selected and placed in a mathematics curriculum. They believe that the mathematics selected must meet the criterion of usefulness in society. By satisfying such a condition, the knowledge can meet the needs of the learner, which, according to Dewey in the 1940s, should be the starting point for any curriculum. Therefore, content that satisfies these needs will also motivate the learner, allowing the teacher to do a better job.[13]

13. John Dewey, *How We Think* (Lexington, Mass.: D.C. Heath, 1953).

EXCURSION
John B. Dewey

The name of John Dewey has become synonymous with the progressive education movement in both the United States and abroad. A philosopher, psychologist, and practicing educator, he was the dean of twentieth-century American educators.

Dewey was born in 1859 in Vermont and died in New York City in 1952. He graduated from the University of Vermont with the A.B. degree (1879) and from Johns Hopkins University with the Ph.D. degree (1884). He taught philosophy at the University of Michigan, the University of Minnesota, the University of Chicago, and Columbia University in New York City.

For Dewey, the everyday world of common experience was all the reality that a person had access to or needed. In his writings, he was careful to make clear the kinds of experiences that were most valuable and useful. An educative experience, according to Dewey, is one in which we make a connection between what we do to things and what happens to them or us as a result; that is, the value of an experience lies in the perception of relationships or continuities among events.

For Dewey, learning arises primarily from the personal experience of grappling with a problem. He argued that schools did not provide genuine learning experiences, only an endless amassing of facts that children were expected to remember. Dewey constantly pointed to the gulf between in-school and out-of-school experiences. Without some connection between the two, he felt genuine learning and growth would be impossible.

Dewey authored many books and articles. Among the most popular were *My Pedagogic Creed* (1897), *How We Think* (1910), *Democracy and Education* 1916), and *Experience and Education* (1938).

The sociological approach allows one to make value judgments about the most important math for each child. Some advocates of this position even propose that the learner take responsibility for determining when specific math content should be learned. A passage from A.S. Neill's *Summerhill* illustrates this view:

> We have no new methods of teaching, because we do not consider that teaching in itself matters very much. Whether a school has or has not a special method for teaching long division is of no importance except to those who want to learn it. And the child who *wants* to learn long division *will* learn it no matter how it is taught.[14]

Overall, those who advocate the sociological viewpoint see the child as the center of the curriculum; whatever is included should reflect his or her emotional, educational, and societal needs.

14. A. S. Neill, *Summerhill* (New York: Hart, 1960) p. 5.

Other Positions

Two additional theoretical positions that surfaced during recent years are information-processing theory and neurologically oriented theory. The impact of these positions on teaching and learning mathematics is still unclear, but both are interesting areas of research and are potentially useful.

Information-processing theory attempts to explain how information is encoded, decoded, and processed in the human brain.[15] Much of the

15. Walter R. Reitman, *Cognition and Thoughts: An Information-Processing Approach* (New York: John Wiley and Sons, 1965).

current research involves developing information-processing models by carefully documenting the length of time needed to process specific bits of information. Some tentative inferences are being made concerning an individual's cognitive structures, elementary psychological processes, and higher-order problem-solving strategies.

The neurologically oriented theories attempt to understand the relative functions of the different regions of the brain. Much of the research involves *hemispheric specialization,* and it attempts to identify the hemisphere of the brain usually responsible for solving specific types of problems. For instance, is the computation of 8 + 4 = □ essentially a right-brain or a left-brain task? Do a child's hemispheres function differently when doing math or reading? Which parts of the brain contribute to a child's inability to perform in mathematics? Such questions must still be answered as researchers build a neurological theory of learning.[16]

Much remains to be studied in the areas of information processing and neurological research. However, we can already speculate about implications such theories may have for teaching.

Cautions

No single theory is comprehensive enough to explain learning and, at the same time, reliably predict the best way to select and organize content and choose a teaching strategy. Few theories address such vital aspects of learning as affective behavior and classroom climate. Hence, you should be cautious about wholeheartedly adopting any one theory. Instead, use ideas expressed in different theories to give you the basic information that can help you make initial, tentative decisions about how you will teach. By the time you finish this book, you should have integrated many theoretical ideas with your own practical experiences, and you will be more articulate concerning your beliefs about teaching and learning.

16. See *Education and the Brain, part 2,* 77th Yearbook of the National Society for the Study of Education (Chicago: National Society for the Study of Education, 1978).

IMPACT OF THEORETICAL POSITIONS ON MATHEMATICS EDUCATION

As we take a closer look at how each theoretical position affects the teaching of mathematics, observe that different answers to the questions posed in Chapter 1 are suggested by the different positions.

What Is Learning?

The beliefs of behaviorists and cognitivists illustrate the varied answers to this question. The common behaviorist definition of learning is that it is any change in behavior[17] or "the relatively permanent modification of behavior as the result of experience."[18] In fact, only when a change in behavior is observed can a teacher assume that learning has taken place. This definition suggests that teachers should induce certain behaviors in students; when these behaviors are demonstrated, we can assume that learning has taken place. This philosophy is reflected by school systems that invest time and money in writing behavioral objectives. Each objective describes a behavior that becomes the object of instruction. An example of a behavioral objective is: Given the statement 3 + 4 = □, the child will state the sum with 100% accuracy.

In direct contrast is the position of cognitivists, who deny that learning is merely a change in behavior. The cognitivist views behaviors as mere *indicators* of learning; learning itself is an internal process that takes place somewhere between the stimulus and the response. Although certain behavior does imply that learning has taken place, it may also have occurred when no overt behavior change can be detected. For example, children who cannot demonstrate their knowledge through acceptable paper-and-pencil work may actually understand the basic concepts

17. Thomas F. Green, "A Topology of the Teaching Concept" in C. J. B. Macmillan and Thomas W. Nelson, *Concepts of Teaching: Philosophical Essays* (Chicago: Rand McNally, 1968), p. 56.

18. John B. Magee, *Philosophical Analysis in Education* (New York: Harper and Row, 1971), p. 71.

involved. A behaviorist would interpret this as evidence the student has not learned because he or she does not display the desired paper-and-pencil behavior. A cognitivist would not form such a conclusion based on paper-and-pencil performance alone.

What Is Learned?

Many learning products can be identified: habits, skills, chains of sequential behavior, cognitions, and attitudes. The specific learning theory you subscribe to will tend to influence your view of *what* is learned. For the cognitivist, cognitions are the objects of instruction, not behavioral chains or habits. Most cognitivists admit that behavioral chains represent some type of learning, but they are not viewed as the primary concern.

What are mathematical cognitions? Consider first a cognitivist's view of mathematics: a system of related ideas or concepts. It is also a way of thinking that produces other related systems. It is precisely the ability to utilize these mathematical interrelationships that the cognitivists seek. As such interrelationships are conceptualized, they are the base from which new mathematics can be learned. Thus, the cognitivist believes that mental schemata are the elements that are learned.

In contrast, assume that the instructional objective for a student is to learn all the basic facts of addition for which five is the sum. From a purely behaviorist position, this objective is straightforward. The student is to respond by saying or writing "five" whenever stimuli such as $1 + 4$, $0 + 5$, $2 + 3$ are given. Through a properly selected teaching sequence (usually involving much drill) this objective can probably be achieved. A cognitivist would want to know if the learner had developed insight into how the facts are interrelated with each other and with other mathematics. Has the child developed schemata that can be applied when finding all the basic addition facts for which the sum is, for example, six or seven? The cognitivist would demand that, in addition to the obvious responses, there must also be some indication that each child has internalized the mathematical relationships.

What Are the Most Valuable Learnings from Mathematics?

The answer to the above question may be more philosophical than psychological. The value of a particular mathematical learning lies on a continuum ranging from that which is valuable for its intrinsic beauty to that which is valuable for its usefulness. During the 1960s, much was said about the value of learning mathematics because of its beauty as a logical art form. Many theorists, such as Zoltan Dienes, suggested that the study of mathematics is so intrinsically rewarding that students should need no outside motivation.

More recent trends have emphasized the learning of basic mathematical skills. In doing this, school systems have developed lists of basic mathematical competencies required of each graduating high school student. Many school systems give mathematics proficiency tests on concepts they view as most important.

The advent of calculators and computer technology has prompted discussion about the nature of basic skills. The emphasis on computational skills, currently a major part of the elementary curriculum, is now recognized as a small part of those skills.

What Factors Affect Learning?

Most of the factors that affect the learning of mathematics fall into four major categories: content factors, instructional factors, assessment factors, and managerial factors.

Content Factors

One of the first questions a teacher must answer is, what content is to be taught? In many instances, teachers simply follow the content outlines of a textbook or curriculum guide. In fact, research has shown that elementary teachers follow the mathematics textbook at least 90% of the time. However, while the textbook provides a good outline for elementary-age children, the teacher must still determine the most appropriate mathematics *for each student.* As a teacher, you must select appropriate content for students with many levels of ability: intellectually gifted, bright, average, and children experiencing extreme difficulty. Clearly, children covering such a range of abilities cannot all benefit from exactly the same content. Such decisions are a major challenge for every teacher.

We want children to see mathematics in the world around them and to be alert to the numerical and spatial aspects of all situations. Therefore, much of the content we select, especially for young children, should help each one appreciate the presence of number and spatial concepts in all that they see and hear.

John Dewey suggested that one criterion in selecting content should be usefulness; that is, the child will be able to use what is being learned in everyday, real-life situations. However, although many students will enter into the world of work, future mathematicians and scientists are in the classroom also. The goals of the latter two groups include advanced study, the immediate usefulness of which may not be readily perceived. Because of such needs, the curriculum should not just be limited to what is considered useful in everyday life.

As teachers we must often develop our own mathematics units, and this requires a good working knowledge of content. We need to have competence in the structure of the real number system, its subsystems, and Euclidean, coordinate, and transformational geometries. A knowledge of computers and the attendant languages, especially BASIC, FORTRAN, and PASCAL, is becoming more important as schools use computer technology in their curricula. Most homes have hand-held calculators, and we must find appropriate ways to use them in the classroom. And efforts must *still* be made to develop basic computational skills and arithmetical knowledge.

As we teach, we need to clearly understand the *sequential* nature of mathematics, choosing content to supplement, replace, and extend the information found in a textbook or curriculum guide. A clear, logical sequence will do much to insure correct learning.

Instructional Factors

These factors relate to the ways in which a teacher designs and executes plans for instruction. Instructional planning is influenced by the teacher's own definition of and beliefs about learning.

Many learning theorists have written about the value of discovery learning versus expository learning. In discovery learning, the teacher provides guidance so that specific mathematical concepts are discovered by the learner. This is in contrast to the expository approach, in which the teacher presents mathematical concepts in a direct way. The discovery approach has much support from theorists such as Piaget, Bruner, Dienes, and Skemp. The limited comparative research indicates that concepts learned through a guided discovery approach are retained longer than those learned through an expository approach. Guided discovery is also helpful in developing problem-solving skills.

Many cognitive theorists argue that activities that move the learner from concrete situations to a more abstract level must be developed. In order to learn a concept, a child should actually manipulate teaching materials. Through reflection on those physical actions, he or she will begin to abstract the concept being modeled.

Advocates of behavioristic learning theory first specify objectives in behavioral terms. Children are then helped to acquire each behavior. Sometimes a behavior is fixed through the use of much repetition. Children who experience difficulty are frequently placed on contingency schedules where they are rewarded, often tangibly, for demonstrating some overt behavior.

Many additional aspects of a teacher's instructional plans must be considered. Activities must be adapted to a range of learning abilities and styles. For instance, some children tend to be visual learners while others tend to be verbal learners. Some learn quickly, almost impulsively, while others who are equally bright learn slowly and methodically. Some children learn best by talking things out, while others do better by using their hands. Some can work accurately when under time pressure, while others fall to pieces. Some children need constant prodding while others do better in a free, relaxed atmosphere.

The varied learning abilities and styles of children suggest different sequences and combinations of manipulative verbal instructions, assignment schedules, and learning conditions. Games and other motivational activities help facilitate learning, and teachers are encouraged to incorporate such activities into their mathematics

Table 2.1
Instructional Factors

Planning
1. Formulating objectives
2. Selecting method
3. Arranging activities
4. Selecting instructional materials

Implementing and facilitating
1. Motivating
2. Introducing
3. Reinforcing
4. Guiding, questioning
5. Soliciting
6. Culminating

Evaluating
1. Appraising
2. Judging
3. Recycling

programs. Correct use of mathematical language, encouraging student discussion, and appropriate uses of feedback and reinforcement have also been found to aid learning. Several instructional factors are listed in Table 2.1.

Assessment Factors

Before appropriate teaching can begin, a teacher should have as much information as possible about content acquisition and emotional and attitudinal strengths and weaknesses of each student. A number of tests are available for an initial diagnosis; several are discussed in Chapter 20. However, assessment is more than an initial diagnosis. It implies a total approach that attempts to determine what and how well each child has learned and what should be taught next.

When planning an assessment program, a behaviorist is primarily concerned with how many items are correct or incorrect. A child's level of performance is based upon his or her recorded responses. Mastery is usually indicated by a percentage correct, often 80 or 90 percent. In contrast, a teacher with a cognitive view is as interested in the process being used as the number correct. In-depth probes, interviews, and other diagnostic procedures are used to determine if a child has conceptualized the interrelationships of concepts and their prerequisites. In general, teaching mathematics involves continuous assessment in relation to grouping for instruction, choosing the next content, and so forth.

Current classroom assessment procedures include standardized and teacher-made tests, structured interviews, formal and informal teacher observations, and evaluations of student products. The value you place on each procedure depends on your view of what is learned and, hence, how to diagnose for that learning.

Managerial Factors

The classroom climate must be conducive to learning. Given the range of cognitive abilities present in a classroom and the choice of instructional procedures available, a major managerial effort must be made to provide the variety of activities appropriate for each learner. In any classroom, a mix of instructional designs must be organized and coordinated, such as large-group instruction, small-group instruction, and independent learning centers. In addition, teachers must coordinate the factors related to content and assessment and prescribe an appropriate mathematics program for each student. They must also be able to maintain an open, warm, and nonthreatening climate that fosters an atmosphere of self-discipline. This may sound like an impossible task, but the authors have observed many teachers who have developed the above skills. Throughout this book, the authors present tips on managing the classroom, ideas which are drawn from their expertise. Your own managerial skills will develop as you gain teaching experience.

THE AUTHORS' BELIEFS ABOUT LEARNING

The authors lean towards the interpretation of cognitive psychologists, that cognitions (rather than behavioral chains or habits) should be the learning products that primarily concern mathematics teachers. However, many other kinds of learning, including the purely behavioral chains emphasized by behaviorists, can be considered.

The authors' beliefs about learning are also reflected in their conception of mathematics:

that mathematics is essentially composed of systems of related concepts, and that it is a way of thinking. Through centuries of experience and reflection upon same, people have noted patterns, seen contrasts, and recognized similarities and stabilities. Gradually and painstakingly, these experiences have been drawn upon to create the systems of ideas we call mathematics. This process continues even today.

As a result, a great variety of overt behaviors or procedures were developed: counting, measuring, computing, and so on. With the help of such procedures, people have been able to solve many everyday problems. However, the essence of mathematics is not merely a set of overt behaviors to be trained into a child. It is more than rattling off 1, 2, 3, and paper-and-pencil manipulation of symbols. Such behaviors are only visible signs of fundamental ideas. Moreover, they are fully useful only to those who have acquired the concepts involved. Children should learn such overt procedures, but the authors believe that the most important objectives of instruction should be the basic, underlying concepts of mathematics: those related to number, numeration, the meaning of operations, shape, measures, and so on.

In addition to the fundamentals, the authors also want children to appreciate the beauty of the mathematical systems that are a part of our culture. These systems, along with art, music, science, and literature, are among humankind's most creative works, and can help us interpret, enrich, and develop our world. Like other creative outlets, mathematics can also give the learner a sense of competence and self-worth.

Many children experience difficulty learning mathematics. The authors believe that diagnosis exposes two types of difficulties: learn*ing* and learn*ed.*

Kirk defines a learn*ing* disability as "a specific retardation or disorder in one or more of the processes of speech, language, perception, behavior, reading, spelling, writing or arithmetic."[19] While there are different definitions of learning disability, all refer to a psychological disorder in certain basic biological or neurological processes. A teacher with special training may infer a possible learning disability from a child's performance or behavior during a thorough diagnosis. As teachers, we must make sure that any such disability, when observed, is actually due to a malfunction of some psychological process. If it is not, it should not be labeled as a learning disability.

Many difficulties are learn*ed* disabilities. Learned disabilities may result when concepts or skills are taught incorrectly or inadequately. They may also result when the child lacks the developmental maturity to insure acquisition of meaningful concepts. For example, children often learn computational procedures that lead to incorrect answers when completing exercises. In *Error Patterns in Computation,* Ashlock gives many examples of children who appear to have no difficulty learning; they have simply learned the wrong algorithm.[20] Erroneous learnings frequently result in low scores that may be interpreted as indications of a learning disability. The authors believe such incorrect behaviors are frequently caused by learn*ed* rather than learning disabilities. It has been their experience that most of the mathematical difficulties elementary school children experience result from learn*ed* disabilities.

Learned disabilities are correctable through appropriate diagnosis and instruction. Accordingly, the authors emphasize diagnosis and focused instruction throughout this book. However, if we keep in mind that children often *acquire* learned disabilities, we can try to prevent them.

How can we ensure that healthy learning will take place? We must encourage each child to explore, to interrelate ideas, to formulate hypotheses, to search for patterns, and to apply what they know to new situations. Mathematical ideas must be understood before we attempt to fix and maintain those concepts through controlled practice. The authors believe that the child who works from his or her strengths and who experiences success will see real progress and will maintain a positive, healthy attitude toward learning mathematics.

19. Samuel A. Kirk, James J. McCarthy, and Winifred D. Kirk, *The Illinois Test of Psycholinguistic Abilities,* rev. ed. (Urbana, Ill.: University of Illinois Press, 1968), p. 398.

20. Robert B. Ashlock, *Error Patterns in Computation,* 3rd ed. (Columbus, Oh.: Charles E. Merrill, 1982).

Consider one final question. *Who* learns mathematics? Your actions in assessing, planning and teaching will reflect your answer. The authors feel that learning is an individual matter and that *each child learns.* Hence, our challenge as teachers is to insure that we facilitate the learning of *each* child.

3

Models to Guide Teaching: An Overview

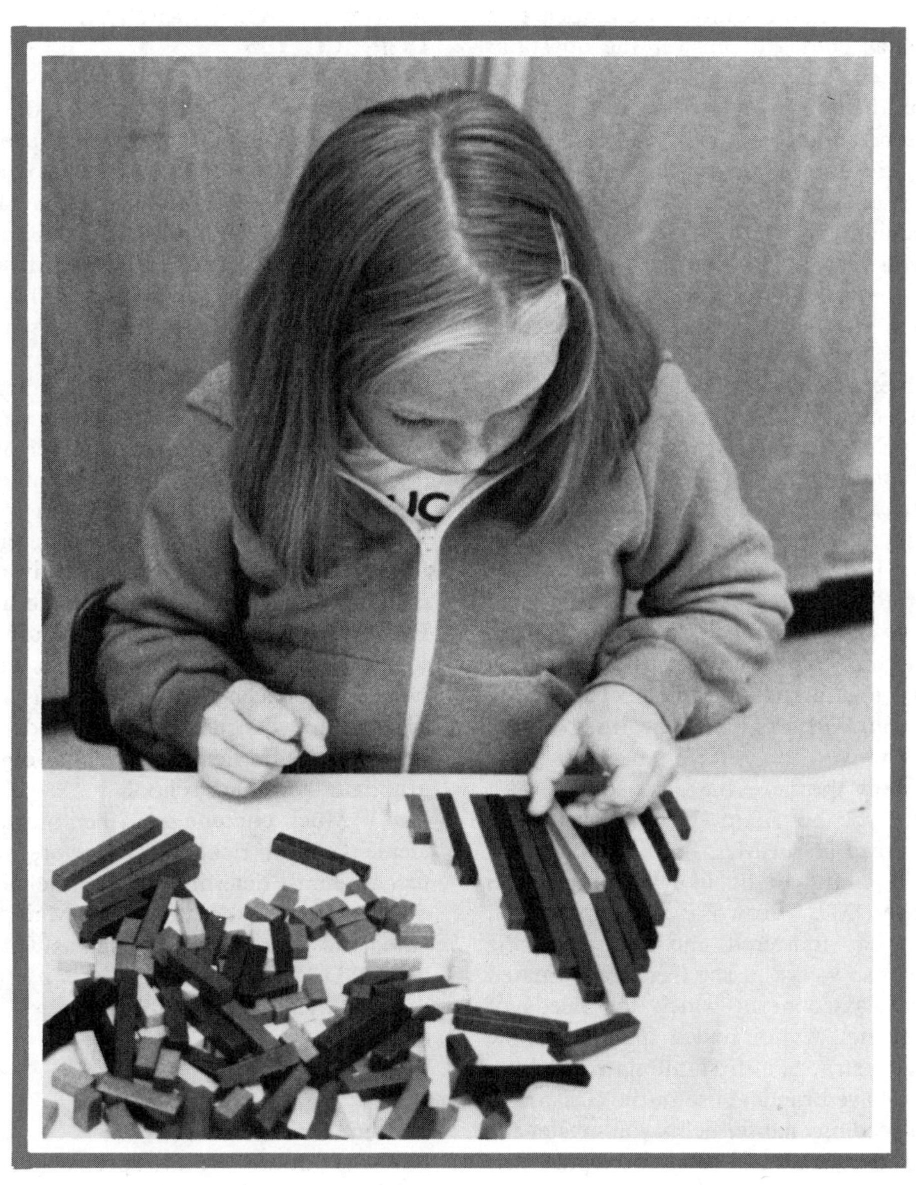

In this chapter the authors introduce models that can be used as guides to answer the questions posed in previous chapters. Each model is developed and used more fully later on in the book. Each has been found to be helpful to teachers in decision making, in planning curriculum and activities for assessment and instruction, and in actually teaching lessons.

As you read about the models, you may ask, "Why do I need to learn theory like this? When will I get to the part about what I am to *do* in class?" Teaching elementary school mathematics is a complex task, much like learning to play a musical instrument. One must learn many different things: which notes to play, where these notes are on the instrument, how fast to play, what rhythm to use, where to put your fingers. When you begin, these things seem impossible to learn individually. The ideas and words are strange; they appear to be unrelated to each other and impossible to simultaneously coordinate. Yet, when you attempt to play a tune you find that all the pieces fit together. It may not be done well the first try, but with practice it begins to sound like music. As you play different songs, you repeatedly apply the same basic concepts.

Before you go on to the next section in this chapter, read about Mrs. Brown's classroom in Appendix A. It will be like listening to an orchestra play. Mrs. Brown is an experienced teacher who has mastered and integrated the basic concepts so well that she has a coordinated mathematics classroom in which the needs of each child are met. As you review the models, the ideas may seem strange at first, unrelated and too theoretical to have practical use in the classroom. But each succeeding chapter helps you master the ideas as you apply them to different topics. The tune changes, but the underlying concepts remain the same, and the process becomes easier. The models even provide help when you are teaching other subject areas.

A MODEL FOR PLANNING CURRICULUM

Mathematics is a major part of the elementary school curriculum. The actual selection of what mathematics to teach is likely to be among your first major tasks. You are already familiar with mathematics from high school and college courses, and you could ask yourself what part of that knowledge should be included in the mathematics curriculum. Furthermore, school systems have adopted textbooks, curriculum guides, and scope and sequence charts that identify important mathematics content. Another factor is the time available for teaching mathematics, requiring that priority be established among content topics. Yet another consideration is basic skill. While the term *basic skills* has different meanings for different people, it almost always includes computational skills. So, you could ask yourself, How much of my curriculum should be devoted to developing basic skills? This question has become increasingly important, as many states and school districts require students to pass mathematics proficiency examinations in order to graduate from high school.

Much content *could* be considered for an elementary curriculum; however, you, as the teacher, must determine the mathematics content most important for each child. Many factors will enter into your decisions, such as what a particular child already knows and how useful a specific mathematical idea or procedure will be for him or her.

Content Objectives

How then do we begin to specify the mathematics curriculum for each child? Initially, it is useful to list major categories of elementary school mathematics: sets, whole numbers, rational numbers, and so on. Next, find out what major ideas, including concepts, principles, and procedures, need to be learned in each of the categories. The following example from whole number numeration helps illustrate this approach.

Suppose you said, "I want my children to realize that in a two-digit number such as 27, the 2 represents two tens and the 7 represents seven ones." This is a reasonable expectation for children in the early grades and an idea on which much teaching time is spent. But we need to analyze the statement according to underlying mathematics concepts.

Obviously the child must recognize the numerals, be able to name the 2 as "two" and the 7 as "seven." Further, the child must have some experience with place value; that is, some knowledge that, in the Hindu-Arabic numeration system, the place in which a digit is written has a value. Consider the following statements.

3.1
Numerals are symbols or sets of symbols used to name or represent numbers.

3.2
A *digit* is a numeral with only one symbol.

3.3
The number to which a digit is assigned is called the *face value* of the digit.

3.4
Each power of 10 has a fixed position or place assigned to it rather than a symbol.

3.5
Horizontally arranged positions are assigned to numbers with respect to a reference called a *decimal point*. When no decimal point is recorded, as is usually the case for whole numbers, it is assumed to be immediately to the right of the numeral.

3.6
The first position to the left of the decimal point in the Hindu-Arabic system is assigned the number 1.

3.7
The second position to the left of the decimal point in the Hindu-Arabic system is assigned the number 10.

3.8
A multidigit numeral names a number that is the sum of the products of each digit's face and place value.

Think back to the statement: "I want my children to realize that in a two-digit number such as 27, the 2 represents two tens and the 7 represents seven ones." The content listed in statements 3.1 through 3.8 is surely part of knowing tens and ones. The authors refer to each statement as a *content objective*. In general, a content objective is a statement about mathematics. The model for curriculum planning presented in this book involves choosing the most appropriate content objectives. Then knowing the content in each statement becomes the object of instruction for each child.

A content objective may take several forms:

1. A definition—Addition is a binary operation that assigns to two numbers, called *addends,* a unique number, called a *sum*.
2. A statement of procedure—Addition may be modeled on the number line.
3. A statement about mathematical relationships or structure—For all whole numbers a and b, $a + b = b + a$.
4. A statement about a verbal or nonverbal association—The sign "+" is read "plus".

These are all mathematical definitions, procedures, relationships, and associations *to be learned.* They are potential objectives of instruction.

The sample content objectives listed in most chapters and summarized in Appendix B do not constitute a complete elementary school mathematics program, but they do represent the major objectives for instruction.

We can now consider the basic question posed earlier: What mathematics is appropriate for each child? By choosing a set of objectives for each child, we take a major step toward individualizing instruction. Also, consider what each child already knows; some children learn content at different levels of sophistication. For example, both of these statements are content objectives:

a. It does not matter in which order you add two whole numbers.
b. For all whole numbers a and b, $a + b = b + a$.

Although both objectives address the principle of commutativity for addition of whole numbers, they are at different levels of sophistication. In choosing an objective, statement **a** may be viewed as appropriate for one child while statement **b** may be more appropriate for another.

In addition to helping us plan individualized curricula, content objectives have other uses. For example, they serve as a basis for developing diagnostic test items. Working from a set of content objectives, we can develop tasks that probe learnings to determine if specific concepts and skills have been learned. Such information is invaluable when planning instruction.

Some teachers use lists of content objectives to identify topics for a mathematics laboratory. They develop activities that focus on each idea stated in an objective. Sometimes independent learning centers are developed that go beyond regular work on a specified topic.

Individual content objectives provide the content outline needed for an *activity plan.* In some instances, a small group of objectives may be more suitable than a single objective. The lesson for one day may include one or more planned activities, or an activity may continue through more than one day's lesson.

Other Objectives

Other kinds of objectives that focus on basic processes or a child's feelings about mathematics require our attention as well. Also, behavioral objectives are sometimes used instead of content objectives.

Basic Process Objectives

Children learn content expressed in a content objective, but also how to organize content so that patterns and rules are discovered, then generalized to related content. In doing this, children use basic conceptualization processes.

Many lists of such processes have been generated. One mathematics program, Developing Mathematical Processes (DMP), lists the following: describing and classifying, comparing and ordering, joining and separating, grouping and partitioning.[1] On the other hand, Bell lists problem solving, proof, and mathematisation. In the latter he includes representation, generalization, and abstraction.[2] Other lists include translating and analyzing.

Affective Objectives

Objectives that state desired attitudes are referred to as *affective objectives.* Feelings about mathematics may be influenced by the immediate and future value of mathematics as perceived by the child, by the level of past success with mathematics, by the teacher's attitude toward mathematics, or by other factors. We need to help each child develop positive feelings toward mathematics.

Objectives concerning content, cognitive processes, and attitudes play important roles in unit and activity planning. Whenever instruction focuses on the development of basic processes, process objectives actually state the substance dealt with during instruction. Similarly, whenever instruction is concerned with developing attitudes or values about mathematics, affective objectives state the content of instruction.

Behavioral Objectives

There is a basic distinction between content objectives and *behavioral objectives.* In a content objective, the object of instruction is knowledge of the procedure, concept, or principle which is defined. Typically, the elements to be learned are themselves complex sets of interrelated ideas. We can infer whether or not a child has acquired these learnings by observing the child's behavior in varied situations. When using a behavioral objective, the object of instruction is usually a specific overt behavior. One objective states an idea, the other describes an action.

---

1. Wisconsin Research and Development Center for Cognitive Learning, *Developing Mathematical Processes* (Chicago: Rand McNally, 1976).
2. Alan W. Bell, "The Learning of Process Aspects of Mathematics," *Educational Studies in Mathematics* 10 (1979): 361–87.

Both of the following objectives are concerned with the product values of digits.

Content Objective
The product of a digit's face value and the value of the place in which it is written may be called the digit's *product value.* (For example, in 75, the product value of the 7 is 7×10 or 70.)

Behavioral Objective
Given a three-digit numeral, the child will be able to identify the product value of each digit in the numeral.

While the content objective states knowledge that a child is to learn, the behavioral objective describes an observable act that a child is to do. A child who has learned to do the task described in the behavioral objective would be able to respond correctly to the following problem:

PROBLEM
Draw a ring around the product value for the 7 in the numeral 75.
7 10 70 75

It is unclear how much we can infer about a child's learning from one specified behavior, which is often at a motor level. Certain stated behaviors are important in mathematics, such as writing the numeral 2, drawing a rectangle, and so on. But we want each child to learn *ideas* also.

A MODEL FOR PLANNING INSTRUCTION

If mathematics is a set of interrelated ideas, then supporting (or at least related) ideas play an important role in learning a mathematical concept. Many learning theorists propose that concept learning begins with an introduction of a concept through some type of exemplification. The learner abstracts the defining attributes of the concept and attaches a label (name), then relates the concept internally to previously learned concepts. There is movement from concrete situations to abstractions.

Activity Type Cycle

The model for instruction presented in this text parallels the learning sequence proposed by cognitive learning theorists, and is referred to as the *Activity Type Cycle.*

Six distinct types of activities are identified for guiding the learning of concepts. Each type serves a different purpose and requires a specific set of techniques. However, not all objectives require each type of activity. The complete Activity Type Cycle is outlined in Figure 3.1.

Diagnosing

Diagnosis involves using a test or other procedure to collect information so that inferences can be made about the presence, correctness, or absence of a concept. It is central to the Activity Type Cycle, using procedures that allow you to infer exactly what a child knows and to determine strengths and weaknesses. Hence, for a given content objective, diagnostic findings suggest the type of activity with which to begin instruction.

Consider this set of content objectives:

3.9
The non-negative rational numbers are defined as the set of numbers that may be named in the form a/b, where a represents a whole number and b represents a counting number.

3.10
A fraction is used to express the number for part of a whole or of a set.

3.11
The denominator below the fraction bar names the fractional part under consideration. It tells how many parts of the same size are in the whole (or how many members are in the set).

3.12
The numerator above the fraction bar tells how many of the parts are being considered.

These content objectives cluster around the meaning of a fraction. Suppose we are faced with these questions: Do these children need

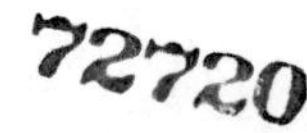

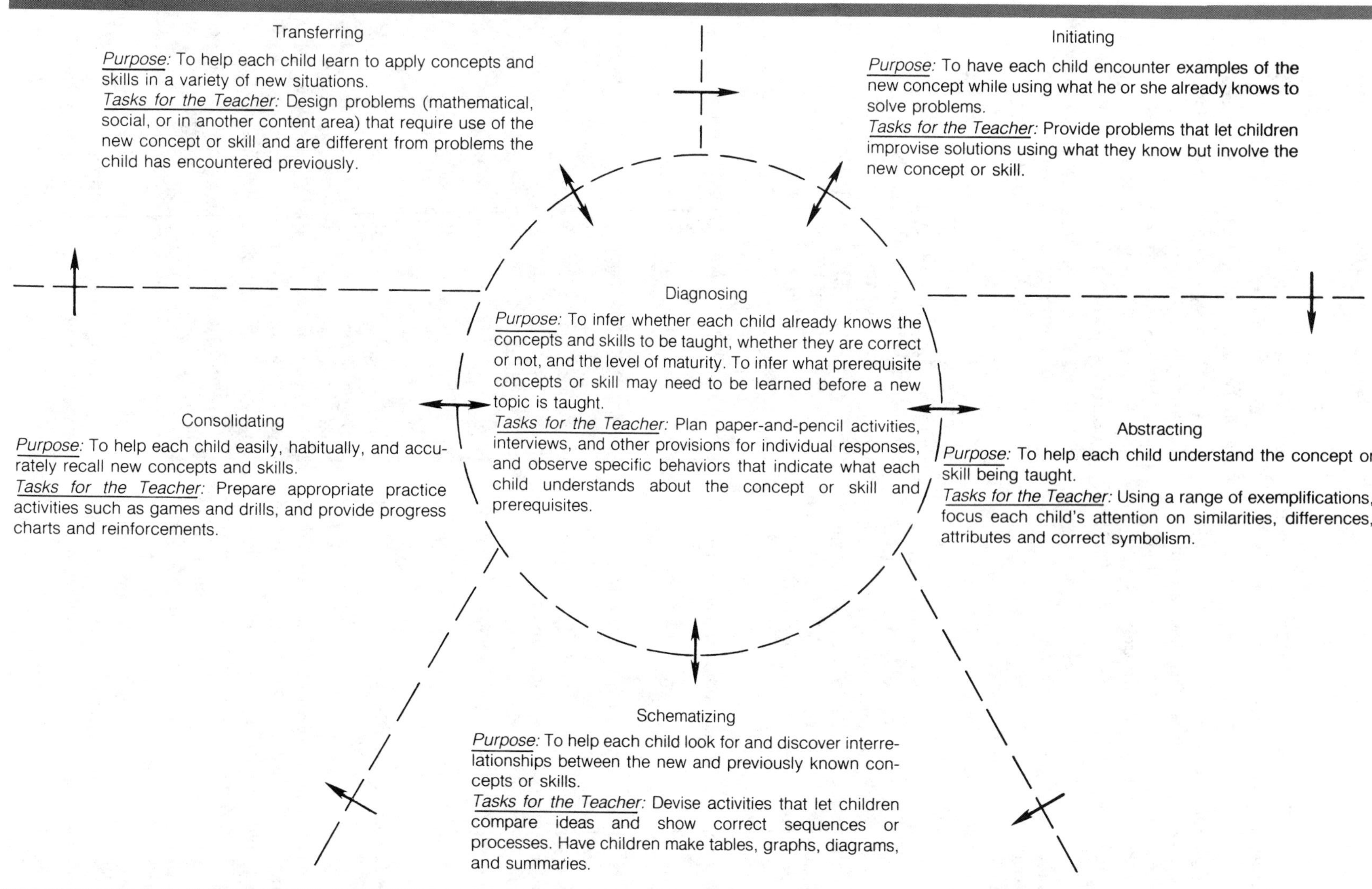

Transferring
Purpose: To help each child learn to apply concepts and skills in a variety of new situations.
Tasks for the Teacher: Design problems (mathematical, social, or in another content area) that require use of the new concept or skill and are different from problems the child has encountered previously.
Initiating
Purpose: To have each child encounter examples of the new concept while using what he or she already knows to solve problems.
Tasks for the Teacher: Provide problems that let children improvise solutions using what they know but involve the new concept or skill.
Diagnosing
Purpose: To infer whether each child already knows the concepts and skills to be taught, whether they are correct or not, and the level of maturity. To infer what prerequisite concepts or skill may need to be learned before a new topic is taught.
Tasks for the Teacher: Plan paper-and-pencil activities, interviews, and other provisions for individual responses, and observe specific behaviors that indicate what each child understands about the concept or skill and prerequisites.
Consolidating
Purpose: To help each child easily, habitually, and accurately recall new concepts and skills.
Tasks for the Teacher: Prepare appropriate practice activities such as games and drills, and provide progress charts and reinforcements.
Abstracting
Purpose: To help each child understand the concept or skill being taught.
Tasks for the Teacher: Using a range of exemplifications, focus each child's attention on similarities, differences, attributes and correct symbolism.
Schematizing
Purpose: To help each child look for and discover interrelationships between the new and previously known concepts or skills.
Tasks for the Teacher: Devise activities that let children compare ideas and show correct sequences or processes. Have children make tables, graphs, diagrams, and summaries.

FIGURE 3.1
Activity Type Cycle: Purposes of Activities and Tasks for the Teacher

instruction on the meaning of a fraction? If so, where should instruction begin? We could carry out a diagnosis consisting of written tests and interview protocols for fractions. Gathering information and analyzing the data will reveal qualitative and quantitative differences in responses. Some children have no workable idea of what a fraction is. Others recognize the numerals but cannot give nonsymbolic representations for fractions. Children in a third group know only one way to use a fraction: as a part of a whole. Still others have many correct fraction concepts yet need to apply what they know.

Obviously, these children should be involved in different types of activities. Some would require an initiating activity; others, a schematizing activity; and still others, a consolidating activity. In each case, the selection is based upon an interpretation of diagnostic findings.

Because diagnosis is a crucial first step in planning instruction for a child, try to determine the appropriate type of activity for each group of learners. Review the purpose of each activity type in Figure 3.1 and jot down your own decisions before reading further. The authors indicate the appropriate activity type for each group of children as they describe the remaining activity types, so you will be able to compare your results with theirs. Diagnosis will be discussed more extensively later in the chapter.

Initiating

The major purpose of an initiating activity is to have students encounter examples of the concept to be learned while using lower-level prerequisite concepts and skills they already know. The problems to be created should meet three criteria. First, the most efficient way to solve the problem would be to use the new learning; but it is not known. Second, the children involved are not likely to be overwhelmed by the problem, and will be able to improvise a solution using what they already know. And third, examples of the new learning are encountered as the problem is being solved.

For children identified as having no concept of a fraction, initiating activities would be most appropriate–specifically, those involving content objectives 3.9 through 3.12. Critical tasks for the teacher are noted in Figure 3.1.

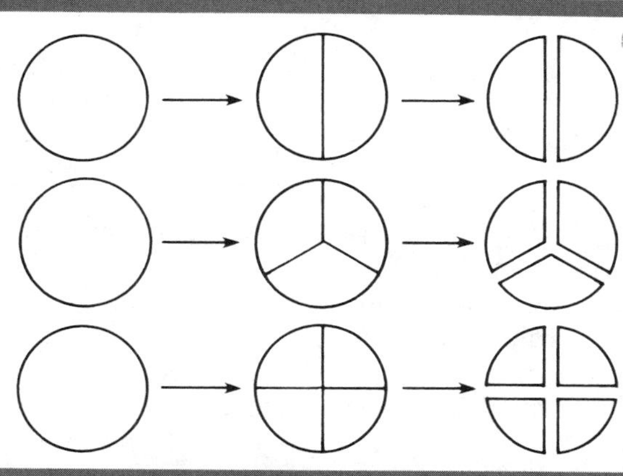

FIGURE 3.2
Unit Regions Being Broken into Equal Parts

Children could be asked to fold or cut geometric shapes into halves, thirds, and fourths as seen in Figure 3.2. In so doing they are confronted with a problem situation. Encourage each child to place pieces on top of each other to show that they really are equal in area; then discuss this concept. This encourages a child to *think* about what he is doing. As the initiating activity continues, you can ask for a specified number of pieces. For example, say, "Show me three of the four equal pieces."

By answering and actually solving problem situations, the child is producing, naming the total number of, and specifically identifying fractional parts. The child may now be ready for an abstracting activity.

Abstracting

Abstracting activities, typically more structured than initiating activities, enable each child to focus on relevant attributes of the examples. This is done to help children understand the concept being defined and learn its name. In abstracting activities, students also learn names for the mathematical symbols associated with concepts.

Again, consider the example involving fraction concepts. During an abstracting activity, students should learn to associate a name with a fraction and to write the appropriate symbol. You may choose to make three columns as in the following, and ask for 2 of 3 equal pieces, then 1 of 3 equal pieces, then 3 out of 3.

Number of Pieces Asked for	*Total Number of Pieces*	*Symbol*
2	3	2/3
1	3	1/3
3	3	3/3

Through questioning, you can help each child focus on invariant aspects of each example. For instance, in writing the symbol *a*/*b*, *b* is the name of the equal parts into which the whole was partitioned. In this case, *a* represents the number of equal parts asked for. This pattern holds in every case. Hence, when a child sees 3/5 he or she should be able to state that this means that 3 of 5 equal parts were asked for. Next, formally assign a name to the parts of the fraction numeral. In the fraction *a*/*b*, *a* is called the numerator, *b* the denominator, and the slash between them the fraction bar.

Fractions also have other meanings, each of which has to be abstracted. Abstracting concepts in mathematics is not limited to abstracting activities as defined, but the cycle of different types of activities is a useful guide for planning instruction.

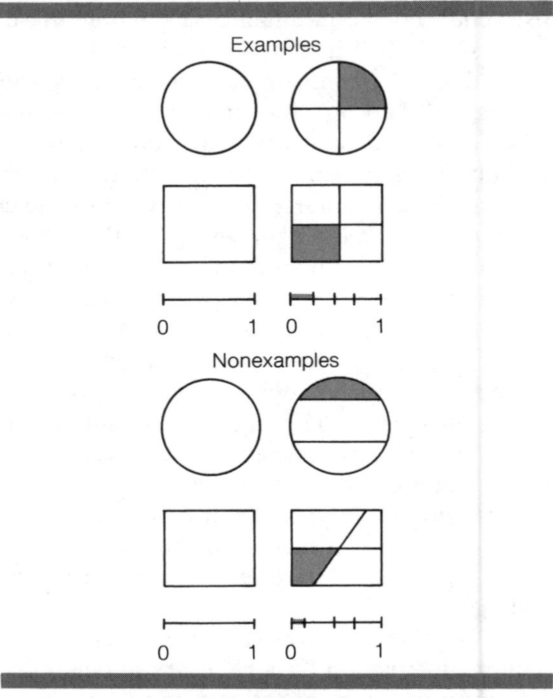

FIGURE 3.3
Examples and Nonexamples of 1/4

Schematizing

A schematizing activity helps each learner focus on existing interrelationships among concepts. Charts, graphs, and tables are often used to display those relationships. General rules are frequently formed from such activities.

Up to this point in the fraction example, children have learned how to write a fraction that tells about part of a unit region. A schematizing activity could highlight a basic relationship involved: although the size or shape of the unit doesn't matter, each part must be equal in area. (At this level, it may be wise to use parts that do have the same shape.) In the activity, each child would continue to write fractions, but many different configurations would be given for units. Figure 3.3 shows some possible configurations. Each child should be led to realize that, when using the "parts of a whole" concept, it is not the number of pieces the unit is cut into that is important; it is the number of pieces having equal area.

Consolidating

Consolidating activities help each child remember already learned concepts and skills until they become accurate and habitual. Games and other practice situations help accomplish this.

The game of Fraction Concentration in Figure 3.4 is an appropriate consolidating activity for the concepts discussed in the illustration. Two sets of cards are used, one showing fractional parts of a unit, and the other, symbols. Fraction Dominoes is another appropriate game to use as a consolidating activity. Following the rules for regular dominoes, each child matches fraction symbols with drawings of fractional parts, as shown in Figure 3.5.

Transferring

In a transferring activity, concepts and processes that have been learned are applied in tasks different from previous tasks. The tasks may involve either horizontal or vertical transfer. Transfer is

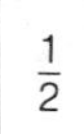

Directions: In this game a match is defined as two cards that name the same fraction: as symbols, as a symbol and an illustration, or as two cards with illustrations. The two cards above illustrate a match.

To play the game, place all cards face down on a table as shown. The first player draws two cards. If they do not match, he returns them to the original position and the next player takes his turn. If the cards do match, he keeps them and draws two new cards. He continues to draw as long as matches are being made. When all cards are matched, each player counts his cards. The player with the greatest number of cards wins the game.

FIGURE 3.4
Fraction Concentration

horizontal when either the social situation or the mathematical tasks involved are at about the same level of difficulty as those used during other types of activity, even though the situation is quite different from those encountered previously. The transfer is *vertical* when the social or mathematical problems are at a higher, more difficult level than those previously used.

A horizontal transfer activity might require a child to consider a set of discrete objects, such as counters or felt pieces, and begin to develop an entirely different meaning for a fraction. A vertical transfer activity might require a child to use a "parts of a whole" meaning to

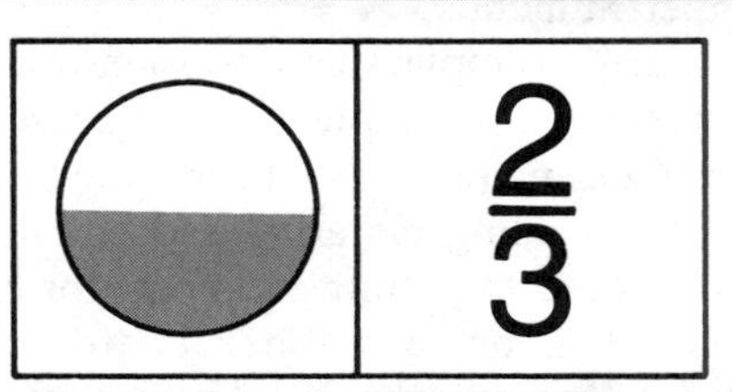

FIGURE 3.5
Fraction Domino

develop an addition algorithm for like fractions, such as 2/3 + 1/3.

Hopefully, from this overview of different types of instructional activities, you can conceptualize the outline of the model. As we plan instruction for different topics in mathematics, the cycle of activity types will be illustrated repeatedly.

Exemplifying Concepts

In using the activity type model, we must consider how concepts are learned, how knowledge of concepts is demonstrated, and how a child interrelates learned concepts in a schema that can then be applied to new learning. Many questions come to mind, such as, How do we exemplify a concept so that a child can identify its defining attributes?

The authors have found it helpful to think of exemplifications as either three-dimensional (3-D) representations, two-dimensional (2-D) representations, or symbols. Models are often three-dimensional representations, whereas pictures and diagrams are two-dimensional representations. Symbols include not only numerals and oral number names, but also descriptions and definitions for the concept.

A nonmathematical example may help clarify the categories. Table 3.1 presents different representations of the concepts "George Washington" and "fire engine."

When we consider a mathematical concept, we can often carry out a similar analysis. The concept "12" is a number idea and, as such, is an abstract idea. However, we can find three-dimensional, two-dimensional, and symbolic representations for 12. Samples are given in Table 3.2.

The authors refer to whatever is used to represent a concept as an *exemplar*. Included are situations, objects, and marks on paper. Thus, we can identify three-dimensional, two-dimensional, and symbolic exemplars.

As teachers, we must choose from each exemplar category those that *correctly* represent the concept. We also need to evaluate exemplars to determine if they are appropriate for particular children.

TABLE 3.1
Exemplification of the Concepts "George Washington" and "Fire Engine"

	Exemplification		
Concept	*3-D Representation*	*2-D Representation*	*Symbol*
George Washington	Statue of George Washington	Pictures of George Washington	The words "George Washington"
Fire Engine	Toy fire engine	Pictures or illustrations of a fire engine	The words "fire engine"

TABLE 3.2
Exemplification of the Concept "12"

	Exemplification		
Concept	*3-D Representation*	*2-D Representation*	*Symbol*
12	Sticks: 1 bundle of ten, 2 loose sticks	Picture of sticks grouped as 1 group of ten and 2 loose sticks	10 + 2, or 12
12	Base 10 blocks: 1 long and 2 units	Drawing of base 10 blocks on chalkboard: 1 long and 2 units	10 + 2, or 12

Physical attributes of three-dimensional representations are of interest, because we want to know what senses will actually be involved. This is important for primarily visual and auditory learners, and for children with similar limitations. Table 3.3 provides a framework for evaluating an exemplar according to sense modality and objectivity[3].

A matrix such as the one in Table 3.3 can be helpful in choosing an exemplar for a specific purpose, for example, in comparing base blocks with a mathematical balance. Both of the latter are useful in a classroom. But what modalities are involved when each is used? A set of base blocks and a mathematical balance are both 3-D visual and 3-D tactile, but the mathematical balance involves more 3-D kinesthetic movement than the base blocks. Using Table 3.3, evaluate a child's workbook, a filmstrip, a calculator, and a set of sandpaper numerals.

Many exemplars can be found in each of the classes; for example, 3-D representations include Cuisenaire rods, bean sticks, pocket charts, chip trading material, and counters. Flat objects that are to be moved about, for example, pieces of felt on a flannel board, are also regarded as three-dimensional representations.

Behavioral Indicators

We cannot see into a child's brain to determine if he has learned, so we look for outward indica-

3. The model as diagrammed is a variation of the system published in A. E. Uprichard, *A Task-Process Integration Model for Diagnosis and Prescription in Mathematics*, 1975, Second National Diagnostic and Remedial Mathematics Conference, Kent, Oh.; and A. E. Uprichard and R. L. Ober, "Functional Analysis of Classroom Task (FACT)," mimeographed (Tampa: University of South Florida, 1971).

TABLE 3.3
Exemplars: Sense Mode X Objectivity

	Objectivity (type)		
Sense Mode	*3-D Representation*	*2-D Representation*	*Symbol*
Visual			
Auditory			
Tactile			
Kinesthetic			
Olfactory			
Taste			

tions by observing what a child does. Before we do that, however, we must know which content objectives the instruction will focus on. We must also have some idea of the specific behaviors that will indicate whether or not a child has learned the content.

The authors refer to these specified behavioral expectations as *behavioral indicators*. Behavioral indicators are statements that describe behaviors that indicate a change in a child's cognitive schema.

Consider the following content objectives related to the concept of twoness.

3.13
Cardinal numbers are whole numbers used to indicate how many.

3.14
A cardinal number is the property common to all sets in a class of equivalent sets.

3.15
Sets such as (#,$), (a,b), (*,X) are representative of the class whose number property is *twoness*.

3.16
The number property of a finite set may be determined by counting.

With this cluster of related content objectives in mind, we can search for behaviors that indicate whether a child is learning the concept of twoness. Because we want the child to exemplify the concept in many ways, we should consider different categories of exemplars.

The three-dimensional representations available for the number two include objects such as counters, balls, and blocks. Among the two-dimensional representations are pictures, diagrams, and sets of two things. Finally, we can consider symbolic representations of twoness, the most obvious of which is the symbol 2.

However, more varied behavioral indicators (BI) can be generated from the translation matrix given in Table 3.4. Behavioral indicators can be written that require a child to consider a particular *stimulus* exemplification, then carry out a response task involving a different exemplification. Consider the following behavioral indicators that are based on the cluster of content objectives 3.13 through 3.16.

BI.1
Given a set of blocks, the child determines a set of two by counting verbally.

BI.2
Given a set of two blocks, the child writes the symbol (digit) for two.

BI.3
Given pictures of animals, the child circles sets of two.

BI.4
Given the symbol 2, the child makes sets of two blocks.

BI.5
Given a set of two blocks, the child draws sets of two on a chalkboard.

These behavioral indicators illustrate specific translations across categories of exemplars as shown in Table 3.4.

Behavioral indicators may involve a translation *within* a class of exemplars. That is, a child may translate to a different exemplar in the same class of exemplars as well as to an exemplar in a different category. Some children will be able to demonstrate all the behaviors we describe, but others may have less understanding and can demonstrate only a few. A child who can illustrate a concept with several exemplars from a given class and can also translate from one class to another has a more adequate understanding of the concept.

Behavioral indicators can also be written for basic learning processes. The following indicators ask a child to interpret, analyze, and solve problems.

BI.6
Analyzes sets of descriptive data to determine mean, median, and mode.

BI.7
Interprets numerical data.

BI.8
Solves word problems using a logical method.

Teaching Moves

Planning is also concerned with exactly *how* teaching is to be organized within an activity. That is, what are we going to do and how are we going to do it? Merrill and Wood refer to this as deciding upon a delivery system.[4] What verbal and nonverbal acts (often called moves) will we make during an instructional activity? What verbal and nonverbal acts will the student make? We need to consider which teacher moves we will use, and how we will sequence them.[5]

Structuring

Structuring moves establish a context for what comes next by beginning or halting interactions

4. M. David Merrill and Norman D. Wood, *Instructional Strategies: A Preliminary Taxonomy*, ERIC Information Analysis Center for Science, Mathematics and Environmental Education (Columbus: Ohio State University, 1974): 4.

5. See Kenneth B. Henderson, "Concepts," *The Teaching of Secondary School Mathematics*, 33rd yearbook (Reston, Va.: National Council of Teachers of Mathematics, 1970): 166–195.

TABLE 3.4
Exemplar Class Translation Matrix

Stimulus Exemplification	*Response Exemplification* 3-D	2-D	Symbolic
3-D		BI-5	BI-1 BI-2
2-D		BI-3	
Symbolic	BI-4		

between teacher and learner and by indicating the nature of the interaction according to time, teaching agent, content topic, cognitive process, rules, or instructional aids. They include decisions about the topics to be taught and the exemplars to be used. The decision to move a group of learners from an initiating to an abstracting activity is also a structuring move. Clearly, structuring moves tend to set parameters for instruction.

Soliciting

Soliciting moves are verbal or nonverbal acts intended to elicit a verbal or a nonverbal response. For instance, when a teacher asks a question, a soliciting move is being used. Gestures, facial expressions, and other nonverbal acts may also be classified as soliciting moves. Examples of verbal soliciting moves include statements, explanations, demonstrations, and illustrations.

Responding

A responding move is a response to a soliciting move. Answering a question may be a response to a verbal solicitation, while curtailing disruptive behavior may be a response to a nonverbal solicitation, possibly a facial expression. Included in soliciting and responding are the moves of *prompting* and *feedback*, which are used to solicit a response through either questioning or nonverbal means. Upon receiving a child's response, the teacher provides some form of feedback to the student.

The pedagogical moves described in this brief overview are illustrated in the activity plans in this book. The moves are often crucial to teaching in general.

Teaching Strategies

A carefully developed sequence of moves is a *teaching strategy*. Careful study of the components of teaching acts can help you develop strategies for specific types of instructional activity and content (e.g., concepts versus principles) and for individual learners.

Some educators take strong positions for and against specific strategies. Two common positions involve *expository* teaching and teaching by *discovery*. Proponents of the discovery method argue that a teacher should lead the child to see the concept or generalization through a combination of structuring, soliciting, discussing, and questioning. The teacher attempts not to tell. On the other hand, proponents of the expository method support a combination of structuring, stating, demonstrating, and illustrating. A child is made aware of the concept or principle being taught through a teacher's lecture followed by examples.

The position you support will depend upon your beliefs about how a child learns and what is learnable. The authors view discovery and expository teaching each as having value with reference to a particular concept or principle. However, the authors favor a discovery-oriented approach in general.

A MODEL FOR PLANNING ASSESSMENT

Determining which behaviors each child can demonstrate and then making inferences about learning is difficult, requiring the use of a variety of assessment procedures. A discussion of an assessment model to help you determine initial strengths and weaknesses of each child and make day-to-day decisions about appropriate instruction follows.

Diagnosis

Diagnosis, a component of the Activity Type Cycle, allows a teacher to infer the presence, correctness, and level of maturity of learned concepts. Diagnosis in some form is necessary prior to, during, and after each instructional activity. One way to proceed is first to determine the broad categories of content known by the child and then how well the content within each of these categories is understood.

Level I Diagnosis

This type of diagnosis is usually made when a change is observed in a child's level of success in mathematics or when there is a pronounced discrepancy in achievement versus aptitude scores. It can also be prompted by psychological test data indicating emotional or social problems. A Level I diagnosis indicates that a child needs a different type of program or teaching procedure. Additional diagnosis is needed to determine the programs or strategies required.

Level II Diagnosis

The purpose of a Level II diagnosis is to provide a more detailed picture of the child's mathematical knowledge. It is impractical to systematically diagnose every single concept; such concepts do not stand alone, but cluster into different but logically related categories. Therefore, *categories* of content in which a child has strengths (i.e., where learnings are present and correct) and weaknesses need to be determined.

One set of categories that can be used is shown in the content taxonomy developed by Wilson, three parts of which are presented in Figure 3.6 (pages 35–36).[6] This particular taxonomy is an outline of the arithmetical component of most elementary school mathematics programs.

One major use of the taxonomy is to categorize the items on a test by keying them to categories in the taxonomy. Thereby we can determine the extent to which all categories are represented on the test, and we can observe strengths and weaknesses in that particular test.

Commercial tests are commonly used in a Level II diagnosis. These are standardized, unlike teacher-made tests, which are usually developed for use in a particular classroom. Frequently used commercial diagnostic tests include the KeyMath Diagnostic Arithmetic Test (grades K–6), the Stanford Diagnostic Mathematics Test (Levels I–IV for grades 1–12), and SRA's "Diagnosis: An Instructional Aid–Mathematics" (Levels A and B for grades 1–8).[7]

The KeyMath test can be used to illustrate procedures involved in a Level II diagnosis. First, each item in the test is categorized according to a content taxonomy, then the items are clustered on a report form as shown in Figure 3.7 (page 37). After the test is administered, correct items are recorded on the report form, creating a profile for the individual child. Content categories of the child's strengths, as well as those in which little appears to be known, will be clearly identified.

The information obtained from a standardized test is unquestionably of value; however, we must be aware that a child's performance is based almost exclusively on symbolic to symbolic translations. The type of qualitative information needed to assess the depth of a child's learning is difficult to obtain from a paper-and-pencil test. Published and teacher-constructed tests cannot measure the full range of exemplifications because they are basically of a written format.

Another concern relates to inferences we make about learning from a norm-referenced score. Instructional groups consisting of children who have comparable scores within a content category are often suggested, but the system of categories used in most commercial tests includes those that are too inclusive to be used effectively in this manner. The score on a subtest is somewhat misleading; we need to know *which* items each child answered correctly.

Because the basic purpose of a Level II diagnosis is to identify content categories of strengths and weaknesses, it is sometimes called an *across-category* diagnosis. On the basis of a Level II diagnosis, we can select categories or major concepts for a more intensive Level III Diagnosis.

Level III Diagnosis

One purpose of a Level III diagnosis is to affirm or modify the inference regarding a child's content strengths and weaknesses obtained from a Level II diagnosis, adding to its specificity. Another purpose is to develop prescriptions regarding exemplars and instructional strategies that might correct the difficulty.

---

6. John W. Wilson, *Some Guides for Elementary School Mathematics* (published privately, 1969).
7. KeyMath Diagnostic Arithmetic Test (Circle Pines, Minn.: American Guidance Service, Inc., 1971); Stanford Diagnostic Mathematics Test (New York: Harcourt Brace Jovanovich, 1976); SRA Diagnosis: An Instructional Aid-Mathematics (Chicago: Science Research Associates, Inc., 1979/80).

III. Sets	*IV. Whole Numbers*	*V. Non-Negative Rationals Fractionals*
A. Concepts & Notation	A. Number & Notation	A. Number & Notation
1. Concepts for	1. Number Concepts	1. Number Concepts
a. Set	a. Meanings & models for	a. Meanings & models for
b. Special kinds	b. Special subsets of	b. Special subsets of
c. Relations	c. Relations among	c. Relations among
	d. Uses of	d. Uses of
2. Notation for	2. Notational Systems for	2. Notational Systems for
	a. Positional	a. Positional
		1. Fraction form
		2. Decimal form
	b. Nonpositional	b. Nonpositional
B. Operations on: Meanings of	B. Operations on: Meanings of	B. Operations on: Meanings of
1. Union	1. Addition	1. Addition
2. Complementation & Comparison	2. Subtraction	2. Subtraction
3. Union & Cartesian	3. Multiplication	3. Multiplication
4. Partition	4. Division	4. Division
5. Intersection		
C. Properties of	C. Properties of	C. Properties of
1. Union	1. Addition	1. Addition
a. Closure	a. Closure	a. Closure
b. Commutative	b. Commutative	b. Commutative
c. Associative	c. Associative	c. Associative
	d. Identity	d. Identity
2. Complementation	2. Subtraction	2. Subtraction
	a. –e.	a. –e.
3. Union & Cartesian	3. Multiplication	3. Multiplication
	a. Closure	a. Closure
	b. Commutative	b. Commutative
	c. Assocative	c. Assocative
	d. Identity	d. Identity
	e. Compensation	e. Inverse
	f. Distributivity	f. Distributivity
4. Partition	4. Division	4. Division
	a. –f.	a. –f.
	D. Computation Procedures for	D. Computational Procedures for
	1. Addition	1. Addition
	a. –(levels)	a. –(levels)
	2. Subtraction	2. Subtraction
	a. –	a. –
	3. Multiplication	3. Multiplication
	a. –	a. –
	4. Division	4. Division
	a. –	a. –

FIGURE 3.6
Content Taxonomy

E. Number sentences with	E. Number sentences with
1. Addition	1. Addition
a. –(types)	a. –(types)
2. Subtraction	2. Subtraction
a. –	a. –
3. Multiplication	3. Multiplication
a. –	a. –
4. Division	4. Division
a. –	a. –
F. Problem Solving with	F. Problem Solving with
1. Addition (types)	1. Addition (types)
a. –d.	a. –d.
2. Subtraction	2. Subtraction
a. –h.	a. –h.
3. Multiplication	3. Multiplication
a. –d.	a. –d.
4. Division	4. Division
a. –e.	a. –e.
G. Measurement	G. Measurement
1. Principles of	1. Principles of
2. Metric Systems	2. Metric Systems
a. Length & Area	a. Length & Area
b. Volume	b. Volume
c. Capacity	c. Capacity
d. Weight	d. Weight
e. Temperature	e. Temperature
3. English systems	3. English system
a. –e.	a. –e.
4. Time	4. Time
5. Other ways numbers are used	5. Other ways numbers are used

FIGURE 3.6
Content Taxonomy (continued)

In order to obtain information needed for a Level III diagnosis, interview protocols for more restricted content categories can be used. These are steps to be followed during an interview, written out for reference; they should focus on a specific taxonomy category. Such protocols are designed to obtain information unobtainable through a written test. A sample protocol for the meaning of addition and subtraction is given in Figure 3.8 (page 38). The use of such protocols is called a *within-category* diagnosis.

The structured interview used in a Level III diagnosis enables a teacher not only to determine the correctness of learning but also to discover misconceptions about concepts, principles, or procedures. A child may miss an entire section on a written test and yet have some knowledge of the categories on the test. The extent of knowledge can often be more adequately determined through an interview. An interview permits a child to respond with exemplars from different classes. Detailed, qualitative data needed to plan for instruction is also gathered.

Other Procedures

Tests and structured interviews are only two methods of assessment. Each method is appropriate for determining the amount and level of a child's knowledge about particular content objec-

III. Sets

A. Concepts and Notation

1. Concepts for
   a. Set–A2, A14 b. c. A7
2. Notation for

B. Operations on–Meanings

1. 2. 3. 4. 5.

C. Properties of–Under

1. 2. 3. 4. 5.

IV. Whole Numbers

A. Number and Notation

1. Number Concepts
   a. A1, A2, A8(9) A9(9), A11(8)
   b.
   c. A5, A10, A11, A12, A15, A18
   d. A13, C11, C13, C16, C17, C18, Sections L, M, N
2. Notational Systems for
   a. Positional Systems–A3, A5, A10, A11, A12, A20, A21, A22, A23
   b. Nonpositional Systems–A16

B. Operations on: Meanings and Symbols

1. Addition–C8, C10
2. Subtraction–C9, C10
3. Multiplication–C10, C12
4. Division

C. Properties

1. Addition
2. Subtraction
3. Multiplication
4. Division

D. Algorithms

1. Addition–D1, D2, D3, D4, D5, D6, D7, D8, D9, D10, D11, H1, H2, H3, H4, H5, H6, H7, H8, H9, H10
2. Subtraction–E1, E2, E3, E4, E5, E6, E7, E8, E9, E10, E11, E12, H4, H5, H6, H7, H8
3. Multiplication–F1, F2, F3, F4, F5, F6, F7, F8, F9, F10, H7, H8, H9, H10
4. Division–G1, G2, G3, G4, G5, G6, G7, G8, G9, H9, H10

E. *Sentence Solving*

1. Addition–I1, I2, I3, I4, I6, I8, I9
2. Subtraction–I3, I4, I5, I7, I8, I9
3. Multiplication–F1, F2, I10, I11
4. Division–I10

F. *Problem Solving*

1. Addition–D1, D2, D3, J1, J2, J3, J5, J9, J13
2. Subtraction–E1, E2, E3, J4, J14, L11, L13
3. Multiplication–J10, J13, J14
4. Division–G1, G2, J5, J8, J13, J14

FIGURE 3.7
KeyMath Test Items Keyed to Content Taxonomy

Category: Meaning of Addition and Subtraction
Exemplar to be used: Math Balance

Planned Procedures:

1. Show the child how the balance works by placing a weight on a peg on one side and asking him where he would place a weight to balance the bar.
2. Place two weights on different pegs on one side, saying, "If these two numbers are each part of another number, can you find the total?" If a verbal response is given, give the child a weight and say, "Show me how you can tell that it is true." Continue the same procedures with other examples of addition.
3. Each time, ask the child to write the addition problem shown by the balance. Ask him or her to identify the numbers that are parts (addends) of the total.
4. Next, place one weight on seven on one side and one weight on four on the other. Say to the child while pointing to the seven peg, "This number is the total, and this number (pointing to the four) is the part. Can you put your weight on the other part to make it balance?" If response is verbal, have the child place the weight on the correct peg to verify the result. Continue with other subtraction problems, stressing part-part-total identification.
5. Ask the child what operation he or she is using to find out where the weight is placed. Ask him to write the problem illustrated by the weights on the balance. Have the child identify the total in the written subtraction problem and then the parts in the same problem. Ask the child to write an addition problem that expresses the same relationship.

FIGURE 3.8
A Sample Structured Interview Protocol for the Meaning of Addition and Subtraction

tives. But instruction and assessment should also focus on objectives other than content. Behavioral indicators can be written for each set of objectives (content, basic cognitive process, and affective).

We can also consider a child's projects and productions and use simulations and role playing, in addition to knowledge displayed during real-life situations. Many different assessment procedures focusing on different objectives must be used. For instance, a child's performance when buying something in a store provides a context for assessment focusing on content objectives, basic process objectives, and affective objectives. The quality of this information will most likely differ from that obtained when a child faces a similar problem on a paper-and-pencil test.

The authors believe that each child learns individually; groups, as such, do not learn. If this is your view, plan assessment procedures that help you evaluate what *each* child knows.

Teaching Whole Number Concepts and Operations

II

4

Developing Early Number Concepts and Notation

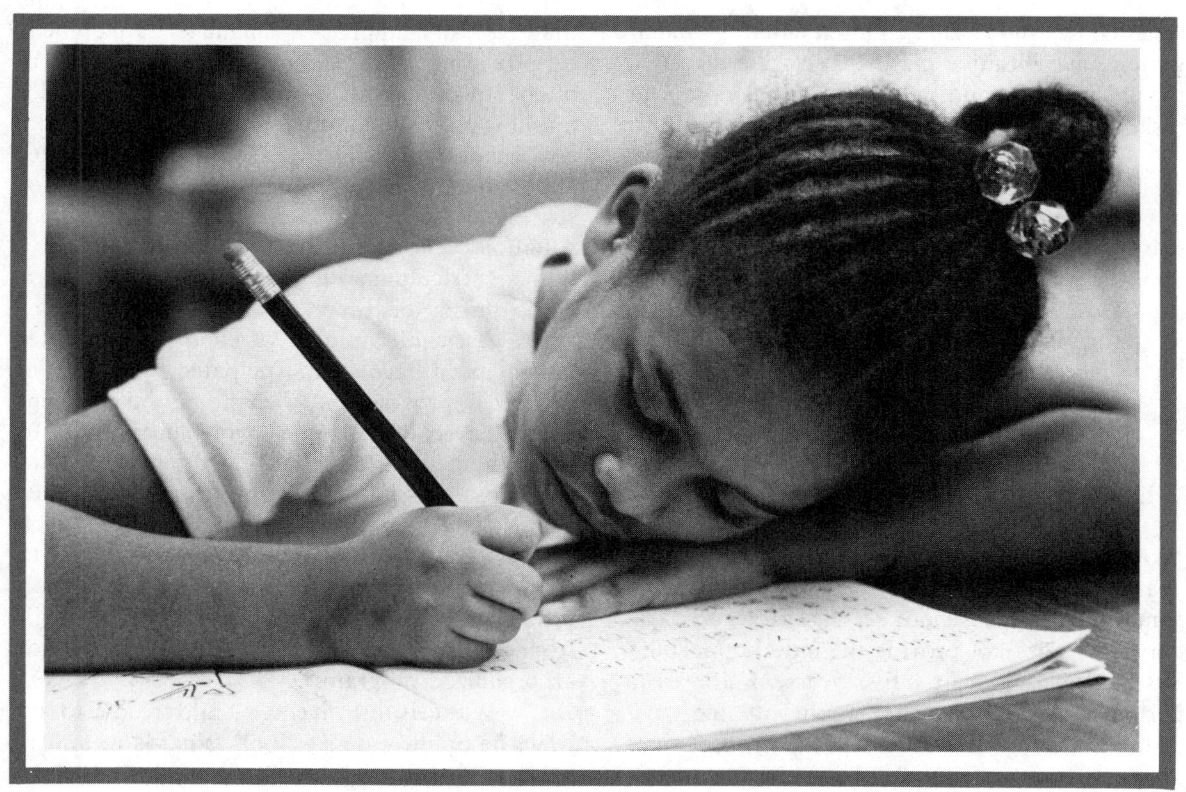

Long before entering school children use mathematical concepts and relationships. Mathematical statements can be heard as they play and talk: "Elaine, are you taller than John?" "I have three balloons: two red and one green." Recent research has shown that children entering kindergarten have intuitive mathematical ideas related to sets, number, and number relationships.[1] Our task is to help them formalize this information.

In order to plan appropriate instruction, we must know what content each child has already mastered, where inadequacies in learning exist, the feelings and attitudes each child has about mathematics, and which cognitive processes are being used. A diagnostic approach is applicable at all levels, kindergarten through high school and beyond.

THE PRESCHOOL CHILD

Most children begin kindergarten at about five years of age; it is often their first experience in a somewhat formal school setting. However, kindergarten does not provide the initial opportunities to learn. The first five years of life are a period of continuous interaction with the environment through play and exploration. Activities are shared with parents, siblings and playmates; often a child explores alone. The child matures greatly during these years in both his physical and mental development. Both positive and negative attitudes are acquired toward adults and other children, toward approaching tasks requiring cooperative efforts, and often toward formal structures such as schools.

A child who comes to school for the first time is already a very complex individual. His mind is not a *tabula rasa,* a blank slate, but instead is filled with ideas generated through continuous interaction with the environment.

Many kindergarten children have already participated in organized preschool programs that typically emphasize language development by providing much verbal interaction. Teachers often stress cognitive development by actually teaching selected concepts and skills from different content areas. The development of attitudes and values is usually emphasized as well, for sharing and cooperation are important in school situations. The fact that there is no national preschool curriculum adds to the diversity of experiences during this time.

A sizable number of children who enter kindergarten have not participated previously in organized programs. A rich home environment helps these children begin school successfully. In such a home a child is challenged to look for relationships, possibly by being asked to place enough chairs at the dinner table so that each person can be seated. The child is expected to share toys and other personal belongings with brothers and sisters. Sadly, some children benefit from neither a stimulating home environment nor an organized program.

What is the average child ready to do when he or she enters school? What is meant by *readiness?* Mowbray and Salisbury define readiness as 'the level of total development that enables a child to learn a behavior, comprehend a concept, or perform in a given way with ease."[2] This implies that readiness occurs on a continuum and is a very personal matter influenced strongly by both biological development and by environmental experiences. Hence, it is unwise to regard *all* five-year-olds as ready to begin rational counting or to learn any specific concept

1. Robert E. Rea and Robert E. Reys, "Mathematical Competencies of Entering Kindergarteners," *The Arithmetic Teacher* 17, no. 4 (January 1970): 65–74.

2. Jean K. Mowbray and Helen Salisbury, *Diagnosing Individual Needs for Early Childhood Education* (Columbus, Oh.: Charles E. Merrill, 1975), p. 83.

or skill. While one might expect certain concepts and skills of a five-year-old, all children do not come to school with the same level of maturity or readiness.

Children possess different levels of readiness in many areas; for mathematics, this includes the ability to process relationships, the range of knowledge, and attitudes towards the subject. The concept of readiness implies optimal time for teaching based on current knowledge, available process skills, or a particular attitude. We must determine what *each* child is ready to learn.

Learning Processes

Young children are naturally curious about their environment. When a child observes objects and events, we can only speculate what is going on in his mind. Certainly he is absorbing much information through the senses. How is this information processed? Is he able to organize this information and then draw inferences from the many observations?

We can begin to answer these questions by considering what is known about how children develop mentally. According to Piaget, children entering kindergarten are typically in the preoperational stage of thought. A child in this stage reasons largely on the basis of how things appear to be. Piaget's classic conservation of number task can be used as an example. In the task, preoperational children believe that the number of counters in a row changes if they are rearranged. Why does a child think this way? The most common explanation of theorists and researchers is that the reasoning is bound to perceptual input, and is based upon how long or how short each row of counters appears to be. However, the child does not consider *both* the length and density of each row. The preoperational child is using the cognitive processes of attending and comparing. Once the attribute is identified (length in this case), comparisons are made. Mathematical relations such as "longer than," "shorter than," "more than," and "less than" are also within the cognitive capability of a preoperational child.

The reasoning displayed in this task is characteristic of the thinking of a child entering school; certain basic processes must be used if concepts are to be formed. The child begins to notice various attributes of objects such as color, size, or length, or she may observe that event A always occurs before event B. As basic attributes are identified, she begins to discriminate among objects by noting similarities and differences. Some objects are red, some blue, some brown, and so on.

The necessary prerequisites for classifying behavior are now present. The child can form groups of objects using the rules "all red," "all square," "all contain five objects." From such beginnings, he or she goes on to reflect upon what has occurred and how the groups were formed. He may wonder if other attributes could be found and forms hypotheses such as, "There are more red things than blue things." Each hypothesis is tested, possibly by matching directly or by counting.

In a further attempt to make sense out of the world, the child begins to notice directly related events; for example, flip a switch and a light comes on or goes off. Reasoning about such cause and effect relationships is basic to mathematical performance. The cognitive process skills developed before and during kindergarten are used over and over again in life, providing a basis for mathematical problem-solving abilities.

As children move through elementary school, they begin to display different learning and cognitive styles. Some children can reason best by being given the individual parts and asked to form a generalization, while others prefer reasoning from the total structure to the individual parts. By identifying different learning styles, we can help insure that each child is given maximum learning opportunities. Children also differ in cognitive tempo, that is, rate and ease of responding. Because a child takes a few seconds longer to process information does not necessarily mean that he or she has a learning disability. Each child should be given enough time to use his or her cognitive processes.

Personal Feelings and Values

Along with basic ways of processing information, children involuntarily bring their individual attitudes, feelings, and recognizable values to school. A value is the amount of importance or usefulness one places on an idea, event, action, or object. Feelings and values are formed during everyday experiences and are influenced by the values and customs of family and society.

Children sometimes come to school with negative attitudes about teachers or other children. They may even have negative attitudes about a school subject such as mathematics. However, attitudes are learned, and we can help change them. Positive feelings towards school, other children, and mathematics can be developed if we plan the right kinds of experiences for each individual. A child who enjoys success with mathematics activities she understands and finds interesting will develop a positive attitude. She will come to value mathematics if it can be used to solve relevant, everyday problems.

Mathematical Knowledge and Skills

Children entering kindergarten have considerable intuitive mathematical knowledge. Numerous studies have shown that five-year-olds can usually count by rote (recite number names) to 20 and can recognize digits and several two-digit numerals. Many preschool children associate their ages with a set of fingers. By the time they come to kindergarten they can often show a set of 1, 2, 3, 4, or 5. Frequently, five-year-olds can name geometric figures such as squares, circles, and triangles and, while they have not formally used a ruler, they can speak of one object being longer or shorter than another. Such children can compare objects and use terms such as "larger than" and "more than." The preoperational child uses language freely, although his logic is somewhat limited.

Thus, a child brings to school a mixture of formal and intuitive mathematical knowledge. He also brings limited skill in organizing and generalizing this knowledge.

PLANNING THE CONTENT

Because there is much mathematics that *can* be taught to children in the kindergarten and primary grades, deciding what *ought* to be taught is a major task of each teacher.

Certain mathematical relationships and concepts are basic to kindergarten programs. Content objectives for many of these are listed in this chapter.

Attribute Discrimination

Objects found in a child's environment possess specific characteristics or attributes. From a very early age, children observe attributes such as size ("My book is bigger than yours"), length ("I want the longest stick"), color ("My coat is blue"), and, for small sets, numerousness ("You have more pieces of candy than I have"). Normally, children are involved in programs before and during kindergarten that emphasize attributes of objects found in their environment and relationships among the attributes.

Some commercial mathematics programs stress attribute identification in kindergarten. Developing Mathematical Processes (DMP) is one such program. Throughout the DMP kindergarten program, the attributes of length, shape, numerousness, and weight are emphasized.

Activities that have children identify the attributes of objects are very important. Early work with attributes should center around easily perceptible characteristics such as color, size, shape, and length. Skill in identifying attributes is necessary in order for a child to classify or sort objects into distinct groups and to distinguish between and among objects. A child is thus helped to organize her environment. Attribute identification and discrimination are basic to concept formation.

Introducing Sets

The child who can identify attributes of objects is ready to group objects into sets or collections based on those attributes. The mathematics to be learned is stated in these content objectives.

4.1
A set is any well-defined collection of real or representative objects or events.

4.2
The objects included in the set are called the elements or members of the set.

4.3
A set may be defined by stating the criterion for membership, for instance, stating the property common to all members.

4.4
The members of the set may or may not have common attributes.

The idea of sets and the mathematics related to a study of set theory became accepted in the elementary school curriculum during the 1960s. The concept of a set is very important for young children because not only does it come naturally and somewhat intuitively but ideas based on sets can be built upon and extended in later years.

Also, remember that *content objectives* presented in the chapters are statements of mathematical content written *for us as teachers.* Although ideas in them are desired learnings, we should not expect a child to state the content as it is set forth in the formal objective. We could expect, however, a child to indicate some knowledge of each idea through varied behaviors.

To construct a set, a child must determine a rule that will clearly identify the objects that do and do not belong in a set. Most sets that young children make are based on a set of common attributes, such as all red triangles, brown bears. As children become more skilled in forming sets and verbalizing reasons why objects are grouped together, they begin to group them on characteristics such as function or location in space. At that point, the objects themselves may not have inherent common attributes.

Consider the following two examples of sets based on function and location in space. In the first example, the silverware constitutes a set because the knives, forks, and spoons are all things you eat with. The second set is described as a set of things on the table.

Set Of Things You Eat With

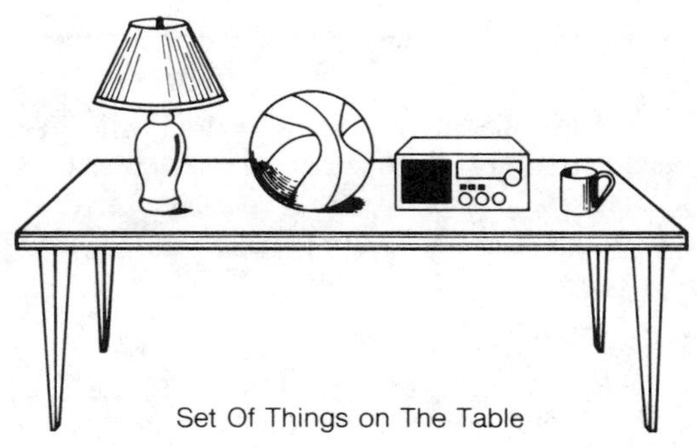
Set Of Things on The Table

4.5
A set with no members is called the *empty* or *null set.*

Teachers often pose statements such as "the set of red elephants" or "the set of five-year-old professional basketball players" to help children develop an intuitive idea of the empty set.

4.6
The union of two sets, A and B, is the set C, such that C contains all elements belonging to A or to B (or to both).

Many operations can be formally defined on sets; among them are *set union* and *set intersection.* Set union is a commonly used model for whole number addition. In the application of this model to addition, the sets must be disjoint, that is, have no common elements. When the sets are disjoint, the number property of the union set is equal to the sum of the number properties of each individual set. A set intersection consists of the common elements of two sets.

Comparing Sets

Sets may be compared on the basis of numerosity, that is, by comparing the number of members in each set. This activity is basic for beginning number concepts.

4.7
If every member of set A can be paired with a distinct member of set B such that each member of set B is also paired with a distinct member of A, sets A and B are in one-to-one correspondence.

4.8
The set that has unmatched elements remaining after the pairing of its elements to the elements of a second set is said to have *more* elements than the second set.

4.9
The set that has a deficiency of elements after the pairing of its elements with the elements of a second set is said to have *fewer* elements than the second set.

4.10
Sets that can be placed in one-to-one correspondence are *equivalent*.

Children regularly encounter situations in which sets are compared. The intuitive notions of one-to-one correspondence, "more than" and "fewer than," must be developed through activities that provide an opportunity for children to actually compare the elements in two sets.

The ideas stated in content objectives 4.7 through 4.10 can be shown as follows:

PROBLEM

Set A: A Set of Cups   Set B: A Set of Glasses

Draw a line from a cup to a glass. Connect only one cup to an individual glass. Is each cup paired with a glass? Is each glass paired with a cup? Since no cups or glasses are unmatched, sets A and B are in one-to-one correspondence and are said to be equivalent.

Consider another example.

PROBLEM

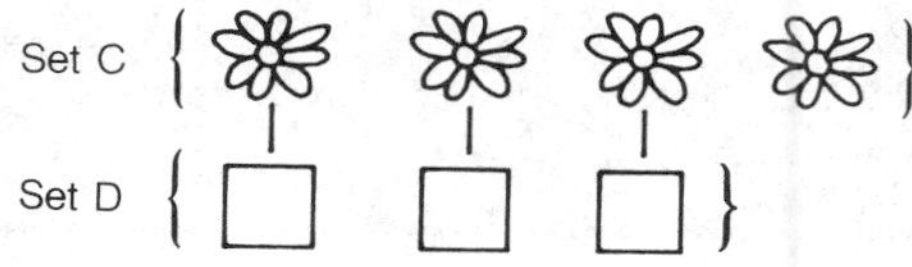

Draw a line from a flower to a square. Continue drawing lines until you have made all possible pairs. Is each square matched with one flower? Yes. Is each flower matched with one square? No. Can the sets be placed in one-to-one correspondence? No, there is not a distinct square for each distinct flower. Thus, set C has more elements than set D, and set D has fewer elements than set C.

The number of elements in set C is *greater than* the number of elements in set D; therefore, $4 > 3$. Similarly, the number of elements in set D is *less than* the number of elements in set C; therefore, $3 < 4$.

Set Relations

Both equivalence and order relations are encountered in prenumber activities.

4.11
A relation is a set of ordered pairs.

4.12
Relations that are reflexive, symmetric, and transitive are *equivalence* relations.

4.13
Relations that are nonreflexive, nonsymmetric, and transitive are *order* relations.

Equivalence relations are used in classifying and in forming groups or sets, often including as many as (numerousness), as long as (length), as big as (size), and same color as (color). Whereas equivalence relations allow a child to classify based on similarities, order relations allow a child to order objects or events based on differences. Common order relations include "greater than, less than" (numerousness), "longer than, shorter than" (length), and "larger than/ smaller than" (size). Situations that require use of order relations allow children to compare objects on the basis of their attributes.

Counting

Several concepts are involved in counting; it forms the basis for arithmetic learning in young children. For example, as children form equivalent sets, the emphasis should be on how many elements are in each set and on the realization that each set has precisely the same number of elements. This is true even when a child cannot count to determine the number. A major step toward developing a concept of number has been taken whenever a child can view each element in a set simply as something to be counted and disregard other nonrelevant attributes such as size or color. When a child can view elements in this way, two sets having four elements each but of different sizes, as in the following illustration, would be counted easily and determined to be equivalent.

> **4.14**
> A *cardinal* number is the property common to all sets in a class of equivalent sets.

Numbers are used in many different ways. When a person uses a number to tell the manyness of a set, the number is used in a *cardinal* sense. When numbers are used to determine relative position of elements within a set, such as first, second, third, or number 1, number 2, number 3, this is an *ordinal* use of number. Sometimes numbers are used in a *nominal* sense simply for identification purposes, such as numbers assigned to players on a basketball team.

> **4.15**
> Sets such as {1}, {1,2}, {1,2,3}, . . . are called counting sets.
>
> **4.16**
> The last number of a counting set is the number property of that set.
>
> **4.17**
> Counting is the process of matching one-to-one the members of a set with the members of one of the counting sets.

Counting has at least two distinct manifestations: rote and rational. Each is necessary as a child begins to find the numerosity of a set.

Rote counting refers to the ability of a child to recite number names in order—1, 2, 3, 4—without relating these names to sets having with the same number of elements. Rote counters can sometimes give the number names up to 100, but know little about associating the number name with a set of objects.

A *rational* counter is able to associate a set with a number name. For instance, the child can show a set having three elements when asked to do so. A rational counter can also determine the numerosity of a set by counting, and can state the number property of the set.

Notation

> **4.18**
> *Numerals* are symbols or sets of symbols used to name or represent numbers.
>
> **4.19**
> A *digit* is a numeral using only one symbol.

A number is an abstract idea, so we use symbols called *numerals* to represent numbers. Many different numerals could be used; for instance, the number four could be represented by 4, IV, quatre, or IIII. Children should be made aware of the number-numeral distinction, but the teacher need not make it a major issue.

A *digit* is a numeral using a single symbol. A multidigit numeral uses more than one symbol, such as 387, which contains the digits 3, 8, and 7.

4.20
Each digit is assigned to or names a unique whole number.

4.21
Each digit has a unique shape.

4.22
The number to which a digit is assigned is called the *face value* of the digit, or the *digit value.*

4.23
The set of digits used in the Hindu-Arabic numeration system is {0,1,2,3,4,5,6,7,8,9}.

4.24
The Hindu-Arabic digits are assigned to the set of whole numbers 0 through 9.

The number property of a set is determined by counting the elements in the set. For sets having 0 through 9 elements, that number property can be named by using one of the 10 digits of the Hindu-Arabic numeration system. Each digit is associated with the numerosity of a set, and this numerosity determines the *face value* of that digit. For instance, the digit 6 has a face value of *six* and names a set having 6 elements. A numerosity greater than 9 can be represented by multidigit numerals.

Ordering the Content

The relative emphasis to be given to a content objective and the order in which objectives are taught can be determined many ways. A purely mathematical analysis of the content can be made, with prerequisite content identified. This method is often followed by textbooks and curriculum guides. Kindergarten materials place heavy emphasis on making sets, doing one-to-one correspondence tasks, and writing numerals to represent the numerosity of sets. This approach is based on the logical relationships that exist among mathematical concepts. Figure 4.1 depicts a content ordering based on mathematical analysis.

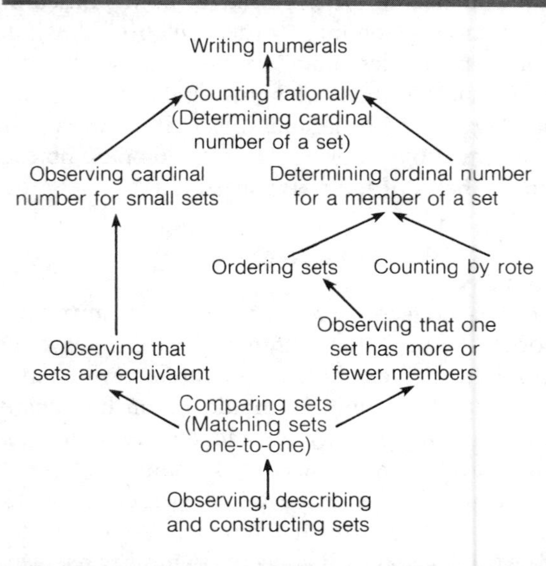

Figure 4.1
A Content Ordering Based on Mathematical Analysis

Another approach is to consider what developmental theory says about how children learn and, in particular, how very young children develop a concept of number. Piaget suggests that number is a synthesis of class and order; that is, the concept of number develops through operations involving both classification (using equivalence relations) and seriation (using order relations).[3] The content stated in content objective 4.16 is attainable only to a child who is able to interrelate both the cardinal (an outgrowth of the classification activities) and the ordinal (an outgrowth of the seriation activities) aspects of

3. Jean Piaget, *The Child's Conception of Number* (New York: Humanities Press, 1952), pp. vii–ix.

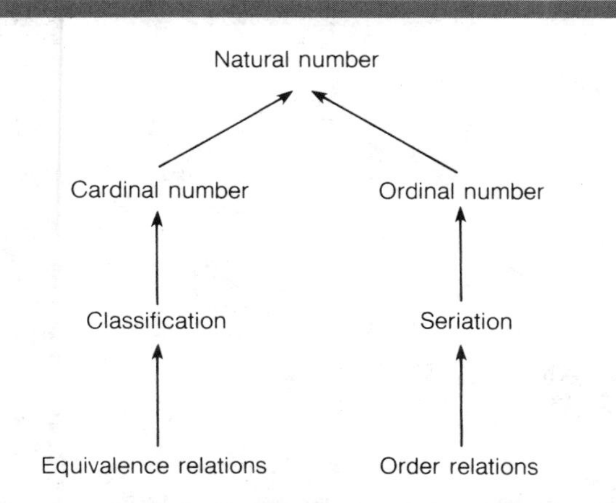

Figure 4.2
A Content Ordering Based on Piaget's Theory

number. An ordering of content based on Piaget's theory appears in Fig. 4.2.

In contrast, Charles Brainerd suggests that young children naturally use ordinal processes, such as ordering blocks or toys, and a beginning number program should build from this naturally occurring strength.[4] Brainerd advocates a lot of counting and less emphasis on sets per se. His research indicates that ordinal number concepts and facility with natural numbers develops in children before they develop a real concept of cardinal number. As yet his research findings have not been applied in a large-scale curriculum project. However, a similar counting approach has been used traditionally by many parents and teachers.[5]

PLANNING INSTRUCTION

In planning number activities for children ages 4–7, keep these two ideas in mind:

4. Charles J. Brainerd, "Mathematical and Behavioral Foundations of Number," *The Journal of Genetic Psychology* 88 (1973): 221–81.

5. For a discussion of alternate orderings, see Chapter 5 in Jon Englehardt, Robert Ashlock, and James Wiebe, *Helping Children Understand and Use Numerals* (Boston: Allyn and Bacon, 1983).

1. Research has shown that children before the age of 7 have somewhat fuzzy ideas about number and number relations, and about the invariant properties of number. Usually they cannot conserve these relations.
2. Number ideas develop slowly. They grow from an experience base; that is, through physical and social experiences each child moves from an intuitive to a more formal meaning of number.[6]

These statements are based on the developmental theory of Jean Piaget, and on the research from his theory. It suggests that, in general, the child before 7 is developing and constantly changing ideas about seemingly simple mathematical relationships. A child constructs mathematical ideas for herself, from experiences, by reflecting upon the contents of those experiences. Such an assumption is basic to a developmental approach to learning and requires planning an early number program that allows each child to use manipulatives and discover relationships. Children also need help in clearly describing the relationships they observe.

The information we gather through diagnostic activities provides a basis for planning instruction. The Activity Type Cycle described in Chapter 3 is one guide. It suggests that we must decide if the appropriate activity is an initiating, abstracting, schematizing, consolidating, or transferring activity. We also need to list behaviors which show if each child is learning (or has learned) the stated content. Exemplars should also be selected.

Discriminating Attributes of Sets

Consider the set of objects shown in Figure 4.3. Children can be asked to describe the objects shown. Help them to focus on attributes such as color, size, shape, and length, as shown in Table 4.1.

6. Barbel Inhelder and Jean Piaget, *The Early Growth of Logic in the Child* (London: Routledge and Kegan Paul, 1964).

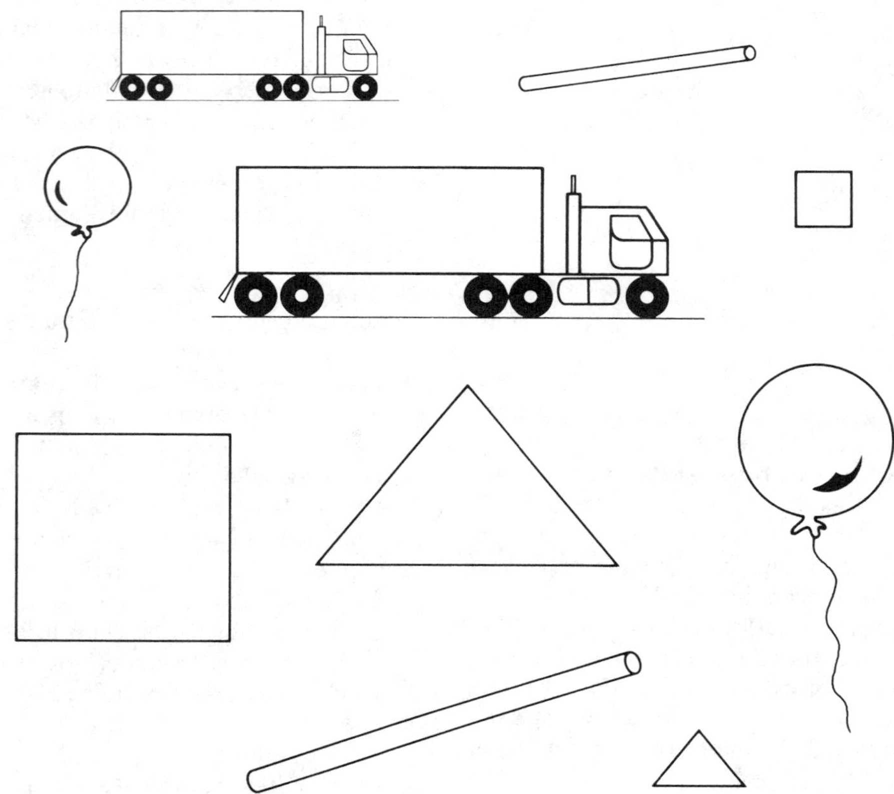

Pointing to the trucks, say, "How are these trucks alike and how are they different?" Continue with other objects, allowing the child an opportunity to discuss as many attributes as possible.

Figure 4.3
Attribute Discrimination

Other objects can be found to highlight attributes such as texture (rough or smooth), weight, numerousness (when the child begins to count), area, volume, capacity, and time. Materials from the child's environment can include fruits, animals, clothing, toys, and games. Collections can be made of objects with a given attribute. Useful commercial materials are also available.

Table 4.1
Attributes of Objects Shown in Figure 4.3

Color	*Shape*	*Size*	*Length*
Red	Triangle	Large	Long
Yellow	Circle	Small	Short
Green	Rectangle		
Orange			
Brown			
Blue			

Instruction will probably begin with an initiating activity, in which each child is asked to use knowledge of lower-level concepts to generate instances of a desired higher-level concept. During activities that focus on objectives 4.1, 4.2, and 4.3, children will use their knowledge of the attributes of familiar objects to form groups that are defined by those attributes; they will also tell why the objects are so arranged.

One favorite exemplar for this activity is a set of three-dimensional attribute blocks. These are available commercially and contain blocks of different shapes, colors, sizes, and thicknesses. Figure 4.4 illustrates a set of attribute blocks.

What behaviors might we look for as indicators that each child is learning as desired? Appropriate behavioral indicators include:

Given a set of attribute blocks, each child will:

- Identify and name the various attributes of each block.
- Sort the blocks on the basis of an attribute.
- For a specified block, name different sets to which it belongs.

A sample activity plan is presented in Figure 4.5. Use it to identify each part of the planning process.

A *variety* of exemplars need to be used—three-dimensional, two-dimensional, and symbolic. In the activity plan in Figure 4.6, the objectives are the same as those listed for the activity in Figure 4.5; however, a two-dimensional exemplar is used.

In addition to forming sets from single attributes, activities should be presented that require the child to consider two attributes simultaneously. A matrix of attribute blocks similar to the example described in Figure 4.7 (page 54) can be presented. In this example, a child must discover the classification rule for the row ("same color as") and for the column ("same shape as") in order to find the correct block for the center cell.

Comparing Sets

When children learn to make and describe sets, they are ready to compare them on the attribute of numerousness. From concepts of set equivalence and one-to-one correspondence, children develop the number ideas of "equals", "more than", and "less than".

But how do we begin to develop the ideas of equivalence and one-to-one correspondence? First experiences with one-to-one correspondence should involve familiar objects. Have very young children arrange a table for a party. Other situations include worksheet activities such as the one

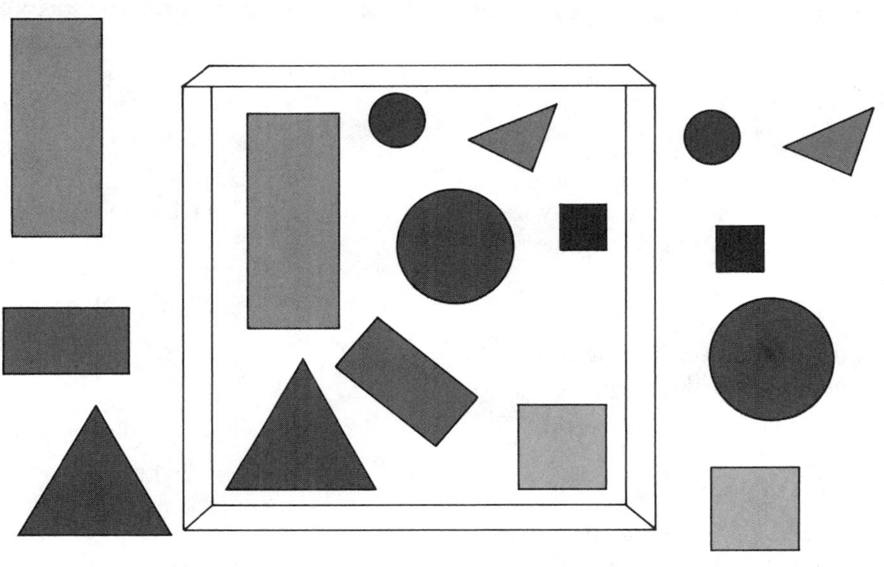

Figure 4.4
Attribute Blocks

ACTIVITY PLAN

Sets: Meanings of and concepts for — ***3-D/Rep→ Symbolic*** — ***Initiating***

Content objectives

4.1, 4.2, 4.3

Exemplars/Materials

Attribute blocks

Behavioral indicators

1. Given a set of attribute blocks, the child will:
   a. Identify the various attributes of each block.
   b. Describe the attributes of each block.
   c. Sort the blocks on the basis of an attribute.
2. Given a set rule, the child will verbally name all members of that set.
3. Given attribute blocks already grouped, the child will determine the set rule.

Procedure

1. Place the attribute blocks on a table or on the floor so all children can reach them.
2. Say, "Look at these blocks and tell me all you can about them."
3. Encourage children to talk about the blocks. Choose particular blocks and encourage discussion on that block; for instance, "It is red, it is big, it is a circle." Then, "It is a big, red, circle."
4. Choose one attribute, such as color, and say, "Let us form a group of red blocks, a group of blue blocks, and a group of yellow blocks." Discuss the blocks which belong to each group.
5. Continue with this activity until each child can name the blocks and display behavior as listed in the behavioral indicators.

Figure 4.5
Sample Plan for an Initiating Activity

illustrated in Figure 4.8. Figure 4.9 contains an abbreviated initiating/abstracting activity for the concept of equivalent sets.

Counting

Both rote counting and rational counting are necessary if the child is to make progress in early number work.

Rote Counting

Teachers have found that rote counting can be taught through nursery rhymes and songs. A sample poem follows.

One, two
    buckle my shoe

Three, four
    close the door

Five, six
    pick-up sticks

Seven, eight
    lay them straight

Nine, ten
    big fat hen.

ACTIVITY PLAN
2-D/Rep→ Symbolic

Sets: Meanings of and concepts for

Initiating

Content objectives

4.1, 4.2, 4.3

Exemplars/Materials

Cards with pictures of animals drawn on them; one set of ten for each child

Behavorial indicators

Given the picture cards, the child will:

1. Describe attributes of each animal.
2. Sort the cards on the basis of an attribute
3. State his or her rule for grouping.

Procedure

1. Give each child a set of picture cards. Ask children to describe the animals on each card.
2. After children have described the animals, ask them to sort the cards so that all animals of the same color are grouped together.
3. Ask children to group the cards together a different way, then have them describe why they grouped them that way.
4. Give each child a chance to explain her grouping.

Figure 4.6
Sample Plan for an Initiating Activity

Rational Counting

Children need *many* activities to develop skill in rational counting. Figure 4.10 (page 56) illustrates an activity in which the child begins to make one-to-one correspondences between number names and members of a set. Examples of the last number named as the number for the set are also encountered.

Red
Red
Red
Yellow
What goes here?
Blue

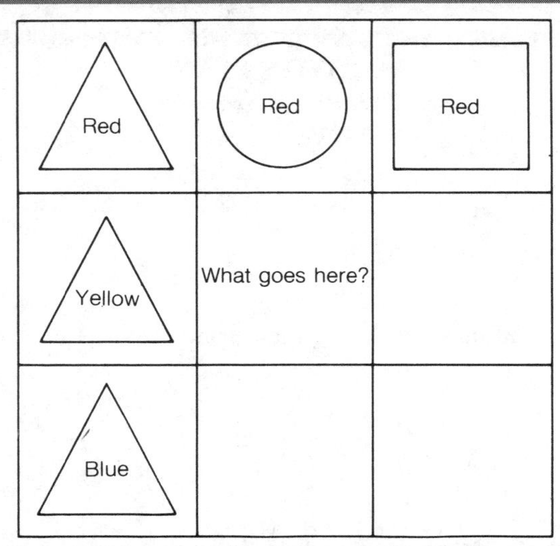

Figure 4.7
Double Classification Matrix

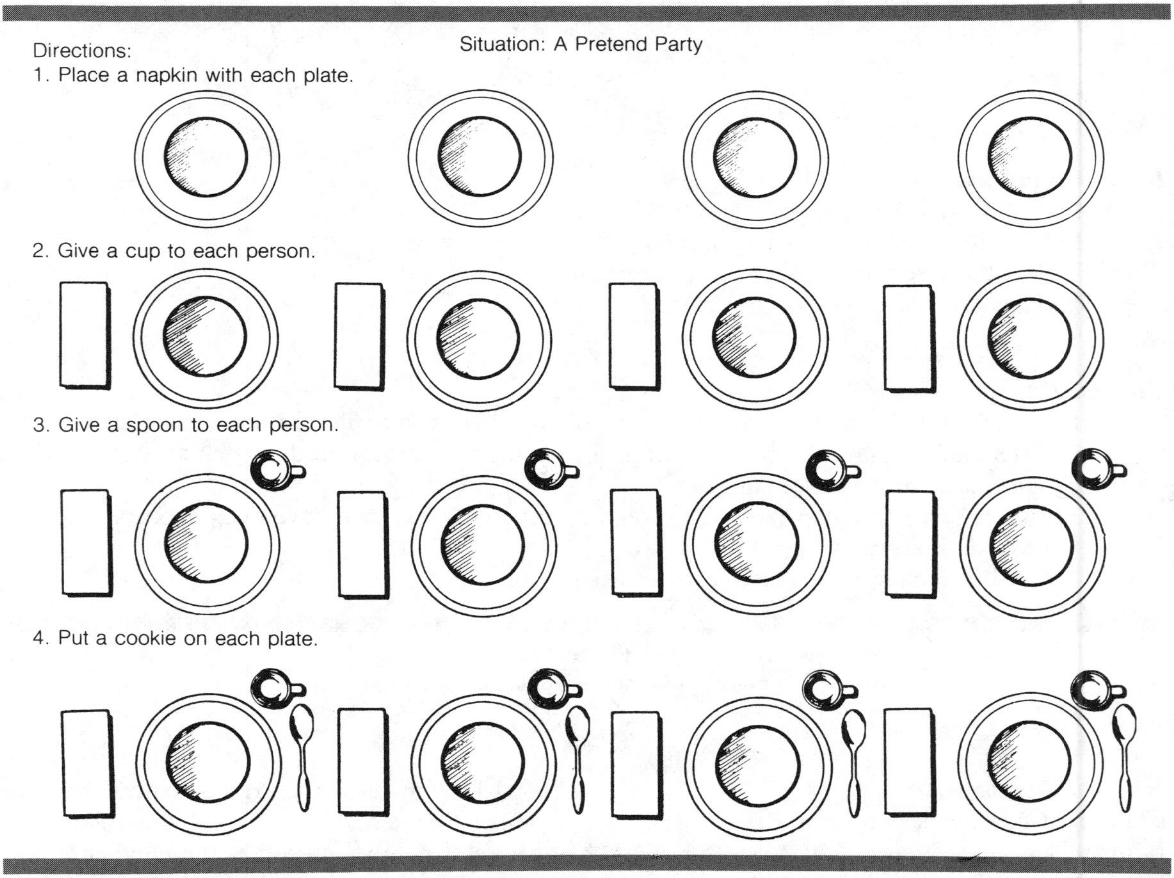

Directions:
Situation: A Pretend Party
1. Place a napkin with each plate.
2. Give a cup to each person.
3. Give a spoon to each person.
4. Put a cookie on each plate.

Figure 4.8
Sample One-To-One Correspondence Worksheet Activity

ACTIVITY PLAN

Sets: Relationships among | *3-D/Rep → Symbolic* | *Initiating/Abstracting*

Content objectives

4.10

Exemplars/Materials

Felt board for the teacher
Felt cutouts of animals: 4 chickens, 4 cows, 4 dogs
Felt cutouts of animal houses: 4 barns, 4 chicken coops, 4 dog houses
Pieces of yarn

Behavioral indicators

Given sets of animals and animal houses, the children can:

1. Match an animal with its house.
2. Determine if sets are in one-to-one correspondence.
3. Determine if sets are equivalent.

Procedure

1. Place a set of felt animals on the felt board. Discuss them.
2. Place the set of animal houses on the felt board. Discuss.
3. Now place only the set of dogs on the board. Also place the set of doghouses on the board. It should look as follows:

4. Ask, "Does each dog have a house?" Have a child take pieces of yarn and connect each dog with its house.
5. Discuss and say, "Each dog has a house. The sets are matched one-to-one. The sets are equivalent."
6. Continue with the other sets of material. Each time, help the child focus on the one-to-one relationship.

Figure 4.9
Sample Initiating/Abstracting Activity for Equivalent Sets

ACTIVITY PLAN

Counting: Rational Counting	*3-D/2-D Rep→Symbolic*	*Initiating*

Content objectives

4.16, 4.17

Exemplars/Materials

Sets of counting blocks, checkers, or other discrete counting material
Table, around which children and teacher are seated

Behavioral indicators

1. Given a set of counters, the child will count the set and express the count verbally.
2. Given a verbal count, each child will construct a set having that numerosity.

Procedure

1. Give a set of counters to each child. Allow a few moments for the children to become familiar with the counters.
2. Place a counter on the table. Have each child place a counter on the table each time you place one and ask the children to count verbally with you.
3. As you place the counter, say, "One." All children should say, "One." Place another counter, say, "Two." Each child should have placed another counter and should say, "Two."
4. Continue counting together up to 10, then say, "We have ten."
5. Make a set of 5. Have the children make a set that is equivalent. Then, by pointing to each counter, determine how many counters are in the set by pointing and saying, "One, two, three, four, five. There are five."
6. Call out the number three. Ask children to make a set having 3 members.
7. Continue with activities as above.

Figure 4.10
Sample Initiating Activity for Beginning Counting

Commercial materials are also available. For example, Stern's pattern boards and adaptations are often used in early counting work as in Figure 4.11. Children can count the holes in each board; they can also fill each hole with a block, counting as each block is put in. When children are ready to count with two-dimensional materials, they will enjoy counting petals in the book *I Can Count the Petals of a Flower.*[7]

Figures 4.12 and 4.13 illustrate abstracting/schematizing and consolidating activities for rational counting. Note that the content objectives are the same as in Figure 4.10.

7. John Wahl and Stacey Wahl, *I Can Count the Petals of a Flower* (Reston, Va.: National Council of Teachers of Mathematics, 1976).

In addition to naming how many members comprise a set, a child should be able to use numbers to determine the relative positions of elements in a set; that is, to make ordinal use of number. Children refer to elements as first, second, third, . . . at an early age; however, few beginning first graders can count using these terms beyond fifth. As children become older, they also use phrases like "number 1, number 2, number 3, . . ." in an ordinal way. Dot-to-dot activities are especially useful in teaching this concept.

Notation

Provide a range of experiences with the digits before asking children to form them with paper

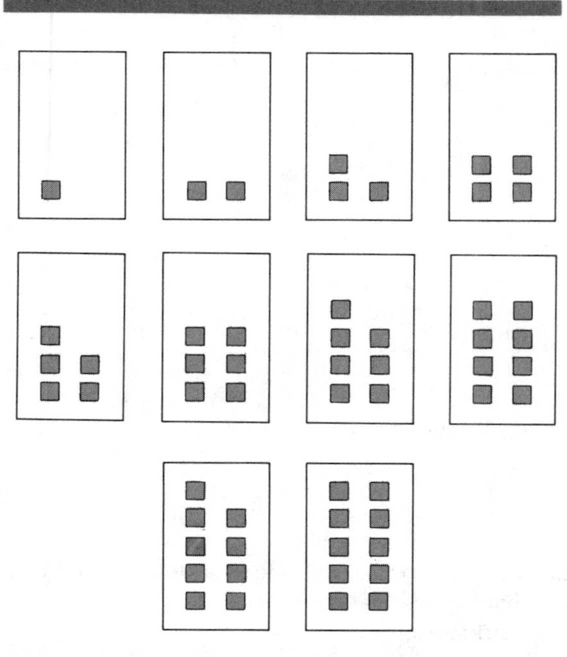

Figure 4.11
Pattern Boards from the Stern Materials

and pencil. Cardboard numerals covered with sandpaper allow both a visual and haptic/tactile modality in perceiving the symbols. Children can also make the symbols out of clay, toothpicks, or yarn as they learn more about the unique shape of each digit. A set of yarn digits is shown in Figure 4.14 (page 60).

After much experience identifying each digit and forming the symbol with materials, children can begin writing the digits with pencil and paper. A favorite method is to copy from a model as illustrated in Figure 4.15 (page 60).

PLANNING ASSESSMENT

Most of what has been said in relation to a child coming to school for the first time is also true of students at all grade levels. As they advance, children come with more and more knowledge in many mathematical areas: number concepts and relationships, geometry and measurement, probability, statistics, logic, and so on. The level of development will vary for each area as the child matures mentally. Therefore, at every level, motor behaviors, attitudes, cognitive processes, and mathematical knowledge must be determined. As we plan instruction that will build on each child's strengths, we must include formal and informal assessment procedures in our programs.

At higher grade levels, inferences regarding a child's reasoning ability can be made from formal tests such as the Cognitive Abilities Test.[8] However, with young children, inferences about cognitive processes can be made only from observations gathered through individual diagnostic procedures such as interviews. This way, a child can explain how he is coming to the solution, and we can determine how he is approaching the task. We can present a variety of situations, tasks that involve many sense modalities for both input and output, and gain insight into a child's cognitive style and preferred input mode. For example, we may find that some children are visual learners while others are oral learners. Or, we may find a child who is unable to process information due to some handicap. Many children arrive at incorrect answers because they learned faulty rules for processing information; they are paying attention to irrelevant attributes. A thorough assessment program will help us identify each child's level of maturity in the cognitive processes.

To assess a child's attitudes and values regarding mathematics, we need to determine how she feels about studying it. What value does she place on it? Formal questionnaires can help gauge a person's feelings, but well-placed questions in a face-to-face situation can determine *why* the child has positive or negative feelings about mathematics. We need to discuss the individual's personal and career goals which interact with the values a child assigns to mathematics.

To assess mathematical knowledge and skills, determine the bits of mathematics content that each child knows and how well this content is understood. Various formal tests, including teacher-made ones, can often be used to identify categories of content in which a special strength or weakness is displayed. Insight into how well mathematics is understood can frequently be

8. Robert L. Thorndike, Elizabeth Hagen, and Irving Lorge, *Cognitive Abilities Test* (Iowa City, Iowa: Test Department, Houghton Mifflin, 1954–74).

ACTIVITY PLAN

Counting: Rational counting | *2-D/Rep→ Symbolic* | *Abstracting/Schematizing*

Content objectives
4.16, 4.17

Exemplars/Materials
Set cards; cards with pictures of stars, dots, etc.
Set of numeral cards with numerals 0–10.

Behavioral indicators
Given the set cards, the child will:
1. Count the elements on the cards.
2. Identify the numeral for the number of the set counted.
3. Place the set cards in order from 0–10.
4. Order the numeral cards from 0–10.

Procedure
1. Give each child a collection of set cards and ask each to proceed one at a time to count the pictures on the cards, then find the numeral that names that amount.
2. Place the correct numeral card with each set card as follows:

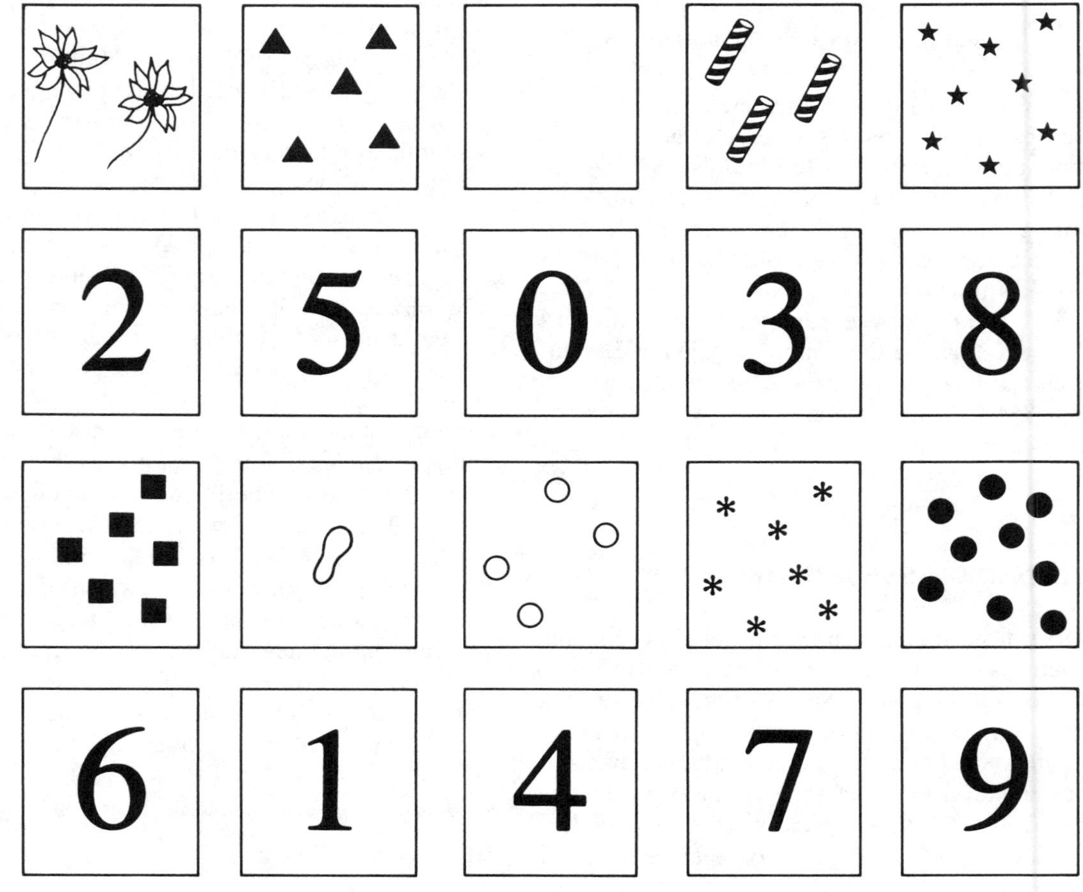

3. Ask each child to find the set card with no pictures. Place on the desk or table.
4. Ask each child to find the set card with one picture. Place it to the right of the first card.
5. Continue with this process until all set cards have been placed in order from 0–10.
6. Have the children repeat the above activity with the numeral cards.

Figure 4.12
A Sample Abstracting/Schematizing Activity for Rational Counting

ACTIVITY PLAN

Counting: Rational counting ***2-D/Rep→ Symbolic*** ***Consolidating***

Content objectives

4.16, 4.17

Exemplars/Materials

Gameboard (as shown)

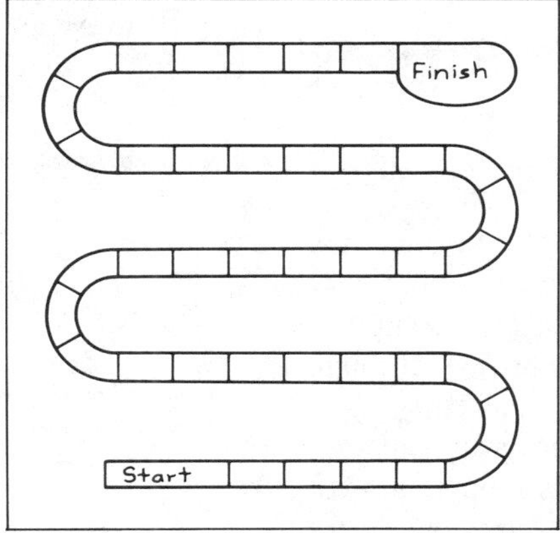

Gameboard markers
3″x 5″ cards with buttons sewn on, numerosity from 0–10
Large paper bag

Behavorial indicators

Each child will:

1. Draw a card from the bag, count the buttons, and state the count.
2. Move as many spaces on the game board as the number of buttons on the card.

Procedure

1. This activity is a game. Each child should place her marker on the start position. In turn, each child reaches into the bag and draws a card. The number of buttons on the card is counted and stated.
2. The child moves her marker as many spaces on the game board as stated for the buttons. The card is returned to the bag.
3. Continue, taking turns and moving the markers. If a child gives an incorrect count, she cannot move on that turn. The first player whose marker reaches or passes finish wins.

Figure 4.13
Sample Consolidating Activity for Rational Counting

Figure 4.14
Set of Yarn Digits

obtained by having the child talk through the task, explaining each step in the reasoning process. Unclear concepts and incorrect computational procedures are often identified through such diagnostic procedures.

A thorough assessment program includes paper-and-pencil tasks with symbolic and two-dimensional representations and interview protocols that allow children to express themselves verbally and with manipulative materials. Such a program permits considerable confidence in the inferences we make concerning the concepts a child knows and the cognitive processes he or she is using.

Paper-and-Pencil Tests

Paper-and-pencil tests, often constructed by teachers, allow us to quickly test for knowledge of concepts. A short inventory test of beginning number concepts is illustrated in Figure 4.16.

Interviews

A special characteristic of the interview is that it provides opportunities for a child to talk through problems with the teacher. A child is able to tell why he or she did a problem a certain way; the teacher obtains insight concerning correct and incorrect concepts, skills, or procedures. Figure 4.17 is an interview protocol for the interrelationship of cardinal and ordinal number concepts. Other interview protocols can be found in the handbook that accompanies this text.

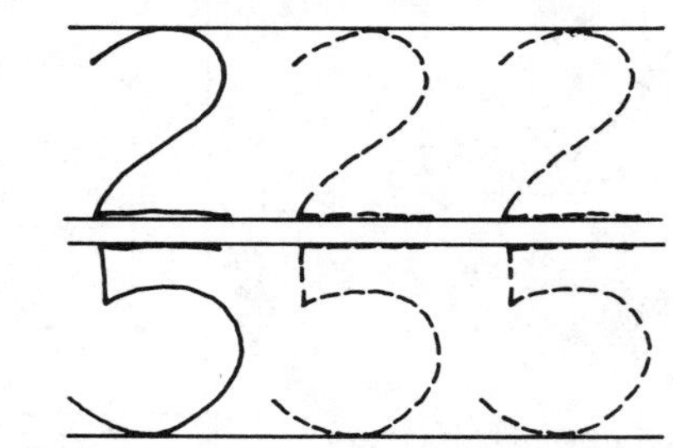

Figure 4.15
Sample Worksheet for Copying Digits

Directions to the teacher:

1. Place nine checkers on the table in a linear sequence as follows:

2. Point to the first one on the left and name it "first". Point to the next one and name it "second", etc. Continue naming each checker until you have reached the "ninth" one.

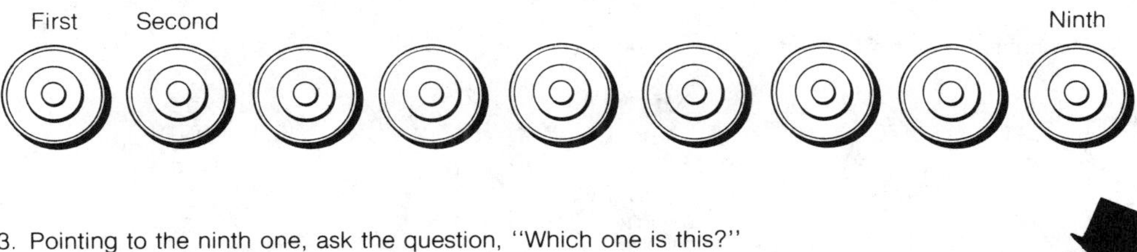

3. Pointing to the ninth one, ask the question, "Which one is this?"

4. Now, using a cloth, cover the first five. Point to the ninth one again and say, "This one is ninth, how many are covered?" "Why?"

5. If child cannot respond, point to the sixth one and say, "This one is sixth, how many are covered and why?"

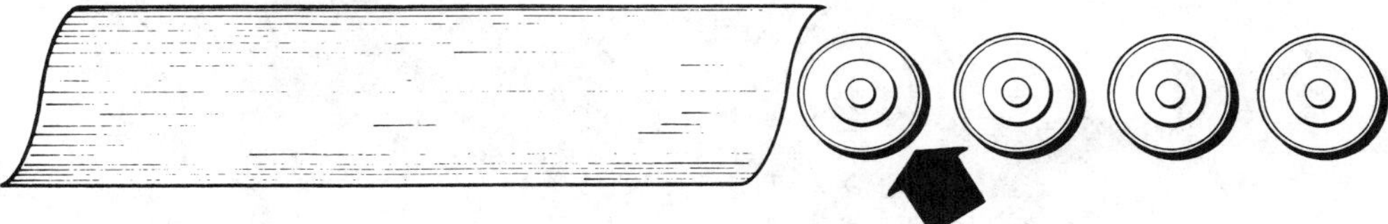

Figure 4.16
Short Beginning Number Concepts Inventory

To the teacher: All directions are given orally.

1. What should go in the missing spaces?

4, __, 6, ___, 8, 9

2. Put an *X* on the set having more.

3. Put the correct symbol in the box; >, =, <.

20 ☐ 20

4. Count the number of boys. Circle the correct numeral

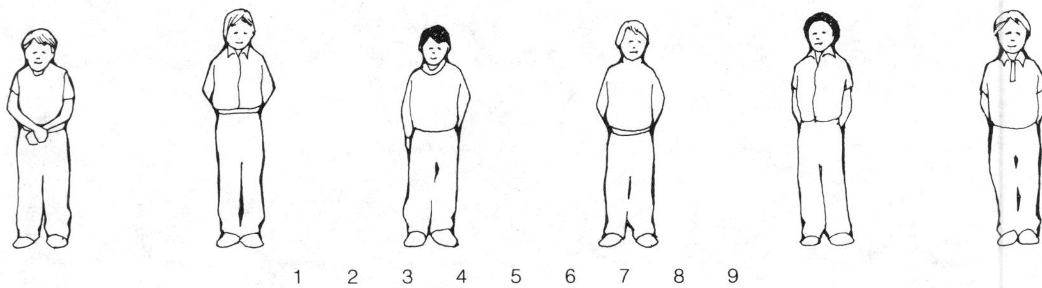

1 2 3 4 5 6 7 8 9

5. Mark the third cat.

6. Mark the set having fewer.

Figure 4.17
Sample Protocol for Cardinal-Ordinal Number Interrelationships

TIPS ON MANAGING THE CLASSROOM

Keep the following in mind as you develop a beginning mathematics program.

1. Mathematical ideas grow slowly, so provide adequate time for children to work with objects before they move to symbols. Each child should be asked to measure, count, make comparisons, and discuss what he or she is doing. In setting up an activity-based program, make sure you have enough manipulative material for each child. Manipulatives that *clearly* exemplify the concept you are developing should be chosen.
2. The attention span of young children is very short, so plan your lessons accordingly and use appealing materials.
3. Allow children to discuss what they are doing. Help each child use words such as *if-then, either-or, not, but,* and *because.*
4. Experiment with different plans for grouping and instruction. Make sure some activities are independent laboratory type, while others require small-group work.
5. Many beginning number concepts such as equivalent sets, counting, and numeral recognition, lend themselves to teaching through games. Make sure children of about the same level of knowledge play together.

5

Introducing Addition and Subtraction: Planning the Content

A child's first experiences with operations on numbers are usually with addition and subtraction of whole numbers. In this chapter the authors are concerned with what addition and subtraction *are* as well as how they are done.

Addition, an operation that needs to be completely understood, involves concepts such as the following:

- Addition is used to find the number of things in the set formed from two disjoint sets.
- The number telling how many things are in one of the two disjoint sets is called an *addend.*
- The symbol "+" is read *plus.*

These are only a few of many addition-related ideas that children need to learn.

Addition is also an operation to be done. When we *do* addition, we execute a procedure; we perform an algorithm to find the sum. It may be mental only, or it may require a paper and pencil.

Sometimes students think of addition as simply putting things together to find how many there are in all. But addition is an operation on *numbers*, not a moving of objects. Many situations requiring the use of addition do not involve moving objects, and some even require the actual removal of objects. If children think addition is taking away, they are soon confused.

TEACHING THE MEANINGS OF THE OPERATIONS

If a child is to be successful in arithmetic and eventually algebra, more is needed than just getting answers right. He needs to understand the *meaning* of addition. Children must be able to choose the correct operation to solve day-to-day problems. They need to understand the meaning of the operation in order to solve different types of open number sentences and story problems.

The child who does not have this understanding survives by learning specific response routines for each situation. Frequently, such a child's learning is so specific that a mere change in the format of the stimulus situation results in the assumption that something completely new needs to be learned. What the child learns is useful only for doing more of the same tasks.

In order to use hand-held calculators, a child must be able to decide which button to push. That decision is based on knowing what the operations mean. The meanings of the operations also help children transfer what they know from one set of numbers to another. The child who is limited to thinking of $15 - 7 = \square$ as taking away 7 later has difficulty using what he knows about subtraction to solve problems like $15 - {}^{-}7 = \square$. The child who is limited to thinking of $4 \times 8 = \square$ as a special addition situation later has difficulty using what he knows to solve problems like $\frac{2}{3} \times \frac{4}{3} = \square$.

PLANNING THE CONTENT

Though textbooks are one of the best instructional aids available, they only provide the framework. As teachers we make the conscious or unconscious decision as to what content will be taught. For example, we may skip certain topics in a book, or bring in additional material. We may even adapt the instructional program to a child with special needs: the gifted, and the learning-disabled or otherwise handicapped.

As each of us plans the specific content to be taught when we introduce addition and subtraction of whole numbers, we need to ask two questions: What content shall I teach? In what order will I teach it?

Selecting the specific content to teach a child will, of course, depend on what the child already knows and can do. Before selecting specific content objectives for one or more children, make sure that they have prerequisite understandings and skills. Keep in mind that all the different models for an operation should not be taught at the same time. Objectives should be selected with each child in mind, remembering that the objectives are designed for *our* use as teachers in planning, and are not worded for children.

Addition of Whole Numbers

Too often, teachers define addition in terms of sets and actions taken with objects when finding the sum. The authors define addition in terms of its function: it is the operation used to find the sum when the addends are known.

Meanings, Models, and Symbols

Among those objectives used in introducing addition are those concerned with meanings, models, and symbols.

> **5.1**
> Addition of whole numbers is an operation that associates each pair of whole numbers with a unique whole number.
> a. The pair of numbers added are called *addends.*
> b. The unique number associated with the addends is called the *sum.*
> c. Addition is used to find the sum when its addends are known.

Each pair of whole numbers is associated with a third whole number through addition. *Three* names exactly 3 and *four* names exactly 4, but *three add four* or *three plus four* is a name for 7, a third unique whole number.

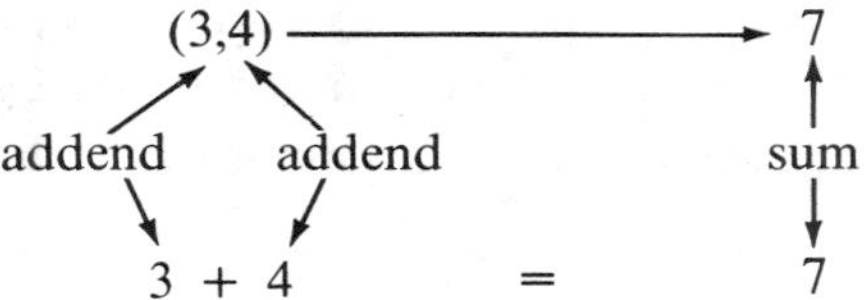

> **5.2**
> The union of disjoint sets is a model for addition on whole numbers.
> a. One addend indicates the number of elements in one set.
> b. The second addend indicates the number of elements in the other set.
> c. The sum is the number of elements in the union of the sets.
> d. Addition is an operation used to find the total number of elements in a union of two disjoint sets when the number of elements in each set is known.

Thus, one way to picture addition for children is with sets of objects. Consider these 7 balls: 4 are grey and 3 are white.

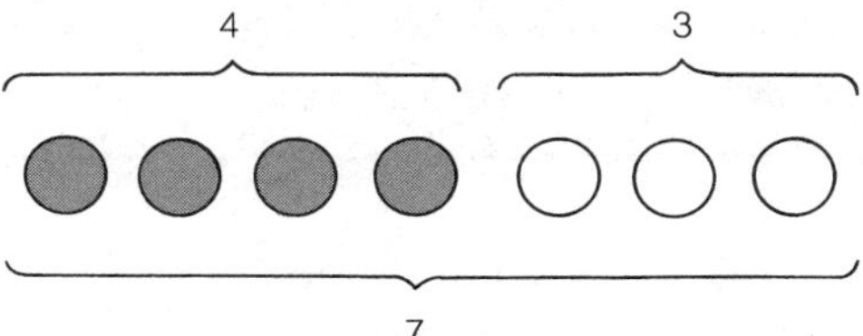

The number in one set (one addend) is 4; the number in the other set (the other addend) is 3. These two numbers are associated with a third number, 7, which tells how many are in union (the sum).

An understanding of addition can be developed in other ways. For instance, the authors partition a set into subsets. The relationships between the subsets and the original set are then developed, as are the relationships between the related sum and addends. This approach is based on the following content objective:

> **5.3**
> A set may be partitioned into two disjoint subsets, even if one or both of the subsets is empty. The subsets and the original set model the relationships between two addends and their sum.

When we relate content objective 5.3 to the previous objective, it becomes clear that whenever a person has two sets, he *has* their union set. The union already exists; no more is made. The primary emphasis of instruction should be on the *relationships* involved. The authors have found that an early emphasis on relationships between two addends and their sum helps children schematize the many addition and subtraction concepts.

> **5.4**
> All possible pairs of addends associated with a given sum can be found by partitioning a set into all possible pairs consisting of a subset and its complement.
> a. The sum indicates the number of elements in the set to be partitioned.
> b. The numbers related to a subset and its complement set form a pair of addends.
> c. All such pairs of addends are the possible pairs of addends for that given sum.

The empty set, with zero members, is also a subset of any set. For the number 7 we observe the partitioning into subsets as follows:

7 = 7 + 0 ● ● ● ● ● ● ● |
7 = 6 + 1 ● ● ● ● ● ● | ●
7 = 5 + 2 ● ● ● ● ● | ● ●
7 = 4 + 3 ● ● ● ● | ● ● ●
7 = 3 + 4 ● ● ● | ● ● ● ●
7 = 2 + 5 ● ● | ● ● ● ● ●
7 = 1 + 6 ● | ● ● ● ● ● ●
7 = 0 + 7 | ● ● ● ● ● ● ●

> **5.5**
> Beginning at the zero point, the union of successive segments of a number line to the right is a model for addition on whole numbers.

Consider the use of the number line to solve 3 + 4 = □.

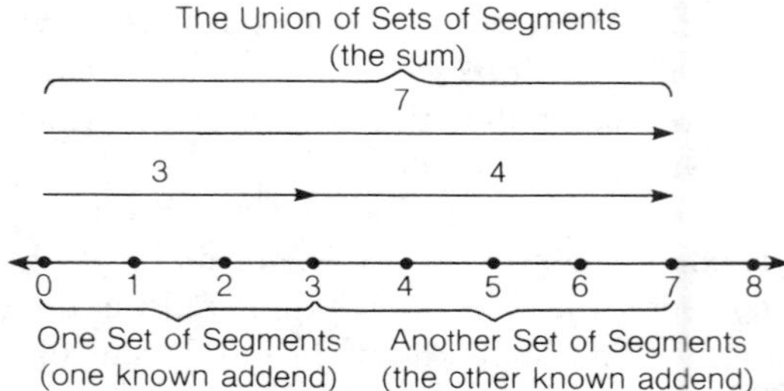

An arrow is used to show each segment. The arrow for the sum starts where the arrow for the first addend starts (zero) and ends where the arrow for the second arrow ends. Seven can be seen as *the same number as* 3 and 4, 2 and 5, and so on.

> **5.6**
> The + (read *plus* or *add*) is a symbol that indicates addition.

Children need to know how to read this symbol and what it indicates. We should not assume every child knows this, and must specifically focus on objective 5.6 during instruction.

> **5.7**
> An *addition phrase* consists of numerals or variables separated by the plus sign.
>
> 3 + 4 53 + □ $a + b$
>
> a. The numerals or variables on each side of the plus sign in a phrase indicate addends.
> b. An addition phrase names a single number, a sum.

A child should understand these ideas in order to fully comprehend addition, rather than thinking of it as putting together.

In 3 + 4 = □, a child knows that 3 and 4 are addends *because* they are on each side of the plus sign. Similarly, in 678 + □ = 952, the variable is an addend because it is on one side of the plus sign.

EXCURSION

Can you tell whether the variable is an addend or a sum in each of the following?

$86 = 47 + \square$ $\square + 965 = 37$

$\frac{2}{4} = \square + \frac{1}{6}$ $^{-}86 = \square + {}^{-}24$

What operation(s) would you use to find the variable in each case?

5.8
A single numeral (digit or multidigit) used to name a sum is called the *simplest name, standard name,* or *standard numeral* for the sum.

In the open number sentence $3 + 4 = \square$, we know that one name for the sum is $3 + 4$, and that the standard name is 7. Similarly, in the open number sentence $278 + 367 = \square$, we know that one name for the sum is $278 + 367$, but we also want to know the standard or simplest name. This can be determined with many different procedures, including paper-and-pencil algorithms. Though we may think of such a procedure as adding, addition is more than that–it is an operation that associates a pair of numbers with a unique number.

5.9
An addition *equation,* or *equality sentence,* consists of addition phrases, variables and/or standard numerals, and an equals sign.

$3 + 4 = 7$ $53 = 17 + 36$ $4 + \square = 13$

a. An addition equation states that numerals and phrases on each side of the equals sign name the same sum.
b. An equation containing a variable is called an *open equation* or *open sentence.*
c. Addition equations may be written in horizontal or vertical form.

$$3 + 5 = \square \quad \text{or} \quad \begin{array}{r} 3 \\ +5 \\ \hline \square \end{array}$$

d. In the vertical form, the bar under the phrase serves as an equals sign.

Recall that the equals relation means *is the same as.* Therefore, to say that $3 + 4 = 7$ is to say that the sum $3 + 4$ *is the same as* the sum 7; they are different names for the same number.

A child who can solve a number sentence like $6 + 7 = \square$ does not always know what to do with the same problem presented in a vertical format. The child does not relate the two unless content objectives 5.9c and 5.9d are taught.

Properties

Other objectives used in introducing addition include those that involve properties.

5.10
Closure–The sum of any two whole numbers is always a whole number.

If a and b name whole numbers, then there is a unique whole number named by c such that $a + b = c$. We say that whole numbers are *closed* under addition.

Content objective 5.10 will seldom be used when introducing addition. It is so self-evident that it is a difficult abstraction for younger children to deal with.

> **5.11**
> *Commutative property*–The order or sequence in which two whole numbers are added has no effect on their sum.

If a and b are whole numbers, then $a + b = b + a$. For example, 3 + 4 and 4 + 3 name the same sum. It is frequently helpful to use the commutative property of the union of disjoint sets to model the commutative property of addition of whole numbers.

Reordered

2 + 3 = 3 + 2

5 = 5

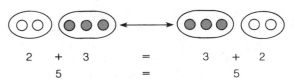

When sets are reordered, the union of sets remains the same; when addends are reordered, the sum remains the same.

> **5.12**
> *Associative property*–The way in which three whole numbers are grouped together in addition has no effect upon their sum.

Addition is a binary operation; that is, it associates exactly two numbers with a third number. Therefore, addends should be grouped in pairs whenever there are three or more. Frequently, parentheses are used to help conceptualize and specify the exact nature of the grouping. In (3 + 4) + 5, the 3 and 4 are grouped together; the other pair of addends consists of the number *3 + 4* and the number 5.

If a, b, and c name three whole numbers, then $(a + b) + c = a + (b + c)$. This is the associative property of addition. As with the closure and commutative properties, it is often helpful to use the analogous property of the union of sets as a model.

> **5.13**
> *Identity element*–Zero added to any whole number yields a sum equal to the original number.

If a names any whole number, then $a + 0 = 0 + a = a$. Zero is called the *identity element* for addition, or the *additive identity* in the set of whole numbers. A child who understands and can already apply this idea grasps nearly 20 of the 100 basic facts of addition.

Subtraction of Whole Numbers

Among the objectives for introducing subtraction are those concerned with meanings, models, and symbols; and with properties of subtraction.

Meanings, Models, and Symbols

> **5.14**
> Subtraction of whole numbers is an operation that associates a pair of whole numbers with a unique whole number.
> a. One of the pair of numbers is a sum (sometimes called a *minuend*) and the other number is one of the addends (sometimes called a *subtrahend*).
> b. The sum's other addend (sometimes called the *remainder* or *difference*) is the third number associated with the sum and one of its addends.
> c. Subtraction is an operation used to find an addend (remainder or difference) when the sum (minuend) and its other addend (subtrahend) are known.

Three names exactly 3, and *seven subtract four* or *seven minus four* is also a name for 3.

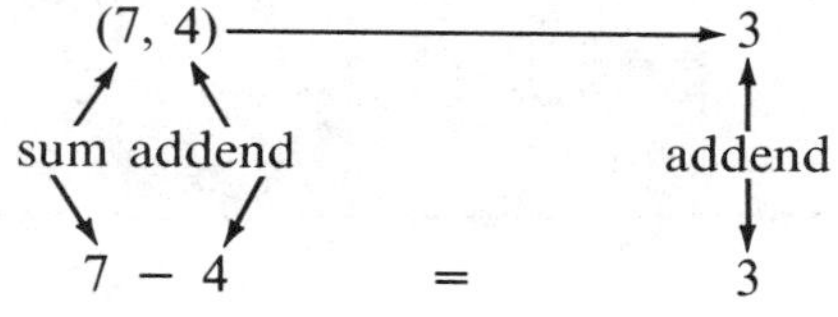

The content objectives that follow focus on ways of modeling the operation and on the symbols used when writing subtraction statements.

5.15
The partition of a universal set into a subset and its complement, or remainder set, is a model for subtraction.

a. The sum (minuend) is the number of elements in the universal set (the starting set).
b. The known addend (subtrahend) is the number of elements in one subset.
c. The other addend (remainder) is the number of elements in the complement, or remainder set.
d. Subtraction is an operation used to find the number of elements in the complement, or remainder set, (one addend) when the number of elements in the universal set (the sum) and number of elements in one of its subsets (known addend) are known.

One way to model subtraction for children is with sets of objects. Consider the set of 7 balls: 4 are grey and 3 are white.

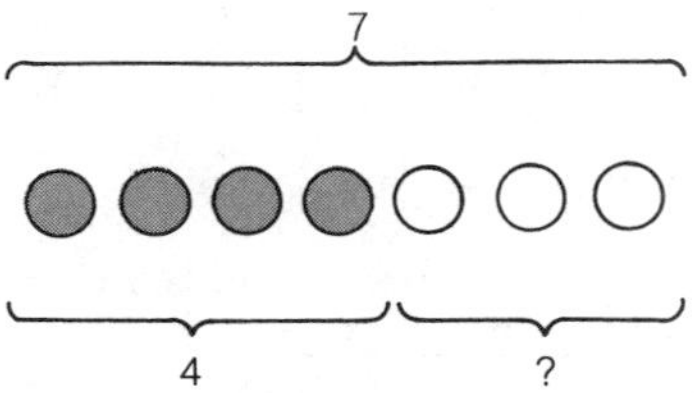

The number in the whole set (the sum) is 7; the number of grey balls (one addend) is 4. These two numbers are associated with a third number, 3, which tells how many white balls there are (the other addend). As with objective 5.3, the emphasis is on the relationships between the sum and its addends.

It is *not* necessary to move objects in order to model the operation of subtraction. Children often come to think of subtraction as taking away, a very inadequate and often misleading way of modeling the operation.

How do children determine an unknown addend if they do not take away part of the objects and count those that remain? If they understand addition and subtraction, they know that in $7 - 4 = \square$, the 4 and the $\square$ are both addends. The missing number is 3 *because* 4 and 3 are addends or parts for 7 (the sum).

5.16
Comparison by matching one-to-one the elements of two disjoint sets is a model for subtraction of whole numbers.

a. The known sum (minuend) is the number property of one of the disjoint sets (the greater set if not equivalent).
b. The known addend (subtrahend) is the number property of the second disjoint set (the lesser set if not equivalent).
c. The unknown addend (difference) is the number property of the subset of unmatched elements (the *difference subset*) remaining in the first set after the elements of the second set are matched one-to-one with elements of the first set.
d. Subtraction of whole numbers is an operation used to find the number associated with the difference between two disjoint sets when the number of members in each of the disjoint sets is known.

Objective 5.16 suggests a very different way to model the operation of subtraction. In contrast to objective 5.15, *disjoint* sets are used. It therefore takes more objects to model the very same number sentence. Consider two sets of balls: 7 grey balls and 4 white.

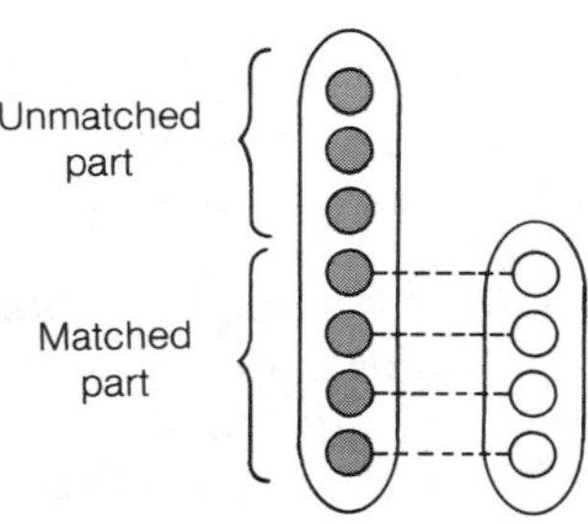

The number in the greater set (the sum) is 7; the number in the lesser set (the known addend) is 4. Subtraction is used to tell us the number of balls in the greater set that are *not* matched with balls in the lesser set: 7 − 4 balls are unmatched. Of course, 7 − 4 is the same as 3, so we can also say that 3 balls are unmatched.

How do children ever find the standard name for a phrase such as 7 − 4 if they do not take away or count unmatched items? Children easily see that the greater set has two parts, a matched part and an unmatched part. The matched part is equivalent to the lesser set (known addend), so that the number of things in the greater set (sum) minus the number in the matched part (known addend) equals the number in the unmatched part (unknown addend). If 7 − 4 = □ is written when comparing two disjoint sets, it is still possible to observe that □ = 3 because 4 and 3 are addends or parts for 7 (the sum).

It is easier for a child to understand the ideas expressed in objective 5.16 if he or she already understands objective 5.15.

5.17
The partition of a line segment from zero on a number line into two successive segments is a model for subtraction on whole numbers.

Consider the following use of a number line to solve 7 − 4 = □.

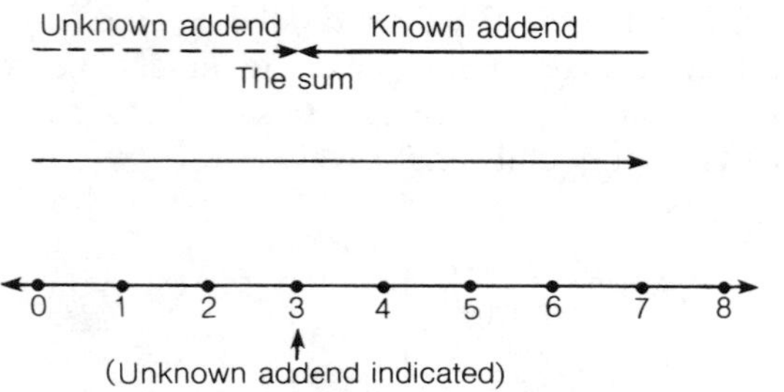

For subtraction, the sum is shown first. The arrow for the unknown addend starts where the arrow for the sum starts (zero) and ends where the arrow for the known addend ends. Three can be seen to be *the same as* (has the same result as) 7 − 4.

In evaluating the different ways of showing subtraction on number lines encountered in textbooks and curriculum guides, we need to discern the methods that will help children understand what subtraction is, rather than just giving them another procedure for getting answers. Unfortunately, some children learn mechanical routines, a specific response to an individual task format.

5.18
Subtraction is the inverse of addition.

a. Addition is used to find a sum when its addends are known; subtraction is used to find one of the addends when the sum and its other addend are known.
b. The inverse of each model for addition is a model for subtraction.

We should introduce addition before subtraction because subtraction is understood in terms of a sum (whole amount) and addends (amount for each part), concepts that a child learns when he is introduced to addition. However, subtraction should be introduced shortly after addition, so the child can more easily understand the inverse relationship. The following scheme may be useful in summarizing this relationship.

addend + addend = sum

sum − addend = addend

4 + 3 = 7

7 − 3 = 4

268 + 475 = 743

743 − 475 = 268

5.19
Elements in each row of an array of objects can be successively partitioned into a subset and its complement. This forms a model for all the possible pairs of addends for a given sum and their inverses for subtraction.

EXCURSION

Can you think of a statement similar to objective 5.18 that relates multiplication and division, a statement involving factors and a product?

For example, for 5:

Subset / Complement

5 + 0 = 5	● ● ● ● ●|	5 − 0 = 5
4 + 1 = 5	● ● ● ●| ●	5 − 1 = 4
3 + 2 = 5	● ● ●| ● ●	5 − 2 = 3
2 + 3 = 5	● ●| ● ● ●	5 − 3 = 2
1 + 4 = 5	●| ● ● ● ●	5 − 4 = 1
0 + 5 = 5	| ● ● ● ● ●	5 − 5 = 0

This model is particularly useful in helping a child see relationships and thereby build a conceptual schema, a mental framework that will facilitate recall and enable him to use this information. Note that objective 5.19 requires that the child already understand the idea of an empty set.

5.20
The − (read "minus" or "subtract") is a symbol that indicates subtraction.

Many children have difficulty solving problems that involve subtraction because they have learned to read this symbol as "take away" and feel that something must be moved aside. An extreme example is the child who believes that 7 − 4 = 7 because, if you have 7 − 4 and take away the 4, only the 7 is left.

5.21
A *subtraction phrase* consists of numerals and variables separated by the − sign.

$7 - 2 \qquad 52 - \square \qquad a - b$

a. The numeral or variable to the left of the minus sign always indicates a sum (minuend).
b. The numeral or variable to the right of the minus sign always indicates one of the addends (subtrahend) of the sum.
c. A subtraction phrase names one number.

A child must understand these ideas if he is to solve open number sentences. For example, given $\square - 8 = 3$, many children will say that

EXCURSION

Can you tell whether the variable is an addend or a sum in each of the following?

$$38 = \square - 36 \qquad \square - {}^{-}97 = 48$$

$$\square - \frac{1}{4} = \frac{1}{8} \qquad {}^{-}24 = \square - {}^{-}36$$

What operation(s) would you use to solve each?

$\square = 5$. But the child who comprehends objective 5.21 will identify the missing number as a sum (because it is to the left of a minus sign) and therefore add the given addends to find the missing sum.

Each child should know that expressions such as "7 − 4" and "632 − 279" name *one* number: an addend is indicated. The answer to a problem such as $632 - 279 = \square$ is merely the more conventional or standard name for that addend.

5.22
A single numeral (digit or multidigit) used to name an addend (remainder, difference) is called the *simplest name, standard name,* or *standard numeral* for the addend.

For the open number sentence $632 - 279 = \square$, one name for the missing addend is $632 - 279$, but the standard name can be found by a calculator, a paper-and-pencil algorithm, or mentally.

5.23
A simple *subtraction equation* or *equality sentence* consists of a subtraction phrase, a standard numeral or variable, and an equals sign.

$$7 - 2 = 5 \quad 8 - 3 = \square \quad 52 - \square = 20$$

a. A subtraction equation states that the numeral and phrase on each side of the equals sign names the same addend (remainder, difference)

b. In a simple subtraction sentence, phrases and standard numerals or variables may be written on either side of the equals sign.

$$7 - 2 = 5 \quad \text{or} \quad 5 = 7 - 2$$
$$13 - \square = 8 \quad \text{or} \quad 8 = 13 - \square$$

c. An equation containing a variable is called an *open equation* or *open sentence.*

d. Subtraction equations may be written in horizontal or vertical form.

$$7 - 2 = \square \quad \text{or} \quad \begin{array}{r} 7 \\ -\ 2 \\ \hline \square \end{array}.$$

e. In the vertical form, the bar under the phrase serves as an equals sign.

To say $7 - 4 = 3$ means that the addend $7 - 4$ *is the same as* the addend 3; they are different names for the same number. Therefore, "Five minus three *is* two" is said instead of "Five minus three are two" because the number 2 is a singular idea.

5.24
For every simple subtraction sentence involving unequal addends, there are three other sentences that state the same relationship among the sum and its addends: one subtraction sentence and two addition sentences

$7 - 2 = 5$	$9 - \square = 6$
$7 - 5 = 2$	$9 - 6 = \square$
$2 + 5 = 7$	$\square + 6 = 9$
$5 + 2 = 7$	$6 + \square = 9$

Objective 5.24 should be used for instruction only after children understand many of the basic ideas for subtraction. However, a child should realize that the relationship between a sum and its addends can be expressed in different formats. Such recognition will help him or her in solving number sentences and other problems.

Properties

It is important for children to realize that subtraction does not always have the same attributes as addition. The objectives that follow focus on properties which may or may *not* hold for the operation of subtraction.

> **5.25**
> *Closure*—It is *not* true that the difference of any two whole numbers is always a whole number.

For example, in $4 - 7 = \square$ the missing addend is not a whole number.

> **5.26**
> *Commutativity*—It is *not* true that the order of two whole numbers has no affect on the difference between the two numbers.

For example, $7 - 4 \neq 4 - 7$.

> **5.27**
> *Associativity*—It is *not* true that the way in which three or more whole numbers are grouped together for subtraction has no affect on the difference.

For example, $(8 - 4) - 1 = 3$ whereas $8 - (4 - 1) = 5$.

> **5.28**
> *Identity element*—Zero subtracted from any sum yields an addend equal to the original sum.

If a names any whole number, then $a - 0 = a$. This is true because the addend a plus the addend 0 equals the sum a. This particular content objective does not need to be stressed with children who understand the relationship between addition and subtraction, and that zero is the identity element for addition.

Basic Facts for Addition and Subtraction

Objectives should focus on definitions, special relationships, and categories of basic facts.

Defining Basic Facts

> **5.29**
> A *basic* (or *primary*) *fact* for addition is an addition equation of the form:
>
> $$a + b = c \quad \text{or} \quad \begin{array}{r} a \\ +\ b \\ \hline c \end{array}$$
>
> Addends have only one digit, and the equation does not contain a variable.

By this definition, each of the following is a basic addition fact:

$$2 + 3 = 5 \qquad 9 + 7 = 16 \qquad \begin{array}{r} 0 \\ +\ 3 \\ \hline 3 \end{array} \qquad \begin{array}{r} 7 \\ +\ 8 \\ \hline 15 \end{array}$$

None of the following are basic addition facts. Can you state a reason in each case?

$$12 + 3 = 15 \qquad 9 + 7 = \square \qquad \begin{array}{r} 10 \\ +\ 6 \\ \hline 16 \end{array} \qquad \begin{array}{r} 3 \\ +\ 4 \\ \hline \end{array}$$

> **5.30**
> A *basic* (or *primary*) *fact* for subtraction is a subtraction equation of the form:
>
> $$x - y = z \quad \text{or} \quad \begin{array}{r} x \\ -\ y \\ \hline z \end{array}$$
>
> Addends have only one digit, and the equation does not contain a variable.

EXCURSION
How many basic addition facts are there?

If a child were to memorize each of the basic addition facts as separate entities, how many facts would he or she have to memorize? Does this incomplete table of sums suggest the number of basic addition facts?

+	0	1	2	3	4	5	6	7	8	9
0	0	1	2	3	4	5	6	7	8	9
1	1	2	3	4	5	6	7	8	9	10
2	2	3	4	5	6	7	8	9	10	11
3	3	4	5	6	7	8	9			
4	4	5	6	7	8					
5	5	6	7							
6	6	7	8							
7	7	8								
8	8	9								
9	9									

Children do not need to memorize each of the basic addition facts. Application of ideas such as the commutative property can cut down the number of facts that must be memorized.

By this definition, each of the following is a basic subtraction fact:

$$5 - 3 = 2 \qquad \begin{array}{r} 3 \\ -3 \\ \hline 0 \end{array} \qquad \begin{array}{r} 15 \\ -8 \\ \hline 7 \end{array}$$

$$16 - 7 = 9$$

None of the following are basic subtraction facts. Can you tell why?

$$15 - 3 = 12 \qquad \begin{array}{r} 16 \\ -\ 6 \\ \hline 10 \end{array} \qquad \begin{array}{r} 8 \\ -5 \\ \hline \end{array}$$

$$9 - 7 = \square$$

Special Relationships

Among the objectives already listed are properties that can cut down on the number of basic addition facts that must be memorized. If doubles (those basic addition facts in which the addends are equal) are momentarily disregarded, application of the commutative property cuts the number of basic addition facts to be learned in half. A child can memorize the sum for an addend pair *regardless of the order* of the addends. A child who can apply the idea of 0 as the identity element has learned almost one-fifth of the 100 basic addition facts.

A number of other special relationships can help children with the basic facts. One, compensation, is a quick, relatively sophisticated way to generate some missing sums.

5.31
Compensation in addition—If the number added to one addend is subtracted from the other addend, the sum is unchanged.

By applying this idea, a child can reason that:

$$9 + 6 = (9 + 1) + (6 - 1) = 10 + 5 = 15$$

$$6 + 8 = (6 - 2) + (8 + 2) = 4 + 10 = 14$$

A child who understands this special relationship need not always make 10, but can use known basic facts to figure out other basic facts. For example, a child who realizes that $8 + 8 = 16$ might reason as follows to solve $7 + 9 = \square$:

EXCURSION

Using compensation in addition, can you mentally calculate the missing sum for each of the following?

198 + 432 = □ *(Think:* 200 + ...*)*

988 + 715 = □

7 + 9 = (7 + 1) + (9 − 1) = 8 + 8 = 16

However, basic addition facts with 8 or 9 as an addend should be consolidated for efficient use. This will usually involve memorization rather than exclusive reliance on self-conscious application of the compensation relationship.

5.32
Compensation in subtraction:
a. If a number is added to both the sum and the known addend, the unknown addend is unchanged.
b. If a number is subtracted from both the sum and the known addend, the unknown addend is unchanged.

By applying these special relationships, a child who knows that 12 − 5 = 7 can reason:

For 11 − 4 = □:

11 − 4 = (11 + 1) − (4 + 1) = 12 − 5 = 7

For 13 − 6 = □:

13 − 6 = (13 − 1) − (6 − 1) = 12 − 5 = 7

Other special relationships can help a child learn the basic addition facts.

5.33
When an addend is increased by 1 and the other addend is unchanged, the sum is increased by 1.

5.34
When an addend is decreased by 1 and the other addend is unchanged, the sum is decreased by 1.

A child who has observed such patterns in tables of basic addition facts or while working with a

EXCURSION

Compensation can be useful in subtracting larger numbers, especially in mental subtraction. Consider 292 − 145 = □. Applying content objective 5.32a, we can reason that (292 + 8) − (145 + 8) = 300 − 153. Application of objective 5.32b further reveals that 300 − 153 = (300 − 150) − (153 − 150) = 150 − 3 = 147. Therefore, 292 − 145 = 147.

The ability to estimate and compute mentally is very important in this era of hand-held calculators, and the principles of compensation should be among the objectives selected for instruction.

Using the principles of compensation, can you mentally calculate the missing addend for each of the following?

204 − 56 = □ *(Think:* 200 − . . . .*)*

277 − 198 = □ *(Think:* . . . . − 200 = *)*

4365 − 1990 = □

EXCURSION

Can you mentally calculate these problems using properties and special relationships?

$25 + 78 = \square$ (*Think:* 25 + 75 = 100. . . .)

$893 + 428 = \square$ (*Think:* 900 + 428. . . .)

$44{,}900 + 25{,}435 = \square$ (*Think:*

math balance can use these ideas to reason as follows:

$5 + 5 = 10$ so $5 + 6 = 11$

$8 + 8 = 16$ so $8 + 7 = 15$

Categories of Basic Addition Facts

The 100 basic addition facts are sometimes divided into three categories: facts with sums less than 10, facts with 10 as the sum, and facts with sums greater than 10.[1] For planning consolidating activities, each category can be listed as a content objective.

> **5.35**
> Basic addition facts with sums less than 10.

After a child understands what addition means and can relate an addition fact to other addition and subtraction facts, skill with these facts needs to be consolidated and maintained.

> **5.36**
> Basic addition facts with 10 as the sum.

1. Donald D. Paige et al., *Elementary Mathematical Methods* (New York: John Wiley and Sons, 1978): 79.

Learning can proceed even though the child has not yet been introduced to two-digit numerals.

This very useful set of facts needs special emphasis. With these addend pairs, a child can figure out sums greater than 10. For example:

$$6 + 7 = \square$$

$$\begin{aligned} 6 + 7 &= 6 + (4 + 3) \\ &= (6 + 4) + 3 \\ &= 10 + 3 \\ &= 13 \end{aligned}$$

A child may not be able to write out this application of the associative principle, but he can think: 6 + *4* = 10, so it is 10, and 3 more equals 13.

> **5.37**
> Basic addition facts with sums greater than 10.

These facts should not be taught until after the child has initiated and abstracted concepts related to two-digit numerals. If facts such as 7 + 5 = 12 are introduced before a child understands the meaning of two-digit numerals, the child may use the 12 as though it were a one-digit numeral. Thus:

$$\begin{array}{r} 47 \\ +\ 35 \\ \hline 712 \end{array}$$

EXCURSION

How can a child who thoroughly knows the family of facts with 10 as a sum use them to solve higher decade problems such as these?

$56 + 7 = \square$ (*Think:* $5\underline{6}$ + $\underline{4}$. . . .)

Premature introduction to two-digit numerals may encourage less sophisticated ways of determining a missing sum. For example, some form of counting is likely to be used for an extended period of time, although the application of the associative principle is often more efficient.

Categories of Basic Subtraction Facts

Each of the 100 basic subtraction facts is an inverse of a basic addition fact. The focus should be on memorizing the addition facts; it is not necessary to actually memorize the basic subtraction facts.

However, it *is* essential that each child understand the meanings of both addition and subtraction, and that children practice using this information to find the missing addend in problems such as:

$$7 - 4 = \square \quad \text{and} \quad \begin{array}{r} 7 \\ -\ 4 \\ \hline \square \end{array}$$

In both cases the missing number is 3, *because 4 and 3 are an addend pair for 7*; that is, 4 + 3 = 7. Older children who have memorized the addition facts may count backwards to find a missing addend. When such children are taught the meanings of addition and subtraction, they can often supply missing addends immediately, even while computing multidigit examples. Corrective instruction for difficulty with basic subtraction facts involves work on the meanings of the operations rather than more practice with the facts.

The 100 basic subtraction facts can be divided into the same three categories as those for addition. Activities should focus on both addition and subtraction facts.

5.38
Basic subtraction facts with sums less than 10.

5.39
Basic subtraction facts with 10 as the sum.

5.40
Basic subtraction facts with sums greater than 10.

Sentence Solving

A child applies previously learned knowledge of the meanings of operations as well as knowledge of basic facts when solving number sentences. Sentence-solving skills should be included when content is planned and objectives are sequenced.

An analogy can be made between solving number sentences and use of sentences in language. Phenix calls our oral and written language behavior "the outer face of language," whereas "the inner face is *meaning*."[2] Transformational grammarians distinguish between surface structure and deep structure: surface structure refers to the form of a sentence, while deep structure is the underlying meaning.[3] The sentence *Marty threw the ball* and the sentence *The ball was thrown by Marty* have different surface structures yet the same deep structure.

Examine each of the following to see if you can find the common meaning or deep structure:

2. Philip H. Phenix, *Realms of Meaning* (New York: McGraw-Hill, 1964): 62–63.
3. Edna DeHaven, *Teaching and Learning the Language Arts* (Boston: Little, Brown, 1979): 66–69.

EXCURSION
Choosing the Operation

Find the deep structure in each of these problems. . . . Decide which operation to use in identifying the missing number.

a. $\square - 13 = 6$ d. $\square - 4\frac{2}{3} = 5$

b. $678 = 295 + \square$ e. $25 = 32 + \square$

c. $\frac{1}{2} = \square - \frac{1}{3}$ f. $14 - \square = {}^{-}23$

If you are not sure, turn to page 80 for more information.

a. $6 + 7 = \square$
b. $7 + 6 = \square$
c. $\square = 6 + 7$
d. $\square = 7 + 6$
e. $\begin{array}{r} 6 \\ +7 \\ \hline \square \end{array}$ f. $\begin{array}{r} 7 \\ +6 \\ \hline \square \end{array}$

g. $\square - 6 = 7$
h. $\square - 7 = 6$
i. $7 = \square - 6$
j. $6 = \square - 7$
k. $\begin{array}{r} \square \\ -6 \\ \hline 7 \end{array}$ l. $\begin{array}{r} \square \\ -7 \\ \hline 6 \end{array}$

Each number sentence has a different surface structure, yet they all have the same deep structure. Two addends are known and a sum must be found in order to solve the problem; and whenever two addends are known, addition is used to find the sum. Therefore, these are all addition situations, regardless of the operation indicated.

5.41

addend + addend = [sum]

Do the following number sentences have a common deep structure?

a. $13 - 6 = \square$
b. $13 - \square = 6$
c. $\square = 13 - 6$
d. $6 = 13 - \square$
e. $\begin{array}{r} 13 \\ -6 \\ \hline \square \end{array}$ f. $\begin{array}{r} 13 \\ -\square \\ \hline 6 \end{array}$
g. $6 + \square = 13$
h. $\square + 6 = 13$
i. $13 = 6 + \square$
j. $13 = \square + 6$
k. $\begin{array}{r} 6 \\ +\square \\ \hline 13 \end{array}$ l. $\begin{array}{r} \square \\ +6 \\ \hline 13 \end{array}$

In each case, the sum and one addend are known, and an addend must be found in order to solve the problem. Whenever the sum and one addend are known, subtraction is used to find the missing addend. Therefore, these are all subtraction situations, regardless of the operation indicated.

5.42

sum − addend = [addend]

Verbal Problem Solving

Verbal, or word, problems—whether oral or written—are frequently used as contexts for writing number sentences as a part of initiating or abstracting activities. Verbal problems can also be used for transferring activities; they provide a context for applying previously learned knowledge and skills. In order to successfully solve verbal problems, a child must understand the operations, know basic facts, and have skill in sentence-solving.

In both sentence and verbal problem-solving, the choice of operations depends upon what is given and what is wanted. For one-step verbal

EXCURSION
Choosing the Operation (Answers)
From page 79

a. Addition (6 + 13 or 13 + 6)
b. Subtraction (678 − 295)
c. Addition ($\frac{1}{2} + \frac{1}{3}$ or $\frac{1}{3} + \frac{1}{2}$)
d. Addition ($4\frac{2}{3} + 5$ or $5 + 4\frac{2}{3}$)
e. Subtraction (25 − 32)
Note: The procedure for subtracting or computing $25 - 32 = {}^{-}7$ also involves subtracting whole numbers ($32 - 25 = 7$).
f. Subtraction ($14 - {}^{-}23$)
Note: The procedure for subtracting or computing $14 - {}^{-}23 = 37$ involves adding whole numbers.

problems involving addition or subtraction, either two addends are given and addition is used to find the sum, or the sum and one addend are given and subtraction is used to find the unknown addend. In sentence solving, known and unknown numbers can be identified by observing the placement of numerals and variables to the left and to the right of the sign of operation. This is not possible with a verbal problem; it is necessary to consider the model for addition or subtraction reflected within the problem. Consider the following example.

PROBLEM

> Joyce had some pennies. She spent 5 pennies on a piece of candy, and she has 7 pennies left. How many pennies did she have to begin with?

To solve a verbal problem such as this, one should think of a whole set and its parts. What is known? The size of each of the parts. What is unknown? The number in the whole set. Both addends are known; the sum is unknown. Obviously, addition is used to find the unknown sum. An open number sentence can be written to show the number of pennies Joyce had initially; that is, $5 + 7 = \square$.

Unfortunately, many children write the number sentence $7 - 5 = \square$ and say that the answer is 2. They have somehow learned to look for so-called key words, using them as the basis for deciding the operation. In this case, the word *left* would seem to be a cue to subtract. Obviously, such cues are unreliable; mechanistic procedures such as using key words ignore what the operations mean in terms of addends and sums. Eventually key words leave the child confused.

Inside the back cover of this book is a list of verbal problem types for each of the four operations. On the addition list, for one-step verbal problems involving addends and a sum, at least four different types of verbal problems call for the use of addition to find the unknown sum. Eight different types of verbal problems are included for subtraction.

A variety of types of verbal problems should be presented for addition and subtraction. The analysis of problems should focus on identifying the parts and whole (addends and sum), whether known or unknown. This variety will help a child schematize information and develop skill in analyzing verbal problems of all types. In contrast, intensive practice with only one type of verbal problem may encourage children to learn specific routines (such as word cues) that produce correct solutions only part of the time. Other ways of varying a child's experiences with verbal problems are presented in Chapter 11.

Make sure that a variety of different types of verbal problems are included in the textual material you select for instruction. The text should probably be supplemented with problems of your own. Also, analyze the tests on problem-solving that are administered to your class. Do the tests assess skill in solving a variety of problems? The list of verbal problem types inside the back cover will be a useful aid in your own analysis of textbooks and tests, and when planning supplementary experiences.

ORDERING THE CONTENT

There is no fixed order for introducing subject matter components when teaching basic facts for addition and subtraction. However, the content should not be taught at random. Consider the logical relationships among various subject matter components: number concepts and number sentences, the meaning of an operation and the symbol denoting it, the meaning of subtraction and the meaning of addition, verbal problem-solving skill and the meaning of the operations, and so forth. The components are sometimes interdependent.

Initially, you should use sums of less than 10. A child can learn a great deal about meanings of the operations, basic facts, sentence solving, and problem solving while working exclusively with one-digit numbers. After we focus on the objectives with one-digit numbers (or while we are working with sentence solving and problem solving), notational concepts for two-digit numerals can be introduced. Children will then understand the addition that is implied in two-digit numerals as they study facts with sums of 10 or more.

6

Introducing Addition and Subtraction: Planning Instruction and Assessment

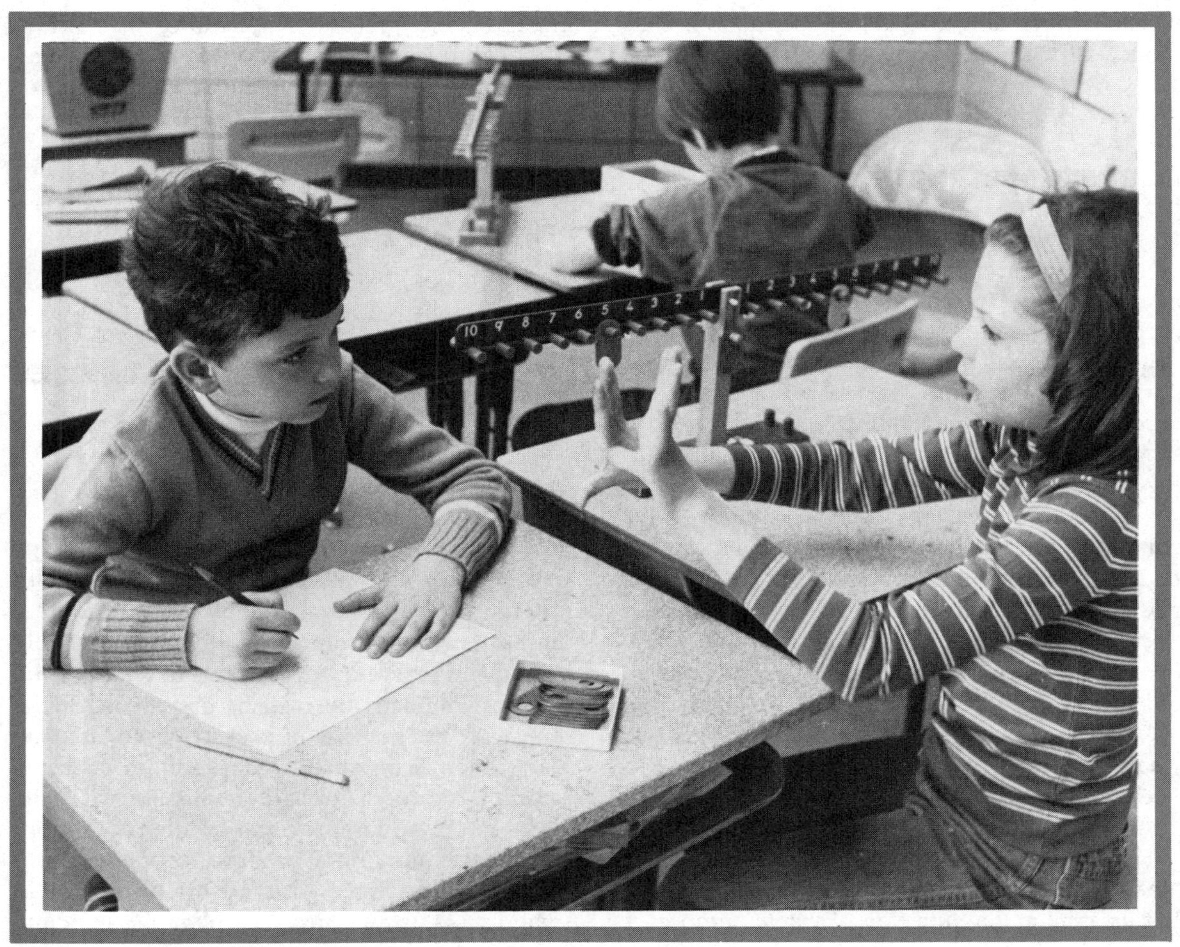

Selecting content objectives, and carefully planning the flow of instructional activities and assessment are equally important. In Chapter 3 the authors introduced the Activity Type Cycle which is a guiding model for instruction. This model will be used frequently as we plan.

PLANNING INSTRUCTION

Instructional activities should accomplish specific purposes according to the objective or objectives selected. Initiating, abstracting, schematizing, consolidating, transferring, and diagnostic activities should be planned with the Activity Type Cycle in mind. Often an activity will accomplish more than one purpose, so we can speak of an initiating/abstracting or an abstracting/schematizing activity.

The authors have selected some of the content objectives listed in the previous chapter to illustrate how to plan lessons for introducing addition and subtraction. Initially, we shall focus on the questions we need to answer, and many instructional activities will be described for given objectives.

The first question to ask is the same one we asked in preceding pages: What content shall I teach?

Meanings, Models, and Symbols

Let us assume that addition of whole numbers is being introduced. Content objective 5.2, which focuses on the union of disjoint sets as a model for addition, is often taught at this level. First, the number of items in each of two sets is usually recorded, then the number of items in the union set. Ultimately, a number sentence (such as $2 + 3 = 5$) is developed. Often, sets are actually joined. Most teacher's editions of first grade textbooks suggest activities of this type.

Content objective 5.3 will be used to illustrate the process of planning instruction. Although this is a less typical way of introducing addition, it can be a very fruitful approach.

> 5.3
> A set may be partitioned into two disjoint subsets, even if one or both of the subsets is empty. The subsets and the original set model the relationships between two addends and their sum.

Initiating and Abstracting Activities for Addition

Where should we start in the Activity Type Cycle? It is not always appropriate to begin with initiating activities. For example, when introducing symbols, abstracting activities will probably be planned first; and for concepts of relationship, schematizing activities will be prepared. Always remember the purpose of each type of activity.

In this instance, when addition concepts are introduced, each child should use what she already knows to improvise a solution to a problem or task, and, in the process, encounter examples of the new idea without formal terms and symbols. Thus, we need to plan an *initiating* activity. Our choice of exemplars for an *initiating* activity is relatively easy, for content objective 5.3 uses sets. Each child can be given a set of objects, such as a set of 5 blocks or discs. After the objects are passed out, ask the children to find all the ways they can use the set of 5 to make 2 sets.

Write out the different answers on chalkboard or chart paper, much as you would make a record during a language experience for instruction in reading, or as follows:

- Five is two and three.
- Five is four and one.
- Five is two and two and one.
- Five is one and one and one and one and one.
- Five is one and four.
- Five is three and one and one.
- Five is three and two.
- Five is zero and five.
- Five is five and zero.

Children will generate statements that include zero, if they understand the concept of an empty set and zero as the number related to the empty set, and if you refer to the blocks as a *set* of blocks. Each child can also take pencils or crayons and separate the blocks into different sets.

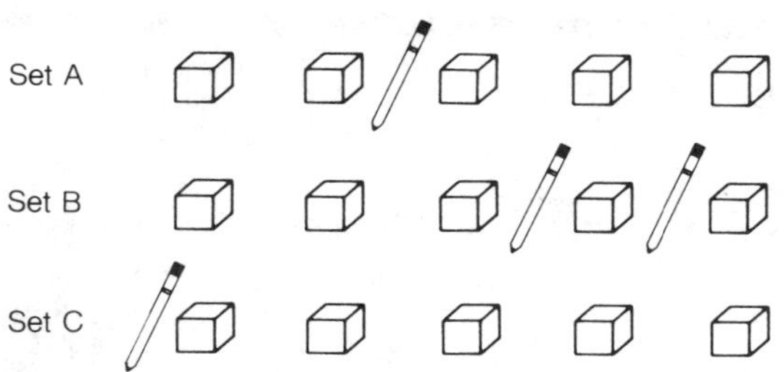

What can we observe that will indicate that a child is learning and is ready to move on to an abstracting activity for the same objectives? What behavioral indicators shall we use? Look for behaviors such as the following:

Given 5 blocks, the child:
1. Separates the blocks into sets.
2. Names the numerousness of each set.

Our plan for an initiating activity can be summarized in a one-page activity plan similar to the plan shown in Figure 6.1.

ACTIVITY PLAN

Addition and Subtraction of Whole Numbers: Meanings, Models, and Symbols | ***3-D/Rep → Symbolic*** | ***Initating***

Content objective

5.3

Exemplars/Materials

Blocks (5 for each child)
Pencils or crayons for partitioning a set
Word sentences written on chalkboard

Behavioral indicators

Given 5 blocks, the child:
1. Separates the blocks into parts or subsets.
2. Names the numerousness of each part or subset.

Procedure

1. Give each child a set of 5 blocks.
2. Say, "Let's find all the different ways we can break this set of 5 blocks into parts or subsets."
3. As needed, help individuals partition or rearrange the set, possibly by using a pencil or crayon to separate the set into subsets.
4. Record what the children observe much as you would record a language experience for instruction in reading. For example, "Five is two and three" and "Five is three and one and one."

Figure 6.1
Sample Activity Plan for an Initating Activity

EXCURSION Instructions for making a flannel-and-chalk board for each child

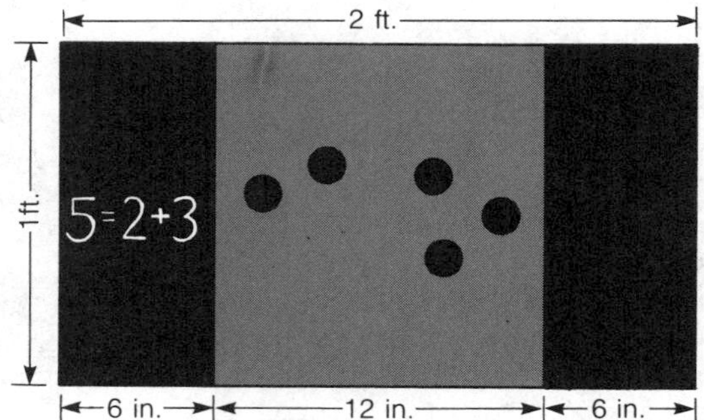

1. For each board you need a 1′ × 2′ piece of masonite (not pegboard). One 4′ × 8′ sheet of masonite can be cut into sixteen 1′ × 2′ sections.
2. Paint each board with chalkboard paint, or with paint used for tennis tables.
3. On the center of the board, glue a one-foot square of flannel or felt.
4. Prepare circular or square felt shapes for use with each board.

The board is a flexible teaching tool, especially useful for initiating, abstracting, and schematizing activities for whole number operations.

A collection of such plans for given content objectives, including different activity types, constitutes a teaching unit, and will be discussed later. In the activity plan in Figure 6.1, certain information is summarized at the very top of the page for future reference: content objective, activity type, and notation of the specific translation involved across classes of exemplars. In a teaching unit, you can include a variety of exemplar class translations for different activities of the same activity type.

Alternative initiating activities can also be considered. Each child should have a set of materials that can be easily rearranged, such as bottle caps, blocks, and plastic discs.

In teaching the concept, each child can be given 5 Unifix cubes as illustrated in Figure 6.2. The cubes can be fastened in a row, then the children can break the set apart in different ways. (Record their responses.)

Children also enjoy the tactile qualities of felt, so we may want to use felt shapes on flannel boards. Each child should have her own small flannel board and a set of shapes to rearrange into subsets.

Cuisenaire rods can also be used. Children should know the value or number name for each rod, when the white rod is assumed to have a value of one. The staircase in Figure 6.3 illustrates relationships among the rods.

Have each child select a 5 (yellow) rod. Then have each make trains of rods that are just as long as the yellow rod. Ask "How many different trains can be made?" (See Figure 6.4.) Record their responses, but write out the number names

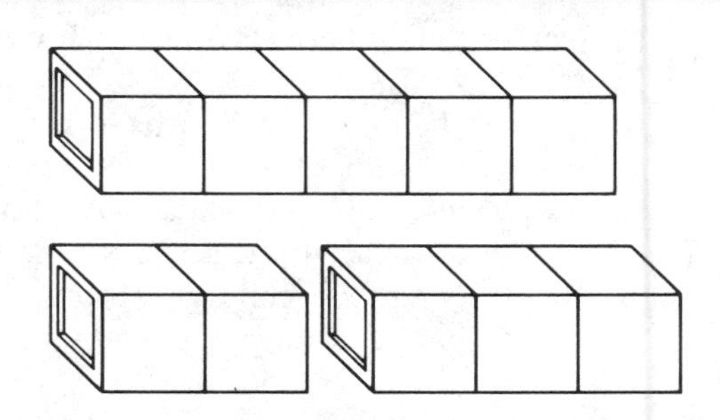

Figure 6.2
Unifix cubes to show that 5 = 2 + 3

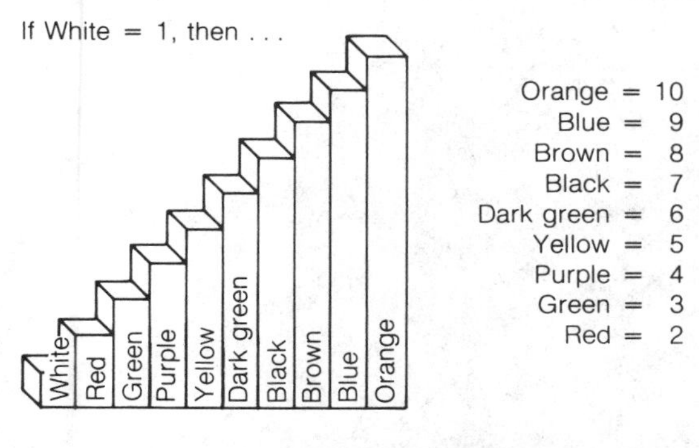

Figure 6.3
A Staircase of Cuisenaire Rods

instead of colors (for example, Five is three and two).

A similar activity involves the Stern blocks and number cases. Give each child a 5 case and ask her to fill it with Stern blocks so that each row is different. You may want to suggest that each row have no more than two blocks, as illustrated in Figure 6.5.

Most children will need several initiating activities before moving on to abstracting activities.

Planning *abstracting* activities, requires focusing on more than one objective, with concern for symbols, models, and the meaning of addition. Abstracting activities should help each child pull out, name, and learn symbols for concepts from specific instances. Hopefully, such activities will also help each child reflect upon the procedure used in abstracting new concepts.

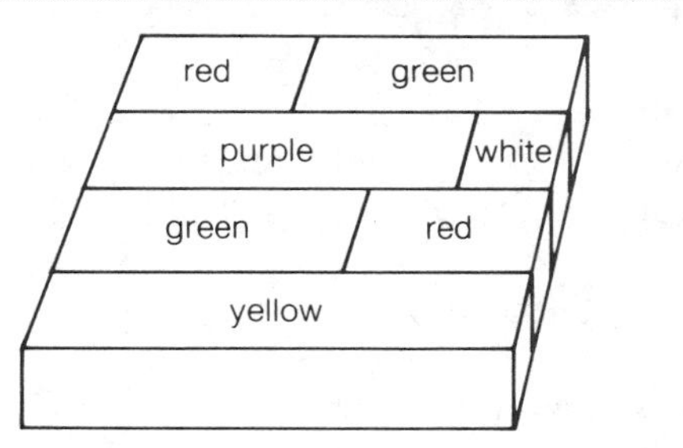

Figure 6.4
Equivalent Trains Made of Cuisenaire Rods

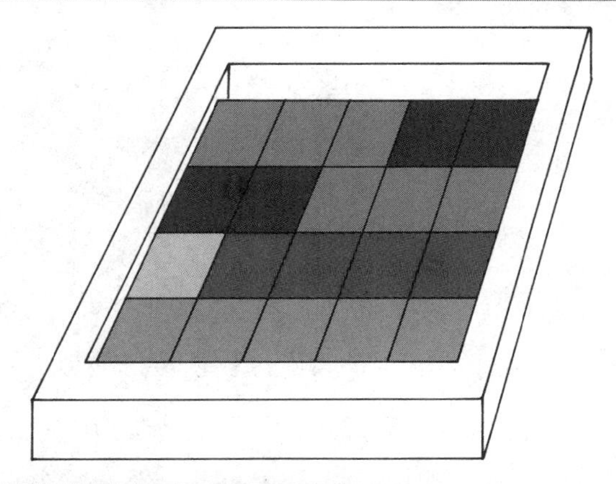

Figure 6.5
Stern Number Case and Blocks for Five

For abstracting activities, we can use the same exemplars as with the initiating activities, as well as others to be described later. For example, felt shapes on a flannel board in connection with a chalkboard can be utilized. Describe a number situation, such as: "Barbara has five pennies. Three are brand new pennies, and two are very old pennies." Interpret the story with felt shapes and a chalkboard record as illustrated in Figure 6.6.

The basic strategy of the activity is to start with an English language sentence, then substitute mathematical symbols as appropriate. For example, a number sentence can be recorded on the chalkboard:

Five is three and two.

Symbols can be introduced through conversation similar to that which follows.

Say: "We already know how to write numerals, so we can use numerals to write this more easily." Erase the number words and substitute numerals.

5 is **3** and **2.**

Say: "We have a special sign we can put in place of *and*. We call it *plus*." Erase *and* and write +. Have the children read the sentence. Say, "The plus sign tells me that the numeral on this side [point] and on this side [point] show how many are in each part."

5 is 3 + 2

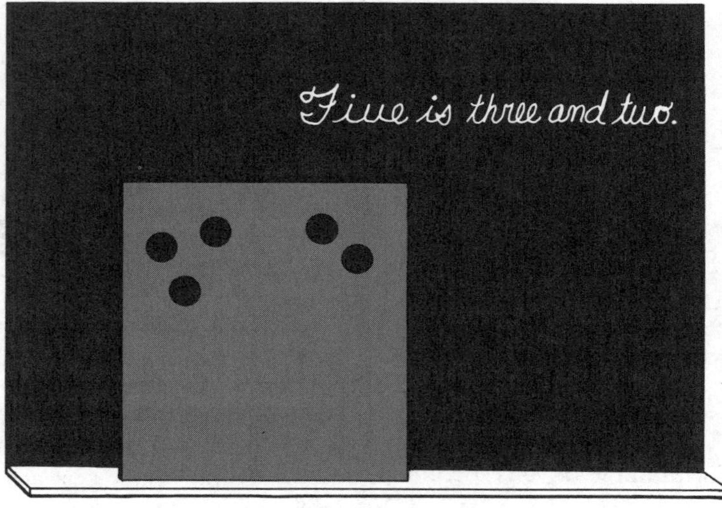

Figure 6.6
Beginning the Lesson

Say, "Another special sign can be substituted for *is*. We call it *equals* and it means *is* or *is the same as*." Erase *is* and write =.

$$5 = 3 + 2$$

Point out that not only is it true that *five* is the same as *three plus two*, but *three plus two* is the same as *five*. Write both sentences.

$$5 = 3 + 2 \qquad 3 + 2 = 5$$

Distribute paper to the children and say, "Let's write down as many names for five as we can." Their suggestions will probably include:

$$5 = 4 + 1$$
$$5 = 3 + 1 + 1$$
$$5 = 2 + 3$$
$$5 = 1 + 1 + 1 + 1 + 1$$

During the activity, introduce the terms *addend* and *sum* gradually, as parallel terms for *number in a part* and for *total amount*. You may respond to a child's comment by saying, "Yes, 3 is the number in one part; it is an addend." Also, continue to ask the children whether a given numeral tells about a part or about the total amount, such as, "Is it an addend or a sum?" If a child responds that it is an addend, then ask the child how she knows. A child should be able to reply that it is next to a plus sign. Move on to other number stories for different sums, with children using materials to interpret the story and recording what they observe with number sentences.

What behaviors indicate that a child understands the concepts? A careful look at the cluster of selected objectives can be helpful. The following are some indicators for the previous activity:

1. Given a set, the child can partition it into two subsets and record what is observed with a number sentence.
2. The child reads + as *plus*.
3. Given an addition phrase, the child states that numerals on either side of the plus sign are addends, or that they tell about parts of the set.
4. The child calls the number for the whole set a *total* or a *sum*.
5. Given number sentences with the sums at both the right and left, the child can identify and name each numeral as an addend or a sum, or as showing how many are in a part or in the whole set.

ACTIVITY PLAN

Addition and Subtraction of Whole Numbers: Meanings, Models, and Symbols

3-D/Rep → Symbolic

Abstracting

Content objectives

5.1, 5.3, 5.6, 5.7, 5.8, and 5.9

Exemplars/Materials

Flannel board and felt pieces
Word sentences on chalkboard
Number sentences on chalkboard and paper

Behavioral indicators

The child:

1. Reads + as *plus*.
2. Given an addition phrase, states that numerals on either side of the plus sign tell about parts of the set (addends).
3. Uses *total* or *sum* to designate the number of the whole set.
4. Given addition number sentences with the sum at the right as well as at the left, identifies and names each numeral as showing how many are in a part or in the total set (an addend or a sum).

Procedure

1. On a flannel board, let 5 felt pieces represent 5 pennies.
2. Regroup pieces as 3 new pennies and 2 old pennies. Write out as: Five is three and two.
3. Introduce the plus sign and substitute, then introduce the equals sign and substitute.
4. Show that the sides of the equation can be reversed.
5. Distribute paper and have each child write as many names for 5 as she can.
6. Use 7 felt pieces to show 7 toy cars. Regroup pieces to show 4 red cars and 3 blue cars. Develop as above.

Figure 6.7
Sample Activity Plan for Abstracting Selected Addition Objectives

The abstracting activity just described is summarized in Figure 6.7. The content area and activity type are noted at the top of the plan. You also find the phrase *3-D/Rep→Symbolic* which should be read as "from three-dimensional representational to symbolic," and refers to the translation that occurs between exemplar classes during the activity.

Several alternative abstracting activities can be planned for one or more content objectives. These vary by using different exemplar class translations and by using different exemplars, even for the same translation.

Exemplars that can be used include Arithmablocks, Hainstock blocks, a math balance, and a number line. A wall of Arithmablocks could be built on the 5 block, as shown in Figure 6.8. Arithmablocks are helpful in initiating or abstracting activities for addition and subtraction because the numerals illustrate that 3 and 2 are the same as 5.

Cuisenaire rods, Arithmablocks, and similar materials model number ideas with units of length; Hainstock blocks, illustrated in Figure 6.9, incorporate a set model, providing a helpful contrast. Hainstock blocks are particularly effec-

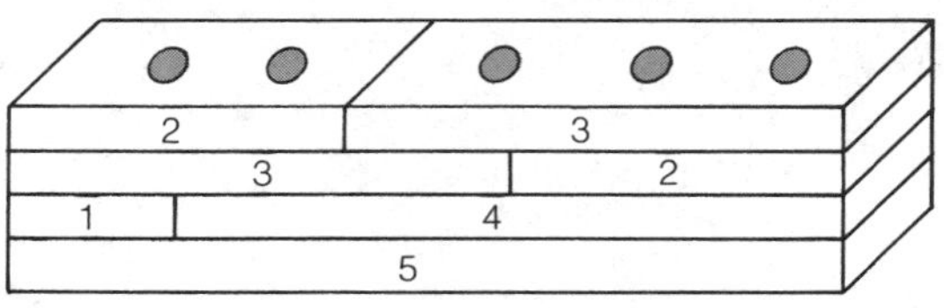

Figure 6.8
Arithmablocks Used for Teaching the Meaning of Addition

tive for exemplifying the basic addition facts with zero as an addend. Children are also curious about the set partitioning results when they shake the blocks.

If Hainstock blocks are not available, one of the following can be used in their place:

- *Divided box with a lid.*[1] Partition a small box with a lid as shown. Place a numeral on the lid and the same number of beans inside the box. Keep the lid on while shaking the box, then remove the lid.

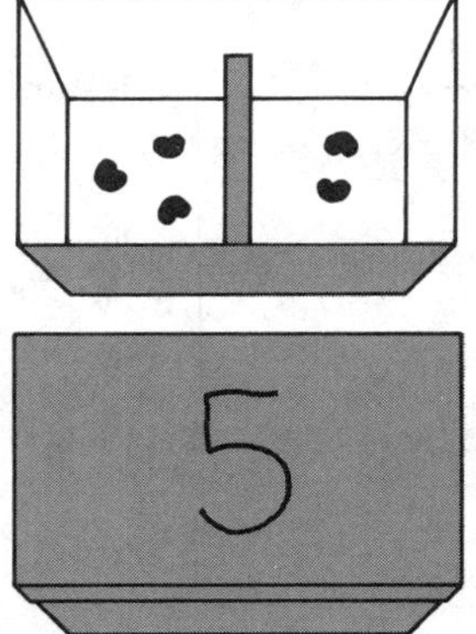

- *Beans sprayed with paint.* Spray beans with paint on one side only. Place a numeral on a cup and the same number of beans inside the cup. When beans are rolled, some beans have the painted side up while others show the unpainted side.
- *Transparent box.* Place dividers in small clear plastic boxes, and show a numeral at the left. Dried beans can be used inside the box.

1. Douglas Cruikshank, David Fitzgerald, and Linda Jensen, *Young Children Learning Mathematics* (Boston: Allyn and Bacon, 1980):83.

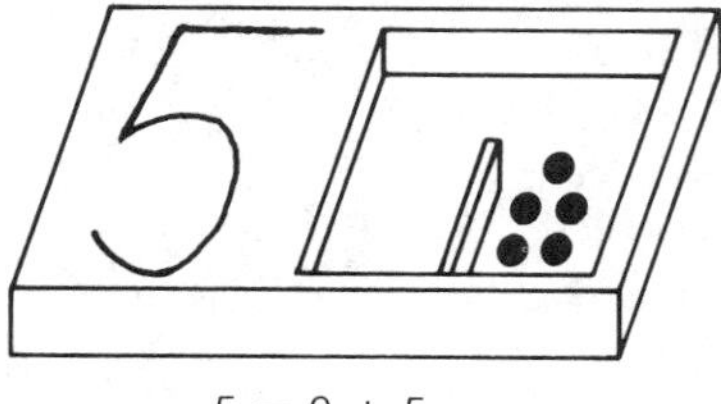

5 = 0 + 5

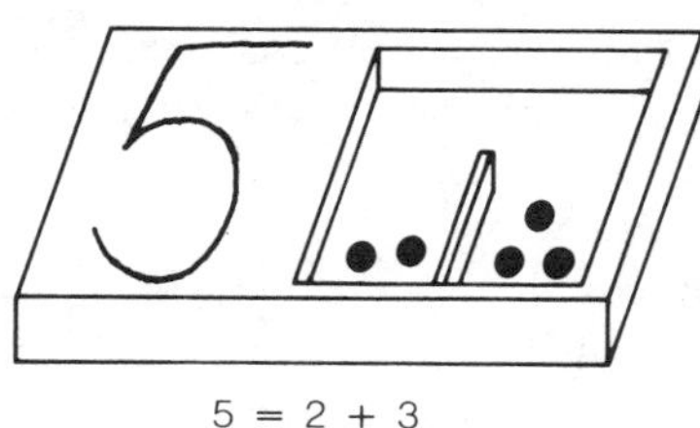

5 = 2 + 3

Figure 6.9
Hainstock Blocks Used for Teaching the Meaning of Addition

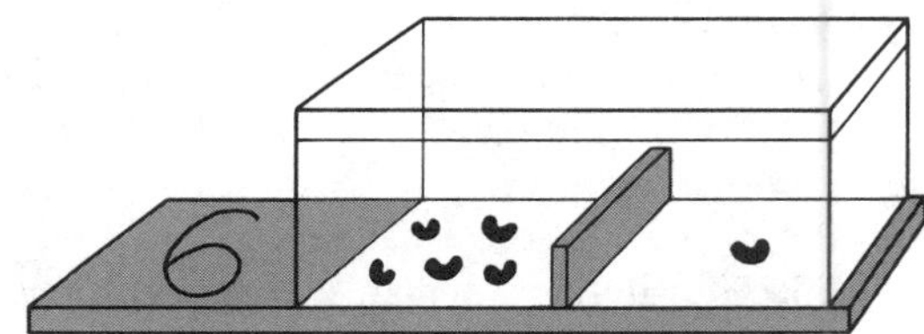

The math balance is another aid to help children abstract the meanings of addition and subtraction. Children enjoy using the balance because of its mechanical nature; it also provides immediate feedback.

During an abstracting activity, place a weight to the left at 5, give the child two weights and ask, "Using both of these on this side [pointing to the other side], can you make it balance?" A number sentence should be written out and discussed as in Figure 6.10. Also, ask, "Are there other ways you can balance the 5 with these two weights?" Be sure to include some activities with the addends shown at the left and the sum at the right.

A math balance is sometimes confused with a number line. Figure 6.11 compares the two. Unlike a number line, the numerals on *both* sides of the fulcrum (0 point) of a math balance

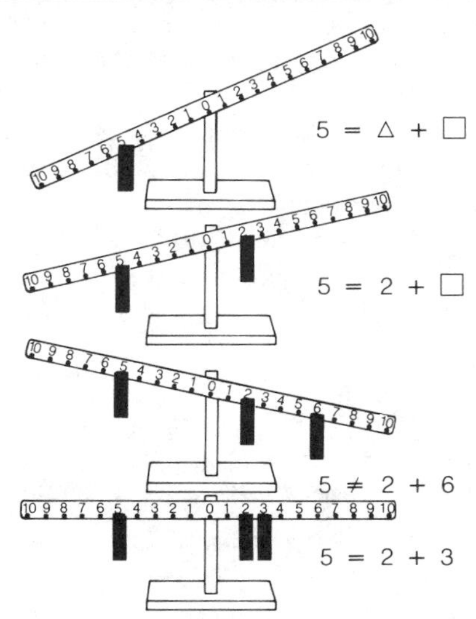

Figure 6.10
Math Balance Used to Find an Addend Pair for the Sum Five

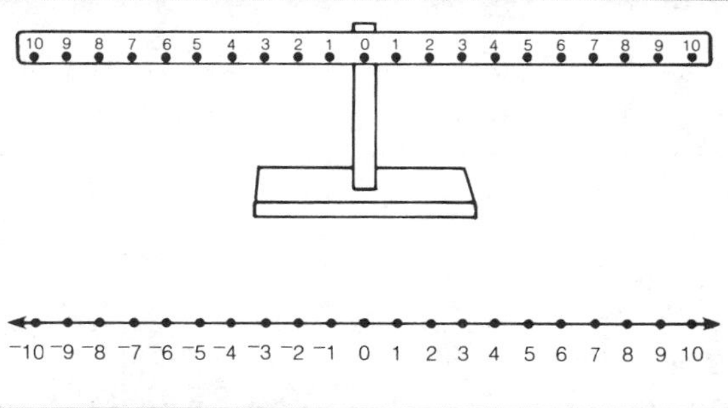

Figure 6.11
Math Balance and Number Line Compared

refer to *whole* numbers. On a number line, the numerals to the left of 0 usually designate negative integers. Throughout the line, numbers are one more to the right, and one less to the left.

The number line should be used for abstracting activities only after a child has successfully translated from three-dimensional representations to symbolic exemplars. On the number line:

1. Space the numerals at equidistant intervals, because a number line is based on a unit of measurement;
2. Place arrows at both ends, only part of the line is shown; and
3. Start at zero when showing operations on numbers.

Figure 6.12 illustrates the use of a number line during an abstracting activity that focuses on content objective 5.5 as well as other objectives. Zero is always the starting point, so begin at zero and show the sum with an arrow to the right. Ask your students to identify the arrows that are joined together for the same result.

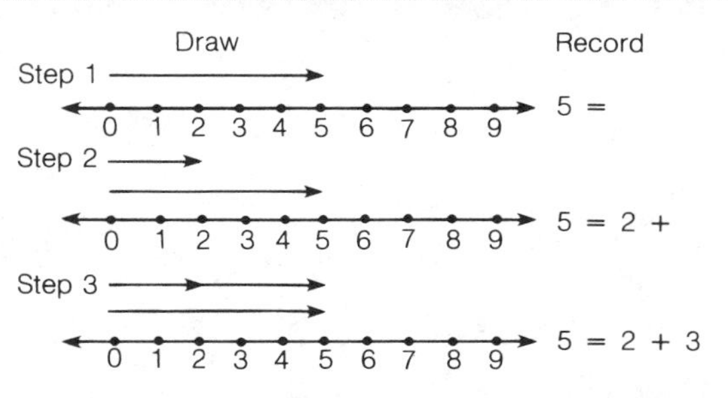

Figure 6.12
Number Line Showing the Meaning of Addition

EXCURSION

Can you use a math balance for other operations? Attempt to show the solution for each of the following by using a math balance.

$\square = 6 + 7$	$4 \times 6 = \square$
$7 - 2 = \square$	$\square = 32 \div 8$
$\square = 25 - 9$	$27 \div 6 = \square$

Diagnostic Activities for Addition

Diagnosis should be made during or after the activities just illustrated, so we can tell what the child understands about the concepts or processes. Diagnosis provides information as to both the correctness and the level of maturity of conceptual learning.

The key to providing effective diagnostic information is well-planned behavioral indicators. For example, each of the behavioral indicators listed in Figure 6.7 (page 89) suggests diagnostic tasks that can be presented to a child if the behavior described has not been observed during other activities.

Table 6.1 lists sample diagnostic tasks for each behavioral indicator. If a child can successfully complete diagnostic tasks such as these, she probably has sufficient understanding of the ideas presented. We can move on to either a more advanced activity type or other content objectives.

Table 6.1
Diagnostic Tasks Suggested by Behavioral Indicators

Behavioral Indicators	*Sample Diagnostic Tasks*
1. Reads + as *plus.*	1a. Have the child read 6 + 7. 1b. Have the child read $\begin{array}{r} 8 \\ +\ 9 \\ \hline 17 \end{array}$
2. Given an addition phrase, states that numerals on either side of the plus sign tell about parts of the set (are addends).	2a. Show the child 6 + 7 and say: What does the 6 tell us? How do you know? What does the 7 tell us? How do you know? 2b. Show the child 8 + □ and say: What does the 8 tell us? How do you know? What does the □ tell us? How do you know?
3. Uses *total* or *sum* to designate the number of the whole set.	3a. Show the child 9 = 5 + 4 and say: The 9 tells us the number in the whole set. What do we call a number for the whole set? 3b. Show the child 8 = 3 + 5 and say: Point to the total. Now point to the sum.
4. Given number sentences with the sum at the right as well as at the left, identifies and names each numeral as showing how many are in a part or in the total set (an addend or a sum).	4a. Show the child 9 = 3 + 6 and 6 + 2 = 8 and say: Point to each numeral in these number sentences. Tell me if it tells how many are in part of a set, or the total set (an addend or a sum). How do you know? 4b. Show the child □ = △ + ◇ and ◇ + △ = □ and say: These are number sentences, but the numerals are missing. Point to each shape. Tell me if the number will tell how many are in part of a set, or the total set (an addend or a sum).

Initiating and Abstracting Activities for Subtraction

At this point in planning instruction, it is best to introduce initiating and abstracting lessons for subtraction rather than schematizing activities for addition alone. Schematizing activities should focus on relationships between addition and subtraction. If you do decide to introduce schematizing activities for addition first, be sure to provide activities for the combination of operations later on.

Problems children can solve by using what they already know about addition can be used for initiating activities for subtraction. At this stage, initiating and abstracting activities are frequently combined.

We shall be concerned with the meaning of subtraction, and with models and symbols as well. Therefore, we will focus on the cluster of content objectives 5.14–15, 5.20–23 that deal with those concepts.

Although most of the exemplars previously described can be used, the felt shapes on a flannel board are convenient for interpreting a simple number situation. The basic strategy of the activity is to start with a language sentence and substitute a number sentence.

1. Say, "Bob has 5 fish in his aquarium." (On flannel board show 5 felt pieces, then cover 2 with your hand.) "Three are goldfish and the others are guppies. How many are guppies? How do we know there are 2 guppies? We know that 3 and 2 are parts for 5." (Uncover the 2 felt pieces.)
2. Say *and write out* on chalkboard, "If Bob has 5 fish and 3 are in one part, then 2 are in the other part."
3. Say, "There is a very short way to write that long sentence." Under the sentence write 5.
4. Say, "Five is the total." "What part did we know about in the first place?" Continue to record: 5   3.
5. Say, "We were looking for the other part, which is 2." Continue to record: 5   3   2.
6. Say, "So we have these numbers: 5, 3, and 2."
7. Continue to record: 5 − 3   2. Point to the minus sign and say, "Put a little line like that. It goes straight across and is called *minus*. Say *minus*."
8. Insert the equals sign: 5 − 3 = 2. Say, "This says '5 minus 3 is 2.' "
9. Say, "What does the minus sign tell us? The 5 in front of it is the total. The 3 after it tells how many are in one of the parts. The total number and the number in one part tells us the number in the other part [point to the 2] because we know that 3 and *2* are 5."
10. Say, "Could we show that with things? How many things do we need?" Point to 5 felt pieces on the flannel board.
11. Say, "How many were in the part I knew about at first?" Move 3 flannel pieces to the left. Say, "How many in the other part?" Move the other 2 flannel pieces to the right.
12. Say, "This is what we did when we were learning addition, but just turned around." Remove the felt pieces.
13. Say, "Suppose you already have 3 pennies, and you want to buy something that costs 7 pennies. What is the total? The 7 is the total; it is all that is needed. And what's the 3? It is part of what's needed. What will you need to know? The other part."
14. Say, "We can write it like this." Record: 7 − 3 = □. Say, "We show the 7 is a total and the 3 is a part of the total by using a minus sign. This will tell us what the other part is."
15. Say, "What is the other part? How do we know the other part is 4? We know the other part is 4 because 3 and *4* is 7." Write: 3 + 4 = 7, then make a 4 in the box. Say, "We can use what we already know to figure out the missing part."
16. Say, "If Mary has only 2 pennies, and she wants to buy something that costs 6 pennies, how many more pennies will she need?" Begin as with 13 above, using the number sentence 6 − 2 = □.

This activity moves from a situation in which subsets already exist to one in which part of a set is needed. Some teachers would question the wisdom of immediately exposing a child to different types of subtraction situations. They prefer to have a child master one type before introduction to another.

However, the authors stress basically one meaning for subtraction—in terms of a sum and a pair of addends—the number of members in a set and in each of the two subsets. Given this emphasis, exposure to varied situations can help a child understand the meaning of subtraction as the same for *all* subtraction situations encountered. Similarly, each symbol used in a number sentence always has the same meaning. There is really less for a child to learn than when each subtraction situation is presented as a separate case to be learned.

What behavioral indicators should we specify for the subtraction initiating/abstracting activity? Examples of indicators might include:

1. Reads – as *minus*.
2. Given a subtraction phrase, states that the numeral to the left of the minus sign tells how many are in the whole set (is the sum), and the numeral to the right of the minus sign tells how many are in one of the parts (is an addend).
3. Given subtraction number sentences, identifies and names each numeral as showing how many are in a part or in the total set (an addend or a sum).

Other activities for initiating and abstracting subtraction should include different exemplar class translations. They should also emphasize vertical as well as horizontal notation, because children use vertical notation when they learn procedures for computing. Make sure each child has many opportunities to identify sums and addends in notation.

What alternative exemplars can be used? A sum can be exemplified with a set of objects: bottle caps, plastic discs, crayons, paper clips. Unifix cubes can be used, either as a set of discrete objects or fastened together. For example, children might work in pairs and take turns doing the following with seven attached Unifix cubes:

- Break your stick into two parts and hide one part behind you.
- Show the other part and see if your partner can tell how many cubes you are hiding.
- Show the hidden part—is your partner right?

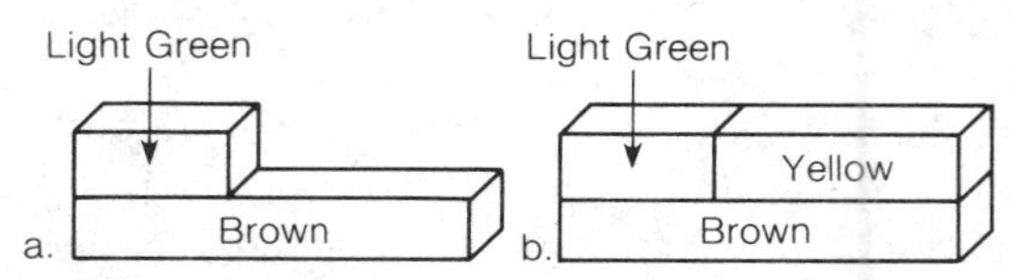

Figure 6.13
Cuisenaire Rods Used for Subtraction

Cuisenaire rods can also be used. A rod (a sum) is selected and an equivalent two-car train of rods is constructed. If a white rod is defined as 1, then Figure 6.13a shows 8 – 3. Why? With Cuisenaire rods, it is always necessary to say what rod is a unit before any rod can indicate a number. The problem posed is: What rod will complete the train so it is just as long? The completed train illustrates 8 – 3 = 5. Stern blocks and Arithmablocks can be similarly used.

Hainstock blocks are especially useful for abstracting activities when teaching subtraction. The numeral on the block tells the total number, even when part of the balls are hidden. Cover one part as shown in Figure 6.14, and pose the question: How many are in the missing part? You may want to write 7 – 3 = □ on the board. A child is able to tell how many are in the missing part when he or she applies the fact that 3 and 4 is 7. The child will be able to verify his answer when you remove your finger.

Figure 6.14
Hainstock Block Used for Subtraction

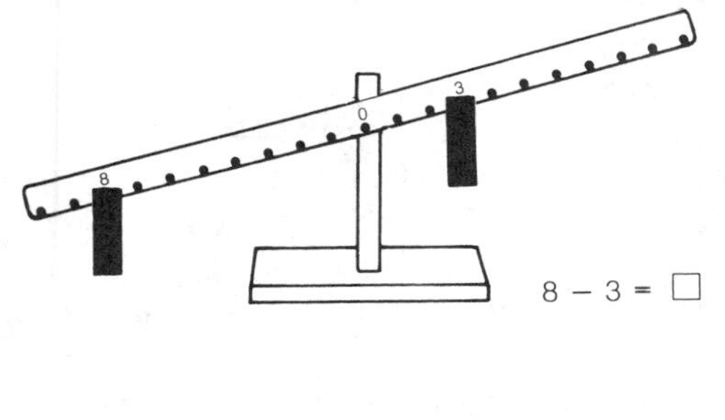

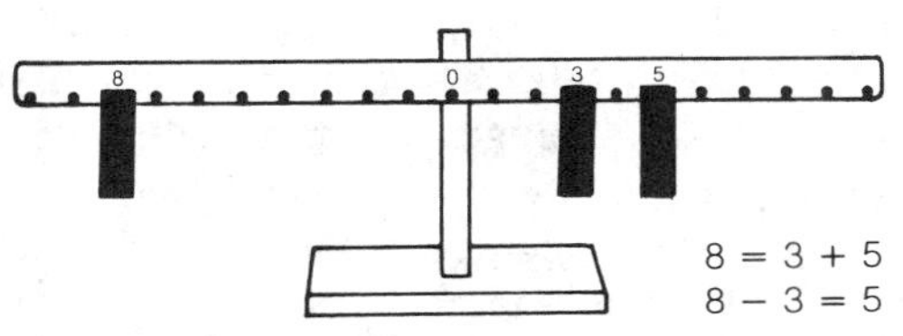

Figure 6.15
Math Balance Used for Subtraction

A math balance can also be helpful in interpreting a subtraction situation or a number sentence. Begin by having the total number shown on the left, and the number for the known part on the right. Later, provide an example with the total on the right. Figure 6.15 shows how $8 - 3 = \square$ has been interpreted on a math balance. In solving the problem, a weight was placed on the 5 peg because 3 and 5 is 8. A math balance illustrates that subtraction tells us the missing addend, rather than showing subtraction as taking away.

A number line should be used in a way similar to the way it is used for addition. Both the known addend and the sum should be shown with arrows. Begin by showing the sum as illustrated in Figure 6.16. The known addend can be illustrated in either of two ways. Figure 6.16a clearly demonstrates that a subtraction situation has the same elements as addition: a sum and two parts. (In addition the sum is unknown; in subtraction one of the addends is unknown.) Figure 6.16b makes it possible to determine the number of the unknown part directly from the number line without counting. Both ways of using the number line can be related to the con-

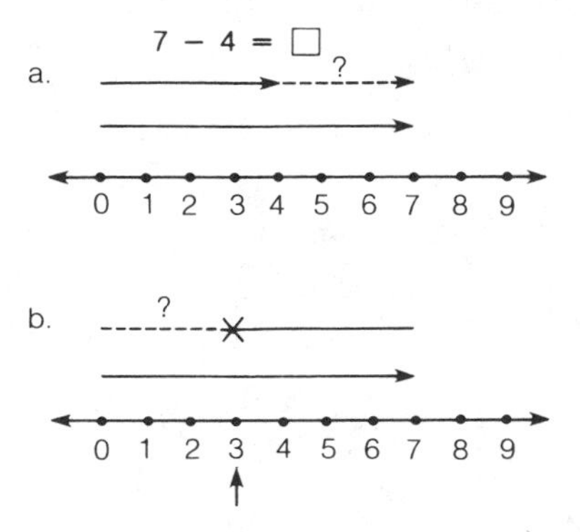

Figure 6.16
Subtraction Illustrated on the Number Line

cept of a sum and two addends. However, in determining the number of the unknown addend, neither counting nor reading numerals on the number line should be emphasized. In Figure 6.16 the missing addend is 3, because 4 and *3* are addends for 7. Make sure that each child is using what she already knows about addition.

Eventually, abstracting activities will be prepared for the following content objective:

> **5.16**
> Comparison by matching one-to-one the elements of two disjoint sets is a model for subtraction of whole numbers.

How can children think about subtracting as finding the number of things in a missing part when they are presented with a situation comparing two disjoint sets? Children can be taught to perceive that the greater set has two parts: one matched and the other unmatched. The matched part is equivalent to the lesser set (known addend) so that the number of things in the greater set (sum) minus the number in the matched part (known addend) equals the number in the unmatched part (unknown addend). Consider the following problem situation:

PROBLEM

Janet's father is serving cupcakes at her party. Seven cupcakes have chocolate icing but only 3 cupcakes have lemon icing. How many more cupcakes have chocolate icing than have lemon icing?

The number in the larger set (sum) is 7, and the number in the matched part (addend) is 3. We know that the number in the unknown part (addend) is 4, because 3 and *4* is 7.

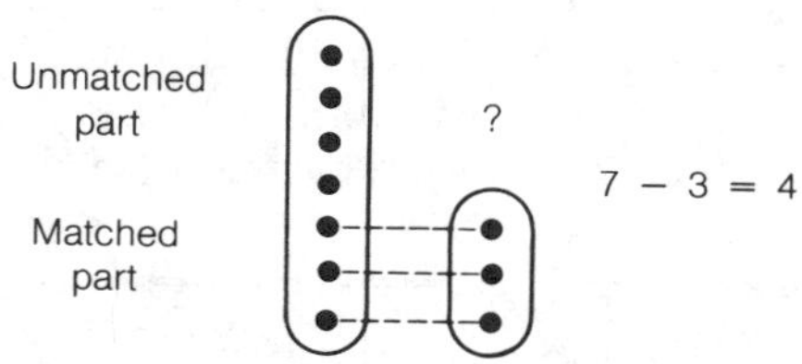

Children need help in using this pattern of thinking over a procedure for counting and matching. When they work with larger numbers, they will not be able to draw sets and count. Instead, they need to be able to identify each number as an addend or a sum.

Schematizing Activities

Once a child abstracts the meanings for addition and subtraction, models those meanings with different exemplars, and uses symbols appropriately, well-designed schematizing activities which help the child perceive different relationships can be provided. We must build the kind of conceptual framework that facilitates both further understanding and recall of information. This way, children will memorize many of the basic facts of arithmetic even though memorization is not emphasized.

Recall that a schematizing activity helps a child identify relationships between two or more concepts that have already been abstracted. It involves looking for patterns, finding interrelationships among concepts, and testing what is observed to make sure it is true. Typically, two or more concepts are juxtaposed in a way that emphasizes their differences and similarities. Sometimes the correct sequencing of concepts is emphasized as a process is studied. Techniques include constructing tables, graphs, summaries, listing or diagramming steps, making flow charts, and constructing proofs.

The same content objectives that were used for initiating and abstracting activities may be used for schematizing activities as well. But other content objectives that specifically state interrelationships can also be included. For example, when planning schematizing activities for the meanings of addition and subtraction, we eventually want to include the following:

5.18
Subtraction is the inverse of addition.

5.19
Elements in each row of an array of objects can be successively partitioned into a subset and its complement. This forms a model for all the possible pairs of addends for a given sum and their inverses for subtraction.

5.24
For every simple subtraction sentence involving unequal addends, there are three other sentences that state the same relationship among the sum and its addends: one subtraction sentence and two addition sentences.

For purposes of illustration, we have chosen the cluster of objectives 5.4, 5.15, 5.18–19. These objectives refer to the partitioning of rows of objects and the inverse relationship between addition and subtraction.

Choosing exemplars can be difficult because the relationship between addition and subtraction can be shown in so many different ways. The flannel-and-chalk board, for instance, allows each child to actively participate. The activity could involve building an array of felt squares, as described in Figure 6.17, with the rows carefully sequenced, allowing the child to observe a variety of relationships. After each row is completed, an addition number sentence is recorded at the left, and a subtraction number sentence at the right. After each row is demonstrated and discussed, children make the same row and number sentences on their individual flannel-and-chalk boards. Later, they are given a blank Parts Page and complete it for 7 as the sum, as shown in Figure 6.18.

A schematizing activity similar to the one just described could be planned by using magnetic discs on a magnetic chalkboard, with number sentences recorded at the side. Unifix cubes can also be used. Two colors are needed, and each row of the array is shown as illustrated in Figure 6.19 (page 100). Number sentence cards can be sorted and placed on either side.

In many schematizing activities, a child constructs a table of basic addition facts. Examples of such tables are illustrated below. While children look for patterns and schematize information they already know, they often discover new ideas.

1. A table of basic addition and subtraction facts for a given sum:

For the sum 5:

$$\begin{array}{ll} 5 + 0 = 5 & 5 - 0 = 5 \\ 4 + 1 = 5 & 5 - 1 = 4 \\ 3 + 2 = 5 & 5 - 2 = 3 \\ 2 + 3 = 5 & 5 - 3 = 2 \\ 1 + 4 = 5 & 5 - 4 = 1 \\ 0 + 5 = 5 & 5 - 5 = 0 \end{array}$$

The facts in this type of table are sometimes called a *fact family*.

2. A table of basic addition and subtraction facts for two given addends:

For the addends 3 and 4:

$$\begin{array}{ll} 3 + 4 = 7 & 7 - 4 = 3 \\ 4 + 3 = 7 & 7 - 3 = 4 \end{array}$$

These basic facts are also sometimes called a fact family, although only a subset of the facts in the first table are included.

3. A table of basic addition facts for a given addend:

For the addend 5:

$$\begin{array}{rrrrrrrrrr} 5 & 5 & 5 & 5 & 5 & 5 & 5 & 5 & 5 & 5 \\ +\underline{0} & +\underline{1} & +\underline{2} & +\underline{3} & +\underline{4} & +\underline{5} & +\underline{6} & +\underline{7} & +\underline{8} & +\underline{9} \\ 5 & 6 & 7 & 8 & 9 & 10 & 11 & 12 & 13 & 14 \end{array}$$

4. A table of basic addition facts related to a given double:

For the double 6 + 6 = 12:

$$\begin{array}{ll} 5 + 6 = 11 & 6 + 5 = 11 \\ 6 + 6 = 12 & 6 + 6 = 12 \\ 7 + 6 = 13 & 6 + 7 = 13 \end{array}$$

5. A table of sums for all basic addition facts:

+	**0**	**1**	**2**	**3**	**4**	**5**	**6**	**7**	**8**	**9**
0	0	1	2	3	4	5	6	7	8	9
1	1	2	3	4	5	6	7	8	9	10
2	2	3	4	5	6	7	8	9	10	11
3	3	4	5	6	7	8	9	10	11	12
4	4	5	6	7	8	9	10	11	12	13
5	5	6	7	8	9	10	11	12	13	14
6	6	7	8	9	10	11	12	13	14	15
7	7	8	9	10	11	12	13	14	15	16
8	8	9	10	11	12	13	14	15	16	17
9	9	10	11	12	13	14	15	16	17	18

The latter table should not be developed until a child has been involved in many abstracting and schematizing activities that focus on the relationship between addition and subtraction, and on properties such as commutativity and an identity element. The table helps a child summarize all of this information. Construct this table *with* children, making sure that each child understands how to read it.

ACTIVITY PLAN

Addition and Subtraction of Whole Numbers: Concepts, Models, and Symbols

3-D/Rep→Symbolic
2-D/Rep→Symbolic

Schematizing

Content objectives

5.4, 5.15, 5.18–19

Exemplars/Materials

Flannel-and-chalk board and felt squares (red and blue)
Number sentences written on chalkboard and paper
Parts Pages and crayons

Behavioral indicators

The child:

1. Partitions a set into all possible pairs of addends.
2. Writes an addition number sentence and a subtraction number sentence for each pair of addends.

Procedure

1. On your flannel-and-chalk board, show a row of 7 red squares across the top. Ask, "What is the largest part of 7 you can get?"
2. Say, "Let's record that with addition and subtraction number sentences." On the chalkboard to the left of the row of squares, write, 7 + 0 = 7 and to the right, 7 − 0 = 7.
3. Have the children make a row of 7 squares across the top of *their* flannel-and-chalk boards, and write the same number sentences.
4. Under the first row, show a row of 6 red squares and 1 blue square. Ask, "What's the next largest part of 7?" Record 6 + 1 = 7 to the left and 7 − 1 = 6 to the right.
5. Continue, and complete the following display:

7+0=7	■■■■■■■	7-0=7
6+1=7	■■■■■■■	7-1=6
5+2=7	■■■■■■■	7-2=5
4+3=7	■■■■■■■	7-3=4
3+4=7	■■■■■■■	7-4=3
2+5=7	■■■■■■■	7-5=2
1+6=7	■■■■■■■	7-6=1
0+7=7	■■■■■■■	7-7=0

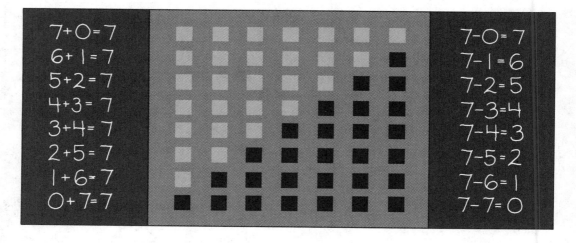

6. As you proceed ask, "Can you figure out what is going to come next?" Discuss any patterns observed.

7. On the flannel-and-chalk board, point to the 6 + 1 = 7 and say, "Is this saying about the same thing as this says over here?" Point to 7 − 1 = 6 at the right. Say, "Yes, 7 is the sum, and 6 and 1 are parts."
8. Repeat step 7 with succeeding rows.
9. Have each child complete a Parts Page for 7 as sum. (This can be part of a succeeding lesson.)

Figure 6.17
Sample Activity Plan for Schematizing Selected Objectives

Another alternative for schematizing addition and subtraction involves dot cards. Have each child write all of the number sentences for the same two addends.

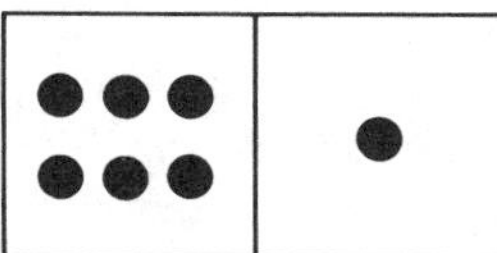

6+1=7
1+6=7
7−1=6
7−6=1

One behavioral indicator for this activity is that, given any one of the addition and subtraction facts for the same two addends, the child can write the other facts. This could be thought of as writing the *whole story*. Whole Stories Pages provide a place to collect stories for a given sum. A form completed for 8 as sum is found in Figure 6.20. Note that both horizontal and vertical notation are incorporated.

Properties

Instruction should also focus on the properties of addition and subtraction. This is best done by including these objectives (5.10–13, 5.25–28) during instruction on the meanings, models, and symbols for the operations. For example, during the abstracting and schematizing activities

+ ADD	Parts of 7	− SUBTRACT
7+0=7		7−0=7
6+1=7		7−1=6
5+2=7		7−2=5
4+3=7		7−3=4
3+4=7		7−4=3
2+5=7		7−5=2
1+6=7		7−6=1
0+7=7		7−7=0

Note: Blank "Parts" Page available in handbook accompanying this text

Figure 6.18
Parts Page Completed for Seven as Sum

5 + 0 = 5 | 5 − 0 = 5
4 + 1 = 5 | 5 − 1 = 4
3 + 2 = 5 | 5 − 2 = 3
2 + 3 = 5 | 5 − 3 = 2
1 + 4 = 5 | 5 − 4 = 1
0 + 5 = 5 | 5 − 5 = 0

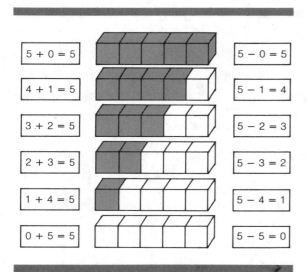

Figure 6.19
Unifix Cubes Used for Schematizing the Meanings for Addition and Subtraction

described in the preceding pages, we can look at a display and ask questions such as: Are there different number sentences with the same addends? Do you see a pattern for zero? You can record the observations of children much as you would in a language-experience approach to reading. The properties or rules could even be temporarily named for the child verbalizing them, and words like *commutative property* could then be introduced as the name we usually use for "Jerry's rule."

Properties of operations were heavily emphasized in elementary mathematics during the late 1960s and early 1970s, with specific focus on the use of such rules in computational procedures. The emphasis greatly subsided during the late 1970s with the increased concern over computational skills. However, properties should not be neglected, for they reduce the memory load involved in mastering the basic facts for arithmetic.

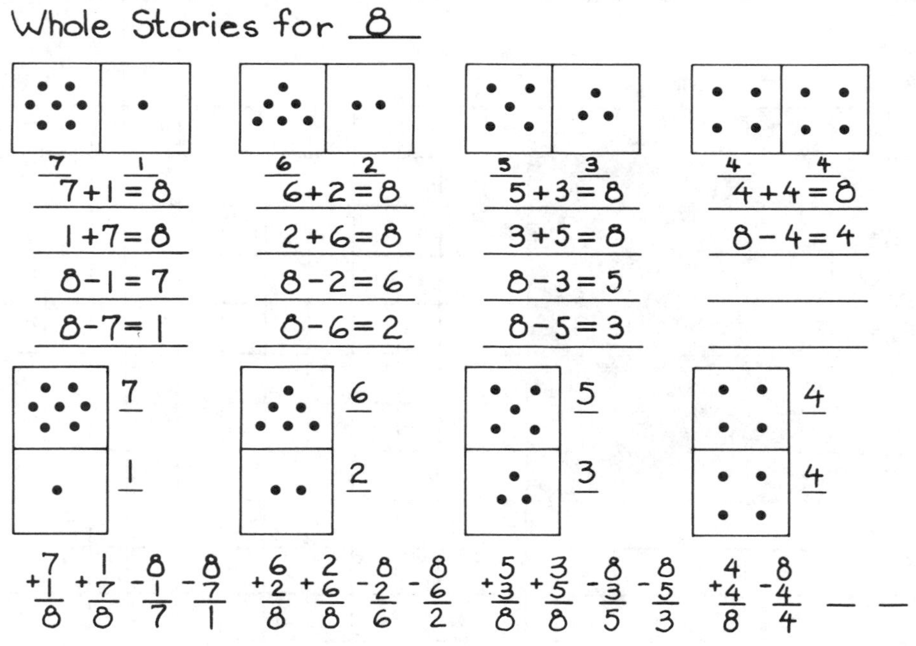

Note: Blank "Parts" Page available in handbook accompanying this text

Figure 6.20
Whole Stories Page for Eight as Sum

EXCURSION

Below is an exercise similar to one found in a fifth grade text published in 1966. Can you give a reason in each case?

$$5 \times 4\tfrac{4}{5} = 5 \times \left(4 + \tfrac{4}{5}\right) \qquad \textit{Why?}$$

$$= (5 \times 4) + 5 \times \tfrac{4}{5} \qquad \textit{Why?}$$

$$= 20 + \left(\frac{5 \times 4}{5}\right) \qquad \textit{Why?}$$

$$= 20 + \frac{20}{5} \qquad \textit{Why?}$$

$$= 24 \qquad \textit{Why?}$$

How can we exemplify the properties of addition? Commutativity is easily shown with a piece of cardboard and clothespins as in Figure 6.21. Also, each child can make a two-color rod of Unifix cubes and turn it around. Number sentences can be recorded for each position in both activities.

For zero as the additive identity, special activities must be structured to help each child discover the rule. A set of Hainstock blocks can be displayed in which each block has all of the little balls on one side of the partition. Ask, "What is the same for every block?" Record a number sentence for each child's block and see if a pattern can be observed. Tables similar to Figure 6.22 are often generated by such activities, and help children observe the special property of zero for addition.

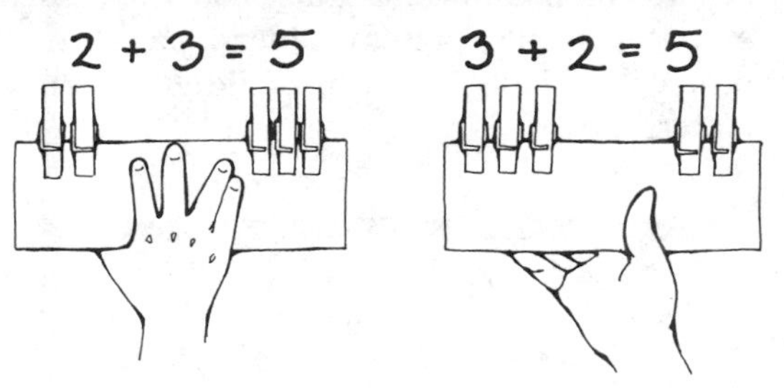

Figure 6.21
Cardboard with Clothespins Showing Commutativity for Addition

Sometimes children memorize the zero facts individually and do not learn the rule that applies wherever zero is an addend. Children who have been exposed to small numbers also assume that the rule applies only to those numbers. Therefore, larger numbers need to be included in some activities.

Associativity for addition is usually taught along with the basic addition facts for sums greater than 10. As can be seen in the illustration for $6 + 7 = \square$, the second addend is renamed as a sum in the process of making a 10.

$$\begin{aligned} 6 + 7 &= 6 + (4 + 3) \\ &= (6 + 4) + 3 \\ &= 10 + 3 \\ &= 13 \end{aligned}$$

With a young child, the illustrated process need not be recorded; that can come later. Instead, we can focus on already familiar materials and concepts to work out solutions to problems with larger numbers. For example, when a train of Cuisenaire rods is used to show a sum over 10, the name for the sum can be determined by

$1 + 0 = 1$	$0 + 1 = 1$
$2 + 0 = 2$	$0 + 2 = 2$
$3 + 0 = 3$	$0 + 3 = 3$
$4 + 0 = 4$	$0 + 4 = 4$
$5 + 0 = 5$	$0 + 5 = 5$

Figure 6.22
Table Pointing to the Zero Property for Addition

exchanging the rods for a train of equal length that includes a 10 rod.

Because addition is an operation associating exactly two numbers with a sum, three addends are grouped or clustered two at a time. Associativity is concerned with such *grouping*, whereas commutativity is concerned with the *order* of two addends. Not surprisingly, younger children sometimes confuse the two properties. But precision in identifying the various applications and names is not as important as understanding the general rule that incorporates both commutativity and associativity: addends can be rearranged in any way without affecting the sum.

Basic Addition and Subtraction Facts

Included in the objectives listed in the previous chapter are the basic facts for addition and subtraction: what they are, principles that make it possible to cut down on the number of facts to be memorized, and categories of facts to be considered for instructional purposes.

We have already given considerable attention to instruction in major concepts such as commutativity and the zero property for addition. These two ideas, when learned and applied, reduce the number of basic addition facts to be memorized from 100 to 45. And, if we regard adding 1 as a trivial counting task, the number of facts is further reduced to 36, which include many that a child is already likely to know. So, if we make individual prescriptions concerning mastery of basic addition facts, children are more likely to regard the task as feasible. If children record progress in recalling the specific basic facts targeted for them, they will be further encouraged.

Abstracting and Schematizing Facts with 10 or More as Sum

After we provide instruction on the basic addition and subtraction facts with sums less than 10, we progress to facts with 10 as the sum. These are important because they help a child learn the basic facts for sums greater than 10, along with computing larger numbers. An example of the latter is $84 + 9 = \square$, which can be solved by thinking $4 + 6 = 10$, therefore $84 + 6 = 90$ and $84 + 9 = 93$.

Ways in which children can exemplify basic addition facts with 10 as the sum include the following:

1. *Cuisenaire rods.* Have each child take an orange rod for 10, then make as many two-car trains as she can that are just as long as the orange rod.
2. *Stern blocks.* Have a child fill the number case for 10 showing each addend pair, then have children write all the number sentences shown with the blocks.
3. *Math balance.* Have each child place a weight on the 10 peg to the left, and see how many different ways it can be balanced by placing two pegs on the right side. Each way should be recorded with a number sentence.
4. *Row of discs.* Have each child make a row of 10 discs and partition them in different ways with a popsicle stick or a crayon as illustrated in Figure 6.23. A number sentence should be written for each partitioning. You may want to conduct an oral exercise with a group of children in which you say, "Ten is 4 and [pause] . . . ," "Ten is 8 and [pause] . . . ," etc. When you pause, each child can partition his or her row of discs to find the other addend.
5. *Beads on a wire.* If each child is given 10 beads on a wire or a string, the beads can be partitioned with activities similar to those described above for a row of discs.
6. *Parts Page.* Have each child complete a Parts Page for 10 as the sum.
7. *Worksheet exercises.* Exercises such as those illustrated in Figure 6.24 can be completed after work with three-dimensional exemplars. As abstracting/schematizing activities, such exercises should present addend pairs in an ordered fashion to help children observe patterns.

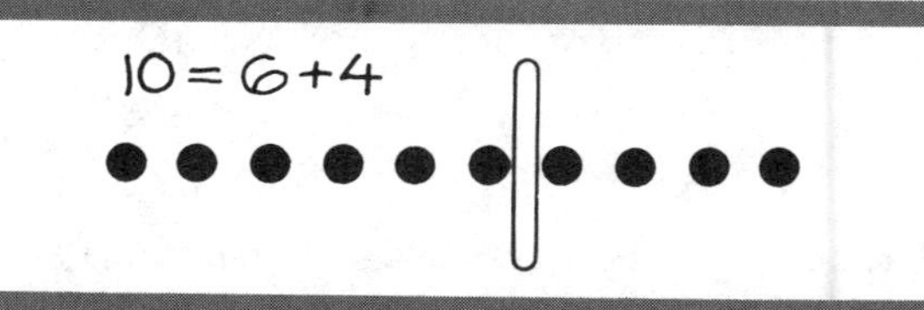

Figure 6.23
Row of Ten Discs Partitioned to Show Addends

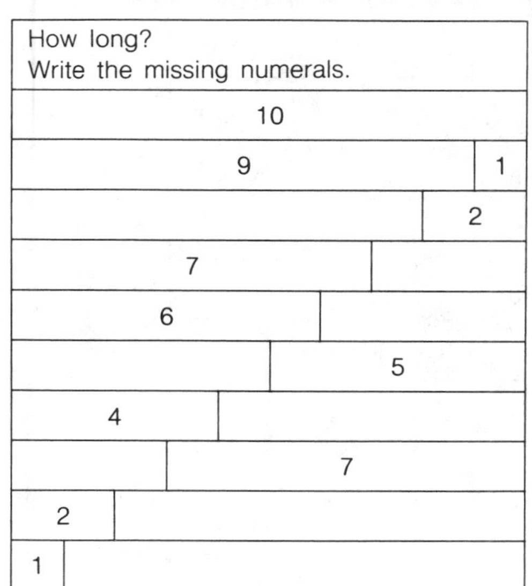

Figure 6.24
Worksheet Activities Involving Addend Pairs for Ten

Only after children have had considerable experience with the addition facts for 10 as the sum should they be involved in abstracting and schematizing activities for basic facts with sums greater than 10. The use of Cuisenaire rods was mentioned in connection with instruction focusing on associativity; Arithmablocks can be used similarly.

Other ways children can exemplify basic addition facts with sums of 11–18 include:

1. *Stern 20-case.* If Stern materials are available, have each child place pairs of rods in the 20-case. Note where the rods cross the 10 line, and describe each pair with number sentences such as 6 + 7 = 10 + 3 and 6 + 7 = 13.

2. *Discs on a lattice.* A lattice for two rows of 10 can be prepared on paper for use with plastic or paper discs, or can be shown on a flannel board with felt discs. Have each child solve an example such as 6 + 7 = □ on the lattice by showing an addend in each row (Figure 6.25a) and completing a row of 10.

3. *Math balance.* Have a child show both addends on one side, then make it balance while using a 10 on the other side. Be sure each child makes a record, such as 6 + 7 = 10 + 3.

4. *Parts Page.* Each child can complete a Parts Page for a given sum greater than 10. Emphasize the making of a 10 by drawing a heavy vertical line on the blank Parts Page to the right of 10 units. Figure 6.26 illustrates a Parts Page completed for the basic facts that have 13 as the sum.

Consolidating Activities for Basic Addition and Subtraction Facts

After children have been involved in abstracting and schematizing activities, we need to decide if we should introduce consolidating activities to help each child acquire and maintain habitual and accurate use of concepts and processes. Activities for mastering the basic facts should focus primarily on information already stored in the memory in an organized way.

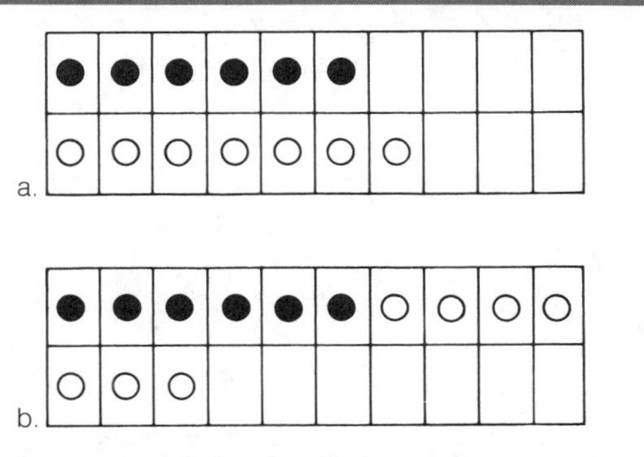

Figure 6.25
Discs on a Lattice for Sums Greater than Ten

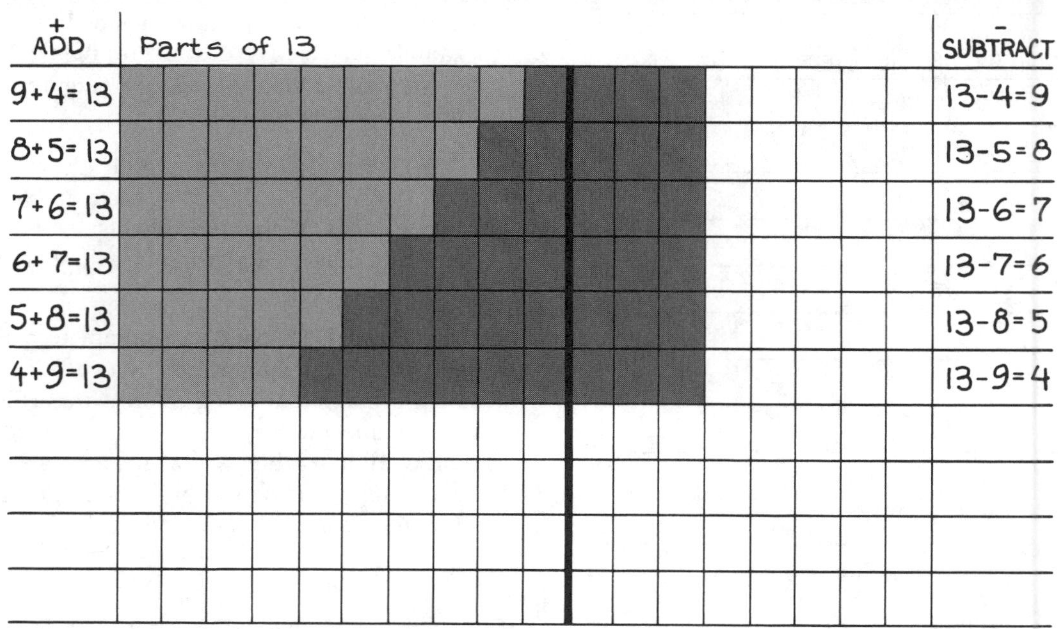

Figure 6.26
Parts Page Used for Basic Facts with Thirteen as Sum

When is a child ready for such practice activities? A child who has been involved in many schematizing activities and has developed a conceptual system for interrelating information is a good candidate. Successful completion of schematizing activities indicates a readiness for consolidating activities.[2]

When possible, consolidating activities for the basic facts of arithmetic should do the following:

1. *Present the facts in random order.* Schematizing activities present the facts in a way that facilitates pattern recognition, but consolidating activities present the facts in random order. A throw of dice, the use of shuffled cards, or a similar mechanism can ensure a random consideration of basic facts.
2. *Provide immediate feedback for the child.* This can usually be accomplished by referring to a fact chart, turning a card over, using a math balance or calculator, or through another child. Also, activities that do not normally provide immediate feedback can frequently be adapted.
3. *Include a time constraint in a game.* Efficient recall needs to be reinforced for the child who already has had considerable success with schematizing activities. A time constraint may encourage more efficient procedures and help the child gain skill and confidence in using direct recall. A game setting is usually preferred, as it creates an informal environment, with less anxiety about making mistakes. The time constraint can be provided by using an egg timer or a stopwatch. In more competitive games, the first child who picks up an object or lays down a row of cubes will score.

---

2. Robert Ashlock and Carolynn Washbon, "Games: Practice Activities for the Basic Facts," *Developing Computational Skills*, 1978 Yearbook of the National Council of Teachers of Mathematics, eds. Marilyn Suydam and Robert Reys (Reston, Va.: National Council of Teachers of Mathematics, 1978):40–41.

4. *Penalize error minimally*. Reinforce *correct* responses and minimize errors. In game situations, avoid backtracking or losing points; instead, errors should be penalized by lack of progress or not gaining points. Each child should also have evidence of long-term improvement. Progress charts are often used in conjunction with practice activities.

The authors feel that most consolidating activities should be games, especially for children who have difficulty in learning arithmetic. Such children are afraid to simply recall the number. Although they have practiced figuring it out, they need to practice *recall*. Games provide a safe environment for this: the teacher always seems to want one specific answer—the right one, but in a game a child expects to lose at least part of the time. It is often important to respond quickly, and a child may try to simply recall the needed numbers. Also, the materials involved in a game tend to increase attention span.[3] Drills can also be used as consolidating activities, but these can serve as evaluations, with the result recorded on a progress chart.

Consolidating activities developed for content objectives will probably specify a set of basic facts for arithmetic. An activity might focus on a set limited to addition facts with 10 as sum, or it might focus on all 100 basic addition facts. Normally, the related subtraction facts are also included so that each child can practice using knowledge of the addition facts to find missing addends. For purposes of illustration, let us assume that all 100 basic facts for addition have been chosen.

Choosing exemplars should be easy, as many games and activities are appropriate. However, most will need to be adapted to meet the criteria listed for consolidating activities.

The game Delivering the Mail described in Figure 6.27 can be used. When children play it, the mail must be thoroughly mixed before sorting begins so that facts will randomly appear. We will know that a child has mastered the basic facts when she sorts the mail correctly and quickly.

3. Robert B. Ashlock, *Error Patterns in Computation*, 3rd ed. (Columbus, Oh.: Charles E. Merrill, 1982):18.

Other games that can be used include Cover-All, Domino Bingo, and Get the Bacon.

Cover-All

Materials: Two dice, chips or other markers, and a number strip with 10 numbered spaces.

Procedure: Each player rolls the dice and uses markers to cover up any two numbers with the same sum as the dice. The first child to cover all spaces is the winner.

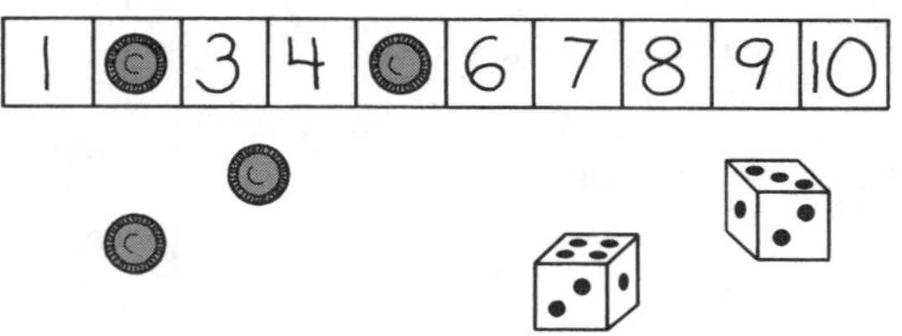

Get the Bacon

Materials: An object such as a bowling pin and two sets of cards with strings to be hung around the neck. An addend phrase, such as 6 + 7, is written on each card. Different colors are used for printing the two sets of cards, and each set has a card for each sum represented. For example, for 14 there is 6 + 8 in one set and 9 + 5 in the other.

Procedure: Assign children to two teams, and have them stand facing each other on parallel lines about twenty feet apart. Place an object (the bacon) in the middle of the space between the teams. Give each player a card with an addend phrase to hang around her neck. When a leader calls out a sum, the children on each team who have the addend phrase for that sum run and try to get the bacon and return it to their own line. A player doing this before being tagged by an opponent scores a point for her own team. If she is tagged, then no point is scored. Have children exchange addend phrase cards frequently. Also make a note of the children who have difficulty with specific facts, for follow-up at a later time.

Variation: The game can be adapted to subtraction facts by giving each child a subtraction phrase and calling out an addend.

ACTIVITY PLAN
Symbolic → Symbolic

Addition and Subtraction of Whole Numbers: Basic Facts

Consolidating

Content objectives

5.35, 5.36, and 5.37

Exemplars/Materials

Cards with addition phrases such as 8 + 4 and 0 + 9
A display board with pockets labelled in order 0 – 18
A timer
A math balance

Behavioral indicators

The child:

1. Sorts the mail correctly.
2. Sorts the mail quickly.

Procedure (for "Delivering the Mail")

Provide cards with addition phrases written on them (the mail), a display board with pockets (mailboxes) labelled with sums, a timer, and a math balance. Each of two children takes turns in starting the timer, delivering as much mail as possible by placing the addition phrase in the pocket showing the sum. After time is up, the children check each pocket to make sure the mail was delivered properly, using a math balance for verification. The number of pieces delivered correctly is the child's score. Basic facts for incorrect deliveries are listed for future work.

Figure 6.27
Sample Activity Plan for a Consolidating Activity

Domino Bingo[4]

Materials: Bingo cards with at least three rows and three columns, and with a numeral (1-12) written at random in each space; a set of dominoes.

Procedure: From a pile of facedown dominoes, each child draws a domino and places it on her card if it matches. If not, the domino is returned to the pile. The winner is the first player to have three dominoes in a straight line.

4	10	7	1
6	2	9	12
11	8	3	5

Variations: Domino cards can be made with addition phrases such as 2 + 4 or 0 + 8, thereby providing experiences with a symbolic→symbolic translation. A larger grid could be prepared, incorporating facts with sums greater than 12.

4. Adapted from Ralph Heimer and Cecil Trueblood, *Strategies for Teaching Children Mathematics* (Reading, Mass.: Addison-Wesley, 1977): 93. Copyright © 1977. Reprinted with permission

For another variation, make domino cards with subtraction phrases such as 9 − 4 and 17 − 8, and use bingo cards showing addends 1–9.

Many activities should include individual practice. Examples are described in the next section.

Worksheets for a Given Sum[5]

Materials: Individual worksheets and pencil, or laminated cards which can be wiped clean with a damp cloth and a felt-tipped pen. If worksheets are used, place a correctly completed worksheet in a folder; if laminated cards, show a correctly completed exercise on the back of each card. Sample exercises appear in Figure 6.28.

Procedure: For consolidation of basic facts with a given number as sum, have the child complete exercises and verify her answers. The child may also keep a record of how long it takes.

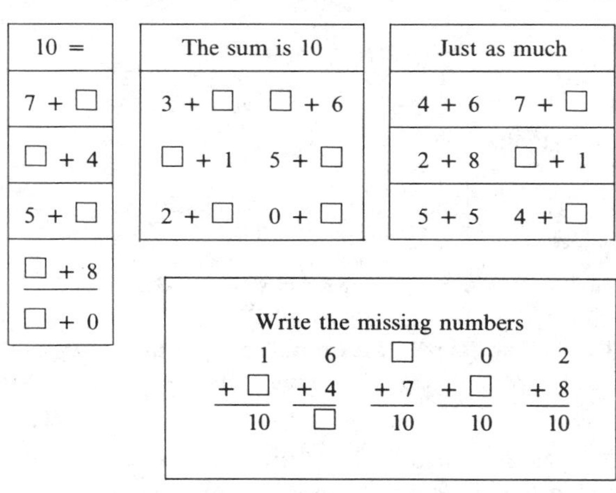

Figure 6.28
Sample Worksheet Exercises for Ten as Sum

Basic Fact Wheels

Different kinds of basic fact wheels can be constructed. Figure 6.29a presents a simple wheel on a worksheet or laminated card to be completed by a child. Addends appear in the two inner rings, with the corresponding sum in the outer ring. Figure 6.29b shows a wheel prepared from cardboard to which a child attaches specially marked clothespins to indicate sums. And Figure 6.29c illustrates fact wheels made of construction paper—a numeral and sign of operation are placed in a pocket in the center. An addend appears in a window as the wheel is rotated, and the child states the missing sum or addend to a partner.

Sum Puzzle

Materials: Worksheets or laminated cards with exercises such as those appearing in Figure 6.30. Include a completed sample and a blank form.

Procedure: Think through the first puzzle with the children. The sum in the lower right should be correct for addition both horizontally and vertically. Have children complete the remaining

5. Adapted in part from Klass Kramer, *Teaching Elementary School Mathematics,* 4th ed. (Boston: Allyn and Bacon, 1978): 242.

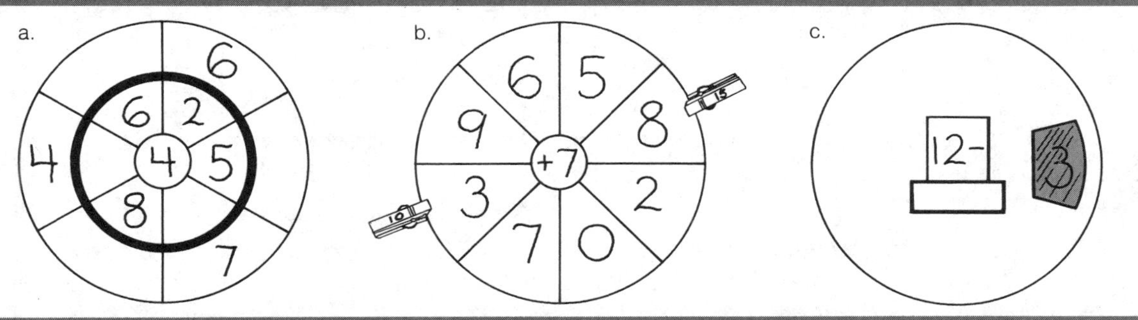

Figure 6.29
Basic Fact Wheels for Consolidating Addition and Subtraction Facts

	addends	addends	sums		addends	addends	sums
addends	5	1	6		5	2	
addends	2	7	9		3		
sums	7	8	15				16

Figure 6.30
Sum Puzzles for Consolidating Addition and Subtraction Facts

puzzles independently. Additional sum puzzles can be made from on forms.

Checking Up with Show-Me Cards

Materials: A set of ten 3″ × 5″ cards for each child, with one of the ten digits (0–9) written on each card; holders for two such cards made from a 5″ × 8″ card; score sheets (see Figure 6.31); and large cards with an addition or a subtraction phrase on each, some written horizontally and others vertically.

Procedure: Give each child a set of digit cards and a holder. Say, "Show me . . . ," and then hold up a phrase card. Each child responds by arranging one- or two-digit cards in the holder and, when you say "Now," holds them up. Each correct child gets an X on the score sheet. Incorrect answerers list the basic fact for further study.

Checking Up with Flash Cards

Materials: Timing device, and flash cards with a missing sum or addend on one side and the complete basic fact on the other. Different sets of cards, with horizontal and vertical formats, should be used.

Procedure: Flash cards are chosen according to the child's needs for mastery and skill maintenance. The cards are placed in a pile, then the timer is started. The missing sum or addend for the top card is whispered aloud, then the card is turned over to verify the solution. Cards for correct responses are placed in one pile, and cards for incorrect responses in another. The timing device is stopped when the child is finished. The time it took to respond to all of the cards and the number of correct responses are recorded on a progress chart. Incorrect responses are listed for further study.

Sentence Solving and Verbal Problem Solving

The key to solving number sentences is understanding the operations—the meanings, models, and symbols for addition and subtraction. For example, a child will know that the open number sentence □ − 8 = 7 presents two addends. Since addition is the operation used to find a missing sum, she will add 8 and 7 to determine the missing number, and will not subtract simply because there is a minus sign.

EXCURSION

Can you complete these sum puzzles?

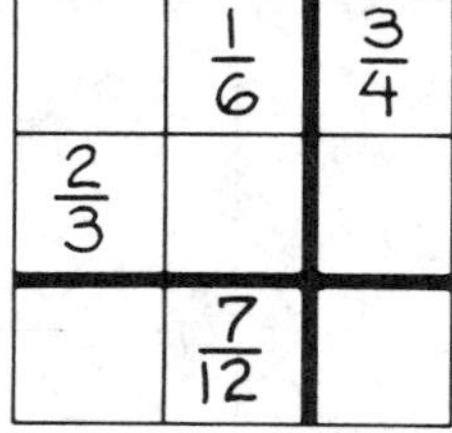

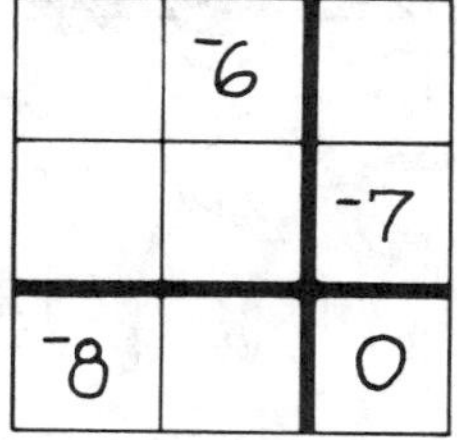

Did you find yourself deciding which operation to use? When would it be appropriate to use hand-held calculators as an aid in completing such puzzles?

Leader

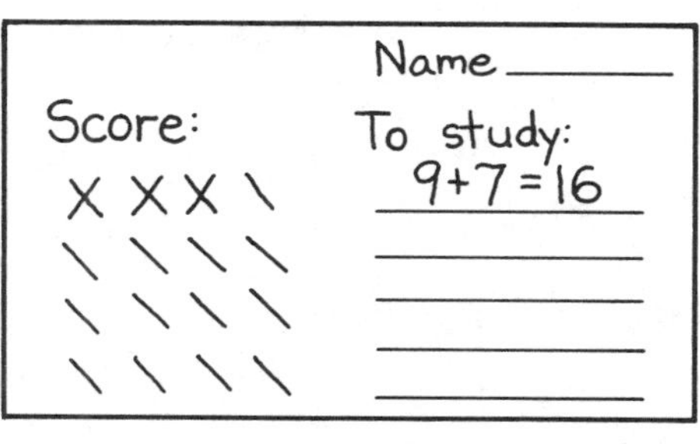

Score Sheet

Child

Figure 6.31
Materials for Checking Up with Show-Me Cards

We shall probably want to cluster content objectives 5.41 and 5.42 with other objectives as we plan instruction focusing on the meanings, models, and symbols for addition and subtraction. Make sure the surface structure of number sentences is sufficiently varied so that each child abstracts the deep structure.

Transferring Activities

Verbal problems should frequently be used in initiating and abstracting activities. They can also be used for transferring activities if they provide a context for applying *previously* learned knowledge and skills. Children who understand the meanings, models, and symbols for addition and subtraction, and who progress well in consolidating the basic facts, are ready for transferring activities.

A transferring activity with verbal problems is likely to focus on content objectives concerned with the meanings of the operations, generalizations that help solve number sentences, and selected types of verbal problems. Assume that we have chosen objectives 5.1 and 5.14, along with the verbal problem types for addition listed inside on the back cover. We must select or create different types of verbal problems, making sure that younger children understand our vocabulary. The activity described in Figure 6.32 is for a group of children who have abstracted and schematized the meanings of addition and subtraction and have already worked with a variety of number sentences. They have also been involved in consolidating activities for the basic facts for addition and subtraction. In the transferring activity, children use what they already know to choose the correct operation and write the solution as a completed number sentence. Notice that all four problem types for addition are included; children will need to focus on the deep structure, the idea that an addend plus an addend equals a sum. A subtraction problem is also included so they will not assume that each equation is addition.

Sometimes transferring activities will involve use of information already learned to solve problems at a more complex level. That is, a transferring activity for one content objective will also be an initiating activity for a higher-level objective.

Consider a group of children who have consolidated the basic facts for addition and subtraction, have abstracted ideas about place value associated with two- and three-digit numerals, and have added with two-digit numerals when no regrouping is required. We could then ask each child to find the answer to the following problem:

ACTIVITY PLAN

Addition of Whole Nos.: Meanings, Models, and Symbols

Symbolic→Symbolic

Transferring

Content objectives

5.1, 5.14, 5.43, 5.44, 5.45, 5.46, and 5.47

Exemplars/Materials

A list of verbal problems, with carefully controlled reading content and spaces for writing number sentences. The first problem has been completed as an example.

A tape recording of the problems.

Problems similar to these:

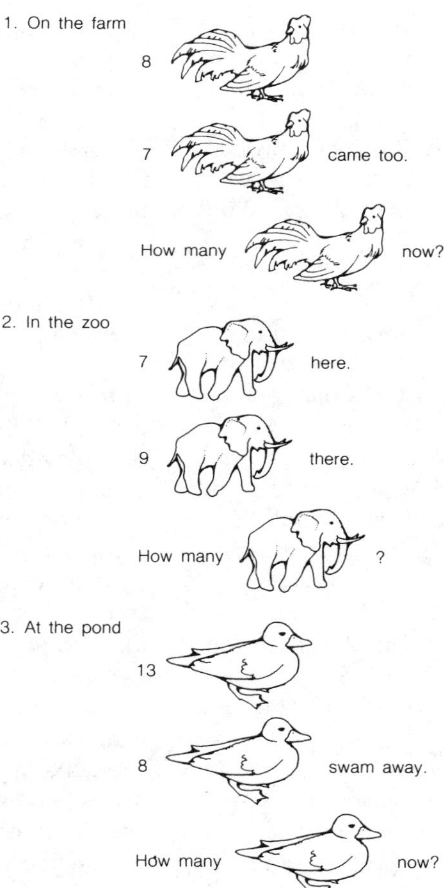

1. On the farm
8
7
came too.
How many
now?
2. In the zoo
7
here.
9
there.
How many
?
3. At the pond
13
8
swam away.
How many
now?

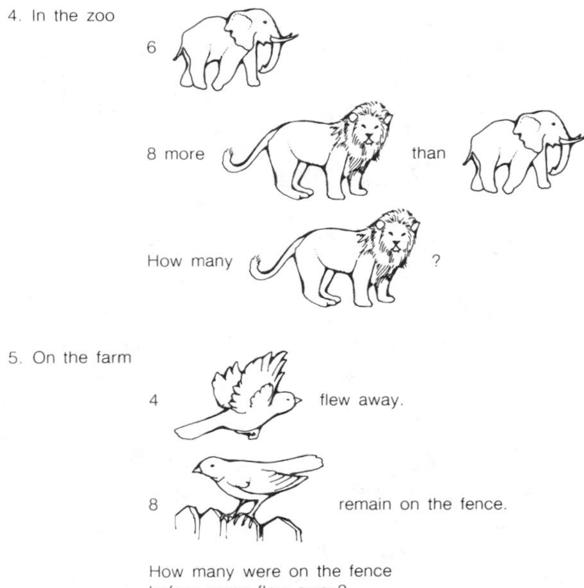
4. In the zoo
6
8 more
than
How many
?
5. On the farm
4
flew away.
8
remain on the fence.
How many were on the fence
before some flew away?

Behavioral indicators

The child:

1. Writes correct number sentences that indicate the operation used to find the unknown.
2. Orally answers the question, telling why addition or subtraction was used.

Procedure

1. Give the list of verbal problems to each child and go over the completed sample together. Do the second problem together.
2. Have each child complete the remaining examples independently, using the tape recording as necessary.
3. Later, after children have completed their work, discuss the problems with the group. Have them answer each question and explain why they decided to add or to subtract.

Figure 6.32
Sample Activity Plan for a Transferring Activity

PROBLEM

Janet has 57 pennies in one box and 29 pennies in another box. How many pennies does she have?

Children are likely to respond differently, some using counting procedures, others using three-dimensional representations, and still others working with paper and pencil. Some children will even be able to find the sum in more than one way.

If we ask children to tell us how they solved the problem, we shall soon find ourselves moving into an abstracting activity for adding two-digit numbers with regrouping. As children solved the problem in their own way, they used the basic facts they knew to solve a problem at a higher level; addition with two-digit numbers with regrouping was also initiated.

Transferring activities do not always lead to higher-level objectives; often the transfer is to other objectives at the same level. For example, having completed a Parts Page for one sum, a child may complete Parts Pages for other sums. A complete set could be made and displayed as an independent activity. A similar activity can be completed for Whole Stories Pages.

Problem solving for children who have not learned to compute with two-digit numbers should be carefully planned; much should be oral. Place emphasis on the varied problem types and on identification of the number in each part (addends) and the total number (sum). Too often,

Table 6.2
List of Verbs for Writing Behavioral Indicators[6]

1. *Identify*. Select (by pointing to, touching, or picking up) the correct object. For example, upon being asked, "Which animal is the frog?" when presented with a set of small animals, the learner is expected to respond by picking up, or clearly pointing to, or touching, the frog; if the learner is asked to "pick up the red triangle" when presented with a set of paper outputs representing different shapes of different colors, he is expected to select the red triangle. This class of performances also includes identifying object properties (such as rough, smooth, straight, and curved) and kinds of changes such as increase and decrease in size. *Selects, points to, touches, picks up,* and *chooses* are behavioral synonyms for identify.
2. *Distinguish*. Identify objects or events from among two or more which are potentially confusable (square, rectangle), or choose between two contrasting identifications (such as right, left).
3. *Construct*. Make a statement or design a procedure which identifies a designated object or set of conditions. Make a drawing or illustration is also included. Example: Given a set of observations ask the individual to "make an explanation of the observations," or, Given a graph with a number of points identified, ask the individual to "construct a prediction." *Make, build, put together, assemble, draw* and *design* are behavioral synonyms for construct.
4. *Name*. Supply the correct name (orally or in written form) for a class of objects or events. Example: "What is this three-dimensional object called?" Response, "A cone." *State, write,* and *read* are behavioral synonyms for this class of performances.
5. *Order*. Arrange two or more objects or events in proper order in accordance with a stated category. For example: "Arrange these moving objects in order of their speeds." *Arrange* is a behavioral synonym for the class of performances called order.
6. *Describe*. Identify and name all the necessary classes of objects, object properties, or event properties, that are relevant to a description which another individual would be able to use to identify the item being described. The learner's description is considered sufficient when there is a probability of approximately one that any other individual reading the description will be able to identify the object or event using only the description. *Report* and *compare* are behavioral synonyms for the class of performances called describe.
7. *Demonstrate*. Perform the operations necessary to carry out a procedure. Example: "Show me how you would tell whether the surface is flat." The response required that the individual use a straight edge to determine whether all points of the straight edge touch the surface, and that this is true in all various directions. *Show* and *carry out this procedure* are behavioral synonyms for this class of performances.
8. *State a Rule*. Make a verbal or written statement which states a rule, a principle, an axiom, or a law including the names of the proper classes of objects or events in their correct order. Example, "What is the rule for finding the area of a circle?" The acceptable response is $A = \pi r.^2$ *State a law, principle, definition,* or *axiom* are behavioral synonyms for state a rule.
9. *Apply a Rule*. Use a principle, a rule, or a law to construct an answer to a question. The acceptable response may be a correct identification, supplying of a name, or the demonstration of a procedure. The task is described in such a way that the individual must apply a rational process to arrive at the answer. *Apply a law, principle,* or *definition* are behavioral synonyms for apply a rule.

6. Henry H. Walbesser, Jr., *An Evaluation Model and Its Application: Second Report*, AAAS Miscellaneous Publication 68–4 (Washington, D.C.: American Association for the Advancement of Science, 1968):7–9. Used with permission. Copyright 1968 by the American Association for Advancement of Science. Reprinted with permission.

verbal problems for younger children are limited to the first addition type and the first subtraction type; as a result, *put together* and *take away* are learned to mean add and subtract, respectively.

PLANNING ASSESSMENT

In Chapter 3, the authors described a model for assessment procedures. For specific objectives, behaviors are described that may indicate a child understands the concept or has acquired the skill. A variety of translations across classes of exemplars must be included among these indicators if we are to assume that the child does understand; the indicators should not be exclusively symbolic to symbolic translations. In addition to paper-and-pencil tests, the tasks should include interviews and problem solving in varied settings. The list of action verbs in Table 6.2 will help us specify behavioral indicators.

The author of this list, Henry Walbesser, designed it so that each verb would suggest a different class of behaviors. Let us plan a few sample assessment activities related to content objectives used in introducing addition and subtraction. One of the content objectives for the abstracting activity illustrated in Figure 6.7 is 5.7a.: The numerals or variables on each side of the plus sign in a phrase indicate addends. What behaviors indicate that the child understands this idea? The list of *action verbs* in Table 6.2 may provide some insights. We can use action verbs to specify indicators for a given objective. For example, *identifying* behaviors for our content objective include:

1. Given number sentences of the forms a + b = c and c = a + b, the child identifies the addends.
2. Given number sentences of the forms a + b = c and c = a + b in which only variables are used, the child identifies the addends.

For each behavior described, ask, How can we elicit this behavior from the child? What tasks will allow us to observe the behavior, and thereby infer that the child understands? For the first behavioral indicator, we might:

1a. Give each child a list of number sentences and the instruction, "Draw a ring around each addend."

$$6 + 0 = 6$$
$$8 + 9 = 17$$
$$9 = 4 + 5$$

1b. Show each child number sentences such as 5 + 7 = 12 and 15 = 9 + 6, and say, "Point to each addend. Can you find them all?"

As children identify addends, we may also ask, "How do you know it is an addend?" We thereby encourage children to actually verbalize the idea that numerals on either side of the plus sign show addends. This is *stating the rule.*

Similar tasks can be easily designed for the second behavioral indicator.

2a. Give each child a list of shapes like the following and say, "These number sentences do not have any numerals. Only the different shapes show us where numerals could be written. Even so, we can tell which are addends and which are sums. Make an X on each addend."

□ + △ = ⬡

⬡ = ▽ + □

2b. Show the child the same type of number sentences made from symbols that can be moved about and say, "Point to each addend. Can you find them all?" After addends are identified for one number sentence, rearrange the symbols and ask, "Now, which are addends? Why?"

Another verb on Walbesser's list is *construct*, which implies something a child could make. We might specify the following behavior:

3. Given cardboard digits, a plus sign, an equals sign, and two numbers specified as addends, the child constructs a number sentence.

This behavioral indicator suggests an assessment task like the following:

3a. Give the child cardboard digits, a plus sign, and an equals sign and say, "Make a number sentence in which the 3 and the 5 are addends." When the number sentence is complete, ask, "How can we be sure the 3 and 5 are addends?"

Note that some assessment tasks can be easily administered to groups, while others are administered more easily to individuals. Tasks similar to 3a can be administered to both: for instance, each child in a group can have appropriate materials, and show the response simultaneously on a desk.

TIPS ON MANAGING THE CLASSROOM

Here are suggestions to help you introduce addition and subtraction.

1. Inventory manipulatives in advance to be sure you have enough on hand.
2. If each child or group of children is to have a specific set of materials (such as 7 discs), count out the materials *before* you start the lesson.
3. Children react best to handling objects with simple geometric shapes such as discs and cubes. Toys may sometimes be used in application settings, but items with minimal irrelevant and potentially distracting attributes are best when introducing concepts.
4. A vertical surface is most visible for demonstrations that involve manipulatives. A flannel or magnetic board or overhead projector can be used. For example, a child can use the projector to rearrange objects, showing others how he or she solved a problem.
5. Select children with comparable ability for a consolidating game. That way, all participants will be stimulated and will experience a measure of success.
6. Independent work is facilitated as children learn strategies. For example, by stressing the meanings of addition and subtraction, properties of the operations, and special relationships, a child typically learns a strategy for figuring out other basic facts. Horizontal transfer activities (e.g., Parts Pages for different sums) can often be used as independent work.

7

Extending Numeration Concepts and Skills

A child learns that the word *five* and the digit 5 can tell how many things are in a set, and that when there is one more in the set the word *six* and the digit 6 can be used to tell how many there are. But a child can learn relatively few such arbitrary associations. In order to deal with larger quantities, a *system* of numeration that allows a child to use just a few digits to record quantities of any size must also be learned.

Our base 10 Hindu-Arabic numeration system is a remarkable invention, but it is not always easy for young children to understand and use. It takes time for a child in the primary grades to learn the system well enough to estimate and solve problems easily. Many ideas are involved that relate to position, addition, and multiplication.

Many errors children make when computing are the result of misunderstanding numeration concepts. Careful instruction in numeration can, therefore, lead to increased success with arithmetic computations. Also, good problem solving requires the ability to estimate and other skills that use numeration concepts. These concepts must be applied even when the student uses hand-held calculators.

PLANNING THE CONTENT

Which numeration concepts are most important? Time is limited, and we must select specific ideas from many possibilities.

Some of the content objectives that can be chosen when teaching children numeration are listed on the following pages. Our choice of objectives will depend largely upon what the child already knows. Make sure that you understand the ideas clearly enough so that you can present them to your students. And, as you read, ask yourself, What behavior would indicate that the child understands the idea?

Digits

The most basic ideas concern the 10 digits used to construct numerals for whole numerals.

> **4.19**
> A *digit* is a numeral using only one symbol.

A single symbol such as 7 is a numeral, but numerals like 254 have more than one digit. A child needs to know what to call a single symbol.

> **4.22**
> The number to which a digit is assigned is called the *face value* of the digit, or the digit value.
>
> **4.23**
> The set of digits used in the Hindu-Arabic numeration system is {0, 1, 2, 3, 4, 5, 6, 7, 8, 9}.
>
> **4.24**
> The Hindu-Arabic digits are assigned to the set of whole numbers 0 through 9.

A child's first label for the numerosity of a set is an oral word. Numerals are used second, and written number words are used third. In order to use a digit for one of the numbers 0 through 9, a child must realize that the numerousness of *many* things can be shown with *one* thing, a digit.

Place Values and Grouping

Historically, people found it easier to count objects if they were grouped. This is reflected in the fact that places within our numerals are assigned values. In our Hindu-Arabic system, there is no digit assigned for the number 10. Instead, a specific place within the numeral is assigned to 10 and to each power thereof. Powers of 10 can be thought of as products that result from using only 10 as a factor 0 or more times.

7.1
A fixed position within the numeral is assigned to each power of 10.

7.2
The power of 10 assigned to each fixed position is called the *place value* for that position.

Sometimes teachers say that a child "does not understand place value," but this phrase does not specify the numeration ideas and content objectives to be learned. What is usually meant is that the child does not apply knowledge about numerals in attempting to estimate, compute, or solve problems.

7.3
The fixed positions to which place values are assigned are arranged horizontally, with values assigned in ascending order from right to left.

Certain characteristics of numerals are not obvious to children. Discuss attributes of the symbols as you introduce them.

He just doesn't understand place value.

7.4
The *base* of a numeration system is a whole number greater than 1 used as a factor to yield place values.

Ten is the base for our Hindu-Arabic numeration system and is used as a factor to generate the powers of 10 to which fixed positions are assigned. But the Hindu-Arabic system can be varied by using a different base. When this is done, the number of digits required is equal to the base.

As objects are counted they can be grouped into sets of the base number, into sets of such sets, then into sets of sets of sets, and so on. For this reason a base is sometimes thought of as a collecting point. In other contexts a set of the base number is exchanged for an object with the value of the base, then a set of these are exchanged for an object with the value of base-X base, and so on. In such situations a base is often conceived as a rule for exchanging.

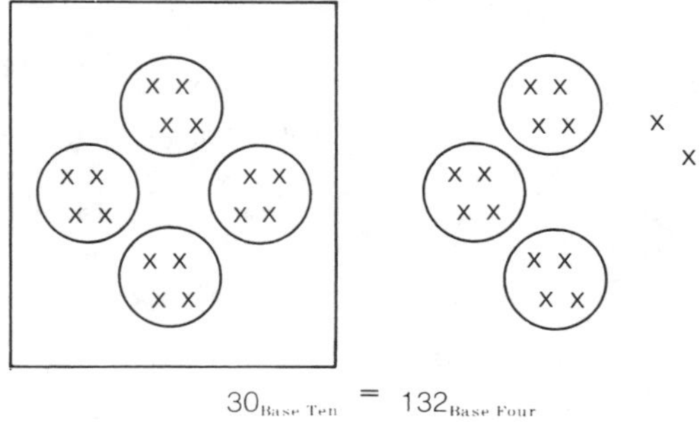

$30_{\text{Base Ten}} = 132_{\text{Base Four}}$

7.5
In our decimal numeration system, each place has a value that is 10 times as great as the place to its right, 100 times as great as the place two positions to its right, and so forth.

7.6
In our decimal numeration system, each place has a value that is 1/10 as great as the place to its left, 1/100 as great as the place two positions to its left, and so forth.

EXCURSION
Nondecimal Numeration

What happens if some number other than 10 is used as the base?

Place Values in Base 5 — Computation to Determine Base 10 Equivalent

Analysis of 4203_{five}	5^3	5^2	5^1	5^0
	$5 \times 5 \times 5$	5×5	5	1
	125	25	5	1
	4	**2**	**0**	**3**

$$= (4 \times 125) + (2 \times 25) + (0 \times 5) + (3 \times 1)$$
$$= 500 + 50 + 0 + 3$$
$$= 553$$

Therefore, $4203_{five} = 553_{ten}$

Basic addition facts in base 5 include:

$$\begin{array}{r}1\\+1\\\hline 2\end{array}\quad\begin{array}{r}2\\+1\\\hline 3\end{array}\quad\begin{array}{r}3\\+1\\\hline 4\end{array}\quad\begin{array}{r}4\\+1\\\hline 10\end{array}\quad\begin{array}{r}2\\+2\\\hline 4\end{array}\quad\begin{array}{r}2\\+3\\\hline 10\end{array}\quad\begin{array}{r}2\\+4\\\hline 11\end{array}\quad\begin{array}{r}3\\+3\\\hline 11\end{array}\quad\begin{array}{r}3\\+4\\\hline 12\end{array}\quad\begin{array}{r}4\\+4\\\hline 13\end{array}$$

Addition computation in base 5 appears as follows:

$$\begin{array}{r}1\,1\\2\,4\,3\\+\ 1\,2\,4\\\hline 4\,2\,2\end{array}$$

Try these on your own, then turn to page 120.

a. 2043_{five} = ________________$_{ten}$

b. ____________$_{five}$ = 182_{ten}

c. $\begin{array}{r}302_{five}\\+\ 143_{five}\\\hline\end{array}$

d. $\begin{array}{r}412_{five}\\-\ 134_{five}\\\hline\end{array}$

Should we teach nondecimal numeration systems?
During the 1960s and 1970s, a number of studies were conducted to determine the impact of instruction in nondecimal bases. The findings did not generally support such instruction. After summarizing the research, Callahan and Glennon concluded the following:

> The evidence would suggest that the teacher can feel quite confident at this point that some supplementary work in other base systems of numeration can be done with no evidence of a decrement in learning in other areas of arithmetic which are judged to be of value. Whether there is any advantage in supplementing instruction with other base numeration instruction over supplementary work with base 10 material is yet unclear.[1]

The current trend is to provide children with multibase blocks or trading chips to give them experience with grouping and help them develop a mental image of the exchange process in computing addition and subtraction. Formal nondecimal notation is sometimes introduced during enrichment activities in later grades.[2]

Multidigit Numerals

The numerals for 10 and for larger numbers are *multidigit numerals*. Several ideas involving place values, multiplication, and addition must be applied in teaching them.

> **7.7**
> In a numeral consisting of more than one digit, only one digit is written in each position to which a place value is assigned.

This content objective is often misunderstood by children. For example, when they are introduced to addition with regrouping, they sometimes proceed as follows:

$$\begin{array}{r} 68 \\ +\,74 \\ \hline 1312 \end{array}$$

Children who use this erroneous procedure frequently assert that, for the above example, there are 12 ones and 13 tens, but they fail to understand that the recorded numeral does not indicate what they are thinking.

Note that 13 tens and 12 ones is a *nonstandard* expression equivalent to 142; such nonstandard expressions are sometimes used in computing. Each child needs to be able to distinguish between standard and nonstandard numerals and realize that the answer to a problem must be expressed with a *standard* numeral.

> **7.8**
> Each digit in a multidigit numeral names a number that is the *product* of its face value and the place value for its position.

This value of a digit is sometimes called its *total value*, but a more appropriate term might be *product value* as it results from multiplication. In the numeral 5768, the product value of the 7 is 7 (face value) × 100 (place value) or 700.

> **7.9**
> A multidigit numeral names a number which is the *sum* of the products of each digit's face value and place value.

This is the meaning of a Hindu-Arabic numeral for a whole number. The numeral 64 means 6 (face value) × 10 (place value) *plus* 4 (face value) × 1 (place value). Multidigit numerals are sums of products. Expanded notation is used to clearly show the addition and multiplication.

$$2845 = (2 \times 1000) + (8 \times 100) + (4 \times 10) + (5 \times 1)$$

Periods

Before a child can learn to read multidigit numerals with more than three digits, he must understand the concept of periods within numerals.

---

1. Leroy G. Callahan and Vincent J. Glennon, *Elementary School Mathematics: A Guide to Current Research*, 4th ed. (Washington, D. C.: Association for Supervision and Curriculum Development, 1975), p. 145.
2. David J. Fuys and Rosamond W. Tischler, *Teaching Mathematics in the Elementary School* (Boston: Little, Brown, 1979), pp. 307–08.

EXCURSION
Nondecimal Numeration (Answers)
From page 118

a. $2043_{five} = 273_{ten}$

b. $1212_{five} = 182_{ten}$

$$\begin{array}{r l}
182 & \\
-125 & = 1 \times 125 \\
\hline
57 & \\
-\ 50 & = 2 \times 25 \\
\hline
7 & \\
-\ 5 & = 1 \times 5 \\
\hline
2 & = 2 \times 1
\end{array}$$

c.
$$\begin{array}{r}
302_{five} \\
+143_{five} \\
\hline
1000_{five}
\end{array}$$

d.
$$\begin{array}{r}
\overset{3}{\cancel{4}}\overset{10}{\cancel{1}}\overset{1}{}2_{five} \\
-134_{five} \\
\hline
223_{five}
\end{array}$$

7.10
A period is any cluster of three adjacent positions such that the first period consists of the positions to which 1, 10, and 100 are assigned, the second period consists of the next three positions to the left, and so forth.

7.11
The names of the periods in our decimal numeration system are, from right to left:

- First period—ones
- Second period—thousands
- Third period—millions
- Fourth period—billions
- Fifth period—trillions

7.12
The names of places within each period in our decimal numeration system are, from right to left: period name, ten period name, and hundred period name.

For example, the names of places within the millions period are millions, 10 millions, and 100 millions (reading from right to left).

In theory the system extends indefinitely, but children do well to learn the following:

4 8 3 ,	6 9 2 ,	0 5 4 ,	3 7 8 ,	2 0 6
Trillions Period	Billions Period	Millions Period	Thousands Period	Ones Period

7.13
The three digits in the ones period are read individually from left to right as the name of the digit (except zero) followed by the place value of the digit (except ones).

EXCURSION

What is the product value for:

The 8 in 28,304?
The 0 in 2,074?
The 5 in 4,276.851? (Can you extend place values into decimals?)

This is the *general* pattern for reading numerals through 999; however, it does not reflect the special names used for teens and multiples of 10. For example, we do not read the numeral 754 as "seven hundred, five tens, four." Instead, the name "fifty" is used for five tens. Similarly, the numeral 413 is not read as "four hundred, one ten, three." The special name "thirteen" is used for one ten and three.

> **7.14**
> The three digits within periods, other than the ones period, are read as if the three digits were in the ones period, followed by the period name.

For example, read 245, 798, 632 as "Two hundred forty five *million,* seven hundred ninety eight *thousand,* six hundred thirty two." A child who understands content objectives 7.10–7.14 should be able to read large numbers with ease.

Nonstandard Forms

The simple standard numeral for a whole number is not always the most useful expression. In paper-and-pencil computation, for example, nonstandard forms are used.

> **7.15**
> Any whole number named by a *standard multidigit numeral* may be expressed in a *nonstandard form* by renaming one of the tens (or hundreds, etc.) as 10 ones (or 10 tens, etc.), and adding these to the number of ones (or tens, etc.) already indicated in the standard numeral.
>
> 367 = (2 hundreds + 1 hundred) + (5 tens + 1 ten) + 7 ones
>
> (2 hundreds) + (10 tens + 5 tens) + (10 ones + 7 ones)
>
> = 2 hundreds + 15 tens + 17 ones

A nonstandard name such as this is often used within the subtraction algorithm so that basic subtraction facts can be used. In the following example, for instance, 367 is renamed as 2 hundreds + 15 tens + 17 ones.

```
  2 15 1
  3 6  7
- 1 9  8
  1 6  9
```

> **7.16**
> Any whole number expressed in a *nonstandard form* may be named by a *standard numeral* by renaming 10 of the ones (or tens, etc.) as 1 ten (or hundred, etc.), and adding this to the number of tens (or hundreds, etc.) expressed in the nonstandard forms.
>
> 2 hundreds + 14 tens + 16 ones =
>
> 2 hundreds + (10 tens + 4 tens) + (10 ones + 6 ones) =
>
> (2 hundreds + 1 hundred) + (4 tens + 1 ten) + 6 ones = 356

Though not actually recorded, the nonstandard numeral 2 hundreds + 14 tens + 16 ones is implied when the sum is computed for the example below. An addition computation procedure enables us to replace the implied nonstandard numeral with the less obvious standard numeral 356.

```
  1 6 9
+ 1 8 7
```

Exponents

As we work with larger numbers, we need to record numbers with more efficient expressions.

> **7.17**
> Each place value in our Hindu-Arabic numeration system can be written in exponential notation.

> **7.18**
> The "3" in 10^3 is called an *exponent* and indicates how many times 10 is used as a factor. It is written above and to the right of 10.

For example, in 10^4 the 4 tells us that 10 is used as a factor 4 times. That is, $10^4 = 10 \times 10 \times 10 \times 10$. Furthermore, $10^1 = 10$ and $10^0 = 1$.

Exponents can also be used with a different number as the base in telling how many times the number is used as a factor. For example, $6^3 = 6 \times 6 \times 6$, $12^1 = 12$, and $12^0 = 1$.

> **7.19**
> We read an expression in exponential form as follows:
> a. 10^2 is read as "ten to the second power" or "ten squared."
> b. 10^3 is read as "ten to the third power" or "ten cubed."
> c. 10^4 is read as "ten to the fourth power", and so on.

Ordering the Content

Certain concepts and skills should be taught before others. It is helpful to think of clusters of content objectives ordered in a logical way. One possible ordering for teaching numeration is pictured in Figure 7.1.

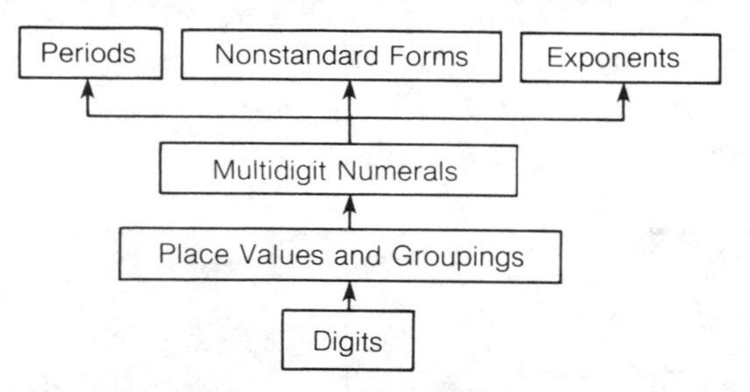

Figure 7.1
Sample Ordering of Objectives for Numeration

Payne and Rathmell suggest that before a child is exposed to two-digit numerals, he should read and write the digits 0–9 and recognize the written words *tens* and *ones.*[3] A helpful article by Ronshausen describes another effective sequence.[4] Often a series of texts is helpful.

PLANNING INSTRUCTION

Digits

Instruction that focuses on single-digit numerals should be designed to help each child build simple associations between ideas of quantity and the symbols used to record quantities. It should also focus on the order of the numbers and numerals.

Let us assume that children are ready for instruction which focuses on the following ideas:

> **4.23**
> The set of ten digits used in our base ten Hindu-Arabic numeration system is {0, 1, 2, 3, 4, 5, 6, 7, 8, 9}.
>
> **4.24**
> The Hindu-Arabic digits are assigned to the set of whole numbers 0–9.

If children are ready for such instruction, they must be able to count a set of objects up to 10 and use oral number words to tell how many are in the set. The abstracting, schematizing, and consolidating activities we design for these objectives should involve each child in matching the appropriate digit to each number. Sets can be used to show ideas of quantity and order, and rods can be used for all numbers except zero. At this level notched rods which can be perceived as a set of units may be helpful.

---

3. Joseph Payne and Edward Rathmell, "Number and Numeration," in Joseph N. Payne, ed., *Mathematics Learning in Early Childhood,* 37th Yearbook of the National Council of Teachers of Mathematics (Reston, Va: National Council of Teachers of Mathematics, 1975), pp. 137–38.
4. Nina L. Ronshausen, "Introducing Place Value," *The Arithmetic Teacher* 25, No. 4 (January 1978): 38–41.

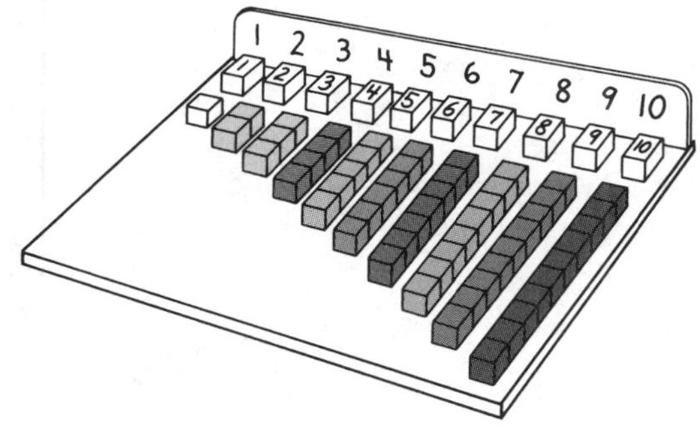

One example of an abstracting/schematizing activity involves the counting board from either the Stern materials or a homemade variation. A child sorts rods for the numbers 1–10 and places them in the order in which they fit on the board. A numeral block is placed as a label above each rod.

In a similar activity involving two-dimensional exemplars (see below), each child orders and matches set pictures with digit cards. A large card with a completed display can be an initial guide, but it should eventually be used only for verification. Additional suggestions for teaching children to write digits are found in Chapter 4.

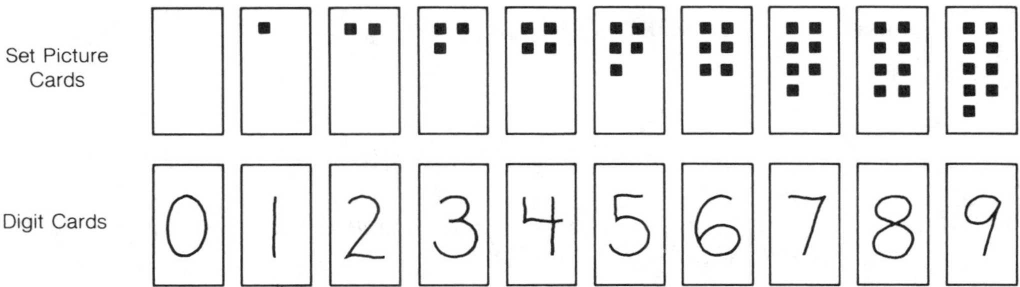

Place Values and Grouping

Children need many experiences counting sets of up to 100 before they work extensively with grouping and with place value. They need to group objects and to count groups before paper-and-pencil activities are emphasized.

Initially, each child should be involved in conversation about numbers and their order. Your questions could include the following: Which number comes next? How do you know? Do we have enough for another group? How many groups do we have? The emphasis should be on correct and intelligent use of oral names for numbers. Numerals, the written number names, are usually introduced in conjunction with counting or grouping activities, often as a record of what a child observes.

A child may learn the sequence of oral number names, but may not be able to relate these names to grouped objects (base representation) or to numerals. As we plan instruction in numeration, children should be provided with many opportunities to translate from one form of representation to another. For two-digit numerals, Payne and Rathmell picture the translations as in Figure 7.2 and note: "If the child can easily shift from one representation to another, he will be able to deal with tasks more flexibly. . . . Similarly, ease in translation is of great use in ordering, in approximating and estimating, and in regrouping and renaming."[5]

5. Payne and Rathmell, "Number and Numeration," p. 138.

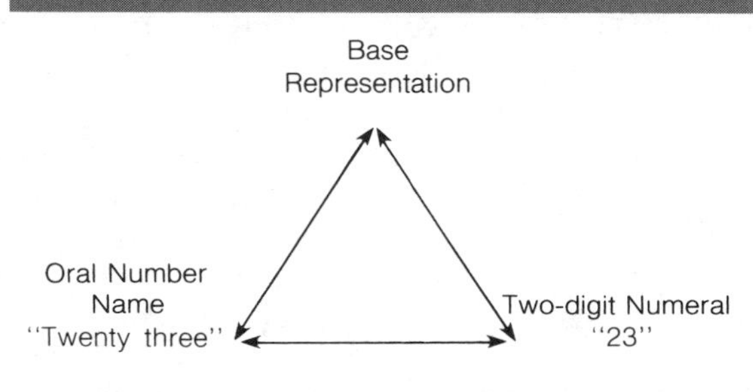

Figure 7.2
Translations Among Representations for Two-Digit Numbers

Initiating and Abstracting Activities

A variety of initiating and abstracting activities can be used in teaching place value and grouping (content objectives 7.1–7.6). A few are discussed here.

After each child has had many varied counting experiences, plan experiences involving counting out sets or groups. Sticks or plastic spoons are often used at first. Figure 7.3 illustrates a plan for an activity in which children learn that bundling sticks in groups of the same size can make it easier to count. After initial work in this activity, then move on to groups of 10.

We can also have each child use collecting places, sheets of paper or box lids arranged in a row horizontally as in Figure 7.4. Counting by ones should proceed very slowly, sometimes using the phrase "And one more is ______." With each succeeding number, one more stick is placed in the collecting place for units; whenever possible, 10 units are bundled and placed in the collecting place for tens. Also, digit cards are selected as labels for each succeeding collection. From time to time, counting should be interrupted to find a corresponding two-digit numeral card and the numeral on a number line.

ACTIVITY PLAN

Numeration for Whole Nos.: Place Value and Grouping	*3-D/Rep →3-D/Rep* *3-D/Rep →Symbolic (oral)*	*Initiating*

Content objectives

7.4

Exemplars/Materials

Sticks for bundling, rubber bands
Chalk and chalkboard

Behavioral indicators

1. Given a set of sticks, the child bundles the sticks into equivalent subsets to make counting easier.
2. Given a set of sticks with as many as possible bundled into equivalent subsets, the child determines the number of sticks in the set.

Procedure

1. Show the children 10 sticks. Say, "How many sticks am I holding?" Then, "How do you know?" (I counted)
2. Say, "We are going to play a game. I will hold the sticks behind my back, then I'll show you some of them and take them away quickly. I want you to tell me how many I show you."
3. Briefly show small numbers of sticks (1, 3, 2). Then show all 10 and say, "Could you count them? Why not?" (There was no time.)
4. As you make two bundles from the 10 sticks, say, "I'm making bundles of 4 sticks each. How many are in each bundle or group? How many groups do I have? Two groups of how many? And how many single sticks are left over?"
5. Say, "Let's play the game again. Try to count how many I show." Show five (one bundle and a single), then other amounts using bundles.
6. Say, "Do all the bundles have the same number of sticks? We see that when we bundle sticks into groups with the same number of sticks in each group, it sometimes makes it easier to count."

7. Say, "What other ways could we bundle sticks besides bundles of 4?" Give each child 13 sticks and say, "Make bundles with your sticks, and make sure you have the same number in each bundle."
8. Record on a chalkboard different ways the sticks are bundled, such as, "Mary's way–5 sticks in each bundle."
9. Play the game again, using sets of sticks bundled by the children. With bundles of 6, show one bundle and four singles. Show three or four different numbers for at least three different bundlings. Make sure bundles of 10 are included.
10. Say, "Bundling the sticks often makes it easier to count, but we have to remember that each group must be ______________." (the same number or size)

Note: With some groups of children you may want to proceed quickly to a similar activity with a larger set of sticks.

Figure 7.3
Sample Activity Plan for an Initiating Activity

EXCURSION
A Word of Caution from Piaget's Research

For some children, the practice of bundling 10 sticks with a rubber band may be confusing. This is especially true if the child does not have the concept of conservation of number equivalence.[6] He may actually think that 1 of the tens is fewer than 10 ones spread out.

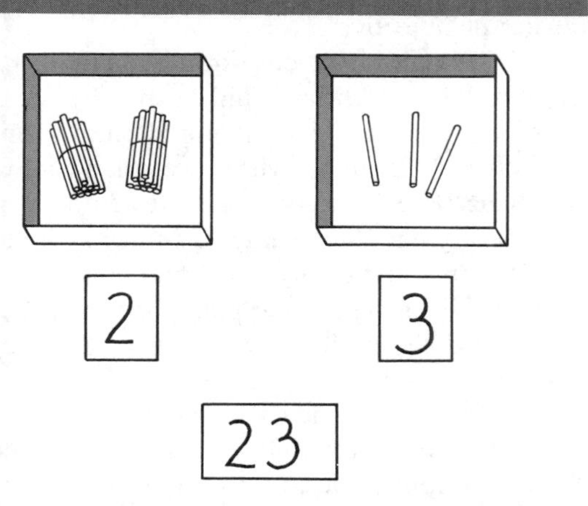

Figure 7.4
Counting with Collecting Places

Abstracting activities should incorporate varied translations among sets, oral number names, and numerals. In her suggested sequence, Ronshausen includes the following:[7]

1. *Counting and writing about sets of 10.* Initially, only two through five sets should be used.
2. *Counting and writing tens and ones.* A prepared form can be used for recording tens and ones, and for writing the corresponding two-digit numeral.

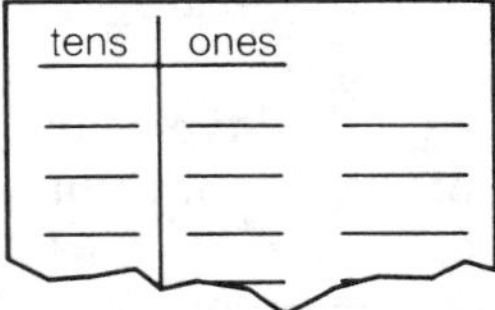

6. Richard W. Copeland, *How Children Learn Mathematics: Teaching Implications of Piaget's Research,* 3rd ed. (New York: Macmillan, 1979), pp. 125–27.
7. Ronshausen, "Introducing Place Value": 38–41.

3. *Interpreting the numeral.* Questions are asked about given numerals.
4. *Matching numeral and set.* The child translates in both directions between numerals and sets which include objects grouped by tens. Multiple-choice arrangements of numerals or sets can be used for a child's response.

Ronshausen's article describes activities in a carefully developed sequence, including suggestions on specific use of language. For example, 30 may mean "three tens," but we want a child to read it as "thirty."

Trading activities have become an increasingly popular way to help each child understand the relation of a given place's value to values assigned to the left and right (content objectives 7.5 and 7.6). Although trading may initially be done with proportional materials reflecting the relative sizes among powers of 10, nonproportional materials should be used before children work exclusively with paper and pencil.

A game that has proportional material emphasizes that trading involves exchanging the item (for example, cardboard or wooden squares) for the same amount of material (a larger length equivalent to the amount traded).

Or, the teacher can use a chip-trading activity. Chip trading is more abstract because a child cannot verify an exchange for the same amount by creating congruent displays of material: one chip is equivalent in value to many only by definition. Base 10 values are often delayed until there is some work with other bases, such as 3 or 5, in which the pattern for trading can be focused upon without extensive counting. Chips are usually differentiated by color, with a different value assigned to each color. Chip-trading materials can be homemade or are available commercially.[8]

Trading games typically involve an exchange sheet or mat for each player similar to the one shown in Figure 7.5. Each space is considered a place for collecting discs of a designated color, thereby helping each child to think of the different values as arranged in a row horizontally (content objective 7.3).

8. See Patricia Davidson et al., *Chip Trading Activities, Book I: Introduction* (Fort Collins, Colo.: Scott Resources, 1975).

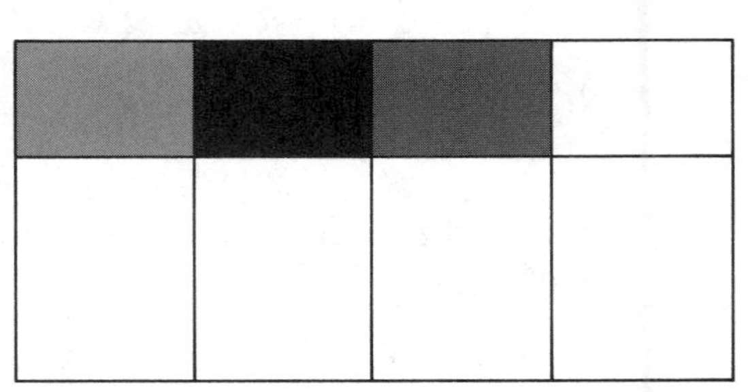

Figure 7.5
Chip Trading Exchange Sheet

Many different chip-trading games can be played.[9] Some involve trading up (exchanging several for one of a higher value) and others trading down (exchanging one for several of a lesser value). One game that involves both concepts is Tax Time. Children roll dice and receive the number of units shown, trading up as appropriate for three turns. On every fourth turn they pay taxes in the amount shown and thereby have to trade down.[10] In games ask from time to time, "Who is ahead? Why?" A child's response will indicate the extent to which he understands the cluster of objectives.

Though initial chip-trading activities may be with bases such as 3 and 5, thereby focusing more clearly on the grouping pattern (content objective 7.4), only when we plan activities involving base 10 do we concentrate on the relation of a given place value to values assigned to the left and to the right (content objectives 7.5 and 7.6). Then we need to plan abstracting activities that help each child relate what is being observed to written symbols. Figure 7.6 (pages 127–28) describes one such activity. Note that, in order to emphasize that values are assigned to fixed positions (places) rather than to colors, the game is repeated with a single color. Observe each child carefully at such a time for any indication of having associated value only with color.

9. Ibid.; See Fuys and Tischler, *Teaching Mathematics in the Elementary School,* pp. 294–308.
10. Davidson et al., *Chip Trading Activities Book I,* p. 21.

ACTIVITY PLAN

Numeration for Whole Nos.: Place Value and Grouping

3-D/Rep →3-D/Rep
3-D/Rep →Symbolic (oral)

Initiating/Abstracting

Content objectives

7.1 through 7.6

Exemplars/Materials

Chips of different colors
Mats for collecting and exchanging (multicolor and one-color)
A pair of dice
A record sheet for each child

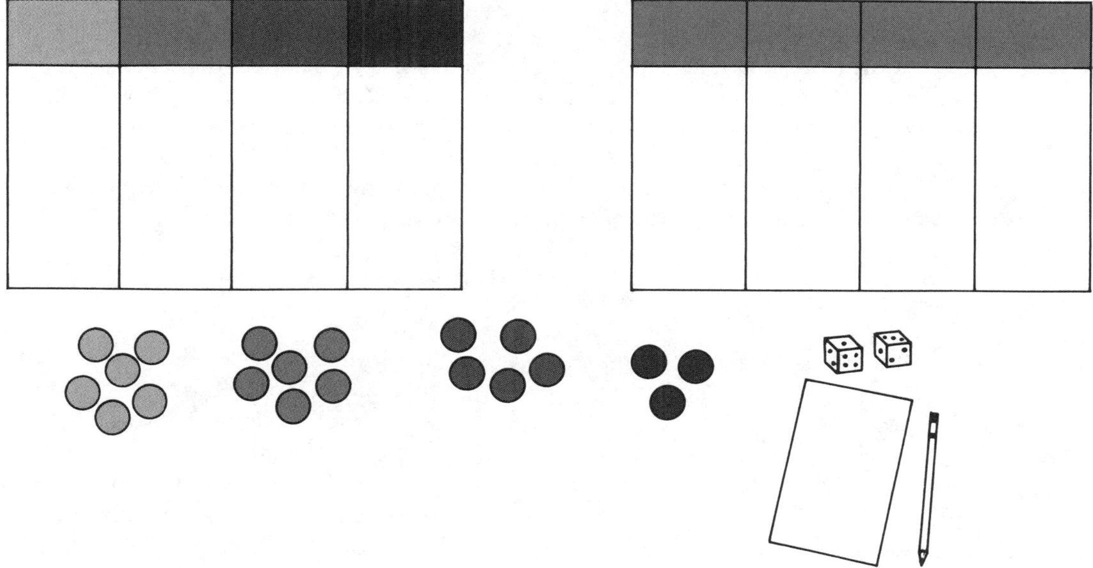

Behavioral indicators

1. In chip-trading games, the child identifies which player is ahead.
2. In a base 10 chip-trading game, the child states the value of a chip of one color in terms of the value of a chip of a different color.
3. Given chips on the mat during a base 10 game, the child writes the numeral for the value displayed.
4. The child plays a base 10 chip-trading game in which value is determined by place only (not by color), and names the value for each place.

Procedure

1. Distribute chips and a mat to each child, and play one game the children have played previously in a base such as 3 or 5; for example, a simple base 3 game in which each child in turn rolls a single die and receives that many chips valued as ones. At every opportunity, 3 chips are traded for 1 chip of the greater value to the immediate left. The winner is the first to receive the chip of the greatest value.
2. During the game, ask or suggest the following:
   a. "Who is ahead? How do you know?"

b. Point to a ones chip. "This is a ones chip. What color chip is next to the left? How many ones is it worth?"

c. Point to a chip. "What color chip is to the right of this chip? It is worth only one-third as much as this chip." Repeat by focusing on different colored chips.

3. Say, "The trading rule we usually use is 10 for 1, so let's play the game again, but trade *10 for 1* whenever we can." The winner will be the first to get a color of chip with greatest value.
4. Make sure each child has enough chips, and proceed with the game using a pair of dice.
5. During the game, intervene from time to time with the same questions as posed in 2 above. The value of each of the chips should be stated in terms of the others. For example, "The green chip is worth 100 yellow chips" or "A yellow chip is worth one tenth as much as a blue chip."
6. Begin to use ones, tens, and hundreds in place of color names.
7. After the game say, "Let's play another game, and each make a record as we play." Give each child a record sheet.
8. Play the base 10 game again. After any one child plays, have *every* child record the new number being shown.
9. Collect the chips and mats and say, "It is fun to see if we can play the same game using only *one* color. Here are mats with no colors on them." (Distribute mats.) "We'll use only the (single color) chips this time. We'll put them on our mats and trade as we did before, except that all the chips will be the same color."
10. Play the game, using record sheets as before. Again, discuss the questions posed in 2 above.

Figure 7.6
Sample Plan for an Initiating/Abstracting Activity

Other trading activities can be planned. These involve counting by ones and trading 10 for 1 whenever possible. A child can use sticks in place value cans, tickets in a place value chart, or an abacus (manual computing device) as shown in Figure 7.7.

From time to time, have each child either select a numeral card or write the numeral shown. Even though the oral number word is spoken as a child counts and shows additional ones, the translation to written symbols needs to be made.

A child can also count using a numeral flip chart (see Figure 7.8). Sections of small spiral notebooks are fastened in a row on a piece of cardboard, and the digits 0–9 are written on the pages of each notebook. As a child counts, the next digit is flipped down to show one more. When a 9 is showing, one more is shown by flipping back to zero and showing one more on the notebook to the left. This particular activity involves only oral and written symbols; therefore it is important that children have previous experience with activities that include 3-D/representational to symbolic translations.

Counting devices that involve ten rows of 10, such as the spool board and counting bead frame pictured in Figure 7.9, can also be incorporated in activity plans. As a child counts by ones and shows 1 more on the device with each number named, emphasize multiples of 10 as they are completed. Also have each child count by tens with the device. When counting by ones, intervene periodically and ask the child to describe the number of tens and ones, then find a numeral card for the number shown, or find the numeral on a number line. Each child should eventually begin with several tens and ones already showing, and count by ones.

Schematizing and Consolidating Activities

One example of a schematizing activity for the numbers 10–20 involves completion of a worksheet similar to the one pictured in Figure 7.10.[11] A child extends the rows of dots one more,

---

11. Adapted from Payne and Rathmell, "Number and Numeration," p. 148.

EXCURSION
Money as an Exemplar

Another context for exchange is learning about money. Children exchange 10 pennies for 1 dime, and 1 dime for 10 pennies. But the equivalence of 10 pennies and a dime is difficult for young children to grasp; a dime is even smaller in size than a penny.

Activities are often planned that involve exchanging dollar bills in different amounts. In using money as a context for exchanging ones and tens, tens and hundreds, and so on, or in teaching a child about money, remember the following:

1. Money values are not the same as base 10 place values. For example, when working with coins, the sequence penny-nickel-quarter illustrates base 5 values. Only selected money values illustrate base 10 place values.
2. Activities sometimes confuse units. Dollar bills are a better exemplar for base 10 place values than coins, because one dollar is maintained as the unit. A penny and a one-dollar bill cannot both serve as the unit at the same time.

records the 10 and an additional 1, then writes the standard numeral. This continues until two tens are recorded.

Many consolidating activities involve worksheets and, because they are consolidating activities, are appropriate only for children who already understand place value ideas. Some activities require interpretations of standard numerals; for example, 64 = ____ tens and ____ ones. However, children soon learn to copy the first digit in the first blank and the next digit in the next blank, forgetting the associated values. Occasionally we should vary the interpretation, such as 64 = ____ ones and ____ tens, in which response blanks are not in the expected order.

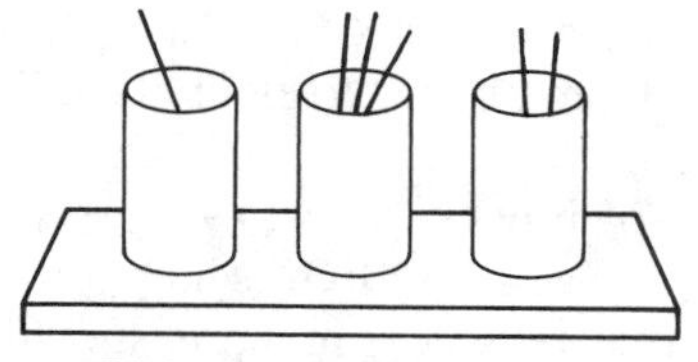
Sticks in Place Value Cans

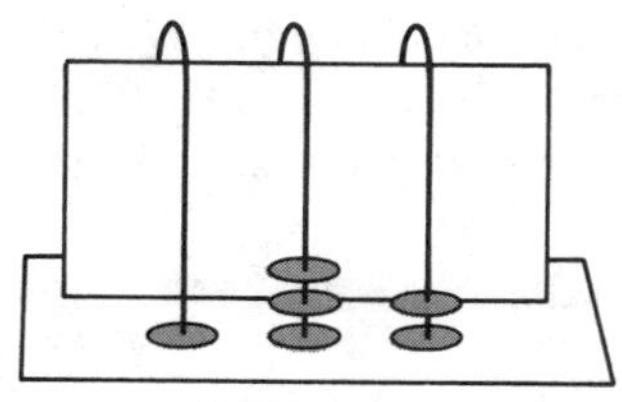
An Abacus

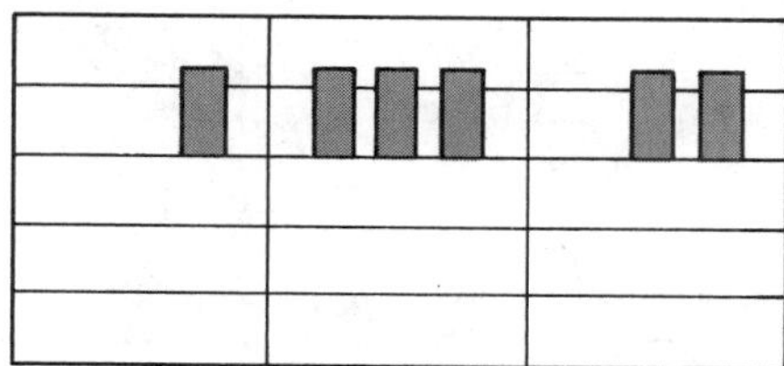
Tickets in a Place Value Chart

Figure 7.7
Devices for Trading 10 for 1

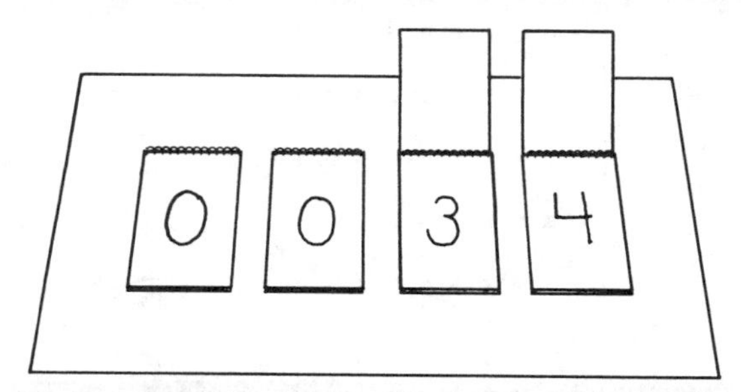

Figure 7.8
Numeral Flip Chart

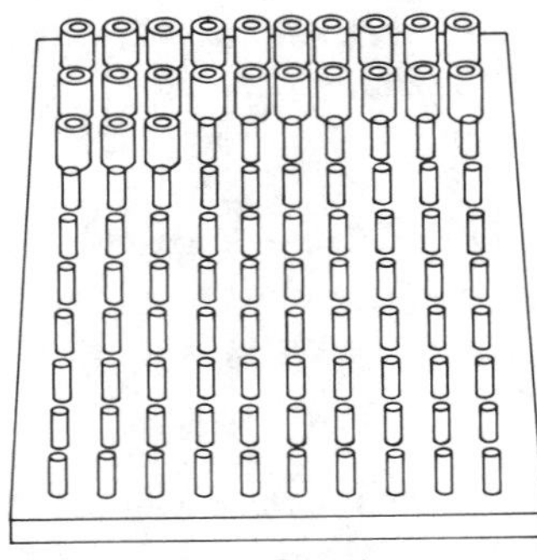

Spool Board

Bead Frame

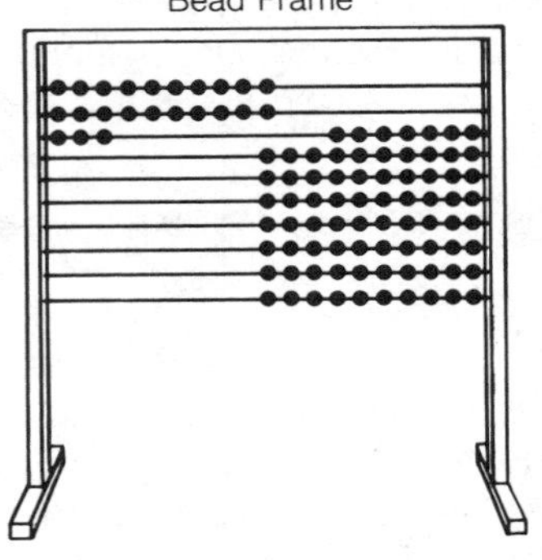

Figure 7.9
Counting Devices

Discussing the examples can also tell us much about what each child understands. Another worksheet is presented in Figure 7.11.

Circle each digit in the tens place.

213 4968 34687
84 609

Circle each digit in the hundreds place.

352 5864 43867
3082 29763

Figure 7.11
Sample Worksheet Problems for Identifying Place Values

Multidigit Numerals

Abstracting and Schematizing Activities

Translation tasks with base 10 blocks or bean sticks (see Figure 7.12) can provide a focus for many abstracting activities. Be sure each child translates from objects to symbols and from symbols to objects. For example, give each child a collection of blocks to sort (9 or less of any one size), then have children select (or write) the numeral that tells how much is shown with their blocks. With base blocks and a set of multidigit numeral cards, say, "Point to the numeral that tells *how much wood* is here. Why did you choose this numeral and not one of the others?" Then have each child also translate from given numer-

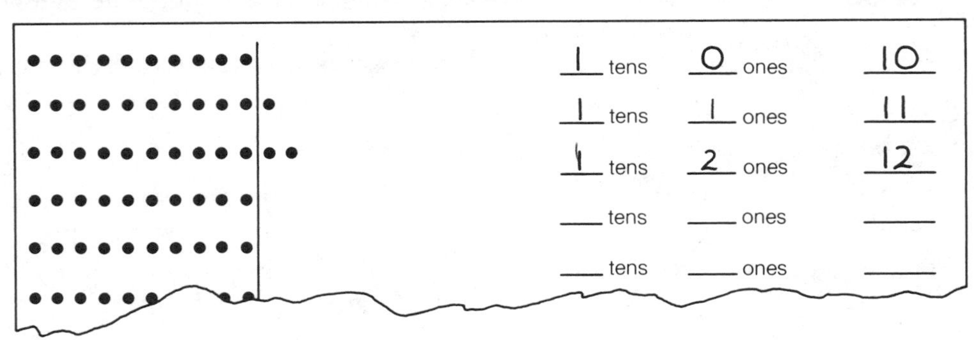

Figure 7.10
Worksheet for a Schematizing Activity

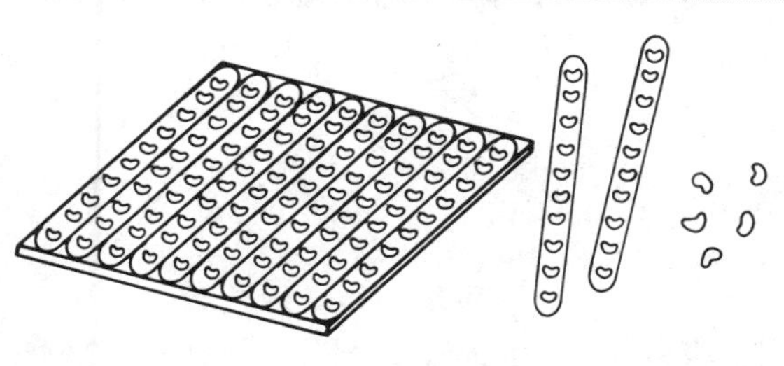

Figure 7.12
Beans and Bean Sticks for Numeration Tasks

als to the blocks or bean sticks. For a given multidigit numeral say, "Make a collection of base blocks with this much wood." Throughout such activities, note that the digit for the number of flats or hundreds tells about *part* of the total amount, and the multidigit numeral itself tells about the *total* amount. We may even choose to introduce expanded notation as a different way of writing the amount of wood in a given collection, as suggested in Figure 7.13.

Other abstracting and/or schematizing activities include:

1. Draw a numeral frame around a multidigit numeral. Have each child point to different places as values are named at random. Note that only one digit appears in each place.

   There were | 6 | 2 | 8 | children.

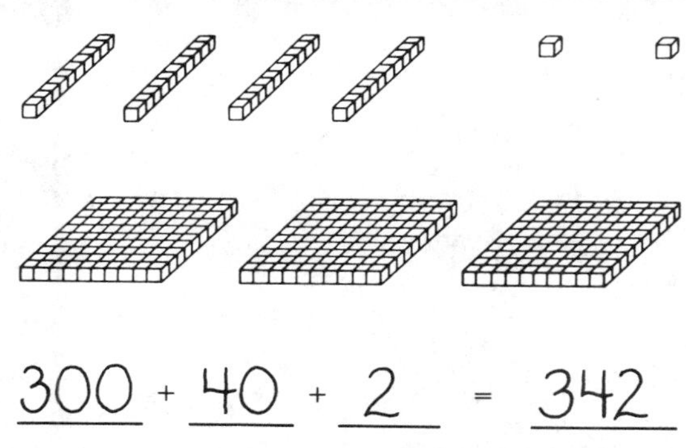

Figure 7.13
Writing a Numeral for a Set of Base 10 Blocks

2. Build numerals on calculators by using only powers of ten and the + and = keys. For example, using only the numbers 1, 10, 100, and 1000, and the + and = keys, show a given four-digit numeral on the calculator. Start with the thousands key, but later try it by starting with the ones key.[12]

3. A wide elastic band attached to two digit cards may help each child understand the term *expanded notation.*

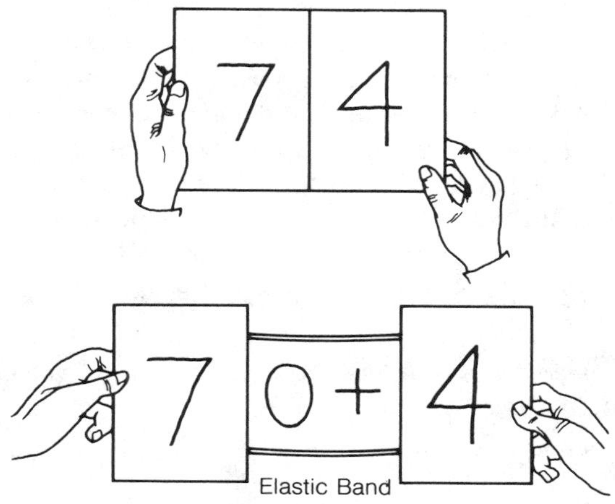

4. Use numeral expanders similarly.[13]

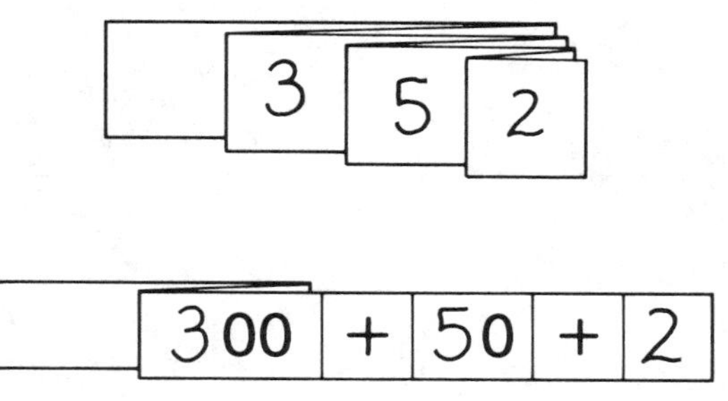

12. Gary Bitter and Jerald Mikesell, *Activites Handbook for Teaching with the Hand-Held Calculator* (Boston: Allyn and Bacon, 1980), p. 28. Reprinted from "Texas Instruments Calculator Inservice Foils," copyright 1977 by Texas Instruments. Reprinted courtesy of Texas Instruments.

13. Adapted from Today's Mathematics, 4th edition, by James W. Heddens, p. 85, © 1980, 1974, 1971, 1968, 1964, Science Research Associates, Inc. Reprinted by permission of the publisher. Also see Shirley Ziesche, "Understanding Place Value," *The Arithmetic Teacher* 17, no. 9 (December 1970): 683–684.

Expanders also allow a child to think of many different names for a number.

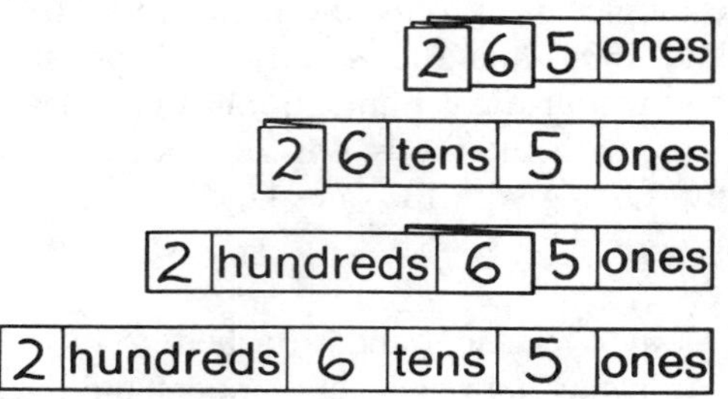

5. Different names for a number can be shown with equivalent collections of base blocks, sets made with the same amount of wood.
6. Provide a child with five large cards, each with a four-digit numeral and pictures of base 10 blocks. Have the child make a collection of base blocks for each card, then order the five collections from least wood to most wood. Later, provide five cards with numerals only, and have the child make and order the collections.[14]

In planning schematizing activities the behavioral indicators we describe will depend on the exemplars. Figure 7.14 describes a schematizing activity in which a child not only relates a standard numeral to expanded notation but also compares different forms of expanded notation.[15] The exemplars are symbols, and the behaviors anticipated involve arranging or writing symbols.

14. Adapted from Mary Laycock, *Base Ten Mathematics: Interludes for Every Math Text* (Hayward, Calif.: Activity Resources, 1977), p. 7.

15. Adapted from Leonard Kennedy, *Models for Mathematics in the Elementary School* (Belmont, Calif.: Wadsworth Publishing, 1967), pp. 33–37.

ACTIVITY PLAN

Numeration for Whole Nos.: Multi-Digit Numerals | *Symbolic → Symbolic* | *Schematizing*

Content objectives

7.8, 7.9

Exemplars/Materials

A display card to be used for reference

Paper and pencil

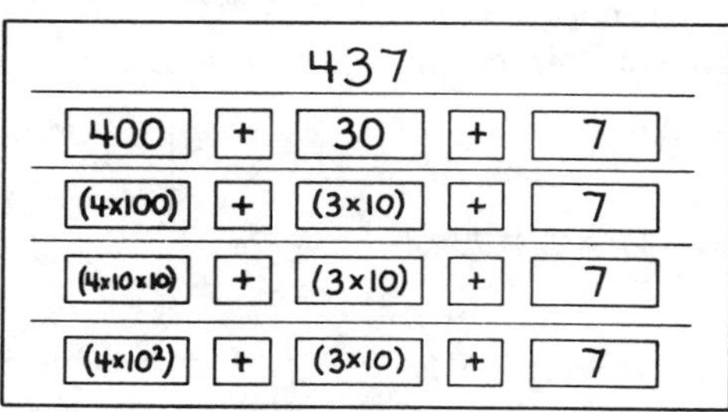

For each child a box with a standard three-digit numeral on the cover, and cards inside for creating a display similar to the reference card. Numerals or phrases should be in one color on 3″ × 5″ cards; the plus signs in another color on 3″ × $2\frac{1}{2}$″ cards.

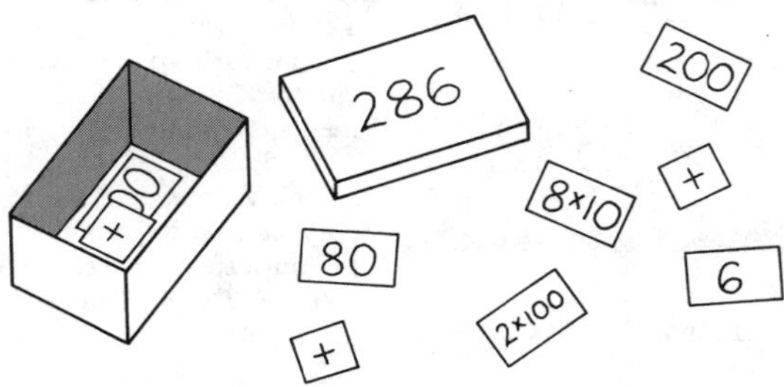

Behavioral indicators

1. For a given three-digit numeral, the child arranges cards to form equivalent expressions using expanded notation.
2. For a given three-digit numeral, the child writes equivalent expressions using expanded notation.

Procedure

1. Provide a box of expanded notation cards for each child, and place a display card where all can see it.
2. Have each child arrange the cards from his box in a display of rows and columns similar to the display card.
3. As the cards are arranged, have each child identify the addends that are shown. Note that each addend is actually a product. Compare the standard numeral on the box lid to the names being constructed with expanded notation.
4. When everyone agrees that the arrangements are correct, ask each child to mix up the cards and return them to the box. Have children exchange boxes and make a display for a different number.
5. When children find the task easy, remove the sample display card and have them continue to make similar displays without a reference card.
6. Have each child make a record of some of the displays by writing a heading such as "Names for 623" at the top of a paper and listing equations that show names using expanded notation:

$$623 = 600 + 20 + 3$$
$$623 = (6 \times 100) + (2 \times 10) + 3$$

Figure 7.14
Sample Plan for a Schematizing Activity

Consolidating and Transferring Activities

Many of the practice activities for numeration will require translation from one class of exemplars to another. For example, if a child is given:

1. A three-digit numeral on the chalkboard, he shows the same numeral on an abacus.
2. Place value cans picturing a standard numeral, he writes and reads that numeral correctly.
3. A three-digit numeral orally, he writes that numeral on the chalkboard.

The Exemplar-Class Translation Matrix (Table 3.4) can help us think of other translations. For example, if a child is given a numeral and shows that numeral with base 10 blocks, he should also be given a set of blocks to sort and then write (or select) the appropriate numeral.

A schematizing/consolidating activity involves a pocket chart and 3″ × 5″ cards with symbols, as illustrated in Figure 7.15. The standard numeral cards are placed at the left, then the remaining cards are arranged to form equations with expanded notation. A key should be available so that children can verify each display. Be sure to include numerals like 333 so children will focus on place value and not just match digits. Also, vary the activity by placing standard numerals at the right or by including more product value cards.

Games can also be included in consolidating activities. Figure 7.16 presents a plan for a card game for children who already understand that a multidigit numeral names a sum of products (content objective 7.9).[16]

Show-me cards are useful in practice activities, especially with larger groups of children. If each child has a show-me card with

---

16. Adapted from Heddens, *Today's Mathematics*, p. 87.

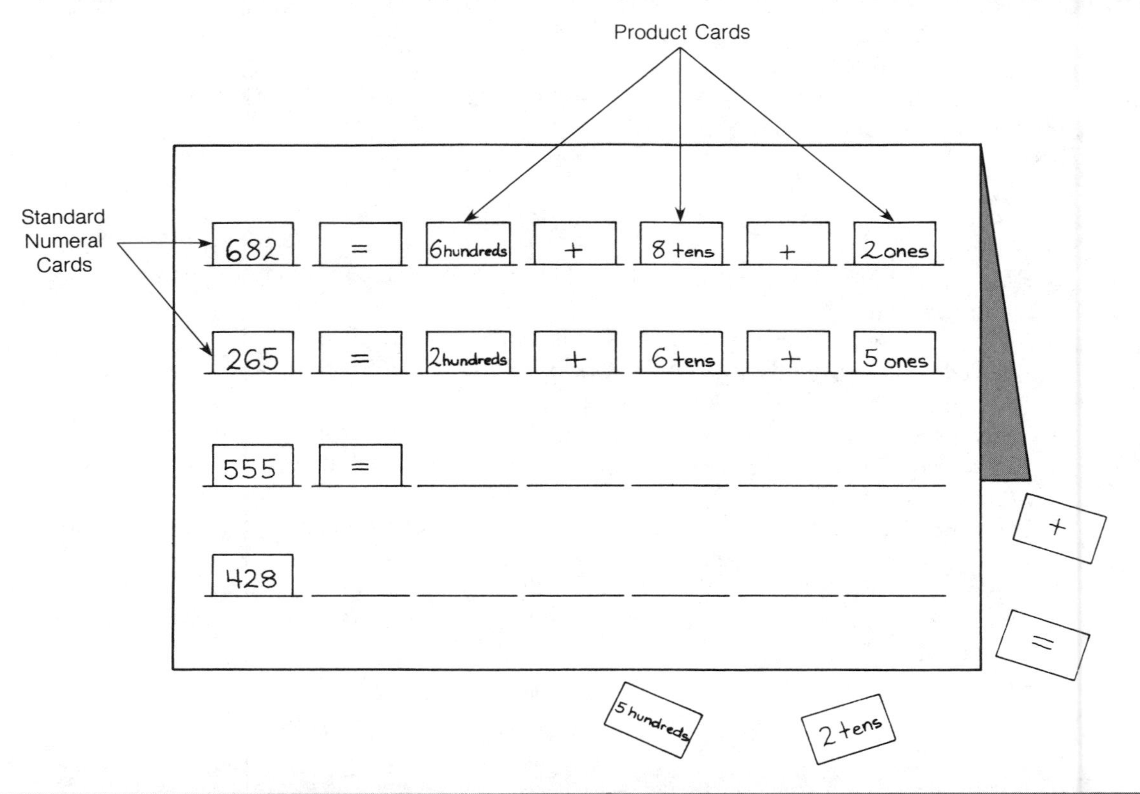

Figure 7.15
Pocket Chart Showing the Meaning of Multidigit Numerals

places for three digits and a set of digit cards, a leader (teacher or child) can say, "Show me 3 in the tens place," and so on. Show-me cards also facilitate translations from three-dimensional representations to symbols. For example, if base blocks (or bundled sticks) are placed on an overhead projector, each child can respond by preparing the standard numeral on a show-me card. (Figure 7.17).

A hand-held calculator can be used for practice in applying the idea that each digit names a number which is the *product* of the digit's face value and its place value (content objective 7.8). One such calculator activity is the game *Wipe Out.*[17] Each child has a calculator and enters the number given by the game leader. Then the leader names a digit that is to be wiped out and replaced with a zero. The product values of other digits are not changed. The activity can be adapted to different grade levels by replacing the digit with a digit other than zero, and by varying the length of the numeral.

In another place value calculator game, players roll a pair of dice and make two-digit numerals. Each time the dice are rolled, the numeral on one die is the number of tens and the numeral on the other die is the number of ones. For example, if a 2 and a 5 are rolled, either 25 or 52 could be made. Each player rolls the dice five times, adding the two-digit numbers on the calculator. The winner is the one who gets closest to but does not go over 222.[18]

17. Wallace Judd, "Instructional Games with Calculators," *The Arithmetic Teacher* 23, no. 7 (November 1976): 516–18.

18. Bill Fisher, "Calculator Games: Combining Skills and Problem-Solving," *The Arithmetic Teacher* 27, no. 4 (December 1979): 40–41.

ACTIVITY PLAN

Numeration for Whole Nos.: Multi-Digit Numerals — *Symbolic→Symbolic* — *Consolidating*

Content objectives

7.8, 7.9

Exemplars/Materials

A set of 4 playing cards for each of 13 different three-digit numerals. For a given numeral, the first card pictures base 10 blocks in the appropriate amount, a second card shows expanded notation, a third contains the standard numeral, and a fourth has number words. Each digit should appear in more than one of the three-digit numerals selected.

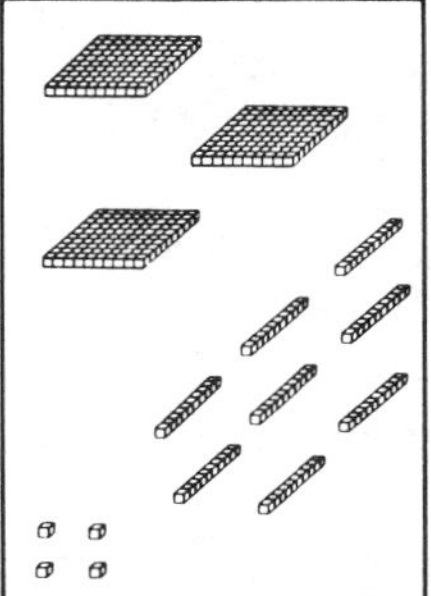

300+80+4

384

three hundreds
eight tens
four ones

Behavioral indicators

The child:

1. Identifies three additional representations for the same number when given any one of the following: a base 10 block card, an expanded notation card, a standard numeral card, or a number word card.
2. Given cards with the same digits, uses place values to explain why one is a correct match and the other is not.

Procedure

1. Make sure children are ready for this game by giving each child 12 cards, a set of 4 cards for each of 3 numerals. Have each child mix his cards thoroughly, then sort them into like piles.
2. Some children will put all expanded notation cards together, others may place all 4 cards for a single number in a pile. Direct attention to the latter.
3. Say, "We are going to play a game in which we put cards together for the same number. There are 4 cards for each number."
4. Play a simple rummy game in which pairs of cards for the same number are formed until one player has no unmatched cards remaining.

Note: The game can be varied by placing addends in unexpected sequences: instead of 300 + 80 + 4, write 80 + 300 + 4 on the card.

Figure 7.16
Sample Plan for a Consolidating Activity

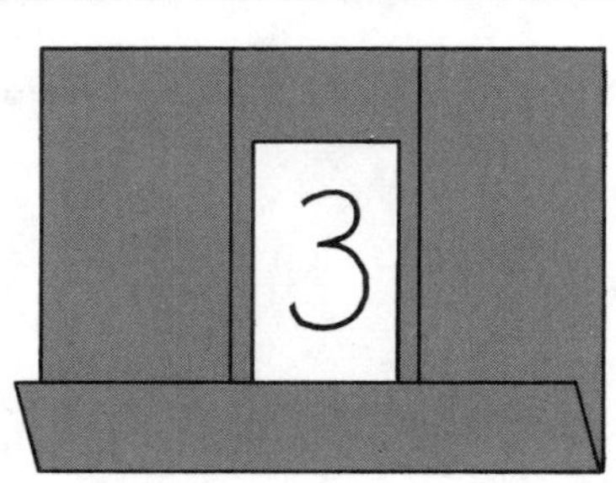

Figure 7.17
Show-Me Card for Individual Response

Another multidigit numeral activity is a trading game in which the winner is the first to have a million dollars. Two cubes are needed: one with 1, 10, 100, 1000, 10,000, and 100,000 on the sides; the other with 1, 5, 10, 50, 100, and 500. Each player is given the total of both cubes in play money. The money is kept on a sorting mat (see Figure 7.18) and 10 of one value is traded for 1 of the next higher value.[19]

Frequently some practice with multidigit numerals can be provided during brief in-between times, while a group is waiting on a bell to ring or a bus to arrive. For one such activity, have children rearrange the digits in a given numeral to make the largest number possible, or the smallest number possible.

Another activity for in-between times emphasizes that a number has *many* names. Give children a number and have them use what they know about place and product values to make different names.

260 = 26 tens
 = 2 hundreds and 60 ones

4000 = 4 thousands
 40 hundreds
 400 tens
 4000 ones

If children have difficulty generating them, use base 10 blocks or bean sticks to determine equivalent names.

Worksheets are limited to two-dimensional representations and symbols, but they can often be used after activities with three-dimensional exemplars. For example, after experiences with sticks in place value cans, children can be asked to select the numeral that goes with a group of mixed-up place value cans (Figure 7.19).[20] Each child can verify his answers by com-

19. Lee Jenkins and Marion Nordberg, Activity Resources Co., Inc., *Place Value and Regrouping Games* (Hayward, Calif.: Activity Resources, 1977), p. 27.

20. Adapted from Bitter and Mikesell, *Activities Handbook for Teaching with a Hand-Held Calculator*, p. 60.

Ones	Hundreds	Tens	Ones	Hundreds	Tens	Ones
Millions	Thousands					

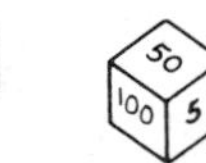

Figure 7.18
Sorting Mat for Play Money and Cubes

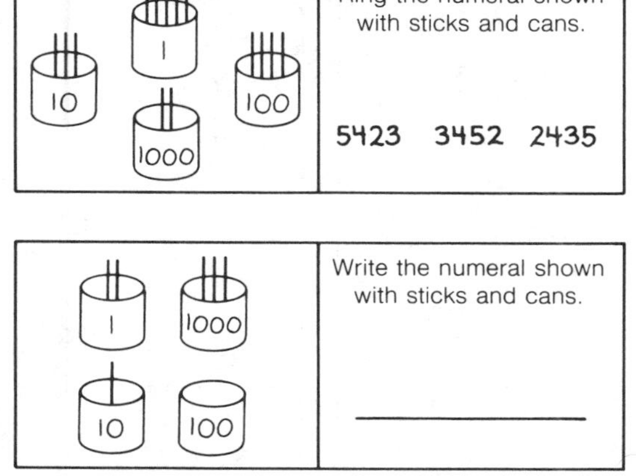

Figure 7.19
Numerals for Mixed-Up Place Value Cans

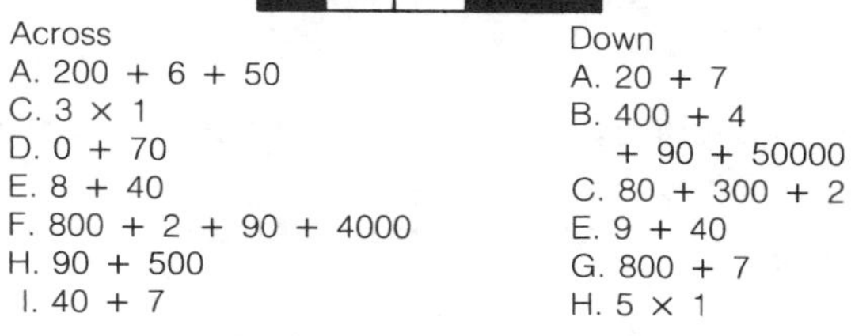

Across
A. 200 + 6 + 50
C. 3 × 1
D. 0 + 70
E. 8 + 40
F. 800 + 2 + 90 + 4000
H. 90 + 500
I. 40 + 7
Down
A. 20 + 7
B. 400 + 4
+ 90 + 50000
C. 80 + 300 + 2
E. 9 + 40
G. 800 + 7
H. 5 × 1

Figure 7.21
Cross-Number Puzzle

paring them with an answer key. Have children write the numerals only after they successfully select the limited choices.

Worksheets can also be used to provide practice writing the numerals that come immediately before or after, as illustrated in Figure 7.20.[21] During such practice, emphasize the fact that only one digit is written in each position to which a place value is assigned (content objective 7.7). If a child has difficulty, show the number with base 10 blocks, then remove or add a unit.

21. Payne and Rathmell, "Number and Numeration," p. 157.

Count by ones.
Write the missing numerals.
a. Before and after
____, 399, ____
____, 420, ____
____, 625, ____
____, 530, ____

b. A chart

341		343				347			
	352				356				
			364					369	

Figure 7.20
Writing Numerals Before and After

Cross-number puzzles can also be used. Each piece of the puzzle is a standard numeral, and equivalent expanded notations are listed under *across* and *down*. When a child is able to complete a puzzle with expanded notation like 7000 + 300 + 5, have the child try a puzzle in which the addends are not in the expected sequence (Figure 7.21).

Many transferring activities for concepts associated with multidigit numerals involve estimation or paper-and-pencil computation. For example, as a child abstracts and schematizes a procedure for addition with regrouping, numeration concepts can be applied. Transferring activities can also involve work with nondecimal number bases.

Periods

For an abstracting/schematizing activity, Persis Herold suggests a periods chart with zeros (Figure 7.22)[22] Have a child place a three-digit numeral card in different periods on the chart and read the numerals that are formed.

Each student can be provided with digit cards and a periods chart similar to the one pic-

22. Persis Herold, *The Math Teaching Handbook* (Newton, Mass.: Selective Educational Equipment, 1978), p. 65.

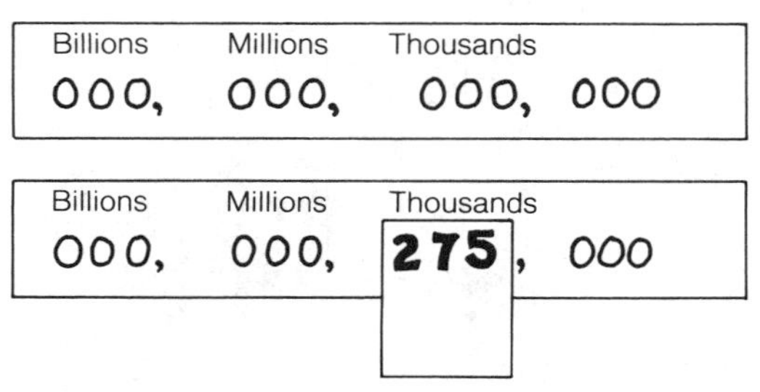

Figure 7.22
Periods Chart with Zeros

tured in Figure 7.23. Record a large number on the chalkboard and have children show the same numeral on their charts. As children read such numerals, emphasize the period names.[23]

For a strictly consolidating activity, pairs of playing cards such as those illustrated in Figure 7.24 can be constructed and used to play a simple rummy game. Pairs of cards for the same number are formed until one player has no unmatched cards remaining.

Nonstandard Forms

As we initiate content objectives 7.15 and 7.16, we need to focus on situations in which a nonstandard form is useful. We can present an addition example that requires regrouping, interpret the problem with wooden base 10 blocks, and then point out that there are more than 9 of one

23. Ralph Heimer and Cecil Trueblood, *Strategies for Teaching Children Mathematics* (Reading, Mass.: Addison-Wesley, 1977), p. 68. Copyright © 1977. Reprinted with permission.

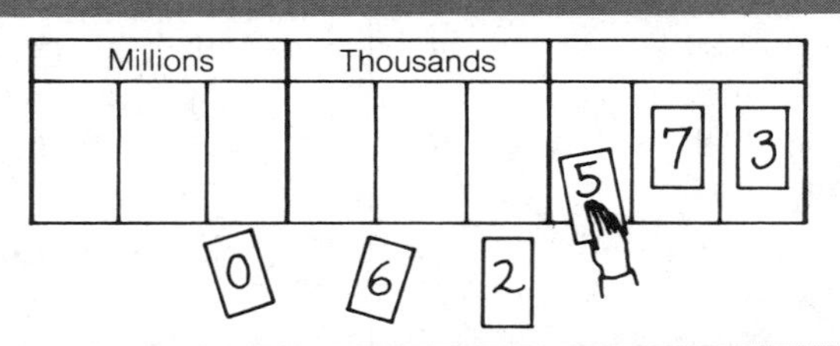

Figure 7.23
Periods Chart with Digit Cards

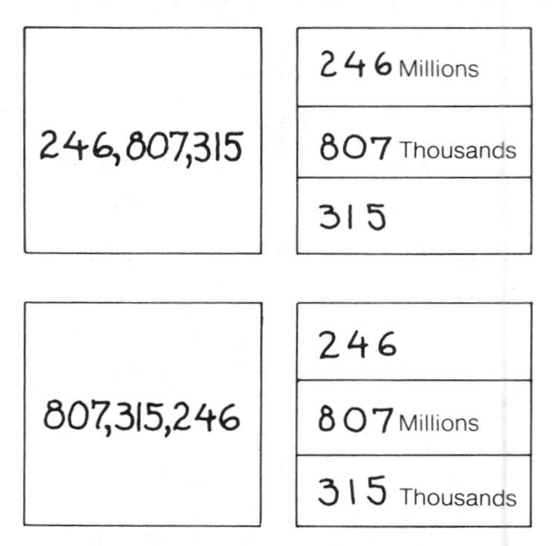

Figure 7.24
Rummy Cards for Periods

size. Because a sum must be written with a standard numeral (with one digit in each place), it is necessary to show the same amount of wood with as few pieces as possible. Children apply what they already know as they decide how to make the exchange.

For an initiating and abstracting activity, present a subtraction example that requires regrouping, interpret the sum (minuend) with base 10 blocks, then point out that there are not enough blocks of one size available to set aside the number of blocks indicated by the known addend. Show the same amount of wood in other, nonstandard ways. Have each child select base 10 blocks for the given sum, and trade to see different ways the same amount of wood can be shown. Record each as follows:

$$\begin{array}{r} 276 \\ -\underline{158} \end{array} \qquad \begin{aligned} 276 &= 2H + 6T + 16U \\ &= 200 + 60 + 16 \end{aligned}$$

Play money can be used for similar activities if 1, 10, and 100 dollar bills are used.

Figure 7.25 illustrates use of a pocket chart and sticks for a schematizing activity. Children should already understand that specific values are associated with each position or pocket; they also need to have previously worked with expanded notation.

ACTIVITY PLAN

Numeration for Whole Nos.: Non-Standard Forms | *Symbolic→3D/Rep Symbolic→Symbolic* | *Schematizing*

Content objectives

7.15, 7.16

Exemplars/Materials

Pocket chart and sticks (one for each child if possible)
Paper and pencil
Chalkboard

Behavioral indicators

1. Given a pocket chart with a nonstandard form of expanded notation, the child trades 10 for 1 to make the standard form.
2. Given a three-digit standard numeral, the child shows equivalent nonstandard forms in a pocket chart.
3. Given a three-digit standard numeral, the child writes equivalent nonstandard forms of expanded notation.

Procedure

1. Say, "We have used a pocket chart before. Can you name the pockets according to their positions?" (ones, tens, hundreds) "Tell me how the pockets are related. How many in the ones pocket at the right is the same amount as 1 in the tens pocket? How many in the tens is the same as 1 in the hundreds?"
2. Say, "We have talked about expanded numerals, and now we will write some."
3. Ask a child to put 1 stick at a time into the ones pocket. As he does, record the total number each time in a column. When 10 appear in a pocket, have him trade the 10 sticks for 1 in the pocket to the left.
4. As you continue, emphasize that:
   a. In the sequence of standard numerals, the digits 0–9 repeat themselves in the ones place.
   b. Expressions like 3 tens + 2 ones and 1 hundred + 0 tens + 0 ones are standard forms of expanded notation.
   c. One ten and ten ones is *nonstandard* notation for 20, because 10 ones can be traded for 1 ten.
   d. Similarly, 10 tens is a nonstandard form of notation.
5. Show 375 in a pocket chart and say, "This shows a standard numeral. Can I trade to make a nonstandard form?" Trade 1 ten for 10 ones, writing "3 hundreds + 6 tens + 15 ones."
6. Write 583 and ask children to show it the standard way in the pocket chart. Write out, "5 hundreds + 8 tens + 3 ones."
7. Say, "Now exchange sticks to show a nonstandard form of notation. Can you do another?" Make a list of nonstandard forms by recording each child's answer.
8. Write 427 and say, "Show me 3 nonstandard forms for this number. You do not have to use the pocket chart; just write them down. Now, read them aloud."
9. Write five three-digit numerals across the chalkboard, then ask five children to write a nonstandard expanded numeral under each.
10. Ask five more children to go to the board to write still another nonstandard form under the numerals.

Figure 7.25
A Sample Plan for a Schematizing Activity

Consolidating activities should involve each child. One such activity requires individual show-me and one-digit cards. Nonstandard expanded numerals are presented on large cards or on the chalkboard, and each child places digits for the equivalent standard numeral in his show-me card.

Games can also be adapted. Figure 7.26 pictures pairs of playing cards which can be made for a rummy game in which pairs of cards for the same number are formed.

A relay race can provide further practice. Divide the group into four teams with the same number of members and have members line up at a starting point. When you call out or write a standard three-digit numeral, a member from each team races to the chalkboard and writes a nonstandard, expanded form. Each team member writes the number in a different form. One member must return to the start line before the next races forward. The first team to finish listing nonstandard numerals correctly and return to the start line receives three bonus points. All teams receive one point for each correct answer written by the time the first one finishes. The first to get 30 points wins.

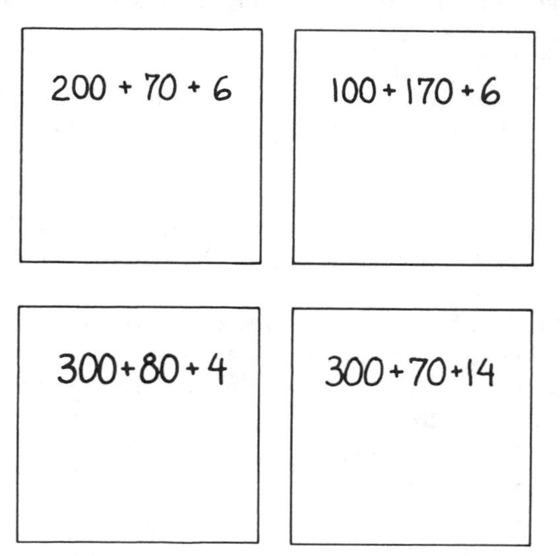

Figure 7.26
Rummy Cards for Nonstandard Forms

We may want to use worksheets for some of the practice activities we plan. Examples of response forms are provided in the manual that accompanies this text.

PLANNING A TEACHING UNIT

The Activity Type Cycle discussed in Chapter 3 can serve as a guide in planning. Furthermore, each activity can be considered as a part of a larger unit that the authors call a *teaching unit*. A teaching unit is a collection of activity plans for a single objective or for a cluster of related objectives, usually the latter, in which activity plans are ordered by type. Typically, more than one plan is included for each type of activity in the unit. Not all types need to be included, but there should be at least three different kinds of activities.

Consider a teaching unit on place values and grouping. For the cluster of objectives 7.1–7.6 we could include Figures 7.2 and 7.5 as a starting point, to develop additional plans.

The number of plans needed for each type varies, and Figure 7.27 notes ways that three different units have been composed. All units include a diagnostic activity to find out if each child understands the concepts and demonstrates the needed skills. Sometimes, as in Teaching Unit B, an abstracting activity starts things off. At other times, as in Teaching Unit C, transferring activities are deleted. The activity in which new concepts and skills are used is actually an initiating activity for a different teaching unit with higher-order objectives.

Sometimes teachers move directly from an abstracting activity to a consolidating activity, and omit schematizing activities. But schematizing activities help children put bits and pieces of ideas together. This facilitates recall and application.

Activity Type	Teaching Unit A	Teaching Unit B	Teaching Unit C
D. Diagnosis	1	1	1
I. Initiating	2		
I/A	1		1
A. Abstracting	2	3	2
A/S	1		
S. Schematizing	3	3	3
S/C			
C. Consolidating	2	2	3
C/T			
T. Transferring	2	2	
Total Activity Plans	14	11	10

Figure 7.27
Number of Activities by Type for Sample Teaching Units

For a given type, activities included can be varied by using different exemplars and by incorporating different translations across classes of exemplars. A teaching unit should contain enough varied activities to be a resource for daily lesson plans. Time for a single activity can vary from an hour to several days. Furthermore, not all activities can be used with each child.

PLANNING ASSESSMENT

When we plan to assess each child's numeration concepts and skills, we focus initially on relevant content objectives. Then we think of one or more behaviors that let us infer a child understands the concept or has acquired the skill. Next, we translate descriptions of behavior into actual tasks that will hopefully elicit the desired behaviors. We thereby infer each child's skill development and comprehension.

To illustrate this process, let us plan sample assessment activities for content objective 7.9 In determining behavioral indicators, consider the matrix of exemplar-class translations described in Chapter 3. Nine different stimulus-response patterns are indicated. Walbesser's list of action verbs in Chapter 6 is also helpful.

Behaviors can include the following:

1. Given a collection of base 10 blocks with nine or less of any one size, and a collection of multidigit numeral cards, the child sorts the blocks and identifies the multidigit numeral card that shows the total number of units (3D →symbolic).

2. Given a multidigit numeral and bundled sticks, the child makes a collection of sticks for the numeral shown with nine or less in each category of sticks: singles, bundles of ten, and so on (symbolic →3D).

Other behavioral indicators can be easily designed if we consider the translation matrix, but the above two illustrate the planning process.

We need to think of more than one assessment task for eliciting each type of behavior. For the first behavior, we could design these assessment tasks:

1a. During an interview with each child, present a collection of base 10 blocks consisting of 4 hundreds, 6 tens, and 3 ones or units mixed up. Also, lay out a numeral card for each of the following: 463, 634, 346, 643, and 346. Say, "Which of these numeral cards tells the *total* number of units in this collection of blocks? Which shows how many?"

1b. Set up a learning center with five stations, each with a different task. Place a box with a collection of blocks as described in 1a and post a list of multidigit numerals at each station. Also provide a response form for each child to copy the numerals.

Provide directions for the center as follows:

- Take one of the sheets and a pencil.
- For each task, sort the blocks in the box.
- How many total units are in the box? Write the answer on the sheet.
- Be sure to put all blocks back in the box.
- When you complete the five tasks, place your paper in the answers sheet box.

For the second behavioral indicator, design similar assessment tasks for interview and learning center settings. Also, pair up children, giving each a supply of prebundled sticks with nine singles, nine bundles of 10, and so on, and a writing pad. Have one of the pair write a multidigit numeral, and the other make a collection of sticks for the numeral shown. Observe the childrens' responses. Then have each pair reverse roles for the next task.

Assessment tasks for numeration are often limited to paper-and-pencil tasks, such as 473 = _____ hundreds + _____ tens + _____ ones, in which children thoughtlessly copy digits in sequence. Assessment tasks planned as described above should show us each child's numeration concepts and skills. We can then plan appropriate instruction.

TIPS ON MANAGING THE CLASSROOM SITUATION

Here are suggestions in teaching children numerals for whole numbers.

1. Increase participation of each child within a group by using show-me cards whenever possible. Each child arranges digit cards in his show-me card, and places it face down until you ask the class to show the cards.

"Show me a three-digit numeral with a zero in the ten's place."

2. Prepare a handy list of appropriate consolidating activities for in-between times. These periods are neither planned or announced, so you have to be ready.
3. Provide immediate feedback, such as answer cards, confirmation by a peer, or a calculator. For example, children can use a calculator to total the numbers recorded upon completion of each set of three exercises. They can then compare the total with your reference list.
4. Introduce choices gradually. Difficulty in learning arithmetic is eased if the number of alternative responses is carefully controlled. Begin by having the child select one of two possible responses. Later proceed to three, then four, then possibly incomplete responses or clues, and finally to open-ended responses such as a box (variable) in which a child writes a numeral. Introduce choices gradually by preparing a worksheet for each level, or constructing a learning center with a series of stations.
5. Help children focus on symbols. Some children become confused and are unable to respond if a text or a worksheet page is crowded with symbols. Each exercise may need to be presented on a separate page or card.

8

Developing Computation Procedures for Addition and Subtraction

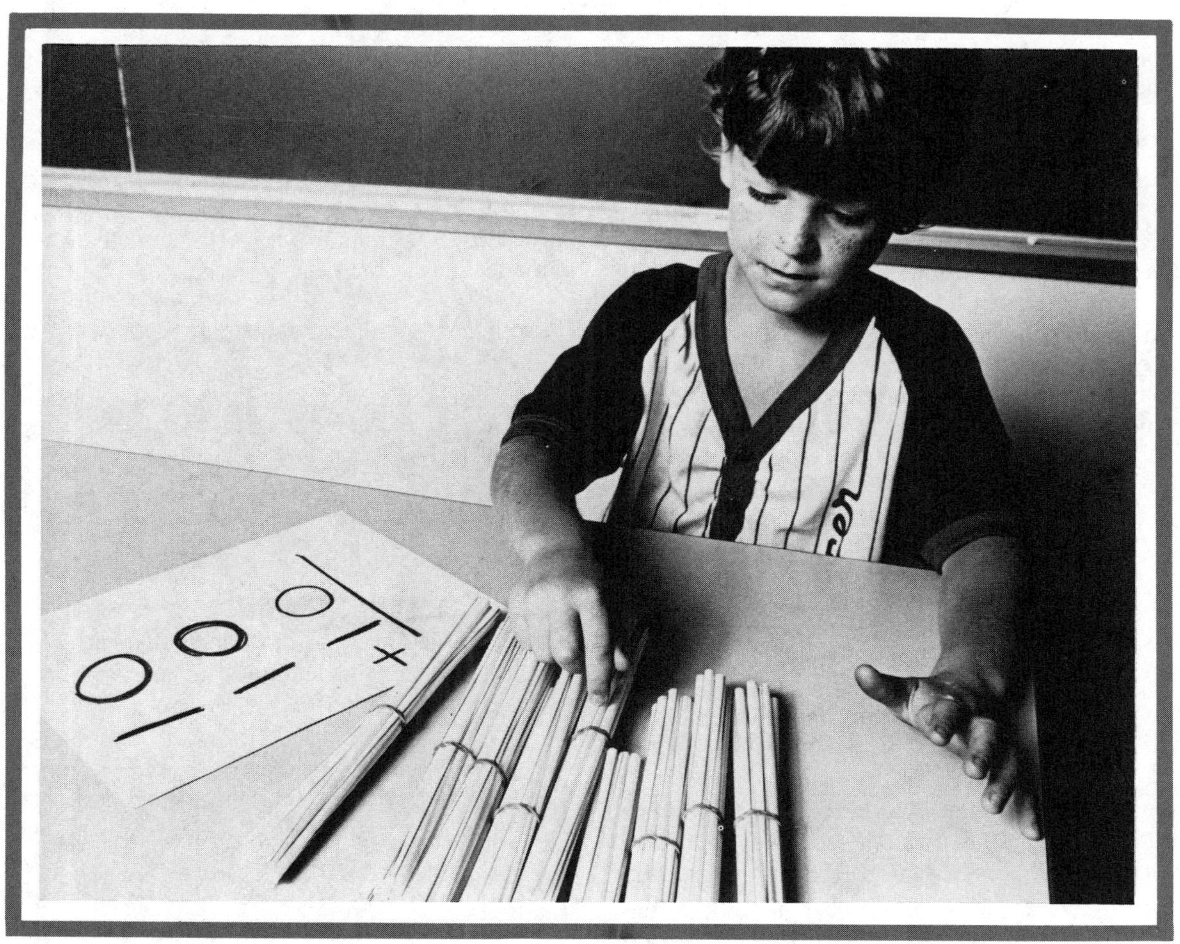

The role of computation is changing. In society, as in technology, calculators and computers are used in situations requiring accurate and rapid computation. Calculators are even used to balance checkbooks.

In light of the decreasing use of paper-and-pencil computation you might wonder about the continued stress on computational skill in elementary schools. Hamrick and McKillip observe that "even though computers and calculators are rapidly changing the way most people deal with arithmetical situations, there are computational skills that continue to be educationally and socially desirable."[1] These authors also believe that development of computational skill is desirable at the elementary level because:

> First, it facilitates the learning of subsequent related topics. Second, computational skill helps pupils to understand both the meaning and the significances of arithmetic operations and to apply these operations appropriately. Third, it facilities an exploration of various topics, generalizations from data, and the recognition of generalizations. Fourth, some aspects of computational skill continue to have considerable social utility.[2]

However, a shift in the emphasis is required. We will need to place more attention on understanding the meanings of the operations and on why an algorithm works to produce a reasonable answer. We should also spend less time dealing with long computations and more time on mental computation and on estimating results.

1. Katherine B. Hamrick et al., "How Computational Skills Contribute to the Meaningful Learning of Arithmetic," *Developing Computational Skills,* 1978 Yearbook of the National Council of Teachers of Mathematics (Reston, Va.: National Council of Teachers of Mathematics, 1978), pp. 1–2.
2. *Ibid.,* p. 3

PLANNING THE CONTENT

As we plan lessons for teaching children to compute accurately and with understanding, we first need to select the specific content to be the focus of the lessons. Many algorithms are used in computing whole numbers; we will discuss both common and alternate forms. In elementary schools we use the word *algorithm* for paper-and-pencil procedures, although the term actually has a broader definition.

> **8.1**
> An algorithm is a process for computing an unknown number by writing numerals and other mathematical symbols in a fixed sequence of steps.

Addition of Whole Numbers

One hundred basic addition facts that each child eventually needs to commit to memory were identified in Chapter 5.

> **8.2**
> Addition is a binary operation that associates a pair of whole numbers with a third whole number called the sum.

Only two numbers can be added at a time. So, to find a sum for three or more numbers, we must add more than once. For example, to add 6 + 8 + 4, think:

$6 + 8 = 14$; $14 + 4 = 18$. The sum is 18.

EXCURSION
Algorithm

Where did the word *algorithm* come from?

ál-go-rithm. Derived "from the name of the Arabic mathematician Al-Khowarizmi, who wrote the first arithmetic book using Hindu-Arabic numerals and place value."[3] Sometimes the word is spelled *algorism.*

EXCURSION
Magic Squares[4]

Numbers can be arranged in squares such that the sums of every row, every column, and both diagonals are equal. These *magic squares* appeared as early as 1000 B.C., and have fascinated mathematicians and children for thousands of years. The earliest known magic square is the Chinese "lo-shu," which has the same number of dots in every row, column and along both diagonals. The magic square on the right appeared in a 1514 engraving by Albrecht Durer. The two central numbers in the bottom row give the date of the picture.

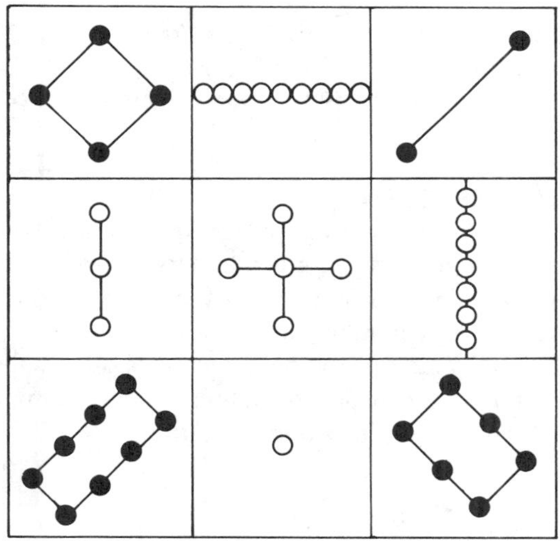

The Chinese lo-shu

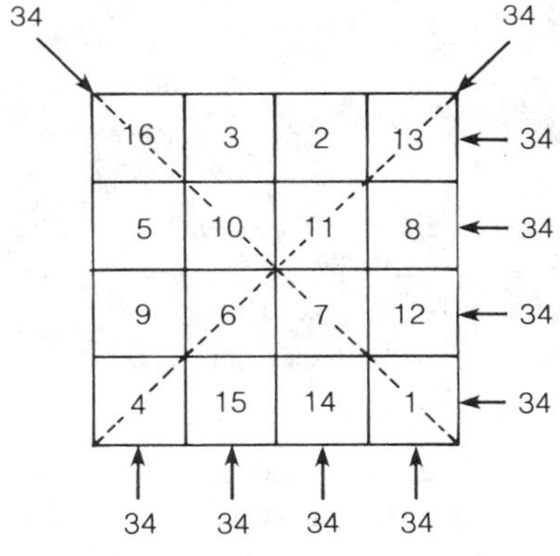

4 × 4 Magic Square

Grouping by Tens

We can simplify column addition by having children learn to group by tens.

8.3
The order in which the addends are associated does not change the sum.

3. Reprinted with permission of Macmillan Publishing Co., Inc. from *Teaching Mathematics to Children,* 2nd ed. (New York: Macmillan, 1973), p. 60. Copyright © 1973 by Esther J. Swenson.

4. Elizabeth Williams and Hilary Shuard, *Elementary Mathematics Today: A Resource for Teachers, Grades 1–8,* 2nd ed. (London: Longman Group Ltd., 1976), pp. 327–28.

```
8                8          8              8
  >15              7        4 >12          4  13
7                  9        5—17          5    (20)
9—24         (10)  3 (10)   7—24     (10) 7
3—27             + 2        6—30          6
+ 2—29           ----       + 7—37       + 7
-----             29        -----        ----
 29                          37           37
```

Vertical addition	Grouping by Tens	Vertical addition	Grouping by Tens

Although grouping by tens often makes addition easier, it does not make it faster.

The Addition Algorithm

These content objectives are important when teaching the traditional addition algorithm.

> **8.4**
> To add whole numbers, add ones with ones, tens with tens, hundreds with hundreds, etc.
>
> **8.5**
> When the sum of the numbers in a column (place) is greater than 9, the sum must be regrouped or renamed.

Computing with addends such as 43 + 35 is simpler than with addends such as 38 + 47. In the first case, no regrouping is needed to find the sum.

$$\begin{array}{rcl} 43 & = & 4 \text{ tens} + 3 \text{ ones} \\ +\ 35 & = & 3 \text{ tens} + 5 \text{ ones} \\ \hline & & 7 \text{ tens} + 8 \text{ ones} \\ & = & 70 + 8 \\ & = & 78 \end{array}$$

Regrouping in addition occurs when the sum of the numbers named in any column is 10 or more. We say that place is *overloaded* and the number named must be regrouped. The traditional name for the regrouping procedures is *carrying*. However, the authors prefer to describe it as *regrouping*, a term that reminds a child of concrete referents used early in instruction. We could also use expanded notation to illustrate the regrouping process.

$$\begin{array}{rcl} 38 & = & 3 \text{ tens} + 8 \text{ ones} \\ +\ 47 & = & 4 \text{ tens} + 7 \text{ ones} \\ \hline & = & 7 \text{ tens} + 15 \text{ ones} \\ & = & 7 \text{ tens} + (1 \text{ ten} + 5 \text{ ones}) \\ & = & (7 \text{ tens} + 1 \text{ ten}) + 5 \text{ ones} \\ & = & 8 \text{ tens} + 5 \text{ ones} \\ & = & 80 + 5 \\ & = & 85 \end{array}$$

We can avoid regrouping by adding partial sums. For example,

```
  375                        375
+ 498                      + 498
   13  (ones)        or       13
  16   (tens)                160
  7    (hundreds)            700
  873                        873
```

Of course, the more familiar forms of the algorithm are usually preferred.

```
                      21
Results of           275     Results of          275
regrouping           368     regrouping          368
recorded           + 493     done mentally     + 493
                    1136                        1136
```

Adding by Endings

Textbooks often refer to mental addition of a one-digit and two-digit numbers as *adding by endings*. The sum may be in the same decade as the two-digit number or in the next decade.

Going into the next decade is often called *bridging* to a new decade. Observe how adding by endings is used in column addition.

```
 6
   >15   Basic fact
 9
 8   23  Bridging the decade
 5   28  Same decade
+7   35  Bridging the decade
 35
```

Compensation

> **8.6**
> If the number added to one addend is subtracted from the other addend, the sum is unchanged.

EXCURSION
Mental Power

Without pencil and paper, use the compensation principle to find the sum for each.

436 + 290 382 + 578

Compensation is a powerful aid to mental arithmetic. If a, b, c, and d are whole numbers and if $a + b = c$, then $(a + d) + (b - d) = c$. To apply this concept when adding 395 + 278, think: (395 + 5) + (278 − 5) = 400 + 273 = 673.

Low Stress Addition

Procedures called *low stress algorithms* increase notation in order to reduce the mental load. Further, they do not intermix fact recall and regrouping. These algorithms are often used to help underachievers. The low stress algorithms developed by Hutchings contain easily written notation for each intermediate step in the computation, allowing a teacher to identify specific errors in a child's knowledge of basic facts.[5]

Examples of the Hutchings algorithm for addition follow. Each sum is recorded with small digits. Units are added to units, then the number of tens is counted.

Checking Addition

The usual procedure for checking is to add in the opposite direction.

Original (↓)	*Check* (↑)
22	22
275	275
863	863
429	429
+ 687	+ 687
2254	2254

Different combinations are added as the check is completed, providing an excellent opportunity to note application of the commutative and associative properties. Grouping into tens while adding a column of numbers is another way to check.

Subtraction of Whole Numbers

Subtraction with regrouping can be difficult for children.

Example 1

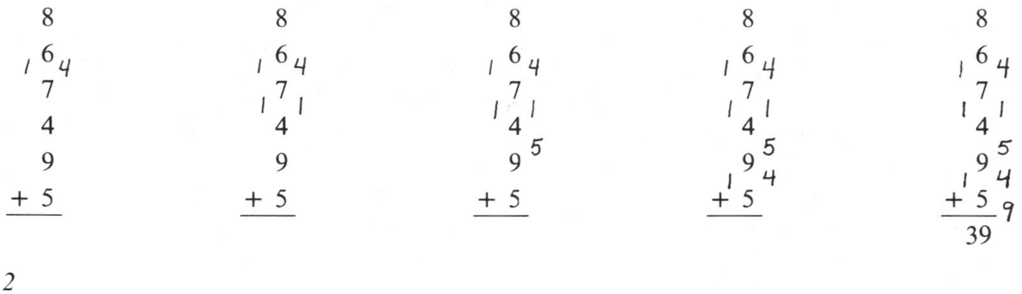

Example 2

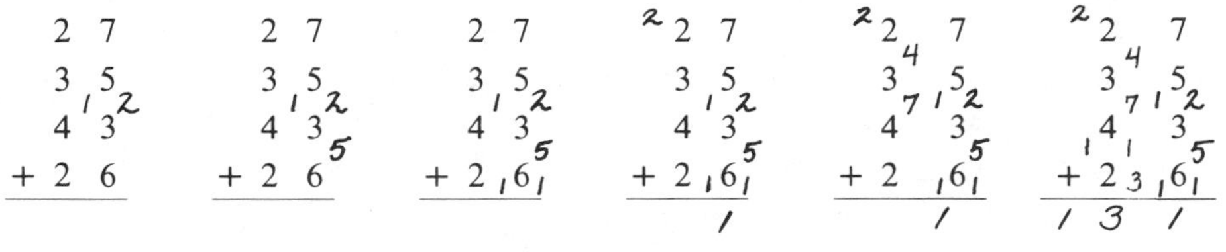

5. Barton Hutchings, "Low Stress Algorithms," *Measurement in School Mathematics: 1976 Yearbook* (Reston, Va.: National Council of Teachers of Mathematics, 1976), pp. 218–39.

EXCURSION
Low Stress Addition

Use the Hutchings low stress algorithm to find the sum for each of these:

```
   8        32       286
   6        49       473
   9        37       268
   3        68       897
 + 7      + 79     + 649
```

Subtraction Algorithms

These content objectives are important in teaching subtraction algorithms.

> **8.7**
> To subtract whole numbers, subtract ones from ones, tens from tens, hundreds from hundreds, etc.
>
> **8.8**
> When the product value of any digit in the subtrahend (known addend) is greater than the product value of the corresponding digit in the minuend (sum), the number named in the minuend must be regrouped (renamed). Subtraction in this case is sometimes called *compound subtraction.*

Subtracting 48 − 35 is easier than subtracting 35 − 28, which requires regrouping. Subtraction without regrouping is usually introduced shortly after a child learns the meaning of subtraction and can use addition facts to name missing one-digit addends.

Two well-known algorithms are used for compound subtraction: the decomposition method and the equal-additions method. Other subtraction algorithms are also available.

The *decomposition method* is widely used in the United States. To subtract 38 from 64, 64 must be regrouped as 5 tens and 14 ones. The 64 is *decomposed* and grouped differently.

$$\begin{array}{rcl} 64 & = & 5 \text{ tens} + 14 \text{ ones} \\ -\ 38 & = & 3 \text{ tens} + \ \ 8 \text{ ones} \\ \hline & & 2 \text{ tens} + \ \ 6 \text{ ones} \\ & = & 20 + 6 = 26 \end{array}$$

The same procedure is involved when the numbers are greater, or when regrouping involves more places. For efficiency, the more familiar forms of the algorithm are preferred rather than those with expanded notation.

Results of regrouping recorded

```
   11
  5 ⁄ 15          5 11 1
  6̸2̸5̸            6̸2̸5̸
- 387     or    - 387
  238             238
```

Results of regrouping done mentally

```
  625
- 387
  238
```

An alternate technique for compound subtraction is the *equal-additions method.* This method, which is more commonly used in Europe, is especially effective if the teacher emphasizes speed. It applies another *compensation* principle.

> **8.9**
> If the same number is added to both the minuend and the subtrahend, the difference between the two remains the same.

If *a, b, c,* and *d* are whole numbers and $a - b = c$, then $(a + d) - (b + d) = c$. For example, when subtracting 83 − 27 we can add 10 ones to 3 ones, and then subtract 13 − 7. But if 10 ones are added to 83, a like amount must be added to 27 so that the difference between 83 and 27 does not change. Adding 1 ten is equivalent to adding 10 ones. If we add 1 ten to the 2 tens of 27, the subtrahend becomes 37. Three tens from 8 tens is 5 tens; thus 83 − 27 = 56.

```
    8 ¹3
 −³2̸  7
    5  6
```

A child can soon learn to extend the procedure to larger numbers.

```
 3 '6 '2 '1
²1 ⁸7 ⁹8  5
-----------
 1  8  3  6
```

A specific variation of this algorithm, sometimes known as the Austrian method, requires thinking in terms of addition facts. This algorithm is especially efficient when the addition of 1 ten, 1 hundred, etc. to the subtrahend is written by indicating a 1 below the given digit. Study the example below and the accompanying thought pattern.

```
  4 2 7 '3
– 2 9 5 8
---------
        5     Eight plus five equals 13.      ┐
                                              │
  4 2 7 '3                                    │
– 2 9 5 8                            Compensate
      1       Add 1 ten.                      ┘
---------
        5

              Six plus one equals seven.
  4 '2 7 '3
– 2 9 5 8
      1       Nine plus three equals 12       ┐
---------                                     │
    3 1 5     (hundreds)                      │
  4 '2 7 '3                          Compensate
– 2 9 5 8                                     │
    1 1       Add 1 thousand.                 ┘
---------
  1 3 1 5     Three plus one equals four.
```

With practice, children can eventually work this algorithm without writing the steps.

As Kennedy notes, "The strength of the equal-additions method of subtraction is the ease and speed with which computation can be done once the process has been mastered. The weakness of this method is that most children find it difficult to understand."[6]

A number of studies have compared the decomposition and equal-addition methods of subtraction. In a 1949 comparison, Brownell found that children taught the decomposition method generally achieved superior results.[7] Thirty years later, Sherrill arrived at a similar conclusion.[8] Both studies stressed the need for children to understand the processes.

Low Stress Subtraction

Like low stress addition, low stress subtraction helps to simplify the regrouping process.[9] The Hutchings low stress subtraction algorithm is different from the conventional procedure in two ways.

1. The regrouped minuend or *rename* is recorded between the minuend and the subtrahend.
2. All renaming is completed before differences are recorded within each column.

If a number recorded in the rename is less than the number below it, a half-space 1 is written at the left and the next number is recorded as one less than the number in the minuend.

---

6. Leonard M. Kennedy, *Guiding Children to Mathematical Discovery,* 3rd ed. (Belmont, Calif.: Wadsworth, 1980), p. 195.
7. William Brownell et al., "Meaningful vs. Mechanical Learning: A Study in Grade III Subtraction," *Duke University Studies in Education* 8 (1949):1–207.
8. James M. Sherrill, "Subtraction: Decomposition Versus Equal Addends," *The Arithmetic Teacher* 27, no.1 (September 1979): 16–17.
9. Hutchings, "Low Stress Algorithms," pp. 218–39.

EXCURSION
Compensation

Applying the property of compensation to subtraction can make mental arithmetic easier. For example, to subtract 28 from 55, proceed as follows:

```
  55  ───→   57
– 28       – 30
           ----
             27
```

Can you tell how the amount added to each was chosen?

PROBLEMS

Example 1

4 3 5	4 3 5	4 3 5
2'5 →	3'2'5 →	3'2'5
−1 6 8	−1 6 8	−1 6 8
		2 6 7

Example 2

7 4 8 6	7 4 8 6	7 4 8 6	7 4 8 6
8 6 →	4 8 6 →	6'4 8 6 →	6'4 8 6
−3 8 5 1	−3 8 5 1	−3 8 5 1	−3 8 5 1
			3 6 3 5

Example 3

6 2 1 3	6 2 1 3	6 2 1 3	6 2 1 3
0'3 →	1'0'3 →	5'1'0'3 →	5'1'0'3
−3 9 7 8	−3 9 7 8	−3 9 7 8	−3 9 7 8
			2 2 3 5

Checking Subtraction

At an early age, children learn that subtraction is the inverse of addition, and that they can check subtraction by adding. The check consists of adding the answer (remainder or difference) to the subtrahend. The sum should be the same as the minuend of the original subtraction.

823	Minuend	245
− 578	Subtrahend	+ 578
245	Remainder	823

Another way to check subtraction is to repeat the subtraction by replacing the subtrahend with the remainder.

823	Minuend →	823
− 578	Subtrahend	− 245
245	Remainder	578

Ordering the Content

Although there is no specific sequence for presenting subject matter components when teaching how to add and subtract whole numbers, dependencies between specific concepts and skills can be observed. By recording such observations in ordered lists, we create guides for selecting content. The guides also help assure that each child has specific concepts and skills needed for new learnings.

An addition or subtraction algorithm applies selected content objectives; concepts must be taught. Also, think of a computational procedure as developing within a sequence of steps or types of examples, such as the following:

Addition

1. Basic facts with sums to 10
2. Zero facts
3. Column of one-place addends with zero, 3 + 0 + 5
4. Column addition of one-place addends, 3 + 1 + 4, unseen addend in a basic fact
5. Adding by endings with no bridging, 15 + 3
6. Adding by endings in column addition, 6 + 5 + 7
7. Column of 4 one-place addends, 1 + 4 + 2 + 4

EXCURSION
Low Stress Subtraction

Use the Hutchings low stress algorithm to find each difference.

45	631	8649
− 27	− 275	− 2978

Note the strong resemblance between this process and the decomposition method. In both cases, the same basic combinations are used, with differences in phraseology and sequence.

---

8. Basic facts with sums from 11 to 18
9. Checking
10. Adding tens to tens, 10 + 20, two-place sum
11. Adding hundreds to hundreds, 500 + 300, three-place sum
12. Addition of 2 two-place addends, two-place sum, no regrouping, 22 + 21
13. Adding 2 three-place addends, no regrouping, three-place sum, 112 + 112
14. Bridging a multiple of 10, 25 + 9
15. Bridging tens in column addition, 6 + 9 + 7 + 9
16. Regrouping tens, two-place sum, 45 + 38
17. Column of 3 two-digit addends, two-place sum, no regrouping, 21 + 33 + 25
18. Column of one and two-digit addends, no regrouping, 12 + 4 + 3
19. Two two-place addends, no regrouping, three-place sum, 45 + 83
20. Regrouping more than 1 ten, 48 + 49 + 28 + 28
21. Regrouping hundreds, 275 + 152
22. Regrouping a ten and a hundred, 399 + 276
23. Regrouping thousands, 3521 + 4803

Subtraction

1. Basic facts with minuends to 10
2. Basic facts with minuends from 11 to 18
3. Facts with zero as remainder
4. Facts with zero as subtrahend
5. Two-place minus one-place, no ones in remainder, 60 − 8
6. Tens minus tens, 80 − 60
7. Subtracting by endings, 28 − 5, no bridging
8. Two-place minus two-place, no regrouping, 38 − 22
9. Checking
10. Regrouping tens, 62 − 38
11. Bridging a multiple of ten, 33 − 5
12. Three-place minus two-place, no regrouping, 365 − 43
13. Hundreds minus hundreds, 900 − 700
14. Three-place minus three-place, no ones in three-place remainder, 387 − 267
15. Three-place minus three-place, three-place remainder, no regrouping, 593 − 471
16. Three-place minus two-place, regrouping a 10, 491 − 98
17. Three-place minus two-place, regrouping a 100, 539 − 83
18. Three-place minus three-place, regrouping a 10, 972 − 347
19. Regrouping hundreds, 628 − 437
20. Regrouping a ten and a hundred, 736 − 587
21. Regrouping a thousand, 3570 − 1850
22. Three-place minus three-place, zero in tens place in minuend, 704 − 375
23. Zero in ones and tens place in minuend, 500 − 284
24. Four-place minus four-place, zeros in minuend, 3000 − 1346

PLANNING INSTRUCTION

Continue to think of the Activity Type Cycle in planning instruction in computational procedures. Also, keep each child's previously learned concepts and skills in mind.

Initiating and Abstracting Activities

A child should understand the meaning of the operations and be able to apply knowledge of basic facts and place value ideas before learning addition and subtraction algorithms. However, some practice with the facts can be provided through algorithms. Use the easier facts when first introducing the algorithms; otherwise, the child's attention may be obscured by the facts that have not been mastered.

Addition

Children often need to add two-digit numbers to find how many are in two sets or the total cost of two items. When problems involve a sum greater than 18, first encourage children to count or group materials to solve the problem. Later as we increase the difficulty of problems, most children will see the need for a more systematic way to find out how many–specifically, the addition algorithm.

As Schminke, Maertens, and Arnold note: "Merely presenting an algorithm to children without helping them to see a need for it is not likely to be effective. Instead children can be provided with a need to develop and use algorithms through carefully structured experiences in problem situations utilizing increasingly greater quantities of concrete materials."[10] An example of a simple problem follows.

PROBLEM

Pencils come in packages of 10 each. In one set there are 23 pencils. In another set there are 15. Name the total number of pencils in the two sets.

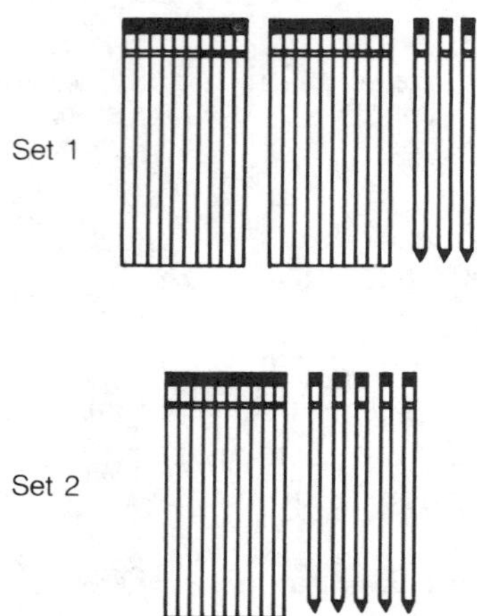

By rearranging the packages and the loose pencils, a child can easily see that there are 3 tens and 8 ones, or 38 pencils altogether. The sum is 38.

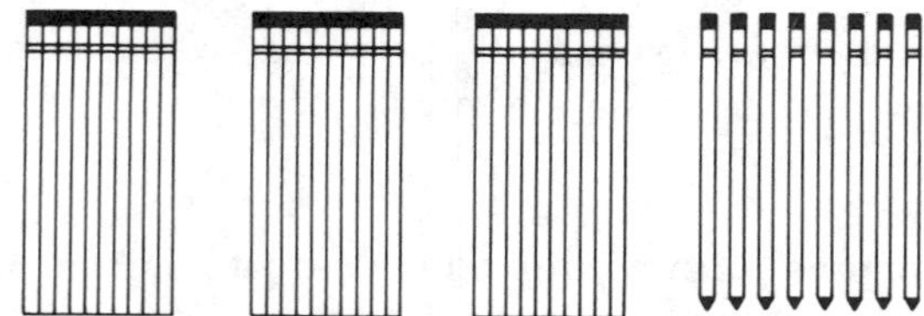

Base 10 blocks can be placed on laminated cards to provide a more structured introduction to the addition algorithm.

Step 1: Show the two addends. What we see is the number for each set; we want to find the total number.

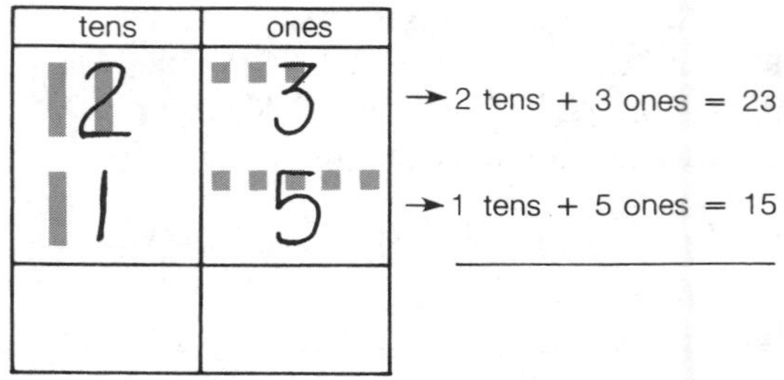

Step 2: Collect the ones together, then the tens. Place them below the bar where the total number is shown.

Tens	Ones	
2	3	→ 2 tens + 3 ones = 23
1	5	→ 1 ten + 5 ones = 15
		3 tens + 8 ones = 38

Straws in bundles can also be used.

Most teachers and textbooks introduce the addition of whole numbers with examples that do not require regrouping or carrying. Some authorities believe children should immediately begin regrouping. Kennedy notes:

> Their argument states that when children add two numbers, such as 23 and 45, using the conventional algorithm form, they can begin their addition in either the tens or the ones place and get the correct sum either way. Thus, they do not learn the value of beginning addition in the ones place. If an addition problem requires regrouping . . . they will learn from the beginning that it is convenient to start the addition in the ones place and then go to the tens place.[11]

Extensive practice with addition without regrouping also allows children to focus on vertical pairs of digits; they forget the place value ideas associated with multidigit numerals. There-

10. From TEACHING THE CHILD MATHEMATICS, Second Edition, by C. W. Schminke, Norbert Maertens and William Arnold. Copyright © 1978 by Holt, Rinehart and Winston, © 1973 by the Dryden Press, Inc. Reprinted by permission of Holt, Rinehart and Winston, CBS College Publishing.

11. Leonard M. Kennedy, *Guiding Children to Mathematical Discovery* (Belmont, Calif.: Wadsworth, 1970), p. 101.

fore, when subtraction with regrouping *is* introduced, children tend to merely compare the two digits, subtracting the small number from the larger.

When children have had sufficient experience with trading activities, as described in the previous chapter, the move to addition with regrouping should be easy. In fact, some children may even ask for "trading problems" because they find them interesting and fun.[12]

Base 10 blocks or similar materials can be used to make the regrouping process meaningful. Introduce the written algorithm as a record of observations made as the blocks are used to find the total number.

Each child should be involved in a sequence of steps similar to the following.

Step 1. Show 25 as 2 tens and 5 ones, and 38 as 3 tens and 8 ones.

Tens	Ones
2	5
3	8

→ 2 tens + 5 ones = 25
→ 3 tens + 8 ones = 38

Step 2. To find the total number, first place all the ones together below the bar.

Tens	Ones
2	5
3	8

→ 2 tens + 5 ones = 25
→ 3 tens + 8 ones = 38
→ 13 ones = 13

In this example there are 13 ones in the ones column, but children have already learned that only one digit is written in each place in a standard numeral. Form a rule with the children such as, *If there are 10 or more, regroup so we can write just one digit.*

After the trade of 10 ones for 1 ten, we may want each child to hold the ten stick in his hand, or place it at the top of the tens column.[13]

Step 3: Exchange 10 ones for 1 ten.

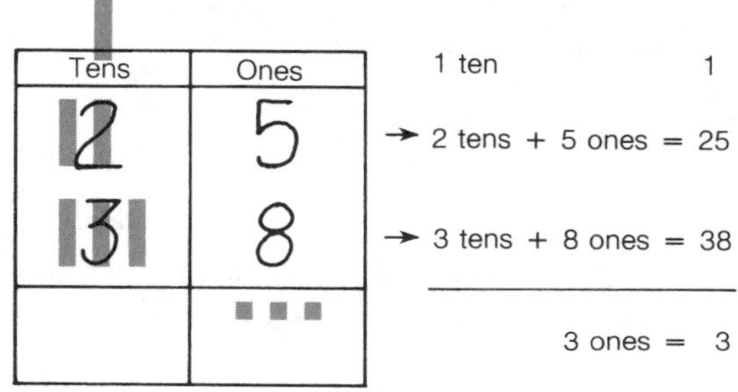

Step 4: Combine all blocks in the tens column, including the 1 ten from the trade.

Tens	Ones
2	5
3	8

1 ten 1
→ 2 tens + 5 ones = 25
→ 3 tens + 8 ones = 38
→ 6 tens + 3 ones = 63

It can now be seen that the total number is 63.

Limit the initial situations to regrouping ones as a ten. As children gain experience abstracting and schematizing the written algorithm, materials can be used to extend regrouping in other columns. As soon as children are comfortable with the trading process using materials such as base 10 blocks or bean sticks, we can provide more abstract experiences with a place value chart.

To find the sum of 47 and 78 on a place value chart, follow these steps.

12. Katherine K. Merseth, "Using Materials and Activities in Teaching Addition and Subtraction Algorithm," *Developing Computational Skills,* 1978 Yearbook of the National Council of Teachers of Mathematics (Reston, Va.: National Council of Teachers of Mathematics, 1978), p. 69.

13. *Ibid.,* p. 70.

Step 1: Show 47 as 4 tens and 7 ones, and 78 as 7 tens and 8 ones. (Note: Numerals are not written on the pockets of place value charts, but they are used here to make the number of sticks in each pocket easily recognizable. Sometimes a teacher will clip numeral cards on pockets during a demonstration.)

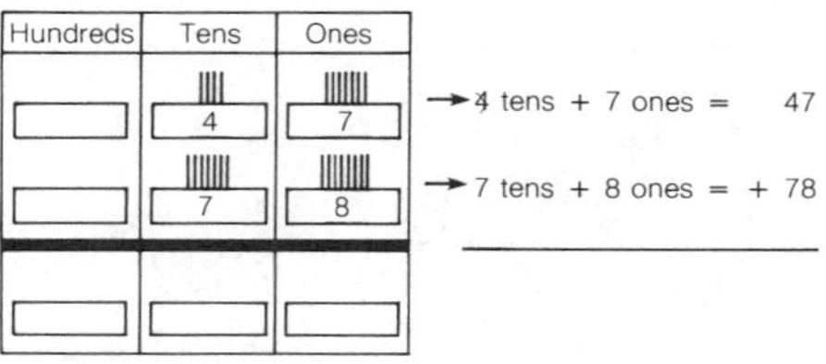

Step 2: Collect the ones below the bar where we show the total number when adding.

Hundreds | Tens | Ones

4 tens + 7 ones = 47
+
7 tens + 8 ones = 78

15 ones = 15

Step 3: Apply the "10 ones to 1 ten" rule and regroup the ones. (The 1 ten received in the exchange can be placed at the top of the tens column if the chart is lying on a table, or it can be fastened at the top if the chart is vertical.)

Hundreds | Tens | Ones

1 ten 1

4 tens + 7 ones = 47
+
7 tens + 8 ones = 78

5 ones = 5

Step 4: Collect the tens, including the 1 ten from the exchange.

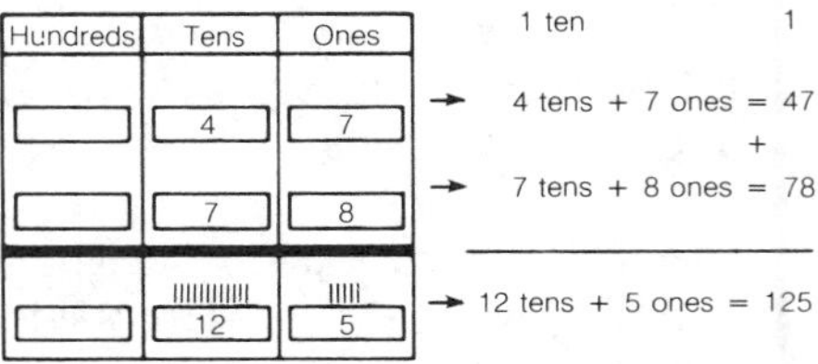

Step 5: Apply the "10 ones to 1 ten" rule and regroup the tens. Place the 1 hundred from the exchange in the hundreds column.

Hundreds | Tens | Ones

1 ten 1

4 tens + 7 ones = 47
+
7 tens + 8 ones = 78

1 hundred + 2 tens + 5 ones = 125

We may also use an abacus and separate the sets with clothespins until ones are collected.

After many experiences using manipulatives to find sums, have each child make a step-by-step written record. With guidance at each step, a child is gradually led to the standard algorithm. A child should be allowed to use manipulative aids as long as needed. However, children do not have to use manipulative aids once they understand both the concepts and the steps in the procedure.[14]

For children having difficulty with the traditional addition algorithm, use one of the nontraditional algorithms described earlier.

Stress the value of accuracy with children as soon as they begin to add and subtract. At first a child should be encouraged to check her addition by using counting devices, markers, or any meaningful method. Later, encourage her to use the more sophisticated checking procedures described earlier.

14. Donald D. Paige et al., *Elementary Mathematics Methods* (New York: John Wiley and Sons, 1978), p. 88.

EXCURSION
A Homemade Place Value Device

This device is quick, easy, and inexpensive to make.

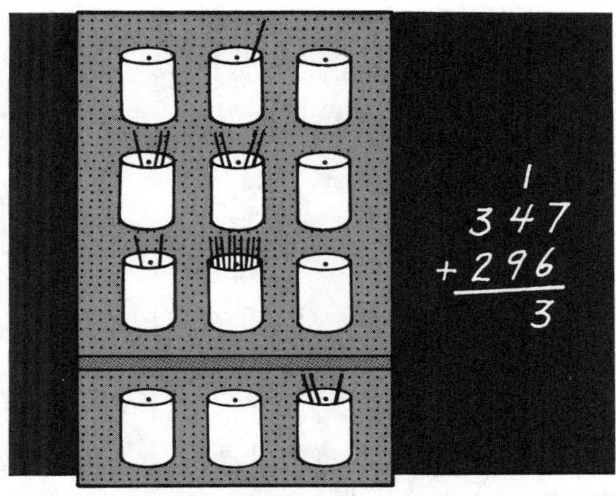

- *Materials needed:*
  16″ × 24″ piece of pegboard (3 pieces can be cut from a 2′ × 4′ sheet)
  12 styrofoam cups or juice cans
  60 plastic coffee stirrers or wooden sticks
  24 brass fasteners
  20″ piece of plastic or masking tape
- *Construction:* Fasten four rows of three cups to the pegboard as pictured. Use two fasteners per cup so the cup will be rigid. Fasteners are easily pushed through the styrofoam. Affix a piece of tape between the two bottom rows.
- *Instructional use:* Sticks are placed in cups as appropriate. Use the device as you would a place value chart. Show each of two given numbers, then regroup, collect, or separate to picture each step of an addition or subtraction algorithm. If the top row is initially left free, it can be used to show regrouping.
- *Variations:* An extra row of tape can be placed across the very top and place values written at the heads of each column. Remove this before children begin to compute with paper and pencil.

Subtraction

Subtraction without regrouping is normally introduced soon after the subtraction facts necessary for their solution.

PROBLEM

> Pencils come in packages of 10 each. Christa has 26 pencils (2 packages and 6 extra pencils). If she gives 12 pencils (1 package and 2 pencils) to her brother, how many will she have then?

We are told the sum and one of the addends. What is the other addend? By setting aside the known part (1 package and 2 loose pencils) the child can see there are 1 ten and 4 ones or 14 pencils in the part remaining.

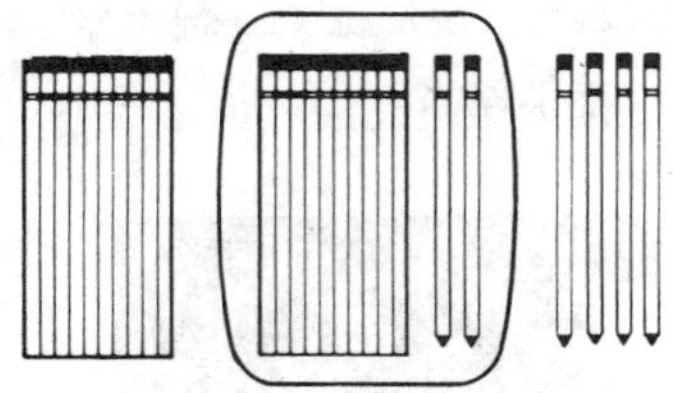

Base 10 blocks and laminated cards can also be used.

PROBLEM

Mrs. Miller had 35 pencils. She gave 21 of them to another teacher. How many does she have left?

Step 1: After writing both numerals, show the total number on the card with base 10 blocks. (Do not show the part we know, which are included in the total.)

Tens	Ones
3	5
2	1

Step 2: The 21 is the known part, so put a one on the 1 and the other ones below the bar where we show the unknown part.

Step 3: Put two of the tens on the 2 and the other ten below the bar.

Tens	Ones
3	5
2	1

The ten and the 4 ones below the bar tell us that Mrs. Miller still has 14 pencils.

Children should have many experiences with exchanging activities before regrouping is introduced in the subtraction algorithm. These experiences should include trading down. Several subtraction algorithms can be used, the two most common being the decomposition and equal-addition procedures. Presently, the *decomposition method* is the most widely used subtraction algorithm. It can be introduced as a written record of observations. When this is done, children regard the algorithm as a reasonable procedure. Furthermore, the close relationship between regrouping in addition and subtraction is easily observed.

Base 10 blocks and a pocket or place value chart can be used for initiating and abstracting activities. Children should follow a procedure similar to this example.

PROBLEM

There were 46 children on the school bus yesterday. Twenty-eight of them were girls. How many boys were on the bus?

Step 1: Show the total number, 46, and 4 tens and 6 ones.

The 28 girls were a part of the 46 children. Take out 28 to see how many boys were in the other part. First, try to take out 8 ones; but a child will quickly see that it is not possible, for there are not enough ones. Therefore, 1 ten must be exchanged for 10 ones, thus renaming 46 as 3 tens + 16 ones.

Step 2: Exchange 1 ten for 10 ones.

Step 3: Move 8 ones down. The 8 ones that remain are placed below a stick or masking tape line.

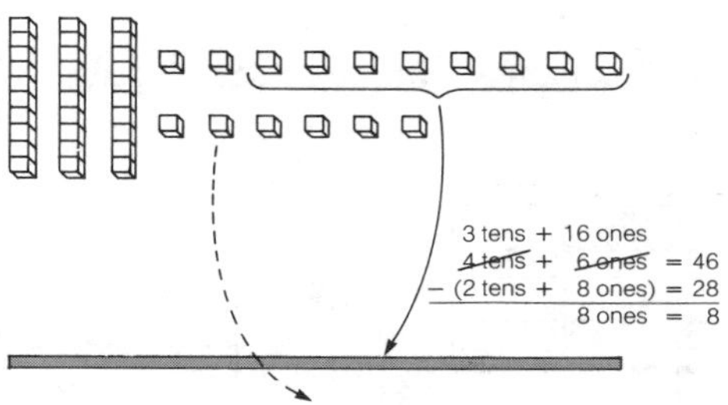

Step 4: Move the 2 tens down. The 1 ten that remains is placed below the line.

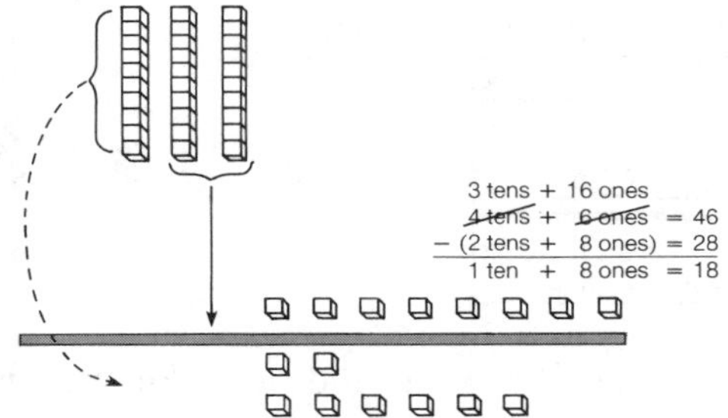

After 28 is taken out, 1 ten and 8 ones are in the other part below the line, telling us that 18 boys were on the bus.

Before children are introduced to the shorter written algorithm, work through the subtraction procedure using a place value chart or an abacus and give children some experience with expanded notation. The chart and the abacus may be more difficult, as they rely on the concept of place having value and thereby more closely exemplify the algorithm.

The following steps are involved in subtracting 186 from 352 using a place value chart.

Step 1: Represent 352, the total amount. Since 6 ones cannot be moved down to show the known part, obtain more ones by exchanging 1 ten for 10 ones.

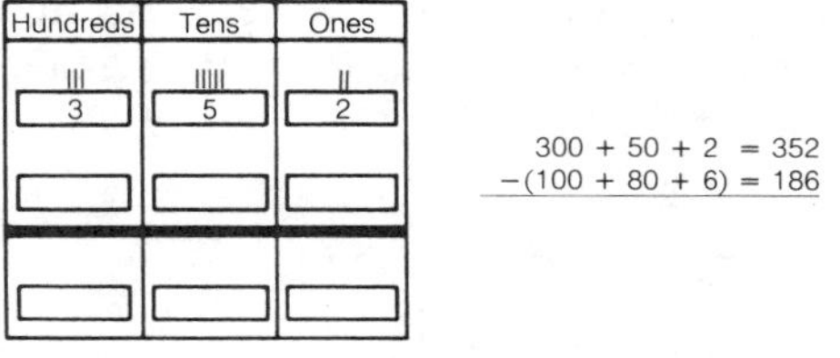

Step 2: Exchange 1 ten for 10 ones.

Hundreds	Tens	Ones
3	4	12

300 + 40 + 12 = 352
−(100 + 80 + 6) = 186

This gives 12 in the ones column and 4 in the tens column. Now subtract 6.

Step 3: Move 6 of the ones down to show the known part and move the remaining 6 ones below the bar.

Hundreds	Tens	Ones
3	4	
		6
		6

300 + 40 + 12 = 352
−(100 + 80 + 6) = 186
6 6

There are not enough tens to subtract 8, so we need to exchange 1 hundred for 10 tens.

Step 4: Exchange 1 hundred for 10 tens.

Hundreds	Tens	Ones
2	14	
		6
		6

200 + 140 + 12 = 352
−(100 + 80 + 6) = 186
6 6

Step 5: Subtract the tens. Move 8 tens down to show the known part and move the remaining 6 tens below the bar.

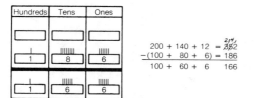

Hundreds	Tens	Ones
2		
	8	6
	6	6

$$\begin{array}{rl} 200 + 140 + 12 & = \overset{2\,14\,1}{\cancel{3}\cancel{5}2} \\ -(100 + 80 + 6) & = 186 \\ \hline 60 + 6 & \quad 66 \end{array}$$

Step 6: Subtract the hundreds. Move 1 hundred to the known part and the remaining hundred below the bar.

Hundreds	Tens	Ones
1	8	6
1	6	6

$$\begin{array}{rl} 200 + 140 + 12 & = \overset{2\,14\,1}{\cancel{3}\cancel{5}2} \\ -(100 + 80 + 6) & = 186 \\ \hline 100 + 60 + 6 & \quad 166 \end{array}$$

The unknown, the remainder below the bar, is 166.

Give special attention to subtraction involving zeros in the minuend. Problems such as the following are difficult for many children.

$$\begin{array}{r} 605 \\ -\ 278 \\ \hline \end{array} \qquad \begin{array}{r} 4002 \\ -\ 1673 \\ \hline \end{array}$$

A place value chart can help a child understand the procedure. For example, to subtract 278 from 605, use the following:

Step 1: Represent 605, the sum.

Hundreds	Tens	Ones
6	0	5

$$\begin{array}{rl} 6 \text{ hundreds} \qquad\qquad + 5 \text{ ones} & = 605 \\ -(2 \text{ hundreds} + 7 \text{ tens} + 8 \text{ ones}) & = 278 \\ \hline \end{array}$$

There are not enough ones to subtract, nor tens to exchange for ones. Therefore, 1 hundred must be converted to tens, and 1 of these tens exchanged for 10 ones.

Step 2: Exchange 1 hundred for 10 tens

Hundreds	Tens	Ones
5	10	5

$$\begin{array}{rl} 5 \text{ hundreds} + 10 \text{ tens} + 5 \text{ ones} & = \overset{5\,10}{\cancel{6}\cancel{0}5} \\ -(2 \text{ hundreds} + 7 \text{ tens} + 8 \text{ ones}) & = 278 \\ \hline \end{array}$$

Step 3: Exchange 1 ten for 10 ones.

Hundreds	Tens	Ones
5	9	15

$$\begin{array}{rl} 5 \text{ hundreds} + 9 \text{ tens} + 15 \text{ ones} & = \overset{5\,9\,1}{\cancel{6}\cancel{0}5} \\ -(2 \text{ hundreds} + 7 \text{ tens} + 8 \text{ ones}) & = 278 \\ \hline \end{array}$$

Now subtract. Show the addend or known part.

Step 4: Subtract the ones, tens, and hundreds. In each case, first show the known addend, then place the remainder below the bar.

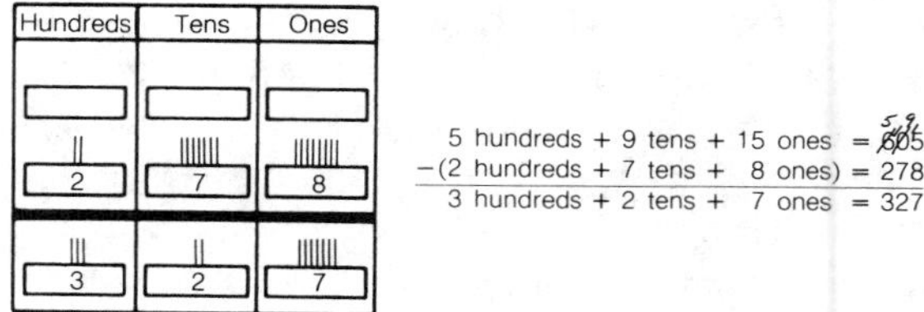

With experience, a child should be able to reason through the same subtraction example more simply by recognizing that 605 is 60 tens and 5 ones, which is the same number as 59 tens and 15 ones. Likewise, for subtracting 4002 − 1673, a child can think of 4002 as 399 tens and 12 ones.

The equal-additions algorithm uses compensation. Specifically, when a child solves a problem that involves regrouping, such as 64 − 27, 10 ones are added to the minuend (64) and one ten is added to the subtrahend (27). This can be illustrated with a place value chart if *disjoint* sets are compared.

PROBLEM

There are 64 fish in a large tank and 27 fish in a small tank. How many more fish are in the large tank?

Step 1: Show *both* the minuend and the subtrahend.

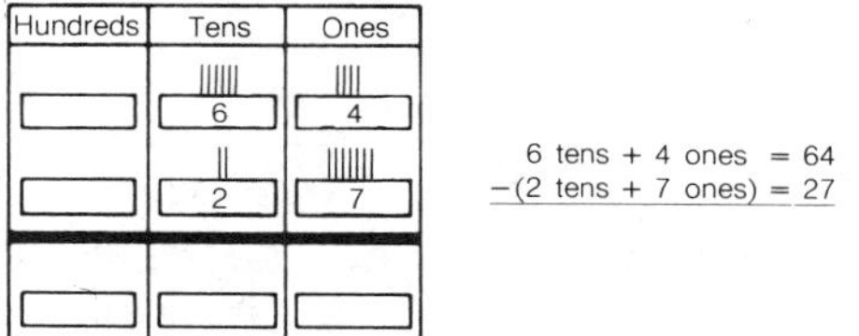

Hundreds
Tens
Ones
6
4
2
7
6 tens + 4 ones = 64
−(2 tens + 7 ones) = 27

In the minuend there is not a one for each one in the subtrahend, so we need to add 10 to both the minuend and the subtrahend.

Step 2: Add 10 ones to the minuend and 1 ten to the subtrahend.

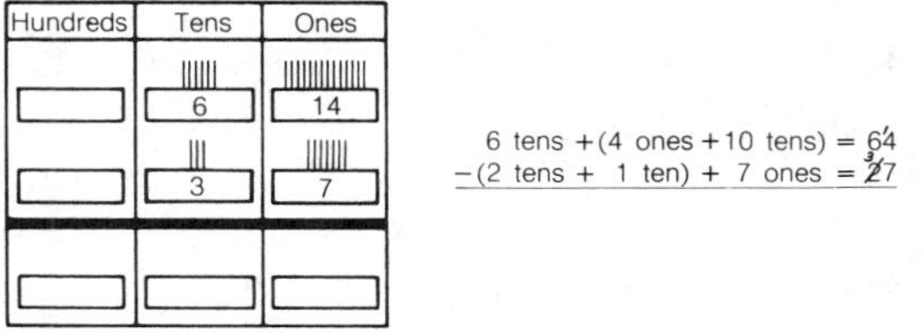

Step 3: Subtract by matching within each place value column to see how many more are in the minuend than in the subtrahend. Move the extras from the minuend to the space below the bar.

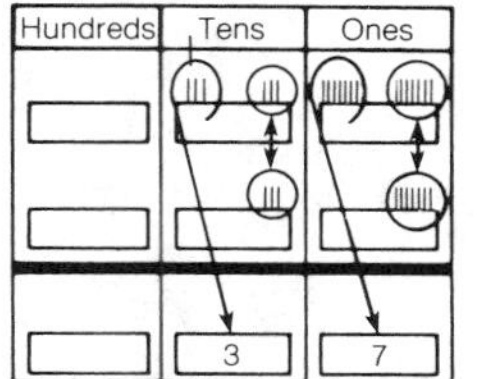

6 tens + 14 ones = 64
−(3 tens + 7 ones)= 27
37

The large tank has 37 more fish than the small tank.

Compensation may be used flexibly, depending upon the particular digits in the problem. It can be an aid to mental computation. For example, to simplify finding the difference between 34 and 18, add 2 to both minuend and subtrahend.

$$\begin{aligned} 34 - 18 &= (34 + 2) - (18 + 2) \\ &= 36 - 20 \\ &= 16 \end{aligned}$$

Schematizing Activities

As children gain experience in adding and subtracting, the bridge from using materials such as base blocks or a place value chart to the written algorithm by itself must be completed. Schematizing activities help a child understand that the steps in the written procedure apply to *any* two whole numbers. The steps represent a general pattern and not just a way to find the sum for specific numbers.

Even when careful groundwork is laid, the move to exclusive use of the written procedure requires careful attention and management. The first step in the transition might be to ask each child to tell what she would do with base 10 blocks as the written record is being made. Then, provide the children with the laminated cards used earlier, without the blocks. For children who have difficulty with the written procedure alone, continue to use more examples with and without blocks, discussing the results and supporting their efforts to move from the concrete to the abstract.

While children are moving to exclusive use of paper-and-pencil procedures, encourage them to record the regrouping. They might write the following:

```
  1 2
  427          3 11 1
   38          4 2 5
 +175         -1 7 9
 ----         ------
  640          2 4 6
```

A child understands the regrouping process more easily when the work is written in full. However, a child who both understands the process and does the operation mentally is performing at a higher level. Remember that learning is a growth process and each child should work at her most successful level. However, encourage the child's progress to a higher and more mature level.

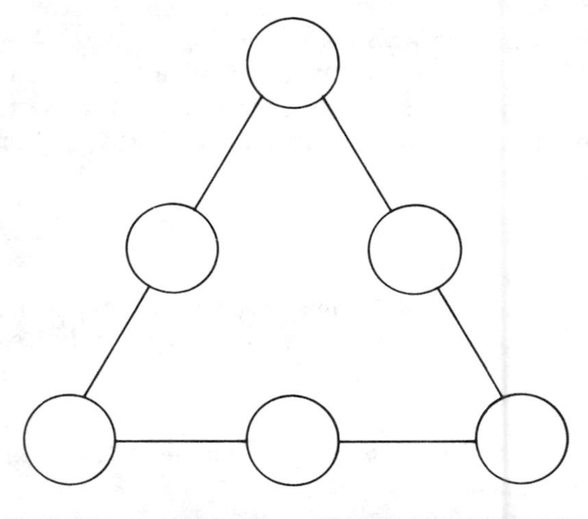

Figure 8.1
Number Puzzle

Consolidating Activities

Consolidating activities are designed to help a child remember new concepts and skills, making them habitual and accurate. Most consolidating activities involve paper and pencil; many will be games.

Configurations of numbers, such as the following, provide both practice with computation and a challenge for each child. The objective is to determine the sum along any straight line constant.

In Figure 8.1, arrange the numbers 7, 14, 21, 28, 35, and 42 so the sum along each side of the triangle is the same. Two possible sums are 63 and 84. For a variation, change the numbers to be used to any six consecutive numbers.

In Figure 8.2, arrange the numbers 25, 26, 27, 28, 29, 30, 31, 32, and 33 so the sum along each line of three is the same. Possible sums are 68 and 90.

EXCURSION
Palindromes

A word or numeral that reads the same forward and backward is a *palindrome,* such as *radar* or 2772. Follow these steps to get a palindrome:

1. Write a numeral with two or more digits	678
2. Reverse the digits	+ 876
3. Add	1554
4. If the sum is not a palindrome, reverse the digits.	+ 4551
5. Add	6105
6. Continue until you get a palindrome	+ 5016
	11121
	+ 12111
The palindrome	23232

Try making a palindrome from these numbers.

16 93 419 2735

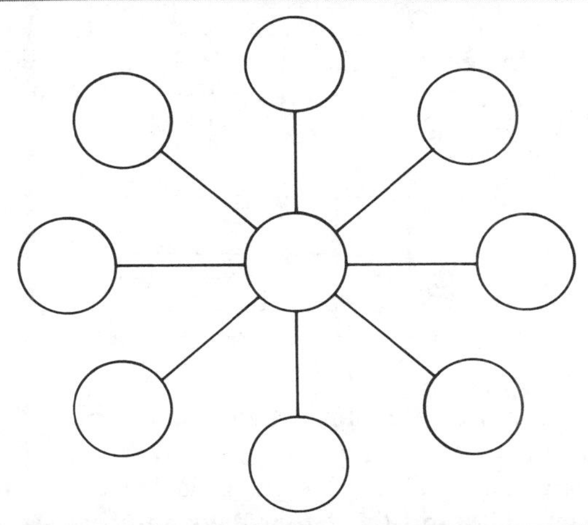

Figure 8.2
Number Puzzle

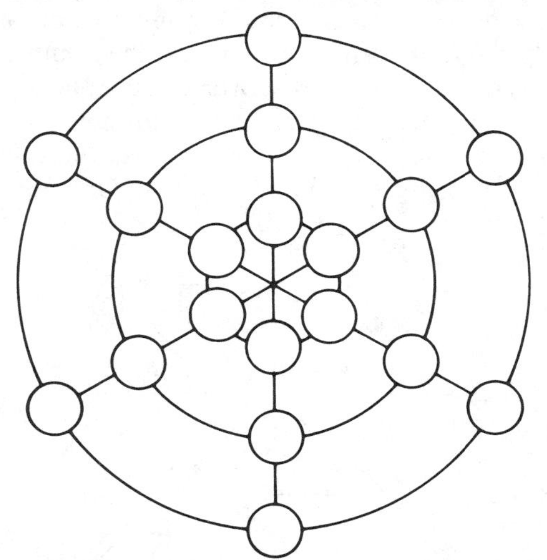

Figure 8.3
Number Puzzle

The gifted child can arrange the numbers from 1 to 18 in Figure 8.3 so that the sum around each circle and along each straight line is the same. Figure 8.4 presents a solution.

Magic squares can also provide an effective drill for addition or for subtraction. Two are illustrated below.

PROBLEM
Find the sum for each row, column, and diagonal.

26	13	12	23
15	20	21	18
19	16	17	22
14	25	24	11

PROBLEM
Complete the following magic square so that sums for all rows, columns, and diagonals are the same.

			28
20		26	
	21	22	27
19	30		16

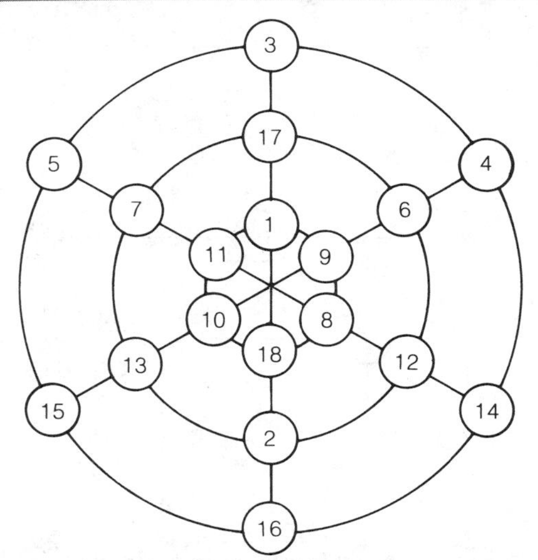

Figure 8.4
Number Puzzle Solution

Activities (see right) that center around finding missing digits provide excellent computation practice, showing the inverse relationship between addition and subtraction.

```
  2 7 6 □
  1 □ 8 4
  5 6 □ 9
+ □ 2 4 3
---------
```

```
  7 □ 2 3
- □ 3 9 □
---------
  5 3 2 7
```

```
  □ □ □ □
- 3 7 2 4
---------
  5 3 1 7
```

Another puzzle that children enjoy is the diffy puzzle.[15] As presented, numbers appear in only the four corners. Begin with the numbers in the outside circles and subtract pairs, placing the results in the inside circles. Continue with the inside square similarly. You can vary the activity by increasing the number of squares in the problem.

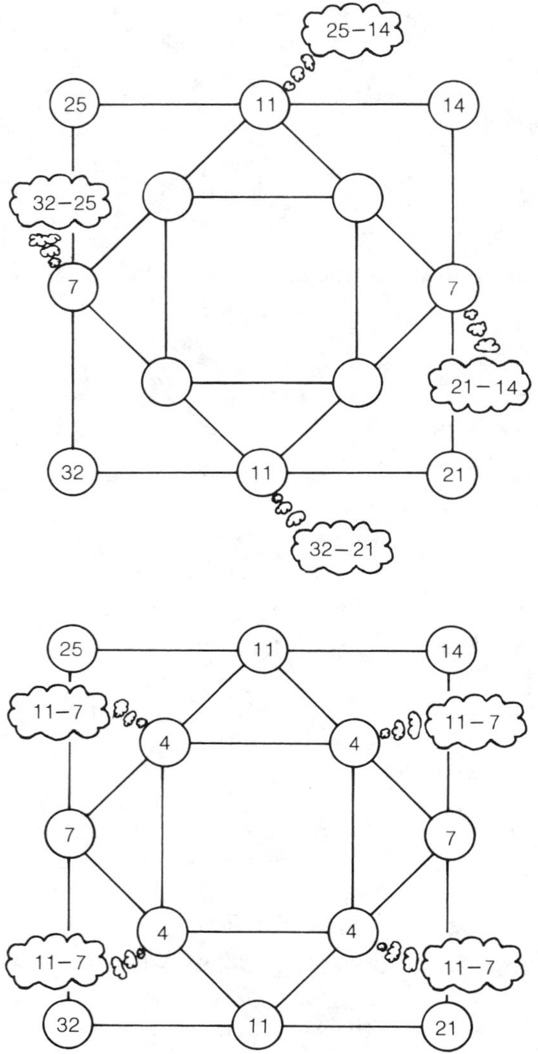

Another game is the greatest sum.[16] Construct 5 sets of ten-digit cards, each set with one card for each of the digits 0–9. Have children draw a table with two rows, each with four cells similar to the one below.

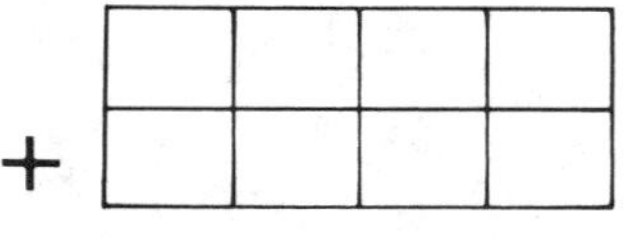

Shuffle the cards, then draw a card and call out the digit. Each child can write it in any of the eight cells. After all eight cells are filled, have each child add the 2 four-digit numbers she has constructed. The player with the greatest sum wins. Vary the game by having the least sum win, or by having children use subtraction.

Games that encourage speed in calculation also provide profitable practice. Have each child write a specified starting number, then ask children to add a two-digit number a certain number of times to the starting number. For example, add 23 five times, starting with the selected number of 45 (see illustration). The first child to complete the five additions with the correct sum wins. This game can be varied by using larger numbers and/or substituting subtraction for addition.

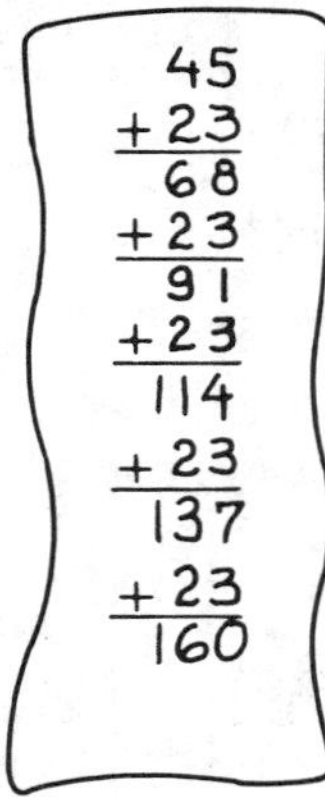

15. Robert Wirtz, *Drill and Practice: Activities Pages with Comment for Teachers* (Washington, D.C.: Curriculum Development Associates 1974), p. 139A.

16. Walter E. Rucker et al., *Heath Mathematics: Teacher's Edition, Level 6* (Lexington, Mass.: D.C. Heath, 1979), p. 17.

EXCURSION
222 Place Value Game[19]

This game is designed for two players with one calculator and a pair of dice. Players take turns rolling the dice, then using the number on one die as the tens place and the other as the ones place to make a two-digit number.

For example, you might roll:

You can make either 35 or 53. The dice are rolled five times. Using a calculator, add the 5 two-digit numbers. The goal of the game is to get as close to 222 as possible. The one who gets closer, without going over, wins.

Use of Calculators

Educators and parents have expressed concern over the widespread effects of calculator use on computation skills. A recent study of 1500 children in Grades 2 through 6 found that children benefitted somewhat from calculator use in mathematics classes. The study also showed that calculators are positive motivators and can be helpful teaching tools.[17] However, long-term effects of calculators are still not known at this time.

A statewide survey of calculator usage in Missouri schools provides these insights.[18]

1. Sixty-eight percent of children, Grades 1 through 12, have at least one calculator available at home.
2. Eighty-five percent of all teachers say that calculators should be available to children in schools.
3. Eighty percent of the teachers felt children should master the basic operations before they use calculators. Some teachers said that certain skills could be developed with the calculator.
4. Over half of the teachers would like to have calculator activities in regular mathematics textbooks.

The authors believe that calculators can be used effectively to support a mathematics program: they can help children understand the fundamentals. The availability of inexpensive hand-held calculators has placed less importance on speed and efficiency in paper-and-pencil computation.

Reasonableness of Results

The National Council of Supervisors of Mathematics' Position Statement on Basic Mathematical Skills (1977) stresses the importance of "alertness to the reasonableness of results" and "estimation and approximation."[20] These skills

17. Grayson Wheatley et al., "Calculators in Elementary Schools," *The Arithmetic Teacher* 27, no. 1 (September 1979): 18–21.
18. Robert E. Reys et al., "Hand Calculators–What's Happening in Schools Today?" *The Arithmetic Teacher* 27, no. 6 (February 1980): 38–43.
19. Bill Fisher, "Calculator Games: Combining Skills and Problem Solving," *The Arithmetic Teacher* 27, no. 4 (December 1979): 40–41.
20. National Council of Supervisors of Mathematics, "Position Paper on Basic Mathematical Skills," January 1977.

are critical in a world of calculators and computers; in determining if a machine-produced result makes sense, or in finding an approximate answer without paper and pencil.

Many everyday problems do not require precise answers. For example:

- How long will it take to walk $5\frac{1}{2}$ miles if I walk a mile in about 15 minutes?
- How many gallons of gasoline can I buy for $10.00?

Children should develop the habit of formally checking computations. But it is even more important for them to be able to estimate a reasonable answer.

Often, reasonableness is taught after the fact, after a child has made a particular error. Rather than expect a child to grasp two very different ideas–correcting the response to a particular question and learning techniques for determining if an answer is reasonable–*plan* instruction in estimating, in making sure answers are reasonable.

Begin by having each child react to two different types of reasonableness:(1) real world situations for which only correct and incorrect results are given, and (2) real world situations for which no numerical information is given.[21] In the first type, each child indicates *reasonable* or *unreasonable,* providing a rationale for her decision. Examples of such situations would be:

- The average fifth grader is 10 years old.
- The cost of a week's groceries to feed a family of four is $2.50.

In the second type, children are asked to indicate what numbers or number range would be reasonable. These include such questions as:

- About how much does a ten-speed bike cost?
- What is the average height of a fifth grader?

Both of these exercises require thinking about real-world settings. The second situation is more demanding in that it requires the mental generation of new material.

In teaching computation you will find many opportunities for illustrating numerical reasonableness:

- When adding 1375 and 982, an answer of 1257 is unreasonable. If whole numbers other than 0 are added, the sum is always greater than any of the addends.
- When subtracting 298 from 1003, an answer of 805 is unreasonable. Since 298 is almost 300, a reasonable answer would be near 700.

PLANNING ASSESSMENT

A child's understanding of algorithms and skill in using them need to be assessed. As with other areas, assessment of addition and subtraction procedures requires planning.

Assessment of paper-and-pencil *procedures* appears to be straightforward: give a child examples to solve. Examples of categories for addition and subtraction were listed earlier. But assessment of paper-and-pencil *skills* is more complex than it appears. A child who misses items on an example-filled page sometimes gives correct answers if the same examples are presented three (or even one) per page. Factors such as a time constraint or the presence of many symbols crowded together can affect skill performance. Test formats may be varied for some children.

Assessment of a child's *understanding* of a computational procedure follows the model presented throughout this text; for this, focus on relevant content objectives. After we decide what we want to assess, determine the behaviors that will allow us to infer that a child understands the concept. When more than one behavior for each content objective is observed, a child probably *does* understand the idea.

Consider content objective 8.4 which was discussed earlier. Desirable behaviors might include:

21. David C. Johnson, "Teaching Estimation and Reasonableness of Results," *The Arithmetic Teacher* 27, no. 1 (September 1979): 34–35.

1. Given numerals on a chalkboard and base 10 blocks, the child uses base blocks to show each amount, then shows addition by combining units with units and longs with longs, etc.
2. Given numerals on a chalkboard and a place value chart, the child uses the chart to show each amount, then shows addition by combining the markers in the ones with markers in the ones, and the markers for the tens with markers for the tens.
3. Given an addition example with numerals aligned vertically, the child adds ones with ones and tens with tens.
4. Given an addition example with numerals written in horizontal form (such as 26 + 51), the child adds ones with ones and tens with tens.

We also need to think of ways we can elicit such behaviors. Can you describe at least one task for each behavior indicator listed above?

9

Introducing Multiplication and Division

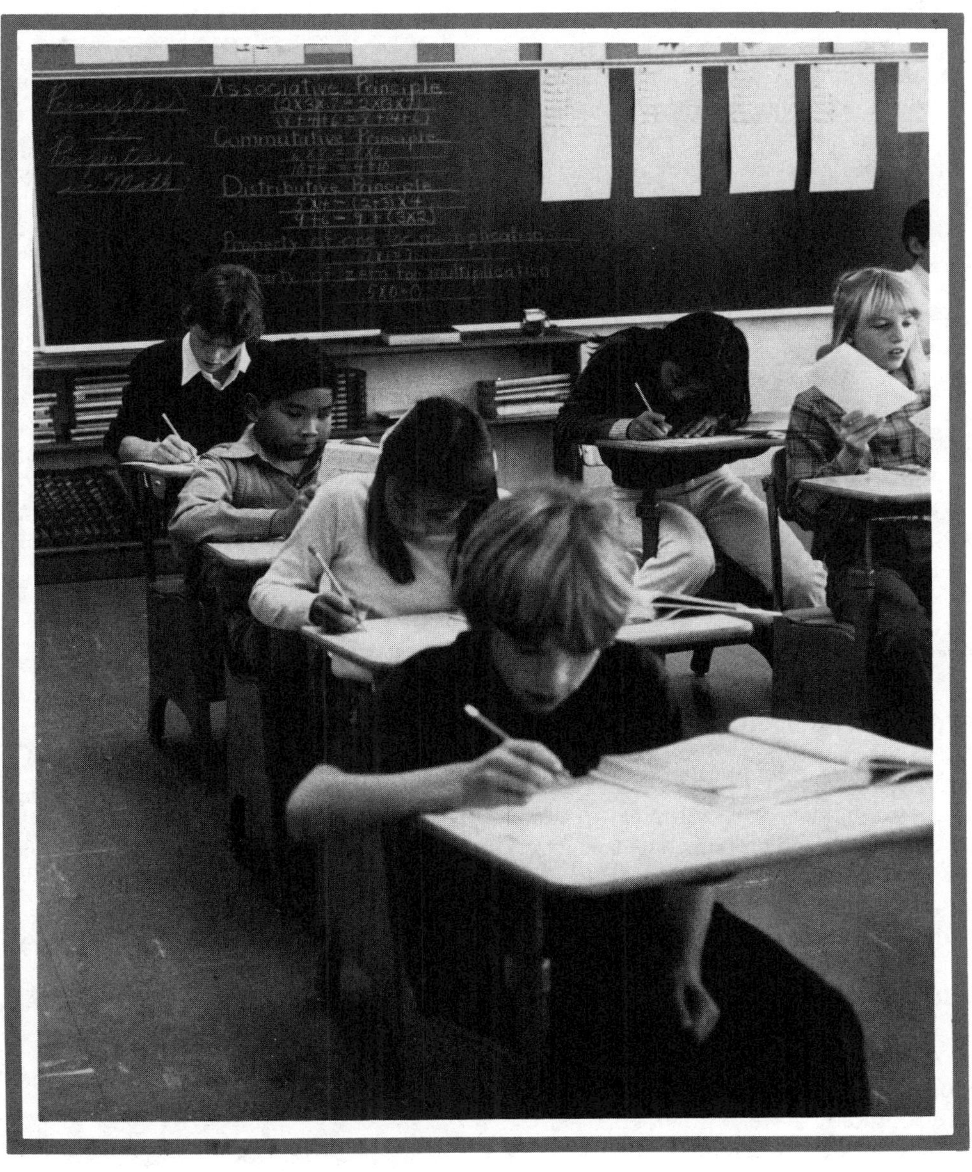

A child first encounters multiplication and division long before he computes paper-and-pencil examples. Early experiences involve questions such as How many shoes are in 3 pairs? and If we are to get the same number, how many can each of us take?

Multiplication and division are operations to be *understood*, and many content objectives are listed in the following pages. In teaching multiplication, we find that some of the ways we commonly model it are not always understood by each child.

Multiplication and division are also operations to be *done*. When we multiply, we perform a procedure to find the result. The procedure may be mental, may involve a calculator, or it may be a paper-and-pencil algorithm.

PLANNING THE CONTENT

The content objectives in the following pages are for our use as we select the content to be taught, and not the words we would expect children to use. Selection of models for multiplication and division should be based upon what each child already knows. Models for both operations should *not* all be taught at one time.

Multiplication of Whole Numbers

Meanings, Models, and Symbols

A child should understand many of the following ideas before beginning systematic practice activities for the multiplication facts.

9.1
Multiplication on whole numbers is an operation that associates a pair of whole numbers with a third unique whole number.
a. The pair of numbers multiplied are called *factors*.
b. The third number associated with the factors is called the *product*.
c. Multiplication is used to find a product when its factors are known.

Each pair of whole numbers is associated with a third whole number through multiplication. *Four* names one number, and *five* names another, but *four multiply five* or *four times five* is a name for a third, unique whole number.

$$(4, 5) \longrightarrow 20$$

Factor ↗ Factor ↑ Product ↑

Factor ↘ Factor ↓ Product ↙

$$4 \times 5 = 20$$

9.2
The *union of equivalent disjoint sets* is a model for multiplication on whole numbers.
a. One factor indicates the number of equivalent sets.
b. The other factor indicates the number of elements in each of the equivalent sets.
c. The product is the number of elements in the union set.
d. Multiplication is an operation used to find the number of elements in the union set when the number of equivalent sets and the number in each equivalent set are known.

One way to model multiplication of whole numbers is with sets of objects. Consider a collection of 12 trophies. One factor is 4, the number of equivalent sets, each on a separate shelf; the other factor is 3, the number of trophies in each set. The product, 12, is the number of trophies in the union set, the total for all sets in the collection. In the United States, the accepted elementary school convention for introducing multiplication expresses the number of sets with the *first* factor.

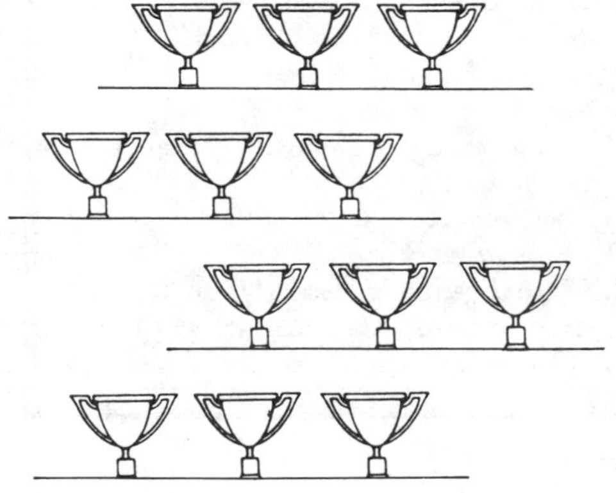

> **9.3**
> *Repeated addition of equal addends* is a model for multiplication on whole numbers.
> a. One factor indicates the number of equal addends.
> b. The other factor indicates the value of each addend.
> c. The product is the sum of the equal addends.

EXCURSION

How adequate is repeated addition of equal addends as a model for multiplication of whole numbers? Interpret these multiplication phrases as addition problems.

$$4 \times 3 \longrightarrow \begin{array}{r} 3 \\ 3 \\ 3 \\ +3 \\ \hline \end{array}$$

$$3 \times 3 \longrightarrow \begin{array}{r} 3 \\ 3 \\ +3 \\ \hline \end{array}$$

$$2 \times 3 \longrightarrow \ ?$$

$$1 \times 3 \longrightarrow \ ?$$

$$0 \times 3 \longrightarrow \ ?$$

Addition is a binary operation that cannot be used to show 1×3 or 0×3. Many children have difficulty with basic facts that have 1 or especially zero as a factor. If children are only taught to model multiplication with repeated addition, the model they have been given does not work. What other models for multiplication can be used to show 1×3 and 0×3?

For example:

$$\begin{array}{r} 3 \\ 3 \\ 3 \\ +\ 3 \\ \hline 12 \end{array} \qquad 4 \times 3 = 12$$

Objectives 9.2 and 9.3 are closely related and frequently clustered together for the same instructional activity.

9.4
The union of successive congruent line segments beginning at the zero point on a number line is a model for multiplication.

Consider the following number line representation for 4 × 3 = 12.

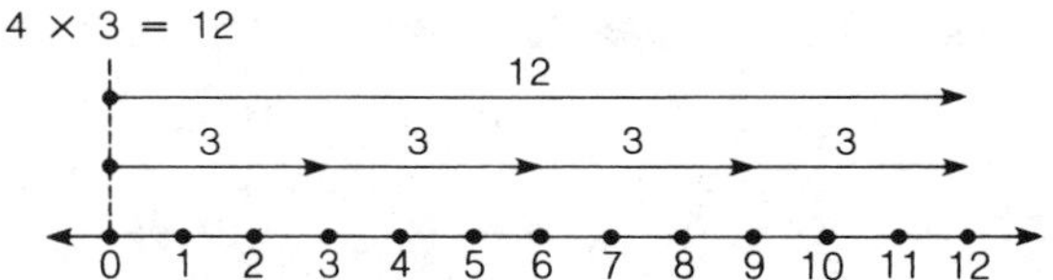

Emphasize zero as the starting point when using a number line. After 4 three-unit vectors are drawn, a parallel vector is drawn starting at zero and extending to 12, thereby demonstrating that 4 × 3 is the same length (quantity) as 12.

Would this model be useful for picturing 0 × 3 = 0?

9.5
The *Cartesian product of two sets* is a model for multiplication of whole numbers.
a. One factor indicates the number of elements in one set.
b. The other factor indicates the number of elements in another set.
c. The product is the *number of ordered pairs* of elements in the Cartesian set.
d. Multiplication is used to find the number of ordered pairs in a Cartesian set when the number of elements in each of two sets is known.

The verbal problem and the following diagram illustrate the Cartesian product of two sets as a model for multiplication.

PROBLEM

At a birthday party, 3 kinds of cake are available as well as 5 kinds of ice cream. If each person is allowed to choose one piece of cake and one dip of ice cream, how many combinations are possible?

The Cartesian product is more difficult for children because, in order to show a number sentence like 3 × 5 = 15, only 8 objects are needed (a set of 3 and a set of 5). Although a child can

Ice Creams

		Vanilla	Chocolate	Strawberry	Peach	Black Walnut
	Chocolate	Vanilla on Chocolate	Chocolate on Chocolate	Strawberry on Chocolate	Peach on Chocolate	Black Walnut on Chocolate
Cakes	Angel Food	Vanilla on Angel Food	Chocolate on Angel Food	Strawberry on Angel Food	Peach on Angel Food	Black Walnut on Angel Food
	Spice	Vanilla on Spice	Chocolate on Spice	Strawberry on Spice	Peach on Spice	Black Walnut on Spice

3 × 5 = 15

count the 15 pairings as objects are matched, the pairings do not simultaneously exist. On the other hand, 15 objects are used to show 3 × 5 as the union of equivalent disjoint sets, and the 15 objects for the total can be seen at one time. Eventually children do need to focus on the Cartesian product, because it is used in everyday applications.

9.6
An *array* is a model for multiplication on whole numbers.
a. One factor indicates the number of rows in the array.
b. The other factor indicates the number of columns in the array.
c. The product is the number of elements in the array.
d. Multiplication is an operation used to find the number of elements in an array when the number of rows and columns are known.

For the number sentence 3 × 4 = 12, construct 3 rows and 4 columns.

3 × 4 = 12

4 columns

3 rows

● ● ● ●
● ● ● ●
● ● ● ●

The array is probably the most useful model in teaching multiplication.

The array helps a child integrate much that has already been learned. It also suggests that multiplication can be visualized as a rectangle, an especially useful procedure when working with larger numbers.

9.7
An *array* is a model which shows the relationships of the models in content objectives 9.2, 9.3, and 9.5.

a. For objective 9.2, all the rows in an array are examples of equivalent disjoint sets.
b. For objective 9.3, the numbers of elements in each row of an array are examples of equal addends.
c. For objective 9.5, each element in an array can represent one of the ordered pairs in the Cartesian set.

Children need to know how to read and understand the meaning of the × symbol. Initiating activities for the meaning of multiplication are appropriate for younger children long before this symbol is introduced, but when it is used with abstracting activities it is usually read *multiply* or *times*. However, children can think of the × as *of this* or *of these* (4 × 6 as *4 of these 6s*.).

9.8
The × is a symbol that indicates multiplication and is read *multiply* or *times*.

9.9
A *multiplication phrase* consists of numerals or variables separated by the × sign, such as 3 × 4, 25 × □, $a \times b$.

a. The numerals or variables on each side of the × in a phrase indicate factors.
b. Multiplication phrases name a single number, a product.

EXCURSION

Often a child is taught to think of the × as *sets of* (2 × 3 as *2 sets of 3*). This is usually satisfactory with whole numbers but when the child begins working with fractions, such as $\frac{2}{3} \times \frac{4}{5}$ it becomes $\frac{2}{3}$ *sets of* $\frac{4}{5}$. The interpretations *of this* and *of these* are satisfactory when working with both whole numbers and fractions.

Content objective 9.9 is especially important for children to understand. In $3 \times 4 = \square$ a child knows the 3 and 4 are factors because they are located on either side of a times sign. Expressions like 3×4 must be understood as a name for one number. In this case, another name for the number is 12. Therefore, $3 \times 4 = 12$.

9.10
Other ways to write multiplication phrases are:
a. Place a dot midway up and between the numerais or variables naming factors. For example, $3 \cdot 4$ means the same as 3×4.
b. Enclose each factor name in a parenthesis and juxtapose; for example, (3)(4) means 3×4.
c. If one or more factors are indicated by variables, simply juxtapose; for example, $3\square$ means $3 \times \square$, ab means $a \times b$.

As different symbols can express the same idea, so one symbol can mean different things. Consider an example of the latter: a short bar (−) may mean minus (the operation subtraction), negative (a signed number), divided by (as in a fraction), or repeated (as in $.3\overline{3}$). Also, the use of the bar in vertical notation for basic facts is often considered to mean equals. Children must understand the symbols used during mathematics instruction.

9.11
A single numeral (one-digit or multidigit) used to name a product is called the *simplest name, standard name*, or *standard numeral.*

The phrase 243×682 is a name for one number, a product. To find the standard name for that number, a calculator or one of the paper-and-pencil algorithms are most likely to be used.

9.12
A multiplication equation consists of multiplication phrases, variables and/or standard numerals separated by an equals sign; for example, $3 \times 4 = 12$, $24 = 4 \times 6$, $\square = 4 \times 5$

a. When a variable such as a $\square$ or a letter is used in an equation, the equation is called an *open equation* or an *open sentence.*
b. A multiplication equation states that the numerals and phrases on each side of the equals sign name the same product.
c. Multiplication equations may be written in horizontal or vertical form, such as

$$3 \times 4 = \square \text{ or } \begin{array}{r} 4 \\ \times\ 3 \\ \hline \square \end{array}$$

d. In the vertical form, the bar under the phrase serves as an equals sign.

Each child should understand the equivalence of horizontal and vertical notation, and that the equals relation means *the same number as.* Read factors from bottom to top when multiplication

EXCURSION

Tell if each variable is a factor or a product.

$\square \times 46 = 38410$ $\quad$ $\frac{3}{4} \times \square = \frac{1}{4}$

$38 = 950 \times \square$ $\quad$ $3024 = \square \times {}^{-}63$

$\frac{1}{9} = \square \times 3$ $\quad$ ${}^{-}30 = {}^{-}12 \times \square$

What operation(s) would you use to solve each?

EXCURSION
Reading in Several Directions

We teach children to read from left to right. But in mathematics we must teach children to read in several directions. Can you think of situations in arithmetic where a child must read print from right to left? from bottom to top? diagonally? After you have noted a few examples, turn to page 174 and compare your list with the authors' examples.

facts are written vertically, as they are usually read in this way during computation. For example:

$$\begin{array}{r} 4 \\ \times\ 3 \\ \hline 12 \end{array}$$ is usually read *3 times 4 equals 12.*

$$\begin{array}{r} 24 \\ \times\ 3 \\ \hline \end{array}$$ Here we also think *3 times 4 equals 12.*

> **9.13**
> A multiplication equation may be read in different ways. For example, $3 \times 4 = 12$ may be read as *three fours equals (is) twelve, three multiply four equals (is) twelve,* or *three times four equals (is) twelve.*

A child can become confused if you read an equation in different ways, so do not teach the ways simultaneously. Make sure, however, that each child does eventually understand that these expressions are equivalent.

Properties

If applied, properties can cut down on the number of basic facts a child needs to learn.

> **9.14**
> *Closure*—The product of any two whole numbers is in every case a whole number.

If a and b name whole numbers, then c is a unique whole number, such that $a \times b = c$. Whole numbers are *closed* under multiplication. Though listed for completeness, this objective is difficult for young children to deal with explicitly.

> **9.15**
> *Commutative*—The order or sequence in which two whole numbers are multiplied has no effect on their product.

If a and b are whole numbers, then $a \times b = b \times a$. The phrases 3×4 and 4×3 name the same product.

> **9.16**
> *Associative*—The way in which three whole numbers are grouped together in multiplication has no affect on their product.

Multiplication is a binary operation; it associates exactly two numbers with a third. Therefore, think of factors as grouped in pairs if there are three or more. If a, b, and c name three whole numbers, then $(a \times b) \times c = a \times (b \times c)$.

> **9.17**
> *Identity*—The product of any whole number and 1 is equal to the given whole number.

If a names any whole number, then $a \times 1 = 1 \times a = a$. One is called the *identity element* for

EXCURSION
Reading in Several Directions
Answers
(From page 173)

These situations require reading in different directions. The arrows indicate the variety of ways.

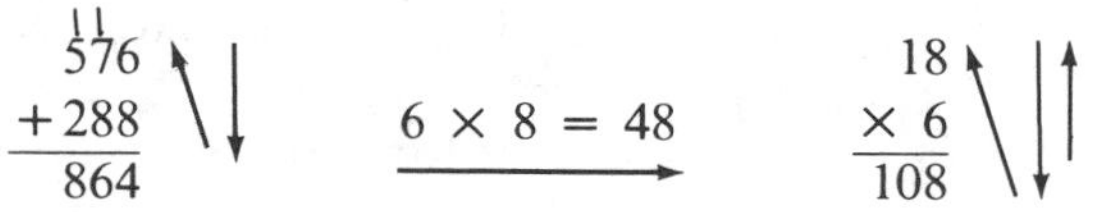

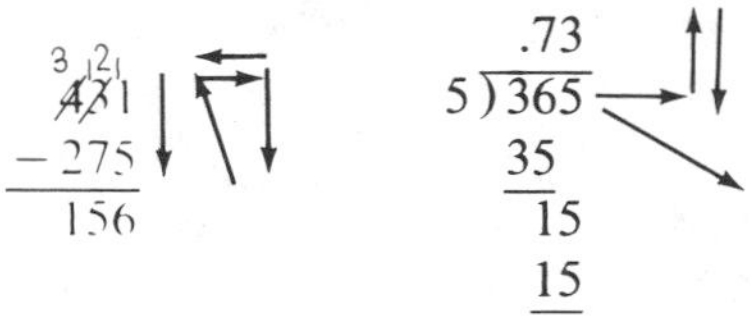

```
  11
  576           6 × 8 = 48        18
+ 288          ----------->     × 6
  864                           108

 3 12
 431              .73
- 275          5)365
  156            35
                 15
                 15
```

Tell the value of the circled digit with a multiplication phrase.

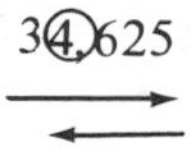

34,625

multiplication, or the *multiplicative identity* for the set of whole numbers. A child who understands and can apply this idea already has use of almost 20% of the 100 basic facts for multiplication.

> **9.18**
> *Distributivity*—Whole numbers under the operation of multiplication are distributive with respect to addition.

Distributivity brings together multiplication and addition. If *a, b,* and *c* represent whole numbers, then $a \times (b + c) = (a \times b) + (a \times c)$. This is sometimes called the *left* distributive property. The *right* distributive property expressed for the same numbers is $(b + c) \times a = (b \times a) + (c \times a)$.

Children should understand distributivity, for it is applied as soon as a child begins to multiply a two-digit number by a single-digit number. The multiplication of larger numbers involves repeated applications of this property.

> **9.19**
> *Zero property*—The product of any whole number and zero is zero.

Children often have difficulty with objective 9.19; they cannot seem to remember if the product is zero or the whole number. A child may be having difficulty conceptualizing multiplication by zero because the models and language used in presenting it do not seem to make sense when the multiplier is zero. Or, the zero property for addition may be confused with the zero property for multiplication.

Division of Whole Numbers

Meanings, Models, and Symbols

> **9.20**
> Division on whole numbers is an operation that associates a pair of whole numbers with a third unique whole number.
> a. One number in the pair is a product (sometimes called a *dividend*) and the other number is one of the product's factors (called a *divisor*).
> b. The third number associated with the product and one of its factors is the product's other factor (called a *quotient*).
> c. Division is an operation used to find one factor (quotient) when the product (dividend) and its other factor (divisor) are known.

In the problem $12 \div 3 = 4$, the 12 is a product and the 3 is one of the factors. The phrase $12 \div 3$ names the other factor, as does the numeral 4. They both name the same number, therefore, $12 \div 3 = 4$.

9.21
Division is not defined for the case in which zero is the divisor.

The phrase $3 \div 0$ is meaningless and it does not even name a number. Many people assume that the quotient is either 3 or 0. But *if* $3 \div 0$ is 3, then $3 \times 0 = 3$, and this is not true. On the other hand *if* $3 \div 0 = 0$, then it would have to be true that $0 \times 0 = 3$, an obviously false statement. No number can be multiplied times 0 to equal 3; therefore, $3 \div 0$ is undefined.

9.22
Division is the inverse of multiplication
a. Multiplication is used to find a product when its factors are known, whereas division is used to find one of the factors when the product and its other factor are known.
b. The inverse of each model for multiplication is a model for division.

These are important ideas for children to understand, for they will apply them often as they solve number sentences and verbal problems.

9.23
The partition of a universal set into equivalent disjoint subsets is a model for division on whole numbers.
a. The product (dividend) is the number of elements in the universal set.
b. One factor is the number of equivalent subsets into which the universal set is partitioned.
c. The other factor is the number of elements in each of the equivalent disjoint subsets.
d. Division is an operation used to find the number of elements in each equivalent subset (quotient) when the number of elements in the universal set (dividend) and the number of equivalent subsets formed (the divisor) are known. (*partitive model*)
e. Division is an operation used to find the number of equivalent subsets contained in a universal set (quotient) when the number of elements in the universal set (dividend) and number of elements in each equivalent subset are known. (*measurement model*)

A division sentence such as $12 \div 3 = 4$, tells us that the 3 and 4 are factors. But if the number sentence is modeled with a set of 12 and equivalent disjoint subsets, the 3 can be interpreted as either the number of subsets (partitive division) or the number of elements in each subset (measurement division). Compare measurement and partitioning division in Figure 9.1

Keep the distinction between measurement and partitive division in mind during introductory instructional activities. Be careful not to randomly switch back and forth between the models and thereby confuse children.

9.24
Repeated subtraction of equal addends is a model for division on whole numbers.
a. The product (dividend) is the sum of equal addends.
b. One factor (divisor) is the value of each addend.
c. The other factor (quotient) is the number of equal addends subtracted from the product.
d. Division is an operation used to find the number of equal addends subtracted from a product when the value of each addend (divisor) and the product (dividend) are known.

The division statement $12 \div 3 = 4$ is modeled by subtraction of equal addends as follows:

$$\begin{array}{rl} 12 & \\ -\ 3 & \checkmark \\ \hline 9 & \\ -\ 3 & \checkmark \\ \hline 6 & \\ -\ 3 & \checkmark \\ \hline 3 & \\ -\ 3 & \checkmark \\ \hline 0 & \end{array}$$

MEASUREMENT	PARTITIVE
Verbal Problem	*Verbal Problem*
If from a collection of 12 books, 3 are given to each of as many different children as possible, how many children will receive a set of books?	If 12 books are placed on 3 shelves so there is the same number on each, how many books will be on each shelf?
Set Representation	*Set Representation*

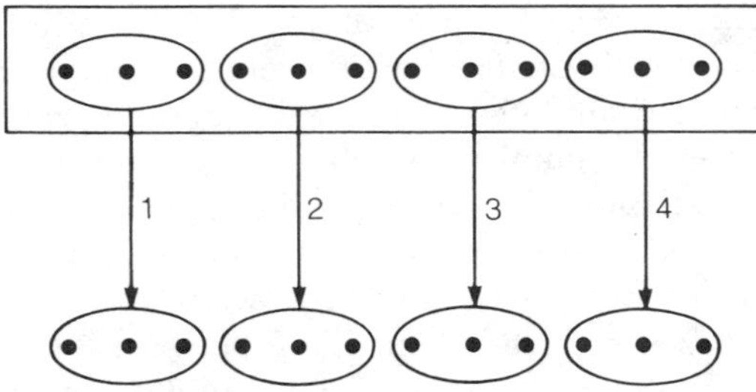

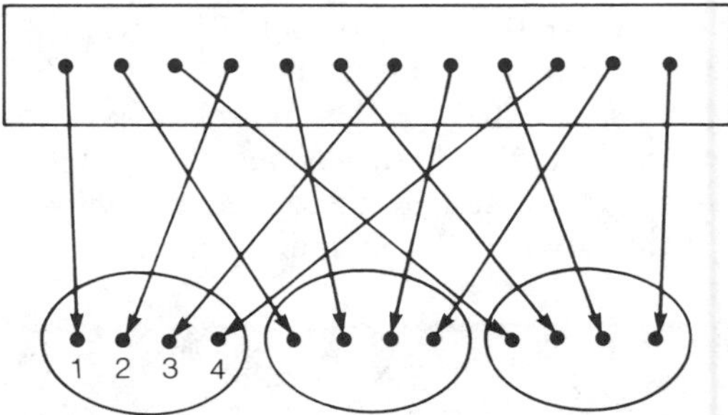

Figure 9.1
Measurement and Partitive Division Compared for 12 ÷ 3 = 4

The factor 4 is determined by counting to find how many times 3 is subtracted. There is a close relationship between repeated subtraction of equal addends and the partitioning of a set into equivalent disjoint subsets. However, only measurement division enjoys this relationship, specifically, objective 9.23e.

Two pairs of models can now be used to help a child understand division as the inverse of multiplication. Objectives 9.2 and 9.23 are one pair, the other pair includes objectives 9.3 and 9.24. In each pair what is shown for division appears to undo what is shown for multiplication.

> **9.25**
> The partition of a line into successive congruent line segments of a number line is a model for division on whole numbers.

A number line can be used to model division when the known factor is conceived as the length of each vector or arrow (measurement model), but the number line is not useful if the known factor is the number of arrows (partitive model). When a number line is used, any accompanying story problem should likewise reflect a measurement interpretation of division.

The number sentence 15 ÷ 3 = 5 is often shown as below. In general, work proceeds from the number line up, with the given product shown first. As many arrows as possible are drawn for the known factor, and the missing factor is determined by counting arrows. Note that the arrow for 15 and the joined arrows for 5 × 3 are the same length; they both show the same number.

In the case of uneven division (such as 16 ÷ 3 = □), the remainder can be interpreted as the number of units left over (one), or as a part of the next arrow (one-third).

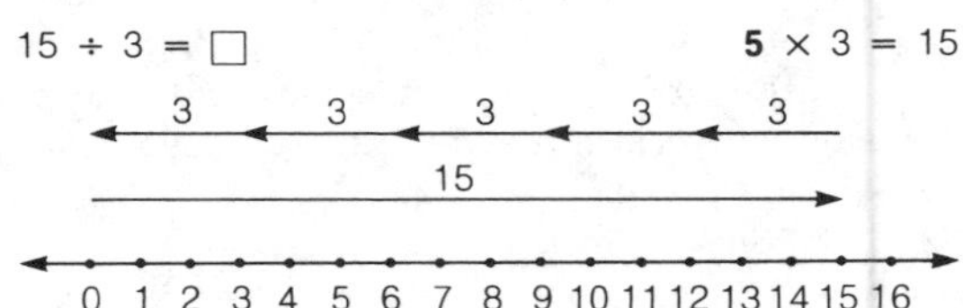

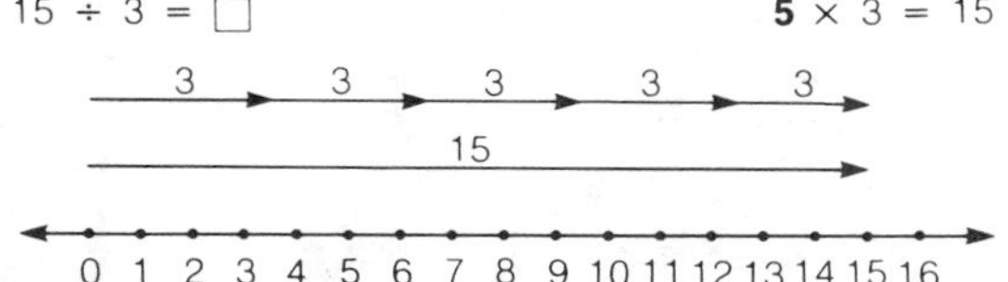

A useful alternative is to show the known factor with arrows to the right as illustrated above.

Whereas the first procedure suggests repeated subtraction of equal addends, this procedure points more directly to multiplication as the inverse of division and allows a child to more easily observe the equivalence of the arrow for the product and the joined arrows for the known factor.

9.26
The partition of the ordered pairs in a Cartesian set into their respective sets of single elements is a model for division on whole numbers.

This model does not point toward computation procedures as directly as other objectives, but children need to understand this idea because it describes a problem setting in which they must choose division as the operation used to find the missing number.

9.27
An array is a model for division on whole numbers.
a. The product (dividend) is the number of elements in an array.
b. One factor (divisor) is the number of rows in the array.
c. The other factor (quotient) is the number of columns in the array.
d. Division is used to find the number of rows (or columns) in an array when the number of elements in the array and the number of columns (or rows) are known.

The array below contains a total of 20 discs, though it is partially hidden by a sheet of

paper. Division will tell us the number of columns.

9.28
An array is a model that shows the relationships of the models in objectives 9.23, 9.24, and 9.26.

Not only does an array show the relationships of certain models for multiplication, it also shows the relationships of the inverses of each model.

9.29
The symbol ÷ indicates division and is read *divided by*.

Children often confuse this symbol with the subtraction symbol, so teach them to look for the two identifying dots.

9.30
A division phrase consists of numerals or variables separated by the ÷ sign, such as 12 ÷ 3, □ ÷ 5, and $a \div b$.
a. The numeral (or variable) to the left of a ÷ sign names a product (dividend).
b. The numeral or variable to the right of a ÷ sign indicates a factor (divisor).
c. A division phrase names one number.

Given the number sentence □ ÷ 8 = 24, many children will say that □ = 3. But a child who understands these ideas will know that the 8 and 24 are factors, and will multiply to find the unknown product.

Each child should know that expressions such as 8 ÷ 2 and 9968 ÷ 28 name *one* number; a factor is indicated. The answer is merely the standard name for that factor.

9.31
The $\overline{)\ \ }$ is also a symbol indicating division.
a. In a division phrase, the numeral for the product (dividend) is written inside the symbol and the numeral for one factor (divisor) is written to the left of the vertical bar. For example:

factor → $3\overline{)12}$ ← product

b. The number named by a division phrase using the $\overline{)\ \ }$ is the *indicated factor* or *quotient*.

The advantage of this particular division symbol is that numerals are arranged to facilitate computation. The standard name for the unknown factor can be determined.

9.32
A single numeral (one-digit or multidigit) used to name the quotient (one factor) is called the *simplest name* or *standard numeral* for the quotient.

We can use different means, such as calculators and paper-and-pencil algorithms of obtaining a standard name.

9.33
A *division equation* consists of phrases, variables, and/or standard numerals separated by an equals sign, such as 12 ÷ 3 = 4, 6 = 24 ÷ 4, 24 ÷ 6 = □, 24 ÷ □ = 6.
a. A division equation states that the phrases and standard numerals on each side of the equals sign name the same factor.
b. In a division equation using the $\overline{)\ \ }$ sign, the horizontal bar serves as an equals sign and the numeral for the quotient is written directly over the horizontal bar.

Both 24 ÷ 6 and 4 name the same number, therefore 24 ÷ 6 = 4. Similarly, $15\overline{)345}$ and 23 name the same number. Therefore:

$$\begin{array}{r} 23 \\ 15\overline{)345} \end{array}$$

Properties

9.34
Closure—Whole numbers are *not* closed under division; that is, a whole number product (dividend) and factor (divisor) will *not always* yield a whole number quotient.

EXCURSION

Tell whether the variable is a factor or a product in each of the following. What operation(s) would you use?

8610 ÷ 35 = □ $\qquad$ $^{-}.375$ ÷ □ = $^{-}.015$

275 = $^{-}6325$ ÷ □ $\qquad$ $\frac{1}{8}$ = 8 ÷ □

For example, in $25 \div 7 = \square$ the missing factor is not a whole number.

> **9.35**
> *Commutative*—Whole numbers are *not* commutative under the operation of division.

For example, $32 \div 8 \neq 8 \div 32$.

> **9.36**
> *Associative*—Whole numbers are *not* associative under the operation of division.

The grouping of three whole numbers does affect their quotient. If a, b, and c name three whole numbers, then $(a \div b) \div c \neq a \div (b \div c)$.

> **9.37**
> *Identity*—A whole number divided by 1 yields a quotient equal to the original number.

If a names any whole number, then $a \div 1 = a$. This is true because the factor a times the factor 1 equals the product a; 1 is the identity element for multiplication. The objective will not need to be stressed with children who understand the relationship between multiplication and division and also understand that one is the identity element for multiplication.

> **9.38**
> *Distributivity*—Whole numbers are right distributive for division with respect to addition.

If a, b, and c name whole numbers, and $c \neq 0$, then $(a + b) \div c = (a \div c) + (b \div c)$, provided that $a \div c$ and $b \div c$ name whole numbers.

Basic Facts

Objectives we choose should include definitions. Children also need to observe special relationships.

Definitions

> **9.39**
> A basic (or primary) fact for multiplication is a multiplication equation of the form
>
> $$a \times b = c \quad \text{or} \quad \begin{array}{r} b \\ \times a \\ \hline c \end{array}$$
>
> that does not contain a variable, in which numerals for the factors have only one digit.

By this definition each of the following is a basic multiplication fact:

$$2 \times 3 = 6 \qquad 4 \times 7 = 28 \qquad \begin{array}{r} 8 \\ \times 0 \\ \hline 0 \end{array} \qquad \begin{array}{r} 7 \\ \times 9 \\ \hline 63 \end{array}$$

None of the following are basic multiplication facts. Can you state a reason in each case?

$$12 \times 3 = 36 \qquad 4 \times 5 = \square \qquad \begin{array}{r} 10 \\ \times 6 \\ \hline 60 \end{array} \qquad \begin{array}{r} 4 \\ \times 3 \\ \hline \end{array}$$

> **9.40**
> A basic (or primary) fact for division is a division equation of the form
>
> $$x \div y = z \quad \text{or} \quad y\overline{)x}^{\,z} \quad (y \neq 0)$$
>
> which does not contain a variable, and in which numerals for the factors have only one digit.

By this definition each of the following is a basic division fact:

$$6 \div 2 = 3 \qquad 64 \div 8 = 8 \qquad \begin{array}{r} 8 \\ 9\overline{)72} \end{array} \qquad \begin{array}{r} 0 \\ 3\overline{)0} \end{array}$$

None of the following are basic division facts. Can you state a reason in each case?

$$36 \div 3 = 12 \qquad 9 \div 3 = \qquad 7\overline{)63} \qquad \begin{array}{r} 3 \\ 10\overline{)30} \end{array}$$

EXCURSION
How Many Multiplication Facts Are There?

If a child were to memorize each of the multiplication facts as separate entities, how many facts would he have to memorize?

X	0	1	2	3	4	5	6	7	8	9
0	0	0	0	0	0	0	0	0	0	0
1	0	1	2	3	4	5	6	7	8	9
2	0	2	4	6	8	10	12	14	16	
3	0	3	6	9	12					
4	0	4	8	12						
5	0	5	10	1						
6	0	6	12	1						
7	0	7	14							
8	0	8	16							
9	0	9								

It is not necessary for a child to memorize each of the basic multiplication facts as separate entities. Application of ideas such as the commutative property can reduce the number to be memorized.

Special Relationships

Children who understand special relationships solve problems more easily, often through mental operations. Along with applying properties, understanding these relationships can cut down on memorization of basic facts.

> **9.41**
> If one factor is multiplied and the other factor divided by the same number, the product is unchanged.

Applying this rule of compensation, a child might reason:

$$
\begin{aligned}
6 \times 15 &= (6 \div 2) \times (15 \times 2) \\
&= 3 \times 30 \\
&= 90
\end{aligned}
$$

> **9.42**
> If the dividend and divisor are both multiplied or divided by the same nonzero whole number, the quotient remains unchanged.

Applying this rule, a child might reason:

$$
\begin{aligned}
200 \div 40 &= (200 \div 10) \div (40 \div 10) \\
&= 20 \div 4 \\
&= 5
\end{aligned}
$$

Basic Multiplication Facts

For instructional purposes, the 100 basic multiplication facts can be divided into categories. The facts can be partitioned in different ways, but the authors recommend the following eight cate-

1. For an alternative, see Ralph T. Heimer and Cecil R. Trueblood, *Strategies for Teaching Children Mathematics* (Reading, Mass.: Addison-Wesley, 1977), p. 152.

EXCURSION

Solve each of the following mentally. Objectives 9.19 and 9.41 suggest ways to proceed.

$390 \times \frac{2}{3} = \square$ (Think: If I multiply $\frac{2}{3}$ by 3. . . .)

$500 \div 25 = \square$ (Think: $500 \div 5 = 100$, and. . . .)

gories.[1] The categories are not disjoint, but they are quite useful for planning instruction.

9.43
Basic multiplication facts with *zero as a factor.*

The zero property for multiplication can be applied during activities designed to consolidate and maintain skill with basic facts.

9.44
Basic multiplication facts with *1 as a factor.*

Similarly, the property of 1 can be applied to 19 basic facts during activities designed to consolidate and maintain skill.

The six remaining categories are described by Rathmell and represent different thinking strategies children use when finding products for basic multiplication facts.[2]

The strategies should be specifically taught.

9.45
Basic multiplication facts for which *skip counting* is a useful strategy.

Skip counting is an effective strategy when one of the factors is 2 or 5.

9.46
Basic multiplication facts for which *repeated addition* is a useful strategy.

As a strategy with basic facts, repeated addition is most effective when a factor is 2, 3, or 4.

9.47
Basic multiplication facts for which *one more* is a useful strategy.

Children can use what they already know to find a missing product. A child can solve $6 \times 8 = \square$ by thinking that $5 \times 8 = 40$, and one more set of 8 would be 48.

9.48
Basic multiplication facts for which *twice as much* is a useful strategy.

When one of the factors is 4, 6, or 8, a child can think of the number which is half of that factor, recall a previously learned basic fact, then double the product.

2. Edward C. Rathmell, "Using Thinking Strategies to Teach the Basic Facts," in *Developing Computational Skills*, ed. Marilyn N. Suydam and Robert E. Reys (Reston, Va.: National Council of Teachers of Mathematics, 1978), pp. 13–38.

9.49
Basic multiplication facts for which *facts of 5* is a useful strategy.

In this additional application of the distributive property, previously learned facts for 5 are used to figure out products which are not immediately recalled. For example, $7 \times 8 = \square$ is solved by thinking $5 \times 8 = 40$ and $2 \times 8 = 16$, so 7×8 is $40 + 16$ or 56. This strategy is especially useful for facts with large factors.

9.50
Basic multiplication facts with *9 as a factor.*

Each successive product has one more 10 and one less unit; also the sum of the digits is 9 for each product.

Basic Division Facts

Each inverse of the 100 basic multiplication facts is a division fact, except the 10 inverses in which the known factor or divisor would be zero, resulting in a total of 90. A child need not memorize the 90 division facts as such; the focus should be on understanding and memorizing the multiplication facts, then using them to solve division examples.

Sentence Solving

Each child should have opportunities to develop skill with sentence solving; to use what he already knows about multiplication and division.

Examine each of the following to see if you can find the meaning they have in common, the deep structure common to every example.

a. $48 \times 12 = \square$

b. $\square = 12 \times 48$

c. $12 \times 48 = \square$

d. $\square = 48 \times 12$

e. $\begin{array}{r} 48 \\ \times 12 \\ \hline \square \end{array}$

f. $\begin{array}{r} 12 \\ \times 48 \\ \hline \square \end{array}$

g. $\square \div 12 = 48$

h. $48 = \square \div 12$

i. $\square \div 48 = 12$

j. $12 = \square \div 48$

k. $12 \overline{)\square}$ with quotient 48

l. $48 \overline{)\square}$ with quotient 12

Each of these 12 number sentences is different in appearance and has a different surface structure. Yet all have the same deep structure because, in order to solve the problem, a product must be found by multiplying two known factors, regardless of the operation indicated.

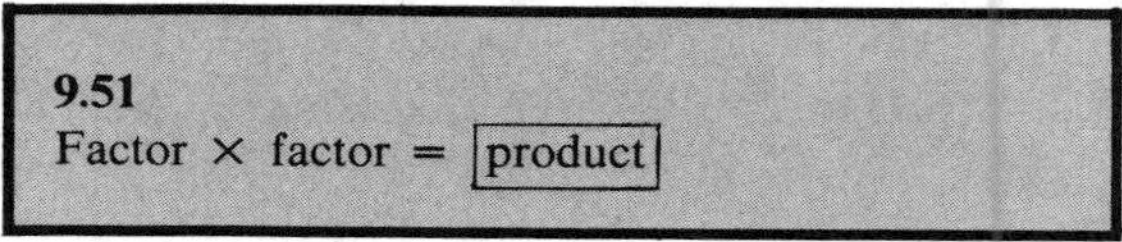

9.51
Factor × factor = [product]

Also consider these number sentences. Can you find a deep structure common to all?

a. $12 \times \square = 48$

b. $\square \times 12 = 48$

c. $48 = 12 \times \square$

d. $48 = \square \times 12$

e. $\begin{array}{r} 12 \\ \times \square \\ \hline 48 \end{array}$

f. $\begin{array}{r} \square \\ \times 12 \\ \hline 48 \end{array}$

g. $48 \div 12 = \square$

h. $48 \div \square = 12$

i. $\square = 48 \div 12$

j. $12 = 48 \div \square$

k. $12 \overline{)48}$ with quotient $\square$

l. $\square \overline{)48}$ with quotient 12

These are all division situations. The product and one factor are known, and a factor must be found in order to solve the problem. Whenever the product and one factor are known, division is used to find the missing factor.

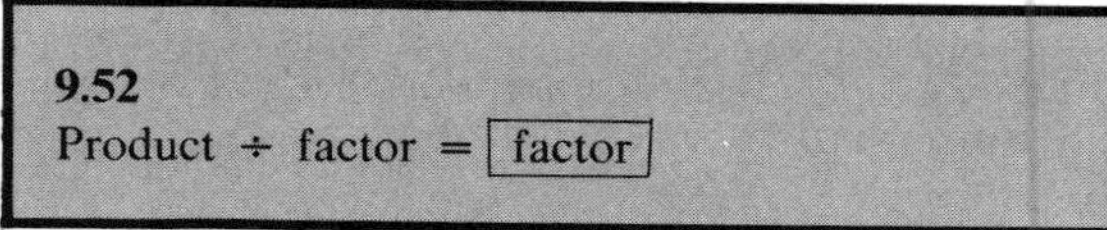

9.52
Product ÷ factor = [factor]

EXCURSION
Choosing the Operation

Find the deep structure in each of these problems. Decide what operation you would use to find the missing number.

a. $\square \div 35 = 5$
b. $24 = 48 \times \square$
(What is the missing number?)
c. $1\frac{1}{2} = \square \div \frac{1}{3}$
d. $\square \div \frac{2}{5} = \frac{3}{4}$
e. $^{-}40 = 5 \times \square$
f. $33 \div \square = {}^{-}3$
(What is the missing number?)

If you are not quite sure, turn to page 184 for more information.

Verbal Problem Solving

When solving number sentences, the placement of numerals and the variable can help you determine if numbers are factors or products, but with a verbal problem, you must consider the model for multiplication or division within the problem itself.

Inside the back cover of this book is a list of simple one-step verbal problems involving factors and a product. Four types of verbal problems are identified for multiplication, and five types are listed for division.

Provide a variety of problem types and, when analyzing problems, focus on identification of factors and the product.

Ordering the Content

How do we create an effective content sequence for multiplication and division? As in addition and subtraction, consider logical relationships between various subject matter components. Sometimes they are dependent. For example, work with meanings of the operations and the introduction of symbols should precede sentence solving. In other cases, they are independent. Models for multiplication and division should be interrelated as they are taught.

Before beginning work with multiplication and division, make sure each child has prerequisite understanding and skill with addition, subtraction, and numeration. The meaning of multiplication and then division should be stressed while models and symbols are introduced. Many of the properties and special relationships can be taught concurrently. We eventually need to make sure that children can use the basic facts in both vertical and horizontal format, and in solving number sentences and verbal problems.

PLANNING INSTRUCTION

Meanings, Models and Symbols

Initiating and Abstracting Activities for Multiplication

Objectives 9.1, 9.2, 9.8, 9.9, and 9.11 can be used to introduce multiplication of whole numbers. These are new ideas for children, even though they already know how to make equivalent sets. Plan an activity that permits each child to improvise a solution and, in the process, encounter instances of the new idea.

When introducing formal terms and symbols, plan an initiating/abstracting activity. For example, when focusing on the equivalent set model for multiplication, use a set of objects,

EXCURSION
Choosing the Operation Answers
(From page 183)

a. Multiplication (35×5 or 5×35)
b. Division ($24 \div 48 = .5$)
c. Multiplication ($\frac{1}{3} \times 1\frac{1}{2}$ or $1\frac{1}{2} \times \frac{1}{3}$)
d. Multiplication ($\frac{2}{5} \times \frac{3}{4}$ or $\frac{3}{4} \times \frac{2}{5}$)
e. Division ($^{-}40 \div 5$)
f. Division ($33 \div {}^{-}3 = {}^{-}11$)

possibly discs, for an exemplar. If each child has 12 discs, he can be asked to use them *all* to make equivalent sets. Try sets of 2 with the group before urging each child to find other ways to make equivalent sets. As children find and report other ways, make a record on the chalkboard as follows:

12 sets of 1	3 sets of 4
6 sets of 2	2 sets of 6
4 sets of 3	1 set of 12

Emphasize that both numbers tell us about sets. One number indicates the size of the set, and the other the number of sets. Then have each child display one arrangement, possibly 4 sets of 3, and introduce the times sign in substituting 4×3 for 4 sets of 3. As 4×3 is also a name for the product, record both $12 = 4 \times 3$ and $4 \times 3 = 12$.

Children can then use discs to make equivalent sets and suggest additional multiplication number sentences. When a number sentence is suggested, have the other children verify the statement with their discs and record the sentence.

Wooden cubes, beans, pennies, or flannel board with felt pieces for objects can be used for instruction with equivalent disjoint sets (see Figure 9.2). Flannel-and-chalk boards described in Chapter 6 can also help the child make sets and record multiplication phrases or sentences.

Or, a strip of tagboard can be folded like an accordion, and equivalent sets of dots colored on each section as shown in Figure 9.3. Different multiplication facts are illustrated as the strip is unfolded.

Both Cuisenaire rods and Stern blocks can be used to suggest equivalent sets; the rods are illustrated in Figure 9.4. Children can make pictures of such displays with crayons on centimeter graph paper. Cuisenaire rods of one color can also be placed along a centimeter ruler.

A math balance can be used when approaching multiplication as repeated addition of equal addends. A set of weights on a single peg can be balanced with a weight or weights to show the standard numeral. Have two records made and displayed: the addition example pointed to by the display on the math balance, and the multiplication sentence (Figure 9.5).

Figure 9.2
Equivalent Sets in an Egg Carton

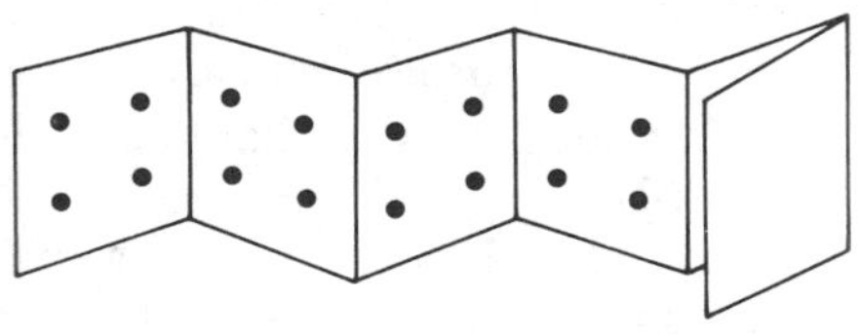

Figure 9.3
Equivalent Sets on a Folded Strip

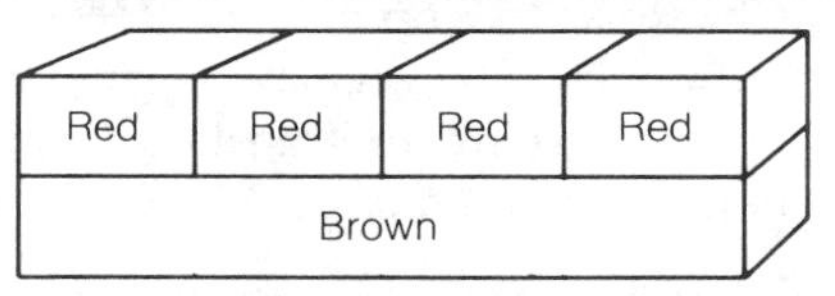

Figure 9.4
Cuisenaire Rods for 4 × 2 = 8

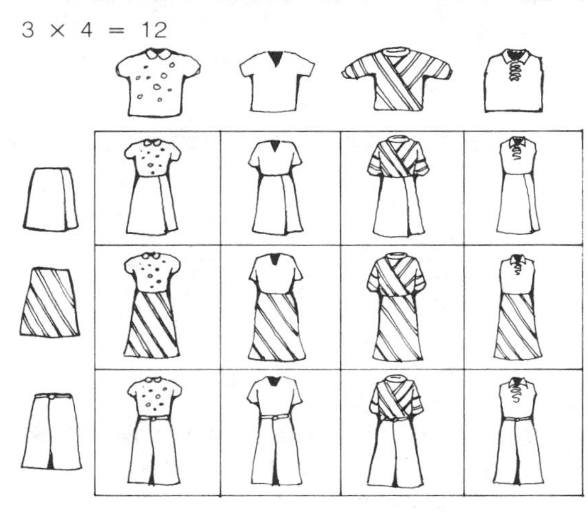

Figure 9.6
A Chart for Showing Multiplication as a Cartesian Product

Above a number line draw a product arrow, then make factor arrows that total the same length. Have each child record a number sentence for the number line display. Next, write a number sentence on the chalkboard and have children create a number line representation. The manual that accompanies this text contains a related worksheet.

Charts can be created to show all possible combinations or pairings for Cartesian products. Figure 9.6 shows the outfits possible given three different skirts and four different blouses. Sticks

Figure 9.5
A Math Balance for Showing Multiplication as Repeated Addition

can also be used to show the number of possible combinations. Have each child arrange sticks as pictured in Figure 9.7a and determine the number of intersections. Figure 9.7b illustrates how crossed lines are often used in a similar manner as a two-dimensional exemplar.

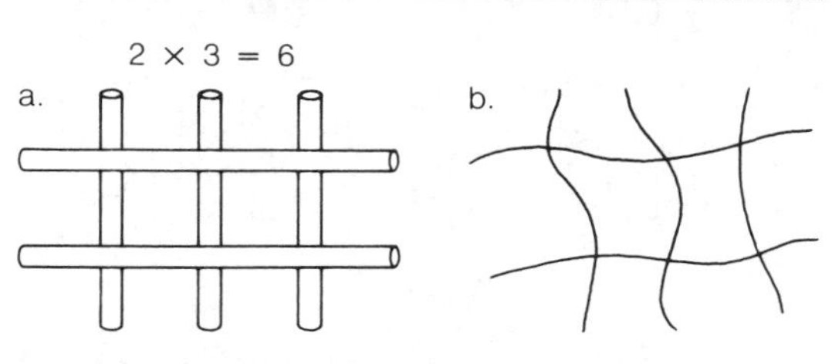

Figure 9.7
Sticks and Lines Showing Multiplication as a Cartesian Product

An initiating activity that focuses on an array is described in Figure 9.8. Children should already know that an array is an arrangement with rows and columns and should be able to make an array of objects. An abstracting activity focusing on symbols and their meaning could follow. Arrays can also be noted in the child's environment: windows, chocolate bars, egg cartons, and store display cards for everything from buttons to Christmas tree ornaments. Record a multiplication phrase for each, as many children need help in learning to translate from three-dimensional or two-dimensional representations to a symbolic statement.

ACTIVITY PLAN

Multiplication of Whole Nos.: Meanings, Models, and Symbols — ***3-D/Rep → Symbolic*** — ***Initiating***

Content objectives

9.1, 9.6

Exemplars/Materials

24 cubes for each child
Word sentences written on chalkboard

Behavioral indicators

Given 24 cubes the child:

1. Arranges all 24 into an array.
2. Names the number of rows and the total number of cubes.

Procedure

1. Give each child a set of 12 cubes.
2. Say, "Make an array with no cubes left over."
3. Record what the children observe much as you would record a language experience for instruction in reading, for example:

   Two rows of six
   Four rows of three
4. Say, "Let's see how many different arrays we can make."
5. When the list is complete (including 12 × 1 and 1 × 12), give each child 12 more cubes and say, "Now let's see how many different arrays we can make using all 24 cubes."

Figure 9.8
Sample Activity Plan for an Initiating Activity

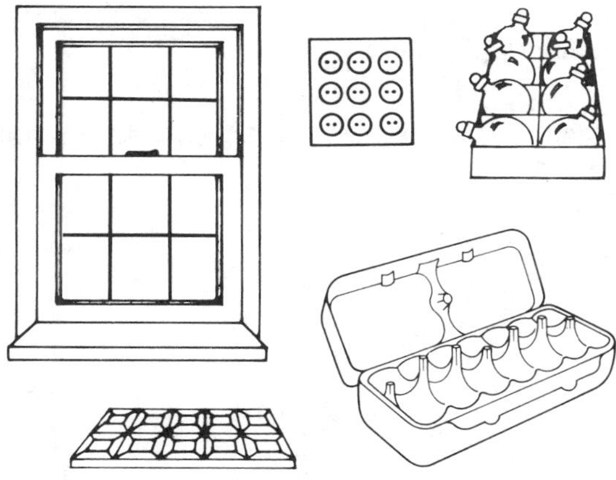

During activities with many three-dimensional exemplars, each child can make a record of what was made or observed. The record can be both two-dimensional (such as arrays colored on graph paper) and symbolic (accompanying multiplication sentences) as illustrated in Figure 9.9.

Cuisenaire Rods are sometimes used as depicted in Figure 9.10a, which suggests an array. At other times factors are shown as in Figure 9.10b; the two factor rods suggest the dimensions of an array.

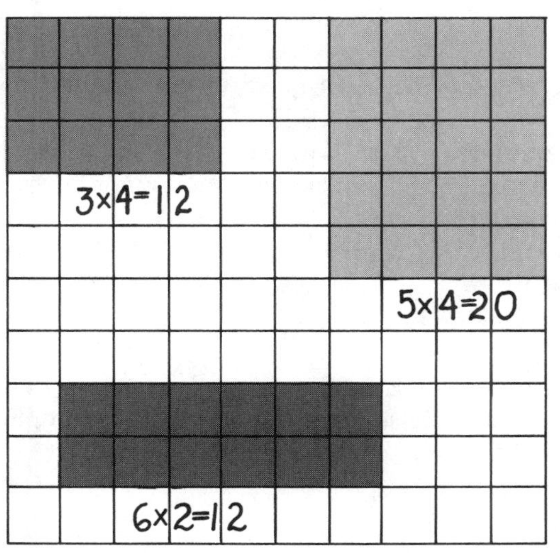

Figure 9.9
Arrays on Graph Paper

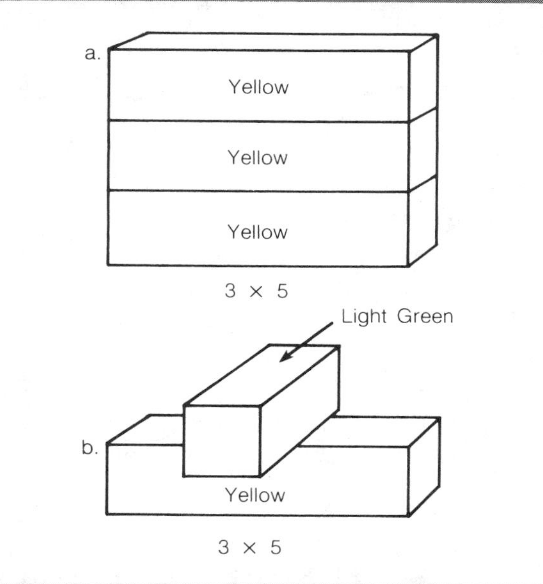

Figure 9.10
Cuisenaire Rods for Multiplication

Initiating and Abstracting Activities for Division

The first activities with division will be both initiating and abstracting, allowing each child to apply what is already known about multiplication. We have chosen content objectives that focus on the measurement model for division and on symbols: 9.23e, 9.29, 9.30, 9.32 and 9.33.

Introduce the activity with a simple number situation: Jill has 15 roses from the garden and wants to put 3 roses in each vase. Record 15 and 3, then give each child 15 cubes. Have children use their cubes to determine the number of vases needed, then record the missing factor (5). Say, "When we know the product and one factor, we can write . . ." Insert a division sign (15 ÷ 3). Teach children how to read the sign, and the fact that division tells us the other factor, so we write 15 ÷ 3 = 5.

Collect 3 cubes from each child, and proceed in a similar fashion with a story and a record for 12 as the product. Then have children use their cubes to complete division number sentences on a brief worksheet. Do the first example together.

Many of the activities described in the following pages point to the inverse relationship between division and multiplication, but a math balance is useful in focusing specifically on this relationship (Figure 9.11).

ACTIVITY PLAN

Multiplication and Division of Whole Nos.: Meanings, Models, and Symbols

Symbolic → 3-D/Rep
Symbolic → Symbolic

Abstracting

Content objectives

9.20 and 9.22

Exemplars/Materials

Math balance and weights for each child in the group
Number sentences
Paper and pencil for each child

Behavioral indicators

Given division number sentences with a missing factor, the child will:

1. Interpret the problem on a math balance.
2. Write the related multiplication sentence stating both factors.

Procedure

1. Briefly review how to solve multiplication problems on a math balance by solving the following:

$2 \times 4 = \square \qquad 3 \times 4 = \square \qquad 4 \times 8 = \square$

Be sure products greater than 10 are expressed with as many tens as possible.

2. On a demonstration balance show $4 \times 8 = 32$ and say, "If one of the factors is missing, we could use what we know about multiplication to figure it out."

3. Remove one of the weights from the 8 peg and write, $32 \div 8 = \square$. Say, "The product is 32 and one of the factors is 8. What is the other factor?" (4) "How do we know?" (We know that *4* $\times$ 8 = 32.)

4. Verify that the missing factor is 4 by placing the fourth weight back on the demonstration balance. Say, "There are *4* eights."

5. Have all balances cleared, distribute paper and pencil, then write $21 \div 7 = \square$. Ask each child to show the product (21) on one side of the balance, then say, "Before you use your balance, think about how many weights you will need to put on the 7. Write down a multiplication number sentence that shows what you think the missing factor is." ($3 \times 7 = 21$)

6. If a cue is needed, say, "How many sevens is 21?"

7. When appropriate say, "How do we know the missing factor is 3? Yes, we know that *3* $\times$ 7 = 21."

8. Have each child verify by placing weights on the 7.

9. Repeat steps 5 through 8 for the following problems. Be sure each child writes a multiplication number sentence before using the balance to verify the missing factor.

$18 \div 6 = \square \qquad 24 \div 6 = \square \qquad 36 \div 9 = \square$

Note: If children are doing well you may also want to present $25 \div 8 = \square$. Later record $(3 \times 8) + 1 = 25$ and $25 \div 8 = 3$ R 1.

Figure 9.11
Sample Activity for Division as Inverse of Multiplication

EXCURSION
Planning Your Moves

The plan in Figure 9.11 has several procedural steps. Review the three teacher moves described in Chapter 3. Then, for each of the following steps in the procedure, record the teacher move or moves involved: structuring, soliciting, or responding.

Step	Teacher Move(s)
2.	______________
3.	______________
4.	______________
5.	______________
6.	______________
7.	______________

When division is modeled with equivalent disjoint subsets and the known factor is interpreted as the number of sets to be formed (partitive division), the resulting action resembles dealing cards: one for each, then go around again. A set of blocks can be distributed evenly into 4 subsets by first designating 4 boxes for the subsets. If square shapes on a flannel board are to be partitioned into 3 subsets, yarn loops can be used to define the boundaries of each subset, as shown in Figure 9.12. A child puts one piece in each loop, then repeats the process until another round is impossible.

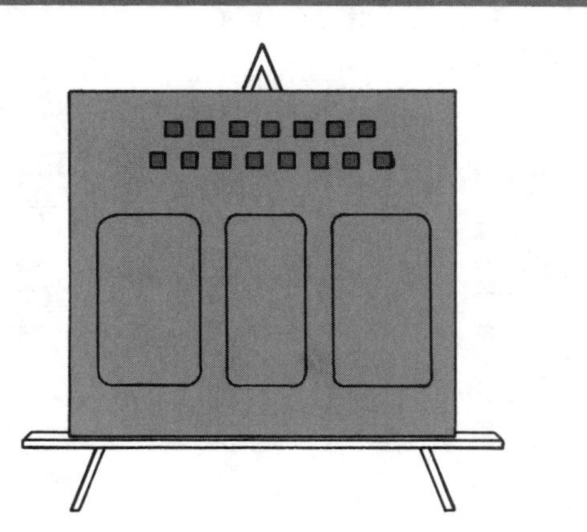

Figure 9.12
Flannel Board and Square Shapes for Partitive Division

When division is modeled with equivalent disjoint subsets and the known factor is interpreted as the number of elements in each subset (measurement division), objects can be rearranged and the number of groups counted, as illustrated in Figure 9.13. Blocks, chips, crayons, cubes, discs, and pennies are but a few examples of items that can be used.

Unifix cubes can be joined to illustrate the product. Ask each child to find how many rods of a shorter length–the known factor–can be made from the long rod. Cuisenaire rods can also picture division, as illustrated in Figure 9.14. For example, if a white rod is 1, a child can easily determine the number of three-unit rods that are equal in length to 15 units. Similarly, it would take 3 two-unit rods and half of another rod to cover a seven-unit rod.

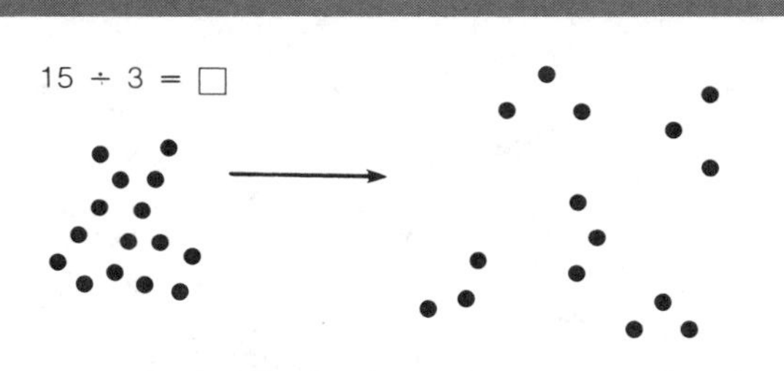

Figure 9.13
Chips Rearranged to Show Division as Equivalent Subsets (Measurement)

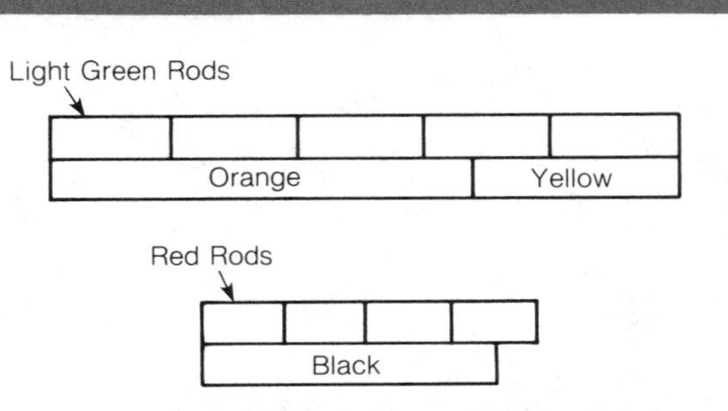

Figure 9.14
Cuisenaire Rods to Show Division as Equivalent Subsets (Measurement)

Textbooks and worksheets are, of course, limited to two-dimensional exemplars. If sets are pictured on a worksheet, subsets can be circled. Sometimes a set picture is used to depict the activity of forming subsets, as shown in Figures 9.15 and 9.16.

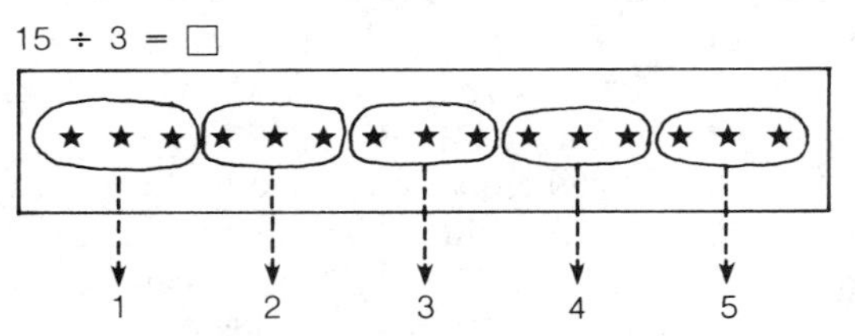

Figure 9.15
Set Picture for Showing Division as Equivalent Subsets (Measurement)

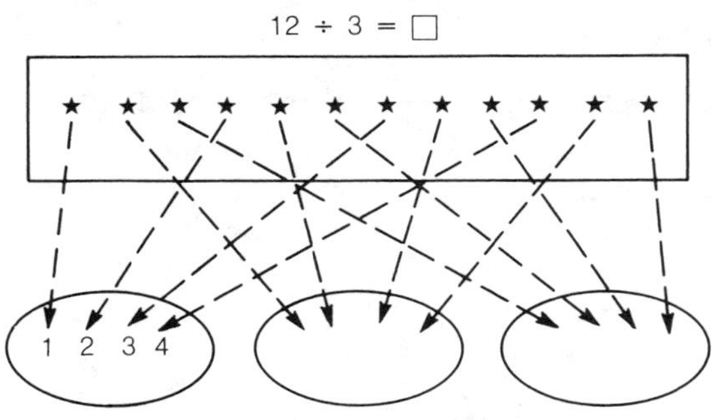

Figure 9.16
Set Picture for Showing Division as Equivalent Subsets (Partitive)

When picturing division with equivalent subsets in measurement situations, record the observations as repeated subtraction. These early paper-and-pencil records lead naturally to certain computational procedures for division. Examples of such records appear in Figure 9.17.

A number line should be used much in the same way it was for multiplication. Begin by showing the product, as illustrated in Figure 9.18. The verbal problem is a measurement situation with the known factor being shown in either of two ways.

Figure 9.18a clearly demonstrates the relation of division to multiplication, whereas Figure 9.18b shows how division can be related to subtraction, a procedure useful for uneven division when the number of objects remaining can be read directly from the number line. The number line should be used carefully when finding the missing factor. Instead of emphasizing counting and reading numerals, stress what each child already knows about multiplication. (In Figure 9.18 the missing factor is 4 because 3 and *4* are factors of 12.)

Arrays can depict both multiplication and division situations, emphasizing the relationship

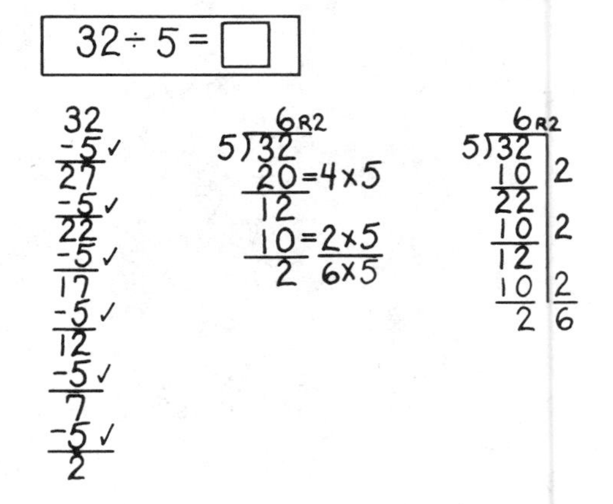

Figure 9.17
Repeated Subtraction Leading to Computational Procedures for Division

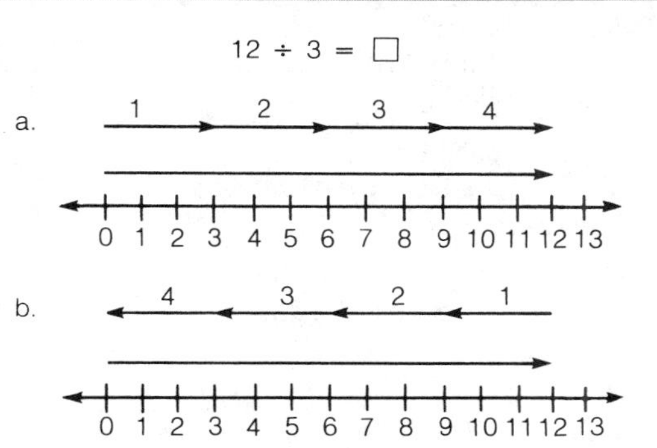

Figure 9.18
Number Line for Showing Division

between the operations. When using an array for division, tell the children the product; part of the array will be covered so that one of the factors is unknown. Have them predict the missing factor for arrays such as those in Figure 9.19.

Activities should include opportunities for children to interpret given number sentences with varied materials, as well as to write or complete number sentences for manipulations or drawings. The ability to translate both ways usually indicates an understanding of the concepts involved.

Diagnosis

Diagnosis is continuous and occurs throughout instruction. Frequently, we should plan specific diagnostic activities before moving from one activity type to another. Sample tasks that are useful to administer after abstracting activities before moving on to schematizing activities are found in Table 9.1. Just as it is desirable to use more than one behavioral indicator for each content objective, it is also best to prepare more than one diagnostic task for each behavioral indicator.

A basic fact family for multiplication and division is usually the set of facts with the same product and factors. For example:

$$3 \times 7 = 21 \qquad 21 \div 7 = 3$$
$$7 \times 3 = 21 \qquad 21 \div 3 = 7$$

In every case 21 is the product and the two factors are 3 and 7. If children color arrays on centimeter graph paper, the arrays can be cut out and displayed. The associated family of facts can be recorded beneath each array. As a cut-out array is turned, the child can easily observe commutativity. An array can be related to both multiplication and division because of the inverse relationship.

Relationship cards are sometimes used. Given one such card (see Figure 9.20), a child writes the complete fact family, then checks the back of the card to see if the correct set of facts has been written.

Schematizing Activities

Once children have abstracted the meanings for multiplication and division, modeled the operations with different exemplars, and used symbols appropriately, schematizing activities can be planned to help them identify interrelationships among concepts. These are among the most important activities we plan.

Each child can be asked to identify all of the factor pairs for a given product, possibly by using number strips, as pictured in Figure 9.21.

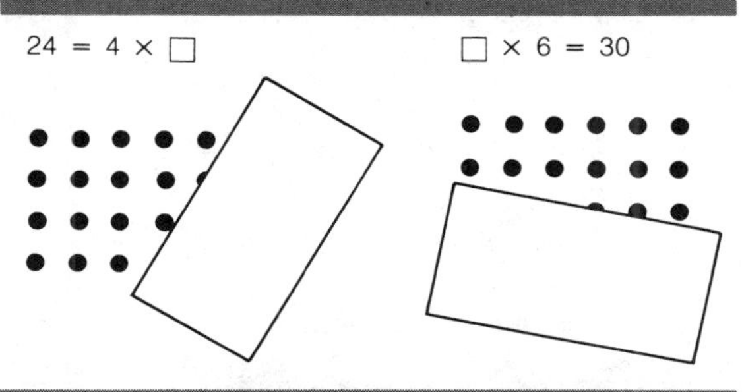

Figure 9.19
Arrays for Showing Division

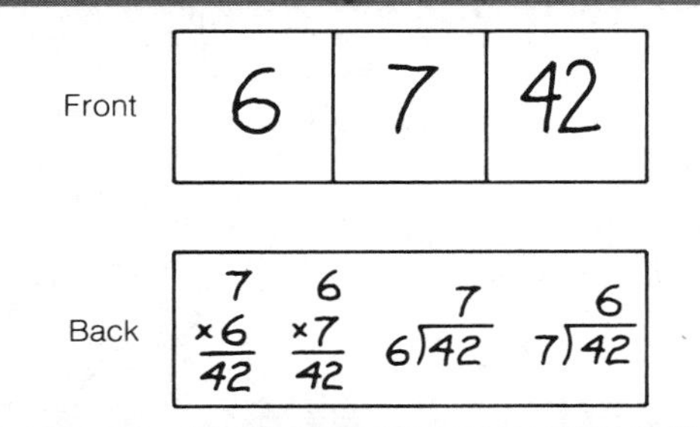

Figure 9.20
Relationship Cards for Multiplication and Division

Table 9.1
Diagnostic Tasks Suggested by Behavioral Indicators

Behavioral Indicators	*Sample Diagnostic Tasks*
1. Given a multiplication phrase and a supply of discs, makes the appropriate array for the phrase.	1a. Give the child about 20 discs, write 3 × 4, and say: "Make an array that shows 3 × 4."
	1b. Give the child about 30 discs, write 7 × 3, and say: "Make an array that shows 7 × 3."
2. Given a multiplication phrase, states that the numerals on either side of the sign name factors.	2a. Show the child 6 × 7 and say: "What does the 6 tell us? How do you know? What does the 7 tell us? How do you know?"
	2b. Show the child 5 × □ and say: "What does the 5 tell us? How do you know? What does the □ tell us? How do you know?"
3. Given a division phrase, states that the numeral to the left of the sign names the total number (product), and the numeral to the right of the sign names a factor.	3a. Show the child 36 ÷ 9 and say: "What does the 36 tell us? How do you know? What does the 9 tell us? How do you know?"
	3b. Show the child □ ÷ 4 and say: "What does the □ tell us? How do you know? What does the 4 tell us? How do you know?"
4. Given multiplication and division number sentences, some with the product at the right and others with the product at the left, identifies and names each number as a factor or a product.	4a. Show the child 8 × 4 = 32 and 56 = 7 × 8 and say: "Point to each numeral in these number sentences and tell me if it is a factor or a product. How do you know?"
	4b. Show the child the following and say: ✧ × □ = △ △ ÷ □ = ✧ △ = □ × ✧ ✧ = △ ÷ □ These are number sentences but the numerals are missing. Point to each shape and tell me if the number will be a factor or a product.

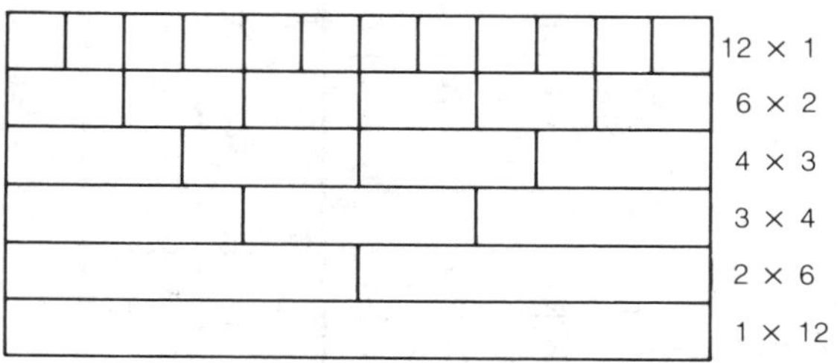

Figure 9.21
Numbers Strips Showing Factor Pairs

Also, a graph of factor pairs for a given product can be constructed, at least for products having several whole number factor pairs. Figure 9.22 shows a graph constructed for the product 12. Ask children to predict if a graph for the product 24 will have the same shape, then let them make one.

Before a child builds a multiplication fact table (Figure 9.23), he should complete a worksheet that relates arrays to basic multiplication and division facts and introduces him to the format of fact tables.

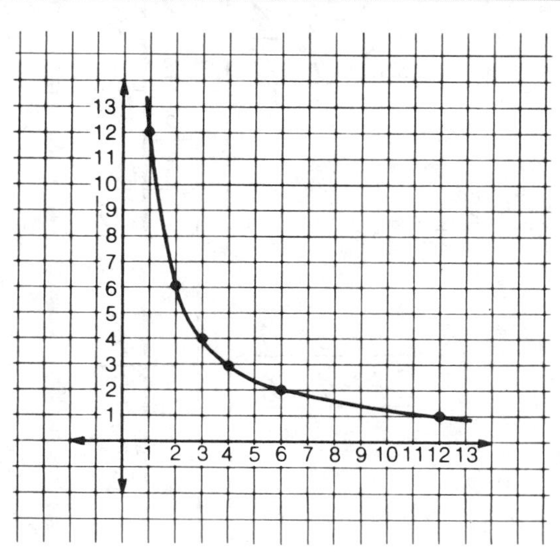

Figure 9.22
Graph of Factor Pairs for 12

×	0	1	2	3	4	5	6	7	8	9
0	0	0	0	0	0	0	0	0	0	0
1	0	1	2	3	4	5	6	7	8	9
2	0	2	4	6	8	10	12	14	16	18
3	0	3	6	9	12	15	18	21	24	27
4	0	4	8	12	16	20	24	28	32	36
5	0	5	10	15	20	25	30	35	40	45
6	0	6	12	18	24	30	36	42	48	54
7	0	7	14	21	28	35	42	49	56	63
8	0	8	16	24	32	40	48	56	64	72
9	0	9	18	27	36	45	54	63	72	81

Figure 9.23
Table for Basic Multiplication Facts

Figure 9.24 presents part of such a worksheet, a two-page response used for basic facts associated with a given factor. A table of products for the factor is in the column of squares at the right. Although the sample only partially illustrates facts for factor 4, worksheets can be completed for factors 1 through 9. The complete two-page response form can be found in the handbook accompanying this text.

Initially, the child should be closely monitored, but as he finishes worksheets for other factors, he can work independently. Later, when building a fact table, the child should write products he already knows, then other facts as they are determined.

Children do not need to make fact tables for division; a multiplication fact table can be used to find a missing factor. Children can locate the known factor at the left, then look across to find the product (or, for uneven division, the largest number less than the product). The missing factor can then be found at the top. Typically, when the teacher sufficiently emphasizes the relationship between multiplication and division, children do not need to use the multiplication fact table in this way. They know, for example,

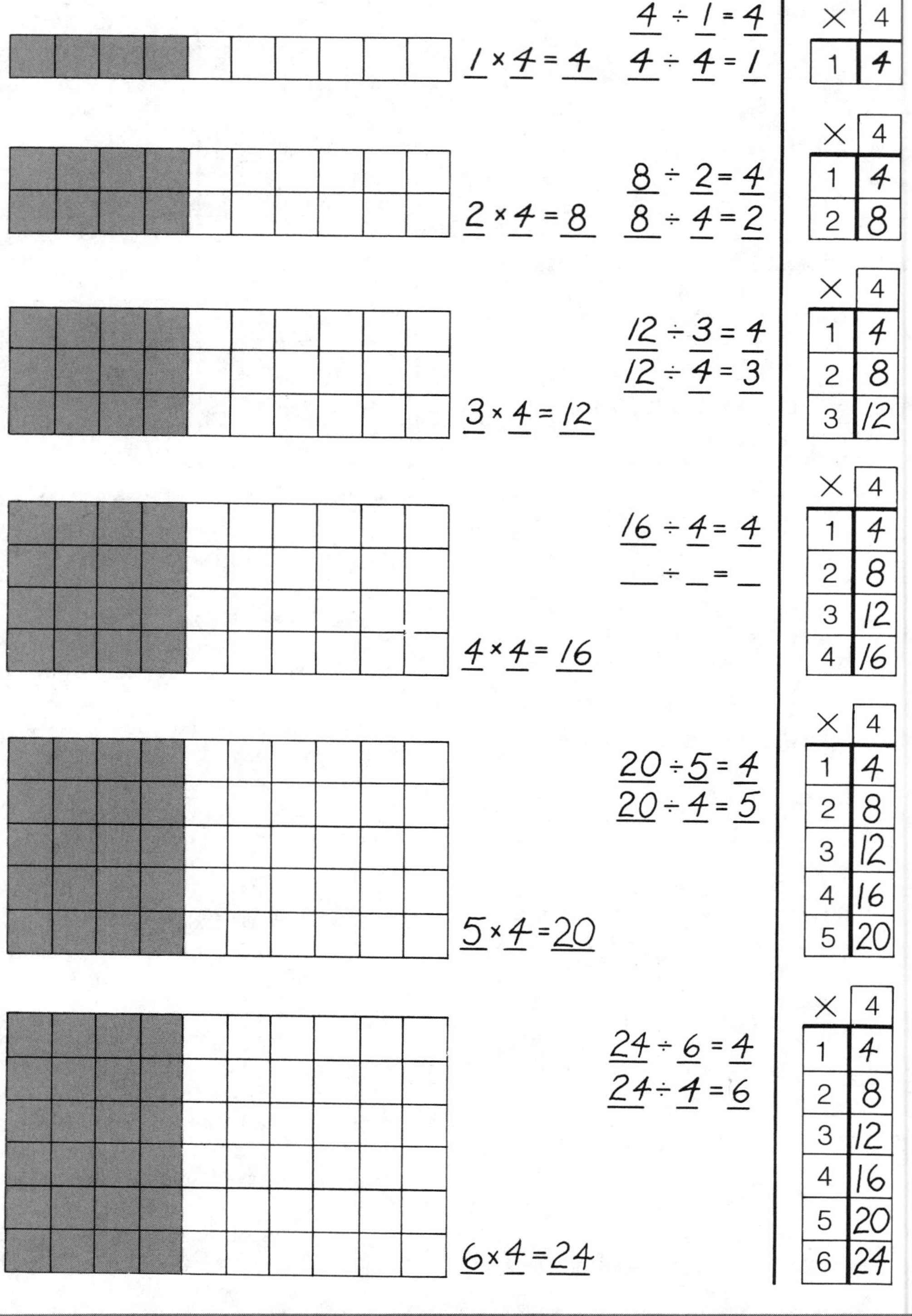

Figure 9.24
Part of Worksheet for Multiplication Facts (Completed)

EXCURSION
Plotting Points on a Graph

Young children can learn to plot points on a Cartesian graph if they play tic-tac-toe.[3] Two teams are necessary, the X's and the O's, with the teacher or recorder plotting number pairs named by individuals on each team as they take turns. In order to win, a team must get four points in a row—horizontally, vertically, or diagonally.

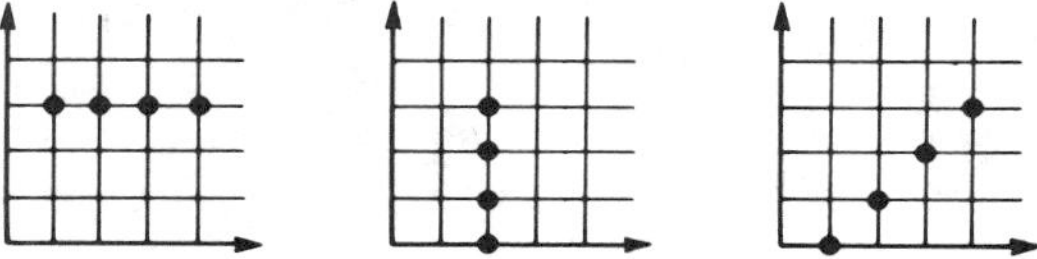

Older children who have studied signed numbers enjoy the challenge of playing this variation of tic-tac-toe on all four quadrants of a Cartesian graph.

that for $35 \div 7 = \square$ the missing factor is 5 because $7 \times 5 = 35$.

Patterns can help a child organize basic facts in his thinking. Figure 9.25 illustrates patterns that emerge for different factors. Multiples of a given number can be highlighted on a chart of whole numbers from 1 to 100 by coloring in each cell where a multiple is indicated.

Factor ladders for a given factor provide another way of picturing some of the patterns.[4] The second factor is shown at the left, and the units digit of the product is indicated at the right. Examples appear in Figure 9.26.

Children can discover many patterns among the products for a given whole number factor. These patterns lead to rules for determining if a number is divisible by a given factor. Rules for divisibility are summarized in Chapter 11.

Properties

Commutativity is easily shown with arrays. A cut-out array can be turned or loops can be drawn on an array in two ways, as in Figure 9.27. Cuisenaire rods can be used similarly, as pictured in Figure 9.28.

Associativity can be observed when a rectangular prism is built from a set of cubical blocks. Figure 9.29 (page 199) shows a set of 24 such blocks used to build a $2 \times 3 \times 4$ prism in two different ways. In Figure 9.29a, four 2×3 layers can be stacked, illustrating $(2 \times 3) \times 4$. In Figure 9.29b, two 3×4 layers are built, illustrating $2 \times (3 \times 4)$. Both constructions contain 24 blocks, therefore, $(2 \times 3) \times 4 = 2 \times (3 \times 4)$.

The identity element can be noted when all products with 1 as a factor are colored in a multiplication fact table. If 1 is a factor, the product is always the same as the other factor. The special zero relationship can be similarly observed: if zero is a factor, the product is always zero. Make sure each child understands and can apply these two rules which account for 36 of the 100 basic multiplication facts.

Distributivity of multiplication over addition is most easily exemplified by having each child partition an array so that he can multiply *in parts*. This can be done by folding a cut-out array or by drawing a line through an array. (See Figure 9.30.) In abstracting activities that focus upon distributivity, children need to make a step-by-step record of what they observe. For example, a

3. Robert B. Davis, "Notes on the Film: A Lesson with Second Graders," The Madison Project, Webster College, St. Louis, Mo., 1962, pp. 7–9.

4. Edward J. Wyatt, "Tables Patterns," *Mathematics Teaching* 84 (September 1978): 58–59.

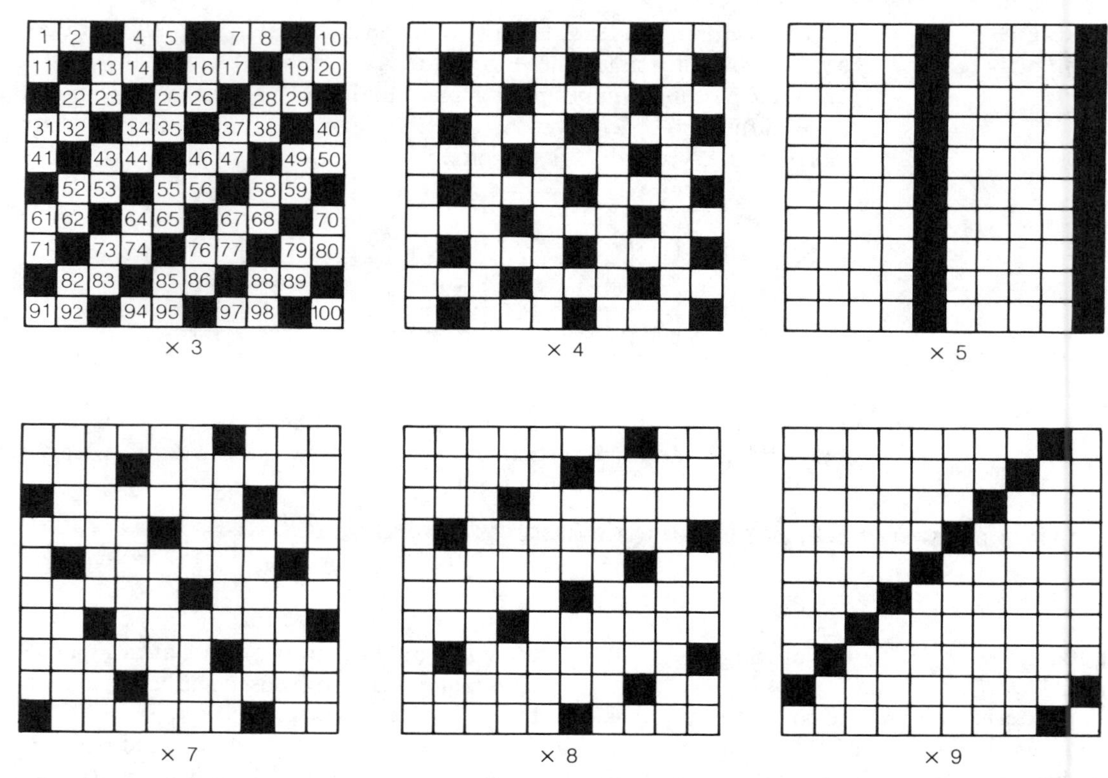

Figure 9.25
Patterns for Factors

3 × 7 cut-out array can be folded so the 7 is renamed as 5 + 2: a 3 × 5 array and a 3 × 2 array result. This record can be made:

$$\begin{aligned} 3 \times 7 &= 3 \times (5 + 2) \\ &= (3 \times 5) + (3 \times 2) \\ &= 15 + 6 \\ &= 21 \end{aligned}$$

A child who understands this property can use it to figure out products for multiplication facts that are not as easily recalled.[5]

5. See Roland F. Gray, "An Experiment in the Teaching of Introductory Multiplication," *The Arithmetic Teacher* 12, no. 3 (March 1965): 199–203.

Basic Facts

Consolidating Activities

Our emphasis has been on helping children understand the relationship between multiplication and division, and on encouraging them to use what they know about multiplication to determine an unknown factor in a division situation. In keeping with this, efforts to help each child memorize the basic facts for multiplication and division should focus primarily upon basic *multiplication* facts.

However, division situations also need to be included in consolidating activities to reinforce each child's skill in applying multiplication. The child needs experience with division symbols, especially for paper-and-pencil computation.

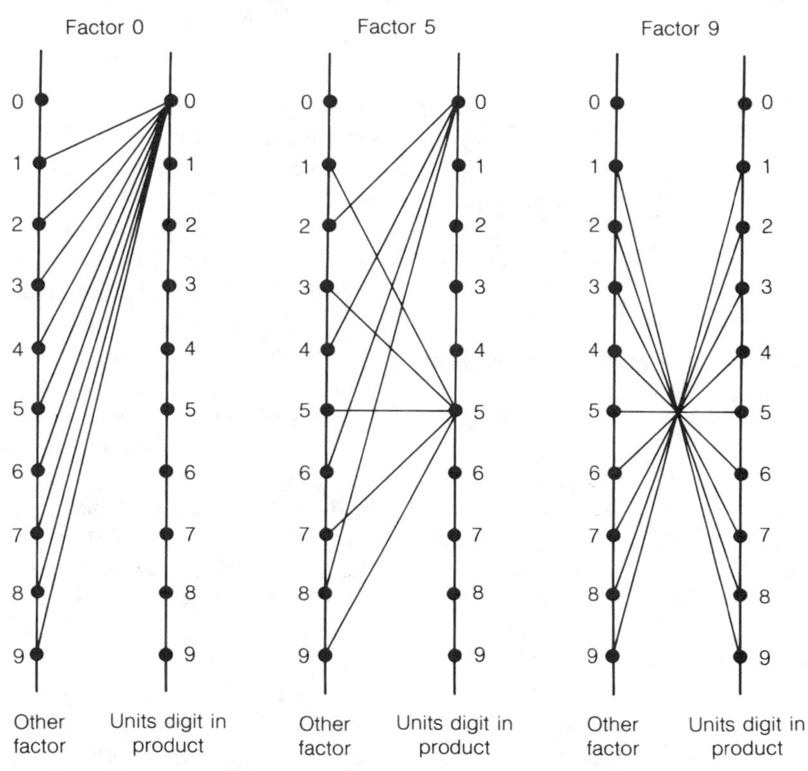

Figure 9.26
Factor Ladders

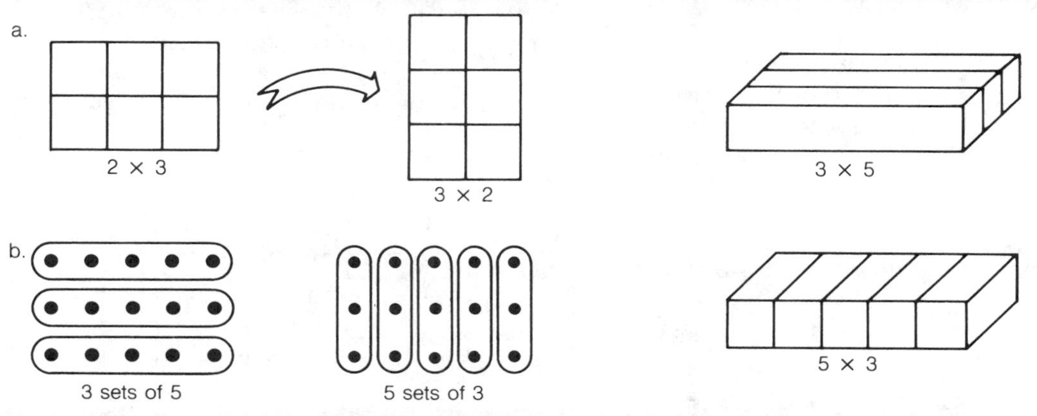

Figure 9.27
Arrays for Commutativity for Multiplication

Figure 9.28
Cuisenaire Rods for Commutativity for Multiplication

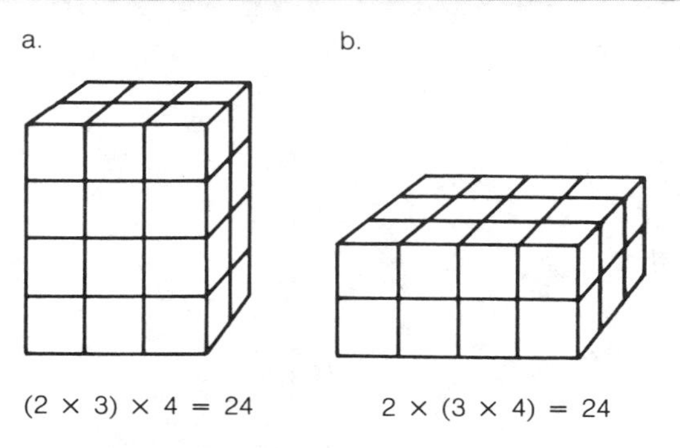

Figure 9.29
Cubes for Associativity for Multiplication

Many of the activities to be described will involve both multiplication and division, but they can also be adapted to help children maintain skill with basic addition and subtraction facts. Some activities actually require all four operations.

In Chapter 6 the authors stated that a child's successful completion of schematizing activities indicates a readiness for consolidating activities. Four desired characteristics of consolidating activities were listed:

1. Present the facts in random order.
2. Provide immediate feedback for the child.
3. Include a time constraint in a game.
4. Penalize error minimally.

Many consolidating activities should be games, although there is a place for drill, primarily as a diagnostic check-up. Individual contracts and progress charts should be used in conjunction with reinforcement activities so that each child can observe his progress.

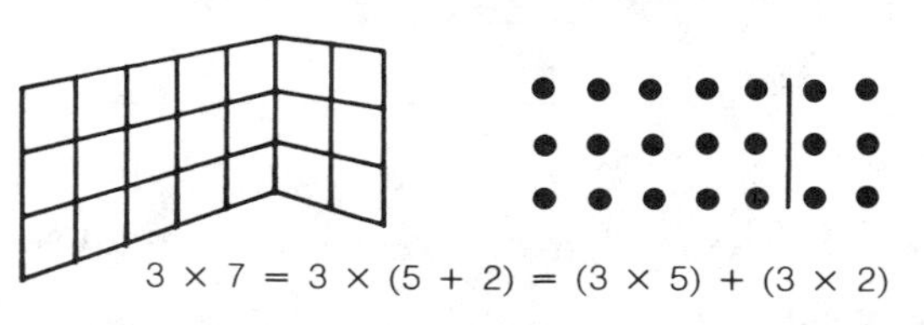

Figure 9.30
Arrays Partitioned to Show Distributivity

Consolidating activities for multiplication should focus on one set of facts as identified in objectives 9.43 through 9.50, reinforcing specific generalizations and strategies. Eventually, activities need to be planned that include most if not all of the 100 basic multiplication facts.

One such activity is Multiplication Tic-Tac-Toe described in Figure 9.31.[6] The game has the four desired characteristics; however, the child never sees the complete written horizontal or vertical number sentence. Make sure that children associate oral statements with written number sentences.

A variation is to require that after the player states the product, he writes the complete basic fact on a card. If the product is correct, the player places a card, rather than an X or an O, on the grid. Players should use different-colored pens. This way, basic facts that are incorrectly or hesitantly recalled can be retained for later review.

Games. As you consider the following examples, think of ways each game can be varied to more completely incorporate the desired characteristics described above.

Multiplication Rummy[7]

Materials: A reference list of 24 basic multiplication facts; 24 three-by-five cards, each with a multiplication phrase from the list (such as 6 × 7); 24 cards, each with a corresponding product; and a supply of cards, each with an equals sign.

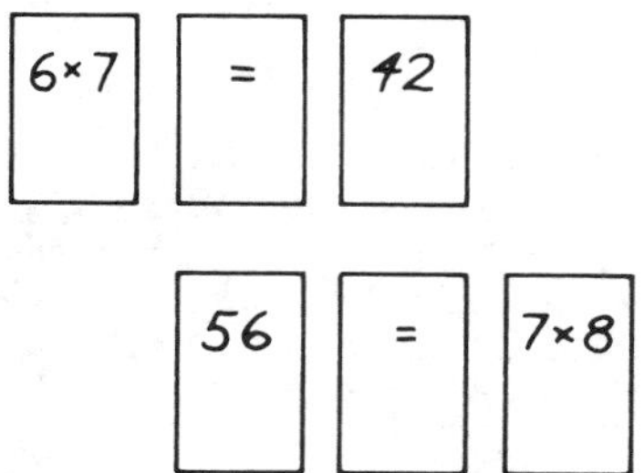

6. Adapted from James W. Heddens, *Today's Mathematics,* 3rd ed. (Chicago: Science Research Associates, 1974), p. 175.
7. Adapted from Leonard M. Kennedy and Ruth L. Michon, *Games for Individualizing Mathematics Learning* (Columbus, Ohio: Charles E. Merrill, 1973), pp. 69–70.

ACTIVITY PLAN

Multiplication of Whole Nos.: Basic Multiplication Facts

Symbolic → Symbolic (oral)

Consolidating

Content objectives

9.45 through 9.50

Exemplars/Materials

Two cubes, one with the digits 3–8 on the six sides, and the other with the digits 4–9
A 10-second timer
Basic multiplication fact reference (such as a table or Multifactor)
Tic-tac-toe grid on paper or chalkboard

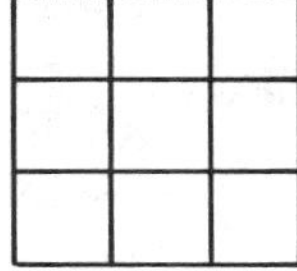

Behavioral indicator

The child states the correct product for given factors within 10 seconds.

Procedures

1. Two children play. One player rolls the two cubes and the other child starts the timer.
2. The player then states the product of the numbers shown. If the product is correctly stated within the 10-second limit (as determined by checking the basic multiplication fact reference), the player places an X or an O on the tic-tac-toe grid. If the product is not stated within 10 seconds or is incorrect, play passes to the other child without marking the grid.
3. The other child plays and the game continues until one child places three X's or three O's in a horizontal, vertical, or diagonal row and thereby wins the game.

Figure 9.31
Sample Activity Plan for Consolidating Basic Multiplication Facts

Procedure:

1. Two to four children can play. A dealer shuffles all 48 cards together and deals seven to each player. Remaining cards are placed in a draw pile, with the top card turned face up to start a discard pile.
2. Children check their cards to identify multiplication phrases and matching products, placing all matching phrases face up and forming number sentences by inserting an equals sign.
3. Each player in turn takes either a card from the draw pile, or the card that is on top of the discard pile. A number sentence is formed and placed face up if possible, then a card is placed face up on top of the discard pile.
4. A player may take a card that is below the top card in the discard pile, but all cards above it must be taken also.
5. A round is over when one player has no cards remaining in his hand. Each player gets one point for each number sentence which has been displayed.
6. The winner of the game is the child with the largest number of points after a predetermined number of rounds.

Factor Finding

Materials: A game board listing numerals for 1–40; 40 markers; and paper and pencil for keeping score.

1	2	3	4	5	6	7	8	9	19
11	12	13	14	15	16	17	18	19	20
21	22	23	24	25	26	27	28	29	30
31	32	33	34	35	36	37	38	39	40

Procedure:

1. Two teams play this strategy game. Team A chooses a number, places a marker on the numeral, and records that many points for itself.
2. Team B responds by identifying all proper factors for the identified number (such as all factors except the number itself), placing a marker on each, and recording as many points as the sum of the factors identified.
3. Team B then chooses a number and Team A responds by choosing proper factors not previously identified. Play continues until all numerals are covered.
4. The team with the most points wins.

Missing Factor Snatch[8]

Materials: A small object such as a beanbag or a tenpin (club) and a play area with parallel lines drawn 10–15 feet apart.

Procedure: This game is played with two teams. From 8–16 players are on each team, and the teams form two lines facing each other. Team members are numbered 2–9 from opposite directions (repeat the numbers if there are more than eight players) then a small object is placed in the middle of the space between the two lines. The teacher, or a pupil acting as the leader, calls out a multiplication sentence with a missing factor (such as "6 times some number is 42") or a division phrase (such as "42 divided by 6 is some number"). Each player with the number that is the missing factor runs out and tries to grab the object and return to his line without being tagged. If a player does so, the team scores two points; if the player is tagged by an opponent, the other team scores one point. Care should be taken to have the competing children as nearly equal in skill level as possible.

There are many sources for games that can be used to reinforce and maintain the basic facts for multiplication and division.[9] Games for addition and subtraction can often be adapted. Games and other activities are described regularly in *The Arithmetic Teacher*. Many commercial games such as Winning Touch, Heads Up, and TUF are also useful.

Paper-and-Pencil Activities. Many consolidating activities are not games, but are useful exercises and provide a measure of reinforcement and skill maintenance. Paper-and-pencil activities are often of this character.

Basic Fact Wheels

Materials: Prepare basic fact wheels similar to Figure 9.32, either on worksheets or laminated cards. Factors appear in the two inner rings and products in the outer ring.

Procedure: Make sure children understand the format of the activity, then have each child independently write in the missing factors and products. Provide a way of verifying responses, and a time constraint if appropriate. Have each child record incorrect responses for further study.

Multiplication Machines[10]

Materials: Worksheets that picture multiplication machines and have tables for recording factors and products. (See the sample in Figure 9.33.)

Procedure: Introduce the concept of a binary machine to the children, noting the name of each machine. A number is put into a machine, and the machine does what its name says. For example, a *Times 6* machine multiplies the number put

8. Adapted from Robert Ashlock and James Humphrey, *Teaching Elementary School Mathematics Through Motor Learning* (Springfield, Ill.: Charles C Thomas, 1976), pp. 39–40.

9. See Robert Ashlock and Carolynn Washbon, "Games: Practice Activities for the Basic Facts," in *Developing Computational Skills*, 1978 Yearbook, ed. Marilyn Suydam and Robert Reys (Reston, Va.: National Council of Teachers of Mathematics, 1978), pp. 39–50.

10. Marvin Karlin, "Machines," *The Arithmetic Teacher* 12, no. 5 (May 1965): 327–34.

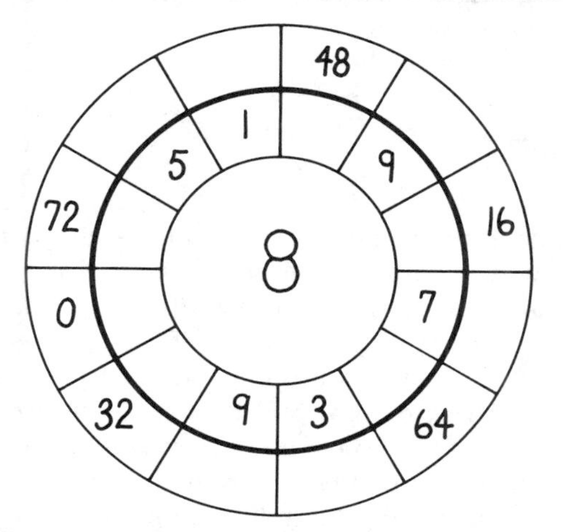

Figure 9.32
Basic Fact Wheel for Multiplication and Division

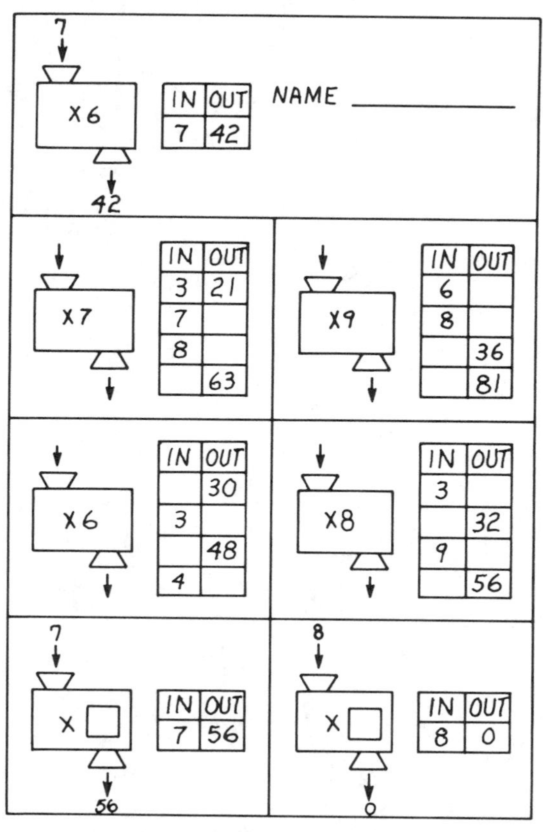

Figure 9.33
Sample Worksheet with Multiplication Machines

into it by 6. The number that comes out of the machine is the result of the operation. The machines shown on the sample worksheet are multiplication machines, and the number that comes out is a product. To find the missing numbers in the tables and on the machine labels, multiplication is sometimes required and division is needed at other times. Have children complete the table for each machine.

Arrow Diagrams

Materials: Worksheets picturing incomplete arrow diagrams similar to those in Figure 9.34.

Procedure: Complete a few arrow diagrams as a group in order to make sure each child understands the format of the exercise. Children draw missing arrows and write missing numerals in the loops provided. Worksheets can then be completed independently.

Calculator Activities. Multiplication can easily be seen as repeated addition of equal addends, and division can be shown as repeated subtraction of equal addends when you use hand-held calculators. They are also helpful for problem solving with larger numbers if the focus is on selecting the appropriate operation rather than computation. Calculators provide a means for verifying correct answers when practicing paper-and-pencil computation or estimation.

When games are played, a calculator can serve as a reference for verifying basic fact statements; it can provide the immediate feedback that is needed. Immerzeel suggests that recall be reinforced by having one child call out a multiplication phrase, and then compute the product on a calculator while another child tries to write down the product before it appears on the calculator.[11] Such activities help children realize that using a calculator is not always the most efficient procedure.

11. George Immerzeel, *Ideas and Activities for Using Calculators in the Classroom* (Dansville, N.Y.: The Instructor Publications, 1976), p. 15.

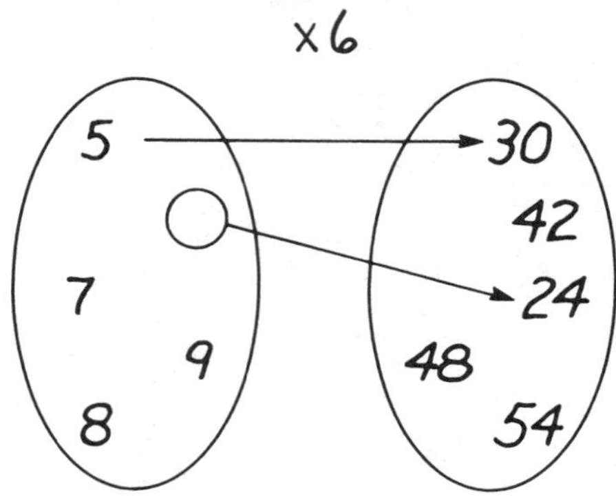

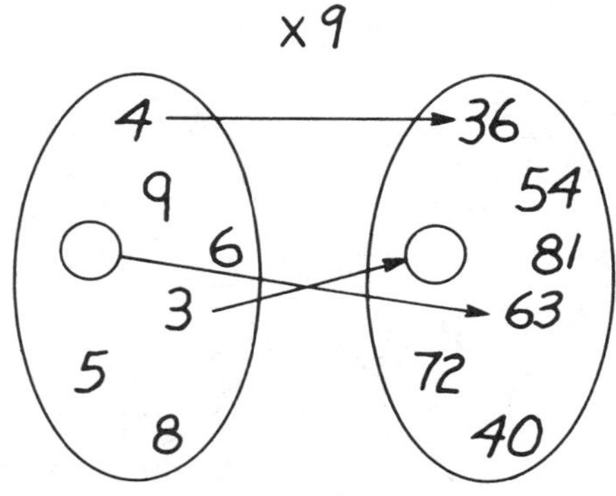

Figure 9.34
Arrow Diagrams for Multiplication and Division

Flash Cards. Some flash cards present multiplication and division phrases horizontally, and others show them as they are used in paper-and-pencil computation. Both kinds of flash cards should be used in games and in diagnostic activities. A child can use cards as follows:

1. Place the set of cards in a pile, then start a timer, possibly a stopwatch.
2. Whisper aloud the missing product or factor for the card on top, then turn over the top card to verify the solution.
3. Place cards for correct responses in one pile and cards for incorrect responses in another.
4. When the products or factors have been stated for all cards in the pile, stop the timing device.
5. Record two scores on a progress chart: the time it took to respond to all of the cards and the number of correct responses.
6. Make a list of basic facts from the cards with incorrect responses for further study.

Other Activities

Bulletin boards and learning centers also enrich early work with multiplication and division. For instance, colorful bulletin boards often focus on models for operations and applications, possibly including practice with basic facts.

One side of a bulletin board can present different arrays, such as egg cartons, button display cards, and arrays cut from graph paper. The other side can list multiplication phrases or basic multiplication facts. A child can match arrays to the appropriate symbolic expressions with pieces of yarn. Children can verify their responses with an answer key picturing the completed bulletin board.

Sentence Solving and Verbal Problem Solving

The key to solving number sentences is understanding the operations; a child who understands the meanings, models, and symbols for multiplication and division can use that information to solve number sentences. Especially important are these generalizations:

Factor × Factor = Product (Objective 9.51)

Product ÷ Factor = Factor (Objective 9.52)

We shall probably cluster these objectives with others when planning instruction. Surface structures among number sentences should be sufficiently varied so that each child abstracts the deep structures: the two content objectives cited above. A few activities follow for helping children solve verbal problems; other suggestions appear in Chapter 11.

Transferring Activities

Verbal problems can be used for transferring activities when a new situation for using previously learned knowledge and skills is introduced. Examples of such activities include Story Prob-

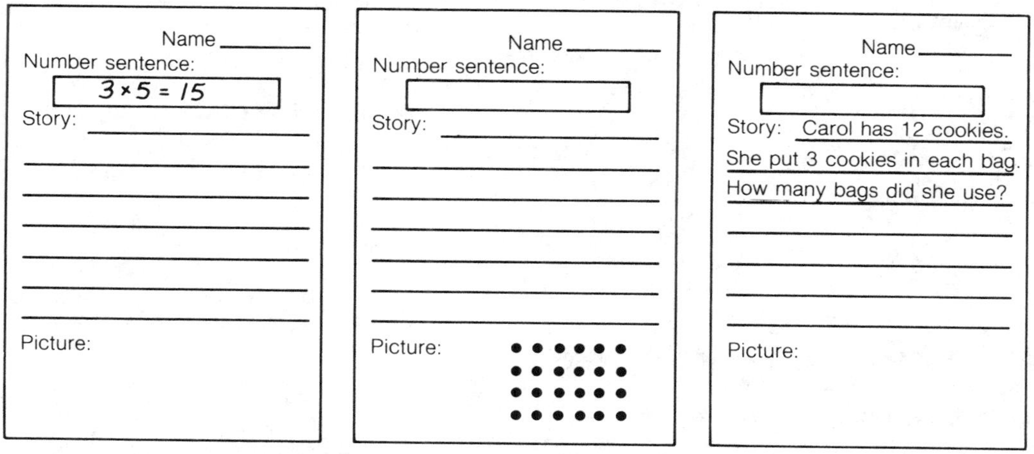

Name ________
Number sentence:
3 × 5 = 15
Story: ________
Picture:

Name ________
Number sentence:
[ ]
Story: ________
Picture: ●●●●●●● / ●●●●●●● / ●●●●●●● / ●●●●●●●

Name ________
Number sentence:
[ ]
Story: Carol has 12 cookies. She put 3 cookies in each bag. How many bags did she use?
Picture:

Figure 9.35
Sample Story Problem Worksheets

lem Worksheets like those in Figure 9.35.[12] Each child is required to translate from a number sentence, verbal problem, or two dimensional representation, to other forms of representation.

Transferring activities in mathematics and other subjects vary greatly and depend somewhat on the topics being studied. For example, concepts of perimeter and area can be initiated as a child uses a geoboard to make an array of square regions for a given multiplication fact. The number of square regions can be recorded, and also the number of unit lengths in the boundary. When this is done for several multiplication facts, relationships between area and perimeter can be explored (see Figure 9.36). A child may discover, for example, that for a given product (area) the factor pair that produces the shortest perimeter is a factor times itself. During such an activity, a child uses what is already known about multiplication to focus on problems in a new and different situation.

A child may learn to make Venn diagrams in working with sets. They can be used to show multiples for each of two or more different numbers. Common multiples are easily noted as intersecting sets, as in Figure 9.37.

12. Adapted from Leonard M. Kennedy, *Models for Mathematics in the Elementary School* (Belmont, Calif.: Wadsworth, 1967), p. 97.

PLANNING ASSESSMENT

Assessment tasks can be planned by considering behaviors that may indicate understanding of an objective. For this, we may want to consult the list of action verbs in Table 6.2.

One important objective for multiplication is objective 9.6, which deals with arrays. In

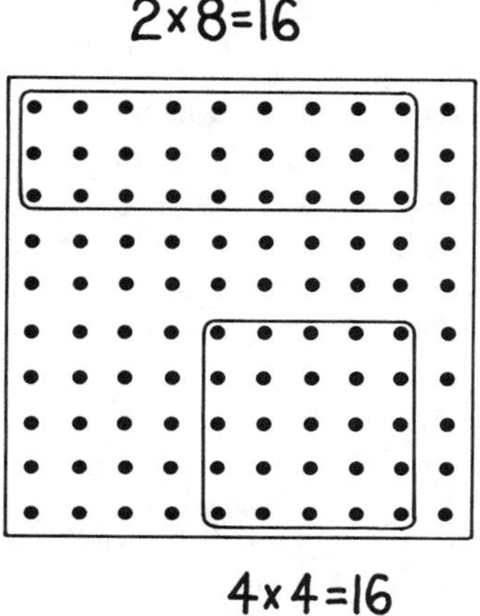

Figure 9.36
Multiplication Facts on a Geoboard

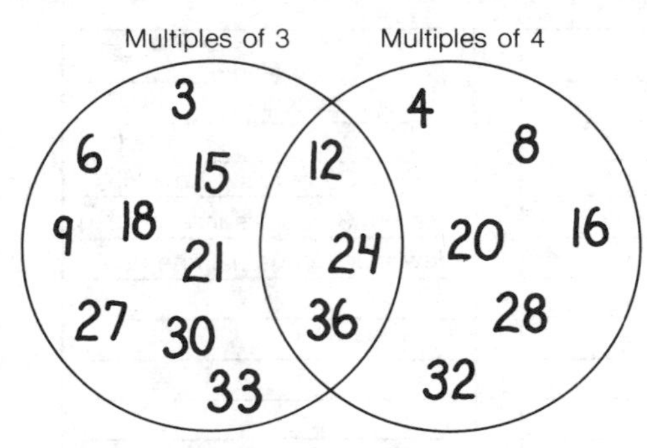

Figure 9.37
Multiples Shown with Venn Diagrams

terms of the action verbs *identify* and *name*, the following behaviors indicate understanding of the ideas:

1. Given an array, *identifies* and *names* the two factors and the product.
2. Given a multiplication number sentence and arrays, the child *identifies* the array that matches the number sentence.

These behaviors may result from the verb *construct*:

3. Given a multiplication number sentence, the child *constructs* a matching array.
4. Given an array, the child *constructs* (or writes) the matching multiplication number sentence.

The last two behaviors constitute an example of going both ways between a number sentence and a model. Children can often do one translation but not the other, indicating that understanding is still incomplete.

How can we elicit such behaviors? We could design the following tasks:

1a. Give each child a cut-out array and say, "What are the factors? What is the product?"

1b. Give each child an exercise like this and say, "Draw a ring around the factors."

2a. Give each child a set of cut-out arrays, then write a multiplication number sentence on the chalkboard and say, "Show me the array that goes with this number sentence."

2b. Give each child an exercise similar to the following and say, "Draw a ring around the array that goes with the number sentence."

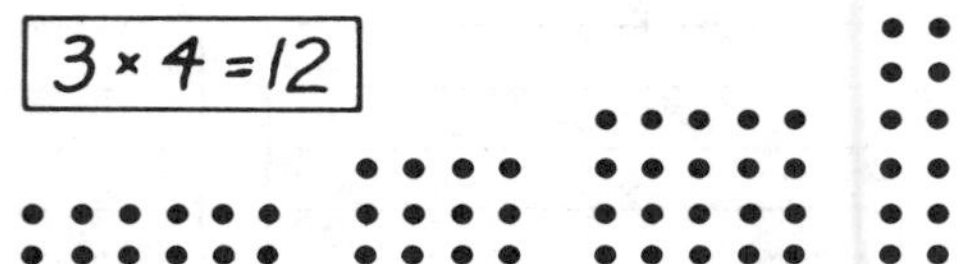

3a. Give each child a set of cubes, then write a multiplication number sentence on the chalkboard and say, "Use your cubes to make an array for this number sentence."

3b. Give each child a paper with multiplication number sentences on the left and say, "Draw an array to go with each number sentence."

4a. Make an array with cubes on the table and ask each child to write a corresponding number sentence.

4b. Give each child a paper with arrays on the left and say, "Across from each array write the number sentence that goes with the array."

When we assess skill with basic multiplication and division facts we must present written symbols such as $9 \times 4 = \square$, and not just say "nine times four." Both horizontal and vertical forms should be presented.

When multiplication and division phrases are presented on individual cards for assessment, place a time limit on each response. Then, as a child approaches complete success with one time interval, introduce a shorter interval. Timing can be done with a timing device in a normal testing situation, and also in a game format. The authors have found that children sometimes have a greater percent of correct responses when playing a game than when participating in a straightforward testing situation.[13]

---

13. See the game Beat the Bell in Robert Ashlock and Carolynn Washbon, "Games: Practice Activities for the Basic Facts," in Suydam and Reys, *Developing Computational Skills*, p. 46.

Skill with a specific set of basic facts should be assessed in brief sessions at regular intervals. Each child should make a record such as a progress graph of his scores.

TIPS ON MANAGING THE CLASSROOM SITUATION

Here are ideas to help you introduce multiplication and division:

1. When you use manipulatives, make sure you have enough on hand; count out the materials *before* you start the lesson.
2. Make certain that two or more children playing a game as a consolidating activity are at approximately the same level of skill development. This way, you assure interest and a measure of success for each.
3. When working with a group of children, frequently use the overhead projector.
   a. The shadows of an array of opaque objects on the stage of the projector will easily be seen by all. Colorful translucent objects can also be used, such as plastic tiles for mosaics.
   b. If you place the array on a clear piece of acetate, it can easily be turned 90 degrees when showing commutativity for multiplication.
   c. Sticks can be crossed to show multiplication as a Cartesian product (Figure 9.7) on the stage of the projector.
   d. When showing division with an array (Figure 9.19), place the array on the stage, then cover part of the array with a piece of paper so that only one factor is visible. Tell children the product and have them predict the other factor. Finally, remove the paper so each child can verify the solution.
4. For number line activities, each child can have a number line covered with clear contact paper on the top of his desk. Arrows that are drawn can be erased, and the number line can be used again.
5. Horizontal transfer activities, such as the worksheet for multiplication facts used with *different* factors (Figure 9.24), can be assigned as independent study while you work with other children on directed activities.

10

Developing Computation Procedures for Multiplication and Division

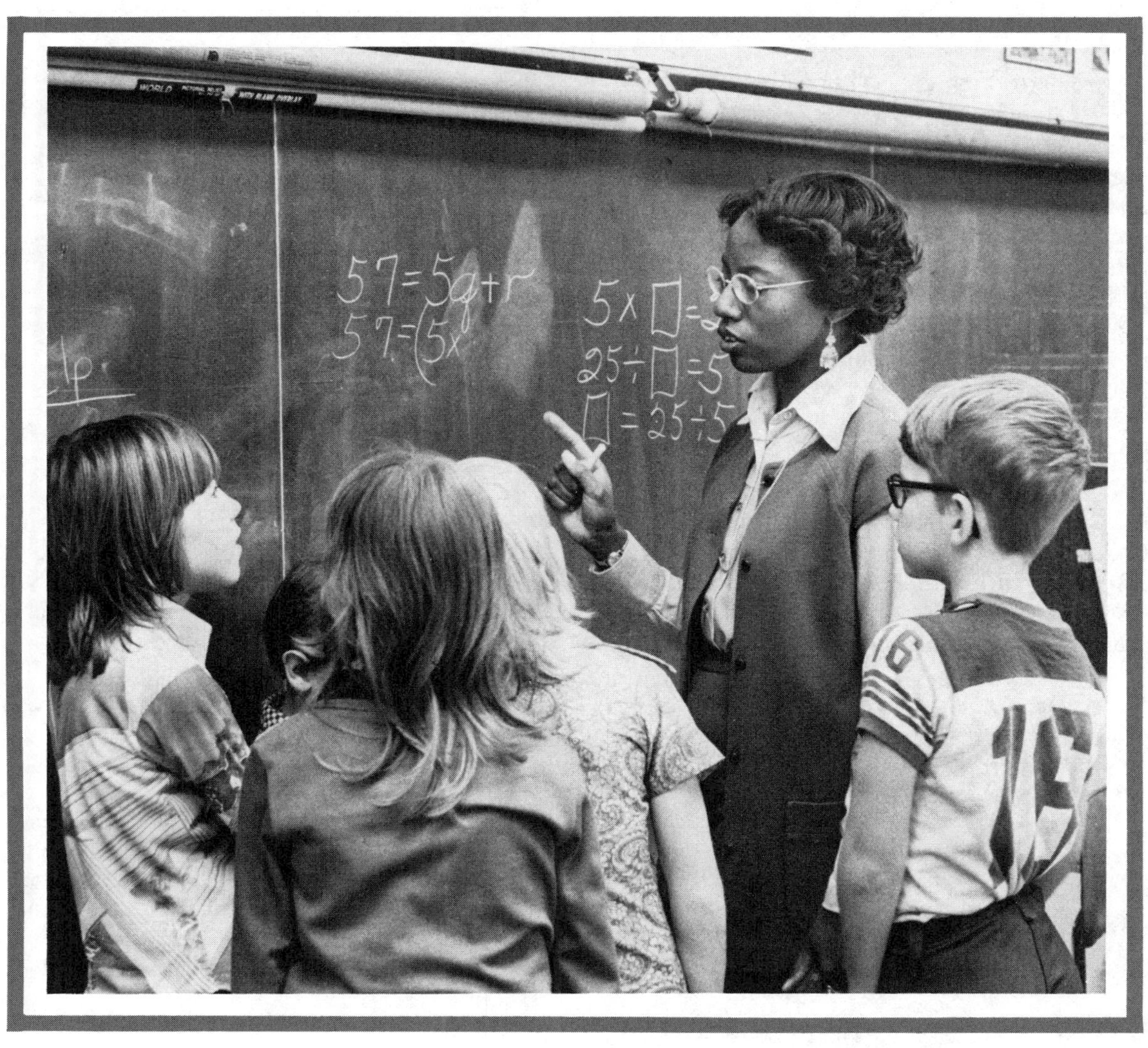

Children should know computational procedures for multiplication and division well enough to compute rapidly and accurately. Yet, as we help them learn, we must continue to stress the meanings and uses of the operations.

PLANNING THE CONTENT

When children need to multiply a two-digit number by any number greater than 1, their knowledge of basic facts is insufficient. But children do use what they already know as they learn procedures for finding products and missing factors.

There are several algorithms for multiplying and dividing whole numbers. The most commonly used algorithms and a few alternative procedures are presented here. Do not try to teach each child everything about these algorithms.

Multiplication of Whole Numbers

Children who know the basic facts, selected properties of multiplication, and numeration ideas typically experience little difficulty with algorithms for multiplication.

Properties of Multiplication

Children use the commutative, associative, and distributive properties when they multiply, as in the following content objective:

> **9.15**
> *Commutative property*—The order or sequence in which two whole numbers are multiplied has no affect on their product.
>
> **9.16**
> *Associative property*—The way in which three whole numbers are grouped together in multiplication has no effect on their product.

> **9.18**
> *Distributive property*—Whole numbers under the operation of multiplication are distributive with respect to addition.

For example, children can use the commutative property to simplify multiplication. Which is easier?

$$\begin{array}{r} 6 \\ \times\ 27 \\ \hline \end{array} \quad \text{or} \quad \begin{array}{r} 27 \\ \times\ 6 \\ \hline \end{array}$$

The associative property is helpful when children want to group factors for convenience.

Left to Right Grouping		Grouping for Convenience
$(27 \times 4) \times 25$	or	$27 \times (4 \times 25)$
$= 108 \times 25$		$= 27 \times 100$
$= 2700$		$= 2700$

Children must understand the distributive property to multiply with two or more digit numerals. This important property will also help a child to mentally compute products. Consider 5 × 27 as an example:

$$\begin{aligned} 5 \times 27 &= 5 \times (20 + 7) \\ &= (5 \times 20) + (5 \times 7) \\ &= 100 + 35 \\ &= 135 \end{aligned}$$

When the factor to be distributed is on the left, as in the example above, the process is an application of *left*-handed distributivity. The example below has the factor to be distributed on the right. This is an application of *right*-handed distributivity.

$$\begin{aligned} 634 \times 4 &= (600 + 30 + 4) \times 4 \\ &= (600 \times 4) + (30 \times 4) + (4 \times 4) \\ &= 2400 + 120 = 16 \\ &= 2536 \end{aligned}$$

EXCURSION
Applying the Distributive Property

With experience, the distributive property can be used for more difficult mental computations.

Thought Process

$$\begin{aligned} 35 \times 102 &= 35 \times (100 + 2) \\ &= (35 \times 100) + (35 \times 2) \\ &= 3500 + 70 \\ &= 3570 \end{aligned}$$

Thought Process

$$\begin{aligned} 45 \times 98 &= 45 \times (100 - 2) \\ &= (45 \times 100) - (45 \times 2) \\ &= 4500 - 90 \\ &= 4410 \end{aligned}$$

Can you find these products mentally?

15×103 $\quad$ 101×92 $\quad$ 18×99

The Traditional Multiplication Algorithm

When teaching the traditional multiplication algorithm, begin with the following content objective:

> **10.1**
> To multiply a number greater than 10, rename the number as a sum. Then, apply the distributive property; multiply each term.

Children who understood expanded notation and regrouping while learning to add and subtract should have little difficulty. We can use expanded notation to illustrate the process.

Expanded Notation Forms

$$\begin{aligned} 4 \times 24 &= 4 \times (20 + 4) \\ &= (4 \times 20) + (4 \times 4) \\ &= 80 + 16 \\ &= 96 \end{aligned}$$

$$\begin{array}{r} 20 + 4 \\ \times \quad 4 \\ \hline 80 + 16 \end{array}$$

$$\begin{aligned} &= 80 + (10 + 6) \\ &= (80 + 10) + 6 \\ &= 90 + 6 \\ &= 96 \end{aligned}$$

Standard Form

$$\begin{array}{r} {}^{1}24 \\ \times\ 4 \\ \hline 96 \end{array}$$

Another important idea results from application of the distributive property.

> **10.2**
> When we multiply a number greater than ten, the product is found in parts called *partial products*. The partial products are added to obtain the final product.

Partial products can be observed in the expanded notation forms above. Repeated applications of the distributive property are required to multiply larger numbers, so there are more partial products. Horizontal notation illustrates this.

$$\begin{aligned} 27 \times 35 &= (20 + 7) \times (30 + 5) \\ &= (20 \times 30) + (20 \times 5) + (7 \times 30) + (7 \times 5) \\ &= 600 + 100 + 210 + 35 \\ &= 945 \end{aligned}$$

Similarly, the following computations show the use of partial products when developing the standard algorithm for multiplying by a two-digit

EXCURSION
Napier's Bones

A set of Napier's Bones can be used for multiplication. They can be made from strips of cardboard or wood as shown in Figure 10.1. Each strip contains multiples of the number shown at the top. The index bone is used as a guide for multiplication.

To find the product of 275 × 36, select the index and the 2, 7, and 5 bones, then place them side by side in that order. Next, find the row opposite 6 on the index and, starting at the right, add diagonally. The first partial product is 1650.

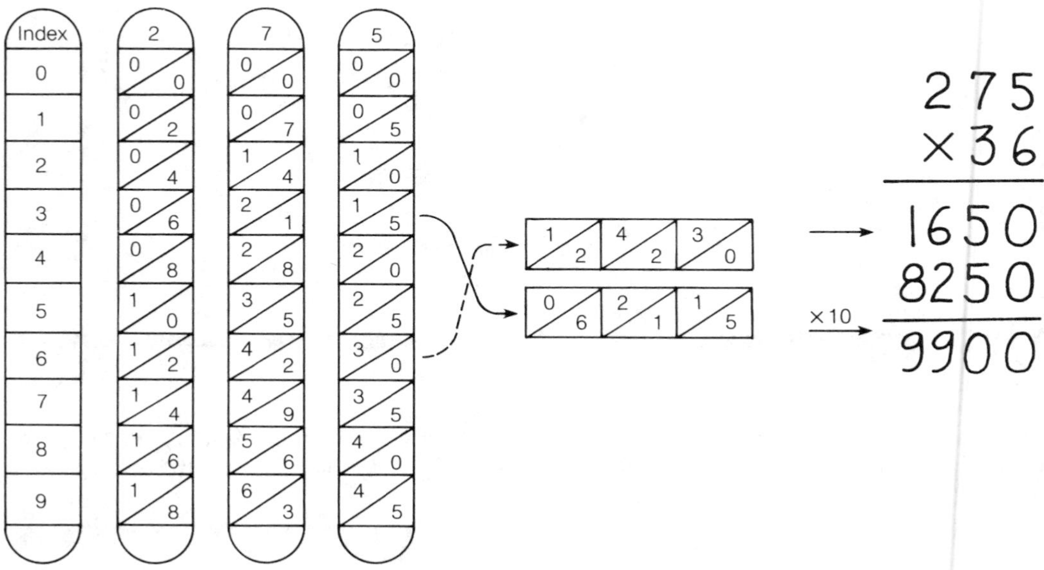

Repeat this procedure for the factor 30. Use 3 on the index, and multiply by 10. Add the partial products to find the product.

Use a set of Napier's Bones to find the product of 483 × 64.

multiplier. In the first example, all of the partial products are shown. In the second, not all possible partial products are written; some are already combined.

	35	
	× 27	
Partial Products	35	←(7 × 5)
	210	←(7 × 30)
	100	←(20 × 5)
	600	←(20 × 30)
	945	←(27 × 35)

35	
× 27	
245	Short Form of Partial Products
700	
945	

In more complex multiplication problems, the principles of place value are applied and the basic algorithm is extended.

368	
× 25	
40	←(5 × 8)
300	←(5 × 60)
1500	←(5 × 300)
160	←(20 × 8)
1200	←(20 × 60)
6000	←(20 × 300)
9200	←(25 × 368)

368
× 25
1840
7360
9200

Checking Multiplication

Each child should establish the habit of estimating answers to make sure each product is reasonable. In order to estimate, a child must know how

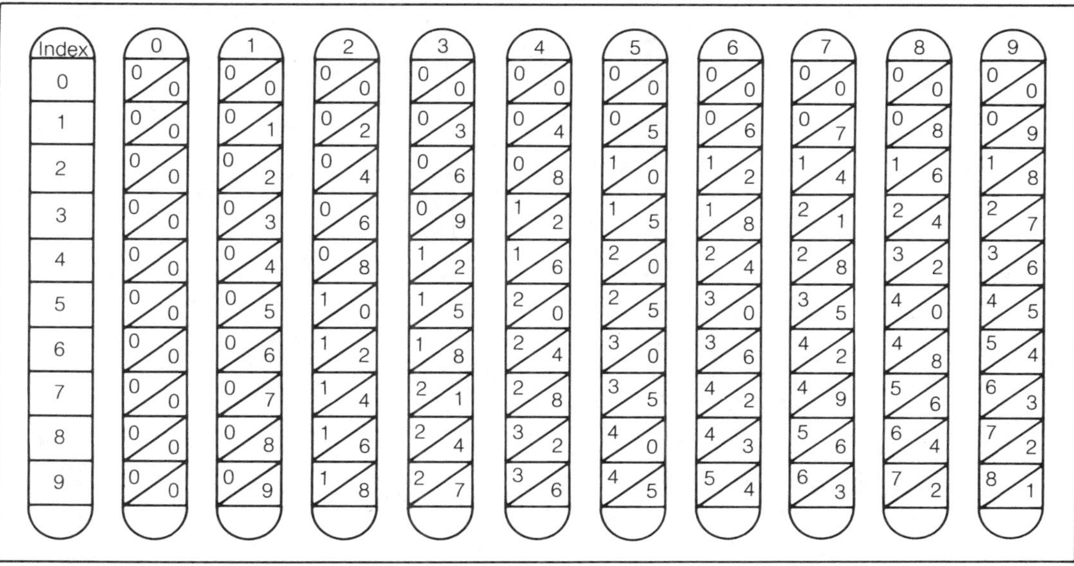

Figure 10.1
Napier's Bones (Rods)

to round numbers to powers of 10 or their multiples. Some of the following procedures can also be understood as applications of commutativity and associativity for multiplication.

10.3
Places to the left of the decimal point are assigned to special products of 10 in a decimal system, and are called *powers of 10.*

10.4
When multiplying by a power of 10 or by a multiple of same, multiply the nonzero numbers and annex as many zeros to the product as there are zeros in the factor.

10.5
The special products of 10 and any other whole number factor are multiples of 10.

10.6
When multiplying two multiples of 10, multiply the nonzero numbers and annex the number of zeros in both factors.

10.7
The estimated product of two factors has upper and lower bounds.
a. The *upper bound* is the product of the two rounded up factors.
b. The *lower bound* is the product of the two rounded down factors.

When we multiply by a power of 10 or by a multiple of a power of 10, the product is easily obtained.

27		27	
× 10	one zero	× 100	two zeros
270	one zero	2700	two zeros
46		46	
× 30	one zero	× 300	two zeros
1380	one zero	13,800	two zeros

Multiplying by multiples of 10 is handled in much the same way.

EXCURSION
Lightning Multiplication

Cross-multiplication, sometimes called the *lightning method*, provides a useful setting for reviewing the multiplication process. The method appeared in Paciali's *Suma* in 1494.[1]

Step 1
2 × 8 = 16
Write 6 ones and save the 1 ten.

```
  38
× 42
   6
```

Step 2
2 × 3 tens = 6 tens
4 tens × 8 = 32 tens
32 tens + 6 tens + 1 tens = 39 tens
Write the 9 in the tens.
Place and save the 3 hundreds.

```
  38
× 42
  96
```

Step 3
4 tens × 3 tens = 12 hundreds
12 hundreds + 3 hundreds = 15 hundreds
Write the 15 hundreds.

```
  38
× 42
1596
```

The lightning method can also be used as an enrichment activity, or for checking computation.

```
  60   one zero         1500   two zeros
× 30   one zero       ×   30   one zero
1800   two zeros      45,000   three zeros
```

Product of 3 × 6   Product of 3 × 15

To estimate the product of 48 and 72, round each factor to the next higher and the next lower multiple of 10.

Rounding Up (higher multiple)		Rounding Down (lower multiple)	
48 →	50	48 →	40
× 72 →	× 80	× 72 →	× 70

The upper bound, 4000, is found by multiplying 80 × 50, and the lower bound, 2800, by multiplying 70 × 40. The product will lie between 2800 and 4000.

A more accurate estimate of the product of 48 and 72 results from rounding each of the two factors to the nearer multiple of 10. Thus

1. From TEACHING THE CHILD MATHEMATICS, Second Edition, by C. W. Schminke, Norbert Maertens and William Arnold, p. 195. Copyright © 1978 by Holt, Rinehart and Winston, © 1973 by the Dryeden Press, Inc. Reprinted by permission of Holt, Rinehart and Winston, CBS College Publishing.

EXCURSION
Estimating Products

Round each factor to the next higher and the next lower multiple of 10. Then find the upper and lower bounds of the products.

24 × 33   27 × 62   67 × 88

In which case is the exact product closer to the upper bound? Lower bound? Midway between the upper and lower bounds?

EXCURSION
Casting Out Nines

Casting out nines can be used to check multiplication. To do this, add across the digits, and cross out (cast out) nines or numbers with a sum of 9. Cast out nines from both of the factors and from the product. Then multiply the excess (the number left) for each factor, and cast out nines again. If the excess is the same for both the product and the result of multiplying the excesses of the factors, the product is probably correct. For example:

```
     475    4 + 7 + 5                 = 16 →    7
   × 291    2 + 9 + 1                 = 12 →  × 3
     475                                      21 → 3
   42750
   95000
 138,225    1 + 3 + 8 + 2 + 2 + 5 = 21 → 3
```

Since the product of the excesses of the factors is the same as the excess in the product, we conclude that the multiplication has been performed correctly. This is not always true. Exceptions include two digits reversed in the product, and a product that is incorrect by an amount that is a multiple of nine.

Find the product of 328 × 48. Then use casting out nines to check your answer.

3500, the product of 50 and 70, is a more accurate estimate.

Estimation serves only as a rough check. Each child should know that the accuracy of a product can be verified by reversing the factors and multiplying. This application of the commutative property requires that a child write different partial products and add different addends, thereby enhancing the likelihood of an accurate check.

```
Original Computation      Check
          27                 48
        × 48               × 27
         216                336
        108                 96
        1296               1296
```

Division of Whole Numbers

Division is often perceived to be the most difficult whole number computation at the elementary level.

Division Algorithms

Several methods for division have been used. The two most widely used are the subtractive and distributive algorithms.

In Chapter 9 we observed that division can be modeled as repeated subtraction. This concept is applied in the *subtractive algorithm*.

> **10.8**
> When dividing by the subtractive method, the divisor or a multiple of same is subtracted from the dividend until the remainder of the subtraction is less than the divisor.
>
> **10.9**
> When dividing larger numbers by the subtractive method, the quotient is found in parts called *partial quotients*. These partial quotients are added to obtain the final quotient.

The subtraction model for division illustrates the close relationship between subtraction and division. To answer the question 24 ÷ 6 = □, we might ask "How many times can we subtract the divisor 6, from the dividend 24?" The answer, or quotient, is 4.

```
 24                      4
- 6  (1)    or      6) 24 |
 18                   - 6 | 1 × 6
- 6  (2)               18 |
 12                   - 6 | 1 × 6
- 6  (3)               12 |
  6                   - 6 | 1 × 6
- 6  (4)                6 |
  0                   - 6 | 1 × 6
                        0 | 4 × 6
```

Instead of subtracting the divisor at each step, subtract *multiples* of the divisor which saves time.

```
      4
 6) 24 |
  - 12 | 2 × 6
    12 |
  - 12 | 2 × 6
     0 | 4 × 6
```

The process for dividing by multidigit divisors is much the same as for dividing by single-digit divisors; however, the ability to multiply by 10 and by *powers* of 10 is a prerequisite for learning to divide by larger numbers. Less writing is required if a child can multiply by a *multiple* of a power of 10.

```
Example 1                        Example 2

      324                              324
24) 7776 |                       24) 7776 |
  - 2400 | 100 × 24                - 7200 | 300
    5376 |                            576 |
  - 2400 | 100 × 24                 - 480 |  20
    2976 |                             96 |
  - 2400 | 100 × 24                 -  96 |   4
     567 |                                | 324
  -  240 |  10 × 24
     336 |
  -  240 |  10 × 24
      96 |
  -   48 |   2 × 24
      48 |
  -   48 |   2 × 24
       0 | 324 × 24
```

The more economical procedure illustrated in Example 2 involves estimating and comparing. A child must first determine the correct power of 10: for instance, *1000* × 24 = 24000 is too great, but *100* × 24 = 2400 is not. Therefore, the child knows that 100 or a multiple of 100 will be used as the first partial quotient. By examining the dividend, she can determine which multiples of 100 can be used. In this case, 300 is the largest possible multiple. In like manner, the child then determines partial quotients for tens and ones.

The *distributive*, or *traditional*, *algorithm* for division relies on the use of the distributive property under division with respect to addition. These content objectives are important when teaching this algorithm.

> **10.10**
> The whole numbers are right distributive for division with respect to addition.

Thus, if *a*, *b*, and *c* name three whole numbers and $c \neq 0$, then $(a + b) \div c = (a \div c) + (b \div c)$, provided that $a \div c$ and $b \div c$ name whole numbers. For example:

$$\begin{aligned} 24 \div 3 &= (18 + 6) \div 3 \\ &= (18 \div 3) + (6 \div 3) \\ &= 6 + 2 \\ &= 8 \end{aligned}$$

> **10.11**
> When dividing by the distributive method, begin at the left of the dividend. To model the procedure, partition the set indicated by the dividend into the number of equivalent subsets indicated by the divisor.

The distributive algorithm is an application of the distributive property. For example:

$$3\overline{)96} \rightarrow 3\overline{)9 \text{ tens} + 6 \text{ ones}}^{\,3 \text{ tens} + 2 \text{ ones}} = 32 \quad \text{or}$$

$$\begin{aligned}(9 \text{ tens} + 6 \text{ ones}) \div 3 &= (9 \text{ tens} \div 3) + (6 \text{ ones} \div 3)\\ &= 3 \text{ tens} + 2 \text{ ones}\\ &= 32\end{aligned}$$

This leads to more efficient forms of notation.

```
  32
   2
  30         32
3)96   and 3)96
```

While developmental forms of the algorithm are being used to help each child understand the procedure, a child may subtract *any* recognized multiple of the divisor.

```
   34
    4
   10
   20
8) 272
 - 160
   112
 -  80
    32
 -  32
     0
```

This developmental form is often called the *pyramid algorithm.* A child does not have to erase even when the quotient figure is underestimated. When she is able to estimate the largest possible figure for each partial quotient, transition to the traditional algorithm is simple. Zeros are omitted and only the final quotient is recorded, one digit at a time.

```
      324
        4
       20
      300                          324
 24) 7776                     24) 7776
    -7200 ← (300 × 24)   or       7200
      576                          576
    - 480 ← (20 × 24)            - 480
       96                           96
    -  96 ← (4 × 24)             -  96
        0                            0
```

The first step in the traditional algorithm is to determine the number of digits in the quotient. To do this, multiply the divisor by powers of 10.

```
    xx                      xxx
6) 276                 28) 8988
6 × 10 = 60            28 × 10 = 280
6 × 100 = 600          28 × 100 = 2800
                       28 × 1000 = 28,000
```

Since 6 × 100 > 276, the quotient will contain 2 digits.

Since 28 × 1000 > 8988, the quotient will have 3 digits.

The second step is to find each digit in the quotient. The procedure is relatively easy for single-digit divisors. Knowledge of basic facts and the ability to multiply with multiples of powers of 10 are used. For example, in 6)276 the quotient is clearly a two-digit number. We observe that 6 × 40 = 240, but 6 × 50 = 300, more than 276. Hence, we write 40 or 4 tens in the quotient. Recall of a basic fact provides the other digit.

```
    46
     6
    40                        46
6) 276                    6) 276
  -240   ← (40 × 6) →        240
    36                        36
  - 36   ← (6 × 6) →        - 36
```

When a child divides by *multi*digit divisors, the process is more difficult, as she must estimate to find the quotient figure. For example, to find the quotient of 28)8988, the divisor is rounded to 30 and the first two digits in 8988 to 90. Then, the child thinks, 90 hundreds divided by 30 equals 3 hundreds. This process is repeated for each of the partial quotients. The final written form looks like this:

```
      321
        1
       20
      300                          321
 28) 8988                     28) 8988
    -8400  ← (300 × 28) →        -8400
      588                          588
    - 560  ← (20 × 8) →          - 560
       28                           28
    -  28  ← (1 × 8) →           -  28
        0                            0
```

In this example, the rounding process yields the correct digit for the quotient. In other examples,

EXCURSION
Subtractive vs. Distributive

The relative effectiveness of these two procedures has been studied over the past 30 years. Van Engen and Gibb reported some advantages to the study of each.[2] The following were among their conclusions:

1. Low-ability children taught the subtractive method had a better understanding of the division process than those taught the distributive process.
2. Use of the subtractive method was more effective in enabling children to transfer to unfamiliar but similar situations.
3. Children taught the distributive method achieved higher problem-solving scores.
4. The two procedures appeared to be equally effective on measures of retention for skill and understanding.

Dilly found significant differences on an application test favoring the use of the subtractive algorithm with fourth graders; however, differences favored the distributive algorithm on the retention test.[3] Kratzer and Willoughby discovered significant differences favoring the standard or distributive algorithm on immediate retention, and delayed retention tests.[4] More recently, Jones found significant differences favoring the use of the distributive algorithm by fifth and sixth grade children.[5]

---

such as the three below, the estimated quotient may not be the real quotient.

13) 416　　24) 1776　　39) 3666

Problems of this type are by far the most difficult. Since the first estimate will not yield the correct quotient, additional trials are required. For example:

Step 1: The first estimate is 40. The quotient is too large.

```
    40
13) 416
    520
```

Step 2: Reduce the estimate by 10. Now the estimate is fine.

```
     30
13)  416
    -390
      26
```

Step 3: Complete the division.

```
     32
      2
     30
13)  416
    -390
      26
    - 26
       0
```

Because there is a wide range of difficulty, we must provide graduated examples.

Division with Remainders

When the divisor is not a factor of the dividend, a remainder results. The remainder is always *less* than the divisor.

2. Henry Van Engen and E. Glenadine Gibb, *General Mental Functions Associated with Division*, Educational Studies, No. 2 (Cedar Falls, Iowa: Iowa State Teachers College, 1956), pp. 87–88.
3. Clyde A. Dilly, "A Comparison of Two Methods of Teaching Long Division" (Ph.D. diss., University of Illinois at Urbana–Champaign, 1970), *Dissertation Abstracts International* 31A (November 1970): 2248.
4. Richard O. Kratzer and Stephen S. Willoughby, "A Comparison of Initially Teaching Division Employing the Distributive and Greenwood Algorithms with the Aid of Manipulative Materials," *Journal for Research in Mathematics Education* 4, no. 4 (November 1973): 197–204.
5. Wilmer L. Jones, "*An Experimental Comparison of the Effects of the Subtractive Algorithm Versus the Distributive Algorithm on Computation and Understanding of Division*" (Ph.D. diss., University of Maryland, 1976).

The way a remainder is interpreted depends on the problem situation.

1. The remainder may be *ignored.*

PROBLEMS

Carol baked 50 cookies. She put them in packages of 6 each. How many packages did she make? Since 2 (the remainder) is not enough to fill another package, the answer is 8 packages. (measurement situation)

Twenty-five children are to play a game. The game requires 3 teams of the same size. What is the size of each team? Each team will contain 8 players. One child will have to sit this game out. (partitive situation)

2. The remainder may be *written as a fraction.*

PROBLEMS

Christa has 21 candy bars. She wishes to divide them equally among 6 friends. How many will each get? Each person will get 6 whole candy bars and one-half of another bar. The answer can be expressed as $6\frac{1}{2}$. (partitive situation)

A bus can only carry 35 children. How many buses are required to carry 95 children? In this situation, 3 buses will be required. (measurement situation)

3. The remainder may be *used to increase the answer.*

PROBLEM

Apples are 3 for 49 cents. What is the cost of 1 apple? Technically speaking, the cost of 1 apple is $16\frac{1}{3}$ cents; however, the seller rounds off to the next highest cent. The cost of one apple is 17 cents. (partitive situation)

We need to provide different types of problem situations regularly, so that each child will understand the ways a remainder can be interpreted.

Checking Division

Kennedy notes that the procedures for checking division are unpopular with children. The procedures often take as much time as the original example. A thorough understanding of the procedure and a good knowledge of basic facts is essential.[6]

One checking procedure is an application of the following content objective.

> **10.12**
> For any division statement, the dividend is equal to the product of the divisor and the quotient, plus the remainder.

To check the answer to $28\overline{)2795}$, this idea can be applied as follows:

```
     97 R26           28   Divisor
28) 2742            × 97   Quotient
    2520             196
     222            2520
     196            2716
      26           + 26    Remainder
                    2742   Dividend
```

Another way to check division is to exchange factors: to use the quotient as a divisor and divide again. The new quotient should be the same as the original divisor, along with any remainder resulting from the original division.

```
      46 R12             37 R12
37) 1714            46) 1714
    148                 138
     234                 334
     222                 322
      12                  12
```

Ordering the Content

In preparing to teach multiplication and division computation, consider the sequence in which the skills and concepts involved should be presented. The subject matter components are not fixed in

6. Leonard M. Kennedy, *Guiding Children to Mathematical Discovery,* 3rd ed. (Belmont, Calif.: Wadsworth, 1980), p. 277.

one specific sequence; however, certain concepts and skills are clearly dependent on others.

Before introducing algorithms, make sure each child understands prerequisite concepts and can do the skills required, such as basic facts, meanings of the operations, place value, and properties of operations. Some concepts and skills may need to be reviewed and reinforced before multistep procedures are taught.

From the following sequential lists, selected skills can be clustered for instruction, for example, multiplication skills 4, 5, and 10. You may discover additional skills or develop your own lists.

Multiplication

1. Basic facts with 1–9 as factors
2. Checking
3. Basic facts with zero as a factor
4. One-place multiplier and 10 as multiplicand (2 × 10)
5. One-place multiplier, tens as multiplicand (4 × 50)
6. One-place multiplier and two-place multiplicand, no renaming (3 × 12)
7. Same, with renaming of tens and two-place product (2 × 46)
8. Same, with renaming and three-place product (4 × 43)
9. Three-place multiplicand, no renaming (3 × 123)
10. Three-place multiplicand, hundreds as multiplicand (4 × 100, 4 × 200, 5 × 500)
11. Zero in three-place multiplicand, no renaming (2 × 404, 2 × 440)
12. Renaming of ones as tens, three-place multiplicand (2 × 236)
13. Renaming of tens as hundreds, three-place multiplicand (3 × 162)
14. Two-place multiplicand, tens as multiplier (30 × 12, 40 × 26)
15. Two-place multiplicand by two-place multiplier, no renaming (13 × 12)
16. Two-place multiplicand by two-place multiplier, with renaming of ones as tens (23 × 24)
17. Omission of zero in ones place of partial products (optional)
18. Three-place multiplicand by two-place multiplier (38 × 244)
19. Three-place multiplicand by three-place multiplier (325 × 647)
20. Same, with zero in the tens place in multiplier (209 × 457)
21. Four-place multiplicand by four-place multiplier (1397 × 5426)

Division

1. Basic facts with 1–9 as factors (such as 27 ÷ 3 = 9)
2. Zero dividend by any number (0 ÷ 2)
3. Tens divided by one-place divisor, no remainder (40 ÷ 2)
4. Two-place dividend by one-place divisor, no renaming or remainder (46 ÷ 2)
5. Remainders in division (47 ÷ 2)
6. Three-place dividend, one-place divisor, two-place quotient (128 ÷ 4)
7. Renaming in division with two-place dividend (75 ÷ 3)
8. Three-place dividend, one-place divisor, three-place quotient (369 ÷ 3)
9. Thousands divided by one-place quotient (8000 ÷ 2)
10. Four-place dividend, four-place quotient (9233 ÷ 4)
11. Zero in ones place in quotient (63 ÷ 6 = 10 R3)
12. Zero in tens place in quotient (324 ÷ 3 = 108)
13. Two zeros in quotient (8016 ÷ 8 = 1002)
14. Tens divided by tens (80 ÷ 20)
15. Three-place dividend, tens divisor, one-place quotient (120 ÷ 20)
16. Two-place dividend, two-place divisor, one-place quotient (82 ÷ 21) and three-place dividend, two-place divisor, one-place quotient (126 ÷ 42)
17. Same, with remainders (83 ÷ 25, 129 ÷ 42)
18. Three-place dividend, two-place divisor and quotient (276 ÷ 12)
19. Four-place dividend, two-place divisor and quotient (1512 ÷ 42)
20. Four-place dividend, two-place divisor, and three-place quotient (4536 ÷ 21)
21. Four- or five-place dividend, zero in tens place in quotient (4692 ÷ 23 = 204)

PLANNING INSTRUCTION

The cycle of instructional activities should be considered in teaching each child to multiply and divide. As we plan we need to know which concepts and skills each child has already learned.

Initiating, Abstracting, and Schematizing Activities

Multiplication and division algorithms can be introduced before all children have completely mastered the basic facts. Practice with the basic facts can be provided through work with the algorithms. However, use easier facts, posting the more difficult ones in conspicuous places; otherwise, attention given to specific facts may distract from learning the computational procedure.

Related multiplications and divisions, such as 3 × 12 and 36 ÷ 3, are often taught consecutively. However, in order to analyze the instructional sequences we will consider activities for teaching multiplication and division algorithms separately.

Multiplication Algorithms

An example with a factor that is a multiple of 10 is usually used in introducing multiplication algorithms. Relate products such as 3 × 40 to basic facts, and have children look for patterns. For example, when multiplying tens by ones, children need to see that the number of ones times the number of tens is a number of tens; basic multiplication facts are used to find the product. This can be suggested with base 10 blocks (see below).

We should develop a similar pattern with examples in vertical form.

$$\begin{array}{r} 4 \\ \times\ 3 \\ \hline \end{array} \qquad \begin{array}{r} 4 \text{ tens} \\ \times \quad 3 \\ \hline 12 \text{ tens} \end{array} \qquad \begin{array}{r} 40 \\ \times\ 3 \\ \hline 120 \end{array}$$

A similar pattern can be shown for multiples of 100.

Present a problem situation that can be demonstrated with three-dimensional materials.

PROBLEM

Cookies are in packages of 10. Each of three children has 12 cookies—a package of 10 and 2 loose cookies. How many cookies do the three children have altogether?

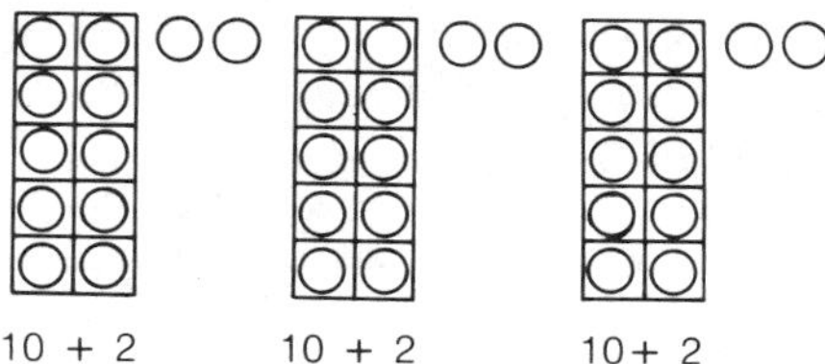

By rearranging the packages and loose cookies, a child can see that 3 sets of 12 are 36. We can record this with vertical notation, or with both horizontal and vertical forms.

$$\begin{array}{rr} 1 \text{ ten} & +\ 2 \text{ ones} \\ \times & 3 \\ \hline 3 \text{ tens} & +\ 6 \text{ ones} \end{array}$$

$$\begin{array}{r} 10 + 2 \\ \times \quad 3 \\ \hline 30 + 6 = 36 \end{array} \qquad \begin{array}{l} 3 \times (10 + 2) \\ (3 \times 10) + (3 \times 2) \\ 30 + 6 \\ 36 \end{array}$$

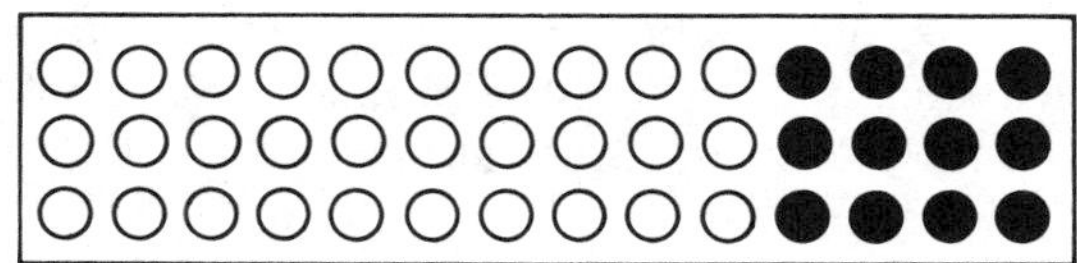

Markers on a magnetic or flannel board can effectively demonstrate the distributive property.

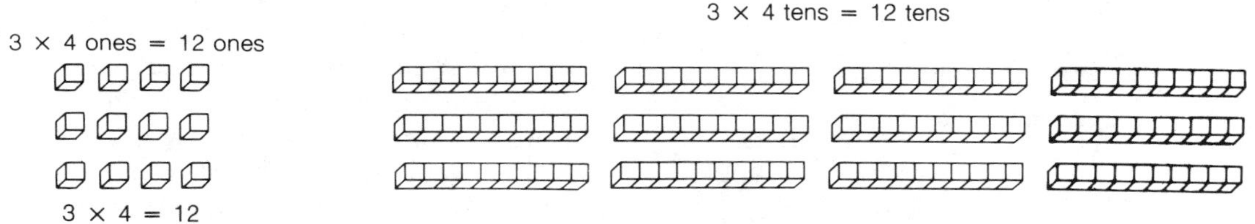

The thought process leads conveniently to multiplication that involves renaming.

$$
\begin{aligned}
3 \times 14 &= 3 \times (10 + 4) \\
&= (3 \times 10) + (3 \times 4) \\
&= 30 + 12 \\
&= 42
\end{aligned}
\qquad
\begin{array}{r}
10 + \;\; 4 \\
\times \qquad 3 \\
\hline
30 + 12 \\
42
\end{array}
\qquad
\begin{array}{r}
14 \\
\times\; 3 \\
\hline
12 \\
30 \\
\hline
42
\end{array}
$$

As we observe the partitioned array and record the multiplication algorithm step by step, we need to be certain that each child understands that he is first multiplying ones and then tens. For example:

	Tens	Ones
	2	8
Multiply the ones.	×	3
(3 × 8 ones = 24)	2	4
Multiply the tens.		
(3 × 2 tens = 60)	6	0
Then add.		
(60 + 24 = 84)	8	4

Children must understand this process to avoid incorrect procedures such as these:

$$
\begin{array}{r}
28 \\
\times\; 3 \\
\hline
624
\end{array}
\qquad
\begin{array}{r}
28 \\
\times\; 3 \\
\hline
24 \\
6 \\
\hline
30
\end{array}
\qquad
\begin{array}{r}
{}^{2}28 \\
\times\; 3 \\
\hline
124
\end{array}
$$

A place value chart or an abacus can be used similarly. To find the product of 2 and 48, note the following sequence:

Step 1. Represent 48 as 4 tens and 8 ones on a place value chart. Show 48 *two* times.

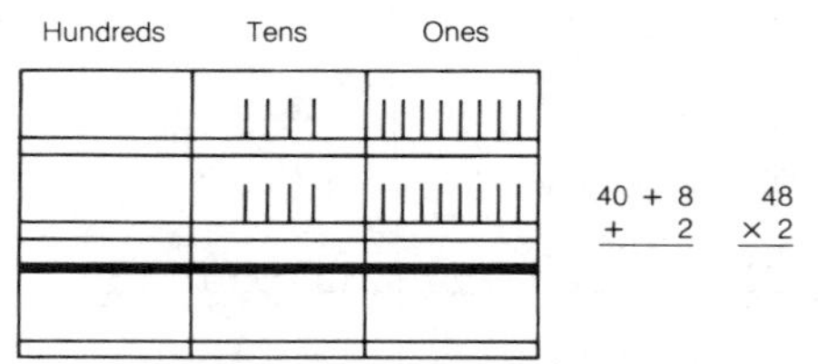

Step 2. To find the product, first multiply the ones by 2.

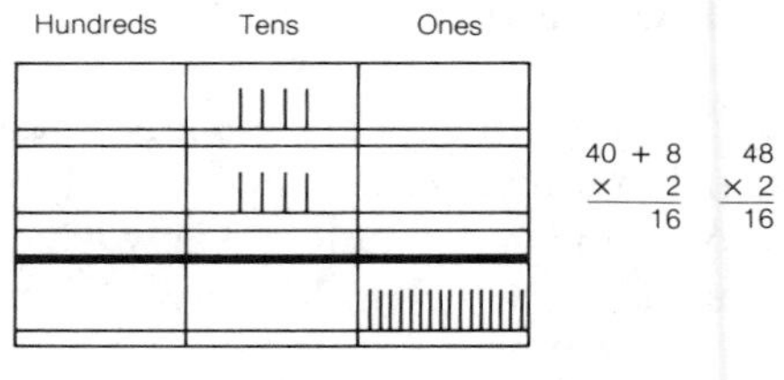

Step 3. Multiply the tens by 2.

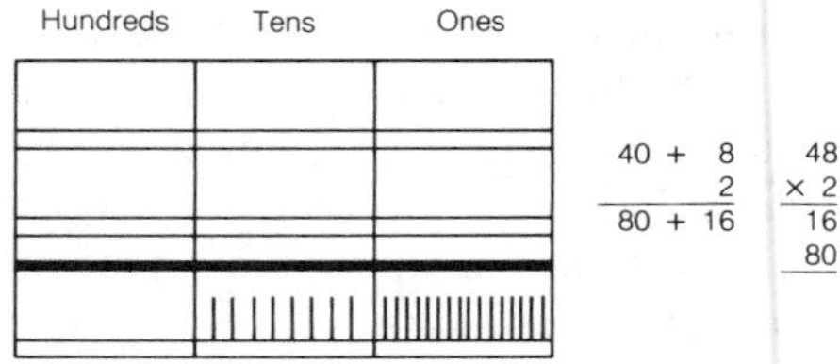

Step 4. Find the standard name for 80 + 16. Exchange 10 ones for 1 ten; rename 8 tens and 16 ones as 9 tens and 6 ones.

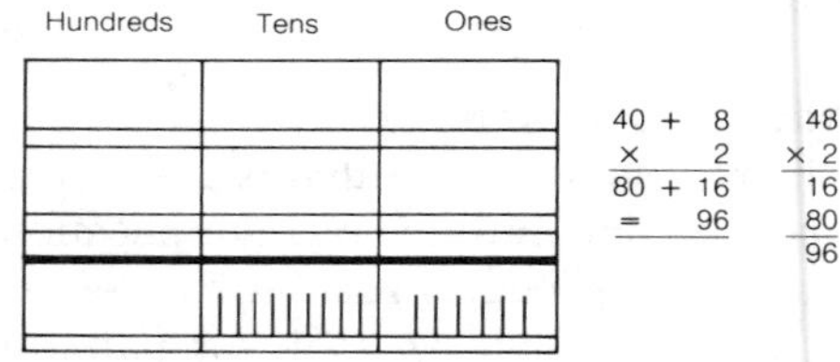

When children can multiply in parts, that is, after they can apply the distributive property, they are ready for the short form of the procedure. In this form, a child must remember numbers of the next higher place value. Initially, children can record the number to be remembered.

$$
\begin{array}{r}
{}^{2} \\
27 \\
\times\; 3 \\
\hline
81
\end{array}
\qquad
\begin{array}{r}
{}^{3} \\
382 \\
\times\; 4 \\
\hline
1582
\end{array}
\qquad
\begin{array}{r}
{}^{4\,3} \\
275 \\
\times\; 6 \\
\hline
1650
\end{array}
$$

Three-place multiplication can be introduced soon after two-place multiplication, since no significant new concepts are involved.

As a first step in the transition to use of the written procedure alone, ask a child to say what she would do with blocks or straws as you make the written record. Materials are present as a reference, but they are not manipulated.

EXCURSION
Multiplication Gotcha[7]

Write each number from 0–9 on two squares of paper. Mix the numbers up in a box.

1. Each player makes a score card (see example). Play 3 games and add the total scores. The player with the highest total wins.
2. A leader draws digits from the box, one at a time. They are to be multiplied as they are drawn to get the total.
3. When 0 is drawn, it is a gotcha. Each player still in the game gets a 0 for that game.
4. A player can go out whenever she chooses. Try to get a big score but go out before 0 is drawn.

One Player's Score

Game 1	Game 2	Game 3
3	2	6
× 8	× 9	× 7
24	18	42
× 5	6	× 8
120 Go out	108	336
	× 0	× 4
	0 Gotcha	1344

Score Card	
1)	120
2)	0
3)	1344
Total:	1465

The distributive property underlies the multiplication algorithm. For many children, distributivity is not likely to be a stable concept until sixth grade or later.[8] Plan carefully in teaching children to multiply in parts.

Examples such as 30 × 60 and 26 × 40 are special types of two-digit multiplication and should be taught prior to the two-digit algorithm, as they do not require use of the usual algorithm. Hazekamp gives the following reasons.[9]

- They can be calculated more efficiently with a one-step procedure than by following all steps in the two-digit algorithm.
- Finding the product of a multiple of 10 and a two-digit number is a skill directly applicable to the two-digit algorithm.
- Pupils recognize the special types more easily if they are taught prior to two-digit algorithm instruction.

Children can apply what they have learned about commutativity and associativity for multiplication: the product 30 × 60 can be thought of as (3 × 10) × (6 × 10), and the factors rearranged as (3 × 6) × (10 × 10), or 18 × 100. When 26 × 40 is factored and rearranged as (4 × 26) × 10, the procedure for multiplying by a one-digit multiplier applies.

Children can see the reasonableness of the general procedure for multiplying two-digit numbers when they consider rectangular regions and an expanded algorithm. The four regions of the rectangle visually represent the four partial products. For a child's first experience with this concept, construct a large array and draw the rectangle around it. This helps her understand that the

7. Adapted from Mervin L. Keedy et al, *General Mathematics* (Menlo Park, Calif.: Addison-Wesley, 1980), p. 67.

8. Douglas H. Crawford, "An Investigation of Age-Grade Trends in Understanding the Field Axioms" (Ph.D. diss., Syracuse University, 1964), *Dissertation Abstracts* 25 (1965): 5728–29.

9. Donald W. Hazekamp, "Teaching Multiplication and Division Algorithms," *Developing Computational Skills,* 1978 Yearbook of the National Council of Teachers of Mathematics (Reston, Va.: National Council of Teachers of Mathematics, 1978), pp. 105–7.

that the key to estimating lies in choosing the part that contributes most to the product. In the following case, 30 × 20 represents the largest part: however, it is obviously less than the exact answer.

Example: 24 × 32

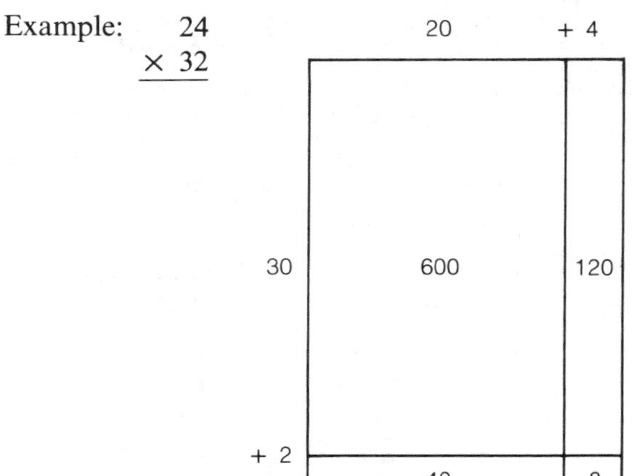

Long Form		Short Form
24		24
× 32		× 32
8	← (2 × 4)	48
40	← (2 × 20)	720
120	← (30 × 4)	768
600	← (30 × 20)	
768	← (32 × 24)	

Show how skills learned earlier are a part of the process. For example, to find the product of 32 × 24 we might record the following steps:

Step 1: Multiply 24 by 2.	24		24
	× 2		× 32
	48	→	48
Step 2: Multiply 24 by 30.	24		
	× 30		
	720	→	720
Step 3: Add the partial products.			768

In earlier exercises minimize difficult multiplication facts and renaming.

Schematizing activities help a child understand not only the steps in the procedure but their sequence, and various relationships between the different parts of the algorithm. These activities also illustrate the general pattern that applies to any two whole numbers. During schematizing activities, ideas already understood are applied; for example, knowledge of the additive identity leads to the conclusion that a row of zeros (a partial product) can be omitted.

Procedures for more complex examples involve additional applications of numeration ideas and extensions of the basic algorithm. Give special emphasis to problems involving zero since they are often a source of error. (Multiplication by zero can be recorded as a complete partial product.) Children who understand place value ideas will probably want to use the second or third of the following forms.[10]

Example 1	Example 2	Example 3
328	328	328
× 403	× 403	× 403
984	984	984
000	131200	1312
131200	132184	132184
131184		

When teaching an algorithm, stress the form and its underlying concepts. When an algorithm is correctly developed, students are usually able to extend the procedure to complex examples.[11]

Division Algorithms

A division procedure is useful any time a product and one factor are known and the unknown factor is required. In teaching division, you need to be familiar with both subtractive and distributive algorithms. Although your textbook or curriculum guide may emphasize one, children may be familiar with the other, having questions or concepts related to same.

When you introduce the subtractive algorithm, children soon learn that they can always find the quotient by subtracting the known factor (divisor) repeatedly from the product (dividend).

10. Donald D. Paige et al., *Elementary Mathematical Methods* (New York: John Wiley, 1978), pp. 149–50.
11. *Ibid.,* p. 150.

EXCURSION
Lattice Multiplication

The lattice method of multiplication was a forerunner to the modern algorithm and requires less mental effort.

To multiply 378 × 43, write the factors on the top and the right edges of a 2 × 3 array of squares with diagonals as in Step 1. Then record the product of each one-digit number along the top with those along the side, as seen in Step 2. Finally, add the partial products along the diagonals. Find the product (Step 3) by reading the digits along the left-hand side and the bottom. The product for 378 × 43 is 16,254.

Note that the product of each pair of digits is written individually with no renaming. The partial products are self-aligning, thus eliminating another common student error.

Find the product for 724 × 38 with this method, and think about ways you can use lattice multiplication in teaching.

Step 1

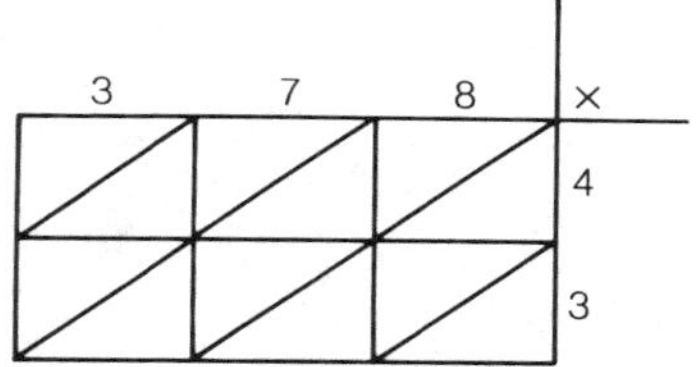

Step 2

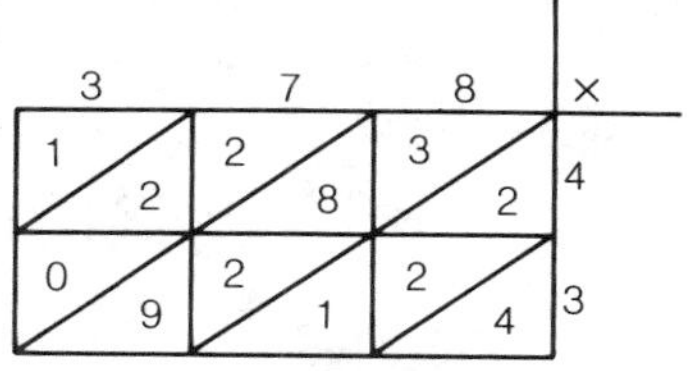

Step 3

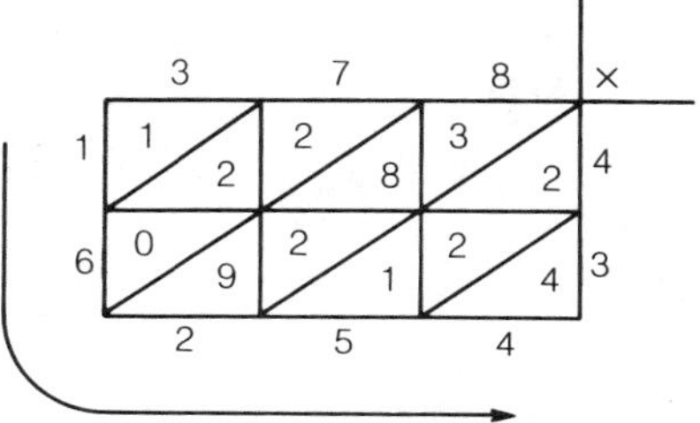

EXCURSION
Low Stress Multiplication

By using a special drop notation, Hutchings has developed a variation of lattice multiplication that does not require a grid.[12] Note that multiplication can be done from right to left or left to right.

Drop Notation	Place Values Aligned (right to left)		Place Values Different Below the Bar (left to right)	
8	85			852
× 3	× 3			× 34
2	21	3 × 5 = **15**	3 × 8 = **24**	**210**
4	**45**			**456**
	255			320
				208
				28968

Inability to perform the required multiplication and subtraction is the chief restriction in doing a particular computation.[13]

Remember to introduce subtractive methods with problem situations in measurement division. Once the child understands that the subtractive algorithm is patterned after this type of division, and gains some independence in working problems, include partitive situations.

PROBLEM

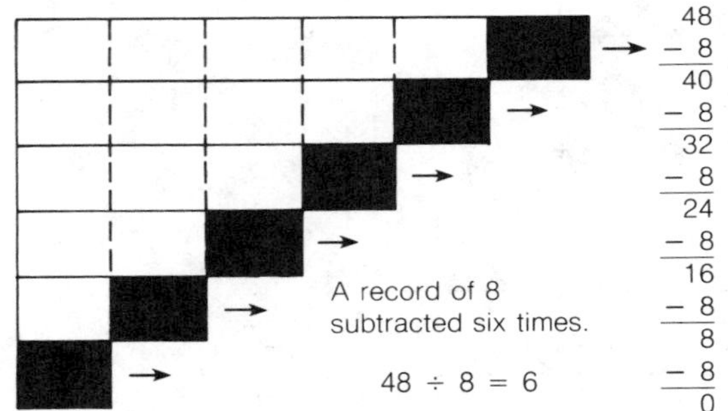

Bill has a piece of ribbon 48 inches long. How many 8-inch prize ribbons can be cut from the longer piece?

After you demonstrate the recording procedure, ask each child to record subtractions for other examples. By using a measurement problem situation, each child is given a rationale for the specific steps involved.

Children can use chips or beans to solve the simple division problems used to introduce the subtractive algorithm. The recording format illustrated below is one way of showing how many sets are removed each time. As children remove sets and record, their solutions will vary.

	Action	*Record*	
Step 1:	Remove 3 sets of 4.	4) 27	
		− 12	3 × 4
		15	

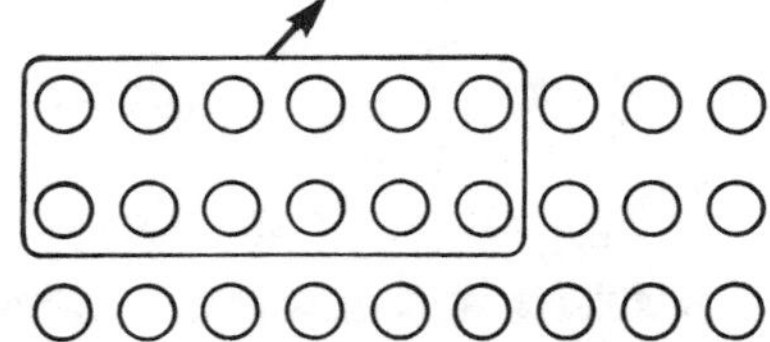

12. Barton Hutchings, "Low-Stress Algorithms," *Measurement in School Mathematics,* 1976 Yearbook of the National Council of Teachers of Mathematics (Washington, D.C.: National Council of Teachers of Mathematics, 1976), pp. 218–39.
13. George F. Green, *Elementary School Mathematics, Activities and Materials* (Boston, Mass.: D.C. Heath, 1974), p. 240.

Step 2: Remove 2 sets of 4.

4) 27	
− 12	3 × 4
15	
− 8	2 × 4

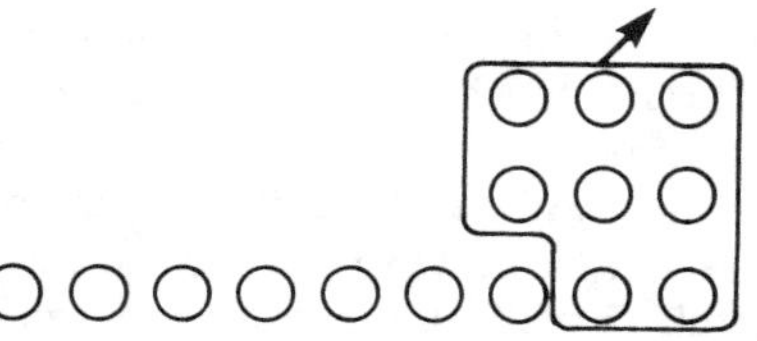

Step 3: Remove 1 set of 4.

6 R3	
) 27	
− 12	3 × 4
15	
− 8	2 × 4
7	
− 4	1 × 4
3	6 × 4

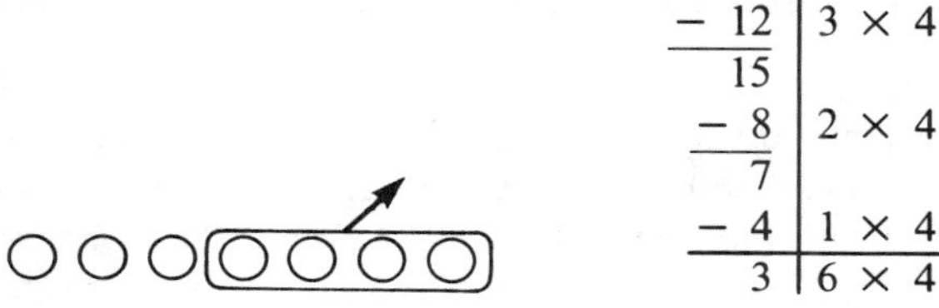

Conclusion: Altogether, six sets of 4 have been removed, and 3 remain. The quotient is 6 R3.

For children who have difficulty writing the complete record, provide partially completed solutions.

?	
4) 35	
−16	?
19	
−12	?
7	
− 4	?
3	?

?	
8) 42	
− ?	2 × 8
26	
− ?	2 × 8
10	
− ?	1 × 8
2	?

?	
6) 50	
−24	?
26	
?	3 × 6
?	
6	?
?	?

Once children are familiar with the mechanics of the algorithm, they are ready for more complex division situations. Tell them that with larger products or dividends, it is easier to remove *ten* sets at a time, or a *hundred* sets if possible. For the division 4)92, one possible solution is illustrated below.

Think — *Record*

Step 1: I can remove 10 sets of 4.

4) 92	
−40	10 × 4
52	

Step 2: I can remove another 10 sets of 4.

4) 92	10 × 4
−40	
52	
−40	10 × 4
12	

Step 3: I cannot remove another set of 10, but I *can* remove 3 sets of 4.

4) 92	
−40	10 × 4
52	
−40	10 × 4
12	
−12	3 × 4
0	23 × 4

Conclusion: 23 sets of 4 have been removed, with no remainder.

Record several solutions on the chalkboard and ask: Which is easiest? Which takes less time? Help children refine their solutions by using larger multiples of the divisor. Subtracting *multiples* of thousands, hundreds, and tens is more efficient and easier. The examples below illustrate stages of maturity through which a child may proceed. Note that as a child progresses, the record is simplified by omitting "times the divisor."

Less Mature — More Mature

24	
5) 120	
− 20	4 × 5
100	
− 20	4 × 5
80	
− 30	6 × 5
50	
− 30	6 × 5
20	
− 20	4 × 5
0	24 × 5

24	
5) 120	
− 50	10
70	
− 50	10
20	
− 20	4
0	24

24	
5) 120	
− 100	20
20	
− 20	4
0	24

To begin instruction with multidigit divisors, we may wish to have each child develop a table of partial quotients:

24	24	24	24	24	24	24	24	24
× 1	× 2	× 3	× 4	× 5	× 6	× 7	× 8	× 9
24	48	72	96	120	144	168	192	216

24	24	24	24	24	24	24	24	24
× 10	× 20	× 30	× 40	× 50	× 60	× 70	× 80	× 90
240	480	720	960	1200	1440	1680	1920	2160

EXCURSION
Representative Materials vs. Abstract Thinking

As the numbers involved in computation get larger, it becomes impractical to use representative materials to demonstrate the procedure. Each child should move to a more abstract level as soon as possible. Weaver states:

> If children always must resort to representative materials for aid in quantitative thinking, . . . [they] will fail to learn in a manner best suited to their increasing maturity and background of experience. . . . Our major purpose is to lead children to higher and higher levels of thinking, so that ultimately they will feel perfectly at home on the highest plane or level. We may resort to lower representative levels when these levels are found to make learning more meaningful. To make this regression unnecessarily is as detrimental to efficient learning as is the frequent practice of forcing children to move from lower to higher levels too rapidly.[14]

Then division problems of this type can be assigned:

24)216   24)288   24)816   24)1392   24)5136

This activity enables the child to directly focus on the selection of partial quotients without the distraction of estimating multiples. Of course, we eventually want each child to solve problems like 24)5136 by finding only the partial quotients needed for division.

You can help a child determine the largest multiple of 1000, 100, or 10 to be subtracted each time by asking how many thousands, tens, and, later, ones can be removed. For example, in the problem 8753 ÷ 36, you might say, "Can we subtract 1000 sets of 36 in this problem?" (No, 1000 × 36 = 36000. This is too large.) Then say "Can we subtract 100 sets of 36?" (Yes, 100 × 36 = 3600. This is less than 8753.) Say, "Can we subtract 200 sets of 36?" (Yes, 200 × 36 = 7200.) "Can we subtract 300 sets of 36?" (No, 300 × 36 = 10800, which is too large.) Then say "Our first partial quotient is 200." And so on.

The *distributive algorithm* is the most widely used procedure after the earlier grades. An understanding of the partitioning concept is especially useful for this algorithm. Problems such as the one below, when illustrated with concrete materials, can help.

PROBLEM
There are 15 baseball cards in the set. If they are to be divided equally among 3 boys, how many will each get?

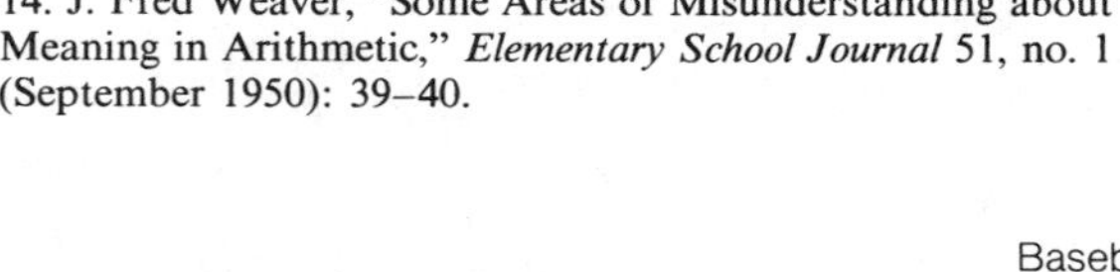

14. J. Fred Weaver, "Some Areas of Misunderstanding about Meaning in Arithmetic," *Elementary School Journal* 51, no. 1 (September 1950): 39–40.

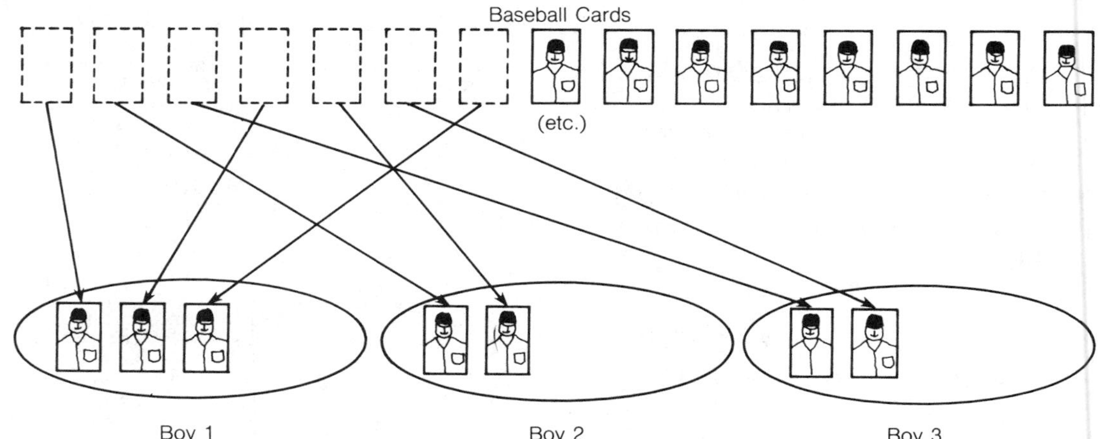

To illustrate the partitioning idea, deal the cards one at a time into three piles. Continue until all cards are given out. Children can easily see that $3\overline{)15}$ is 5. Include situations for which there is a remainder.

The difficulty of a division example usually depends upon the divisor, so one-digit divisors are introduced first. Problems like $4\overline{)80}$, $3\overline{)600}$ and $5\overline{)350}$ can be presented when children are learning to solve corresponding multiplication problems: 20 × 4, 200 × 3, and 70 × 5. Solutions should be based on the relationship between multiplication and division. Emphasize that dividends like 350 can be thought of as 35 tens. In this way, children see the close relationship between the divisor and dividend and the basic facts they already know. Manipulatives used with multiplication examples can be employed again with division.

As the distributive algorithm is introduced, continue to emphasize the partition interpretation of division. For $5\overline{)530}$, encourage each child to think, If 53 tens are to be separated into 5 matching parts, how many tens will be in each part?

Initially, children can use chips or base 10 blocks to act out simple divisions. Each child should gain experience with both concrete materials and in making records.

Action	Record
Step 1: Partition the tens.	$\begin{array}{r} 1 \\ 3\overline{)\,36} \\ -30 \\ \hline 6 \end{array}$ (1 ten in each set)

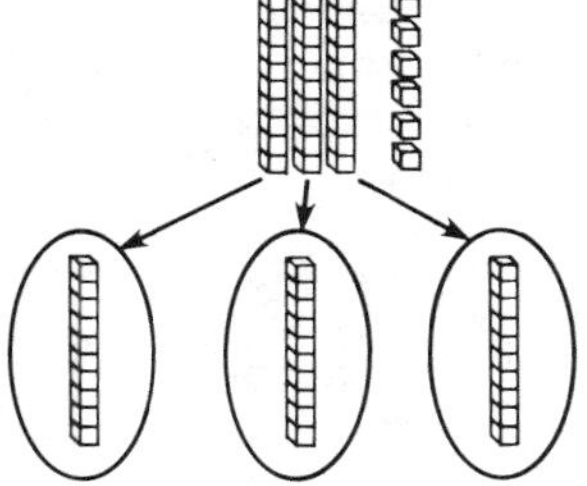

Action	Record
Step 2: Partition the ones.	$\begin{array}{r} 12 \\ 3\overline{)\,36} \\ -30 \\ \hline 6 \\ -\ 6 \\ \hline 0 \end{array}$ (2 ones in each set)

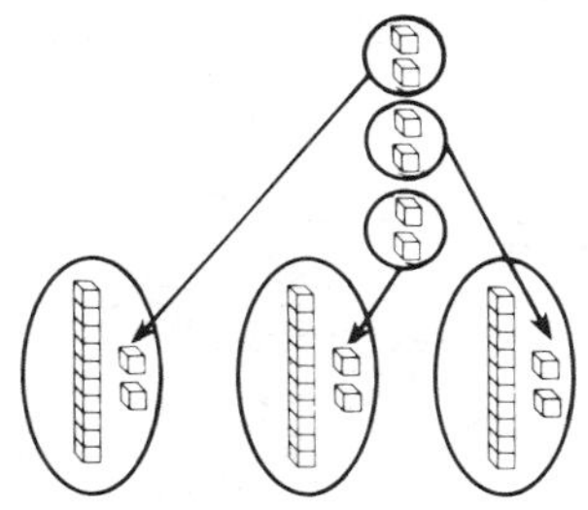

Conclusion: The quotient is 1 ten and 2 ones, or 12.

More difficult divisions that involve renaming may be demonstrated in the same way. In order to point to the algorithm encourage children to deal only one time around. Otherwise, they may just learn a mechanical routine for getting an answer. Ask them to use what they already know about multiplication to decide how many tens to place in each set. For example, for 92 ÷ 4, 2 tens can be given to each of the four sets, but 3 tens would be too many.

Action	*Record*
Step 1: Partition the tens.	$\begin{array}{r} 2 \\ 4\overline{)\,92} \\ -80 \\ \hline \end{array}$ (2 tens in each set)

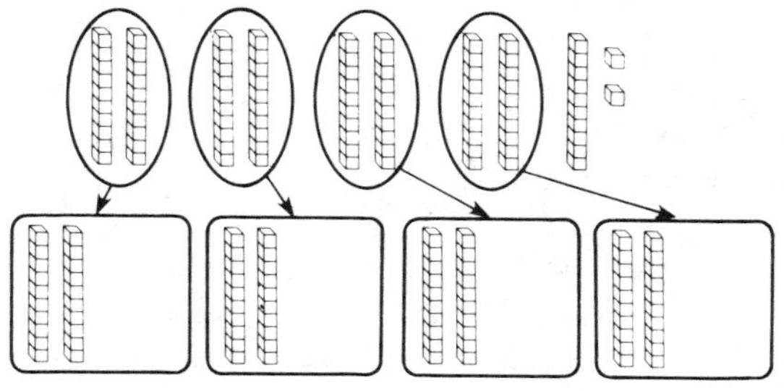

Action	*Record*
Step 2: Exchange 1 ten for 10 ones. Combine the 10 ones with the 2 existing ones to make 12 ones.	$\begin{array}{r} 2 \\ 4\overline{)\,92} \\ -80 \\ \hline 12 \end{array}$

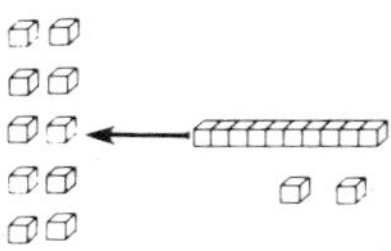

Step 3: Partition the ones.

```
     23  (3 ones in each set)
 4) 92
   -80
    12
   -12
     0
```

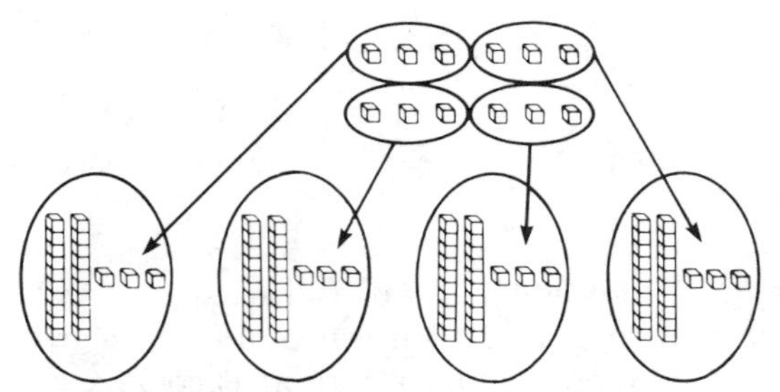

Conclusion: The quotient is 2 tens and 3 ones, or 23.

When children first learn to divide, they may have difficulty writing the complete procedure on their own. Partially completed solutions provide needed guidance.

```
      5 R?          ?3 R?          4?5 R4
 6) 34         4) 135         6) 2794
   - ?           -120           -   ?
     ?              ?             394
                 - 12             360
                    ?               ?
                                    ?
                                    4
```

The pocket chart which incorporates the concept of place having value can also be used. The steps involved in dividing 141 by 3 with the aid of a pocket chart follow. Some experience with expanded notation is usually included in abstracting activities for the division algorithm.

Step 1: Show 141 at the top of the pocket chart where the product (dividend) is shown.

Hundreds Tens Ones

```
3) 1 hundred + 4 tens + 1 one
```

Step 2: One hundred cannot be partitioned into 3 equivalent subsets, so regroup 1 hundred as 10 tens.

Hundreds Tens Ones

```
3) 14 tens + 1 one
```

Step 3: Separate the 14 tens into 3 matching subsets in the pockets below the product. Each subset contains 4 tens. There are 2 tens and 1 one remaining in the product.

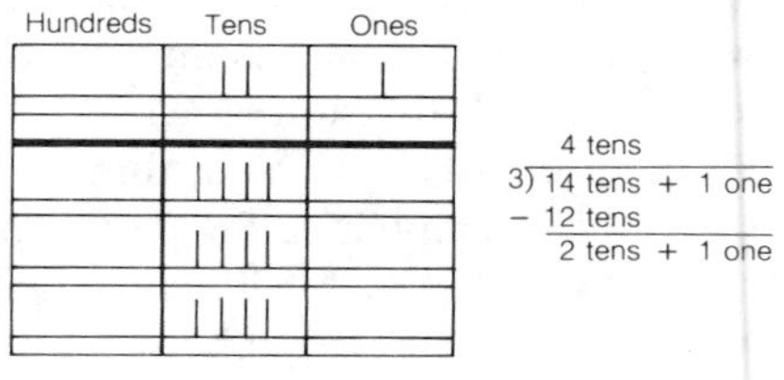

Step 4: Two tens cannot be partitioned into 3 equivalent subsets, so regroup each ten as 10 ones.

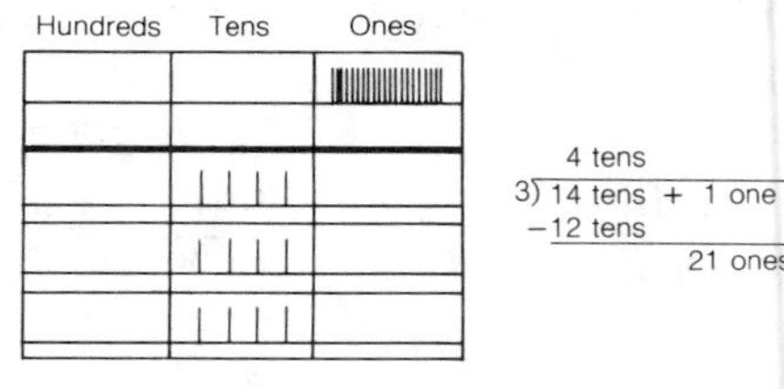

Step 5: Separate the 21 ones into the 3 matching subsets. Each subset receives 7 ones.

Hundreds Tens Ones

```
     4 tens + 7 ones
3) 14 tens + 1 one
  -12 tens
          21 ones
         -21 ones
           0
```

Conclusion: The quotient is 4 tens + 7 ones, or 47.

EXCURSION
Number Detective

Here are two partially completed examples. Can you find the missing digits?

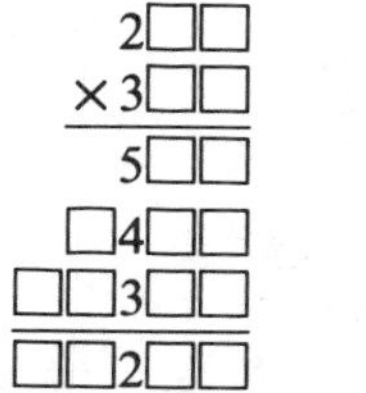

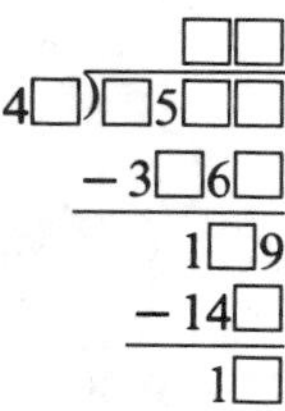

When children are able to divide with the use of manipulatives, have them estimate the product in order to determine the positions of digits in the quotient. Emphasize that the powers of ten are the key. Introduce the following pattern of thinking, as illustrated for 8)4372. Say, "Are there as many as 10 in each of the 8 groups?" (Yes, because 8 × 10 is 80.) Then ask, "Are there as many as 100?" (Yes, 8 × 100 is 800.) Say, "Are there as many as 1,000?" (No, because 8 × 1000 is 8000, and that is larger than the dividend. Since the quotient is greater than 100 but less than 1000, it must contain 3 digits: ones, tens, and hundreds. The first digit to be placed in the quotient will be above the hundreds place in the dividend.)

Organize the work as shown below to help each child make estimates for the quotient.

8 × 10 = 80	- - -	(3 digits in
8 × 100 = 800	8) 4372	quotient)
8 × 1000 = 80000		

EXCURSION
Finding Trial Quotients[15]

A calculator game can be used to help children develop their estimating skills and, at the same time, help them learn to determine trial quotients for division. Use the following or make up your own examples.

23)72,345	63)10,469	603)14,503
909)71,725	7123)18,861	14)90,863

- Start by selecting one of the division examples.
- Have each player write down her estimate of the answer.
- Use the calculator to determine the exact answer.
- Each player scores 1 point if her estimate has the correct number of digits, and 2 points if the estimate has both the correct number of digits and the correct first digit.
- Continue the game until a predetermined score is reached.

15. Adapted from "Check It," in Earl Ockenga, "Calculator Ideas for the Junior High Classroom," *The Arithmetic Teacher* 23, no. 7 (November 1976): 519.

In the same way, quotients for more difficult divisions can be estimated.

28 × 1 = 28
28 × 10 = 280
28 × 100 = 2800

$$\begin{array}{r} \text{- -} \\ 28\overline{)\,1976} \end{array}$$ (2 digits in quotient)

A child who can determine the number of digits in the quotient may find it helpful to mark the place to indicate each digit.

As we focus on the algorithm itself, children need to refine their estimates for the quotient. For example, to find the first trial quotient in the example $4\overline{)145}$, a child thinks, Four times what number is equal to 14 . . . or a little less than 14? (Not 4 "goes into" 14 . . .) Children can also conceal those digits that are irrelevant at a given step in the dividend.

$$\begin{array}{r} 3 \\ 4\overline{)\,145} \\ -120 \\ \hline 25 \end{array}$$

There are two distinct ways to proceed from this point. We can teach children to reason, 3 tens times 4 is 12 tens or 120. They write 120 and subtract it from 145 to find the amount yet to be divided. Or, if we teach the second approach, only the 12 is recorded, then subtracted from 14. In the latter, 5 has to be brought down before the second partial quotient can be found. Green notes that:

> Both methods can be equally reasonable to children with sufficient reference to manipulative activities. The first is mathematically a little more straightforward . . . and is preferred by some on that basis; the second generally involves conscious manipulation with smaller numbers at each step, which is why some prefer it. The first method is sometimes introduced and later replaced by the second.[16]

Before moving on to more difficult divisions, make certain that each child has abstracted and schematized the processes involved in the algorithm. Directions for the algorithm can be outlined in a series of steps as follows:

Step 1: Estimate the quotient.

6 × 10 = 60
6 × 100 = 600
6 × 1000 = 6000

$$\begin{array}{r} \text{- - -} \\ 6\overline{)\,1572} \end{array}$$

The quotient will have 3 digits.

Step 2: Find the first trial quotient. Estimate the quotient by thinking, Six times what number is equal to 15 . . . or a little less than 15?

$$6\overline{)1572}$$

Step 3: Write the hundreds in the quotient. Multiply and subtract.

$$\begin{array}{r} 2 \\ 6\overline{)\,1572} \\ -1200 \\ \hline 372 \end{array}$$

Step 4: Find the second trial quotient. Estimate the quotient by thinking, Six times what number is equal to 37 . . . or a little less than 37?

$$\begin{array}{r} 2 \\ 6\overline{)\,1572} \\ -1200 \\ \hline 372 \end{array}$$

Step 5: Write the tens in the quotient. Multiply and subtract.

$$\begin{array}{r} 26 \\ 6\overline{)\,1572} \\ -1200 \\ \hline 372 \\ -\ 360 \\ \hline 12 \end{array}$$

Step 6: Find the third trial quotient. Estimate the quotient by thinking, Six times what number is equal to 12? Write the ones in the quotient. Multiply and subtract.

---

16. Green, *Elementary School Mathematics*, p. 233.

```
    262
6) 1572
  -1200
    372
  - 360
     12
  -  12
      0
```

A common error in using the distributive algorithm involves proper placement of zeroes in the quotient. (This difficulty is not as common in the subtractive algorithm.) One effective way to avoid this error is to have children estimate the quotient before beginning to divide. They should also get into the habit of checking.

The importance of understanding underlying place value ideas is seen clearly as children work with multidigit divisors. The first two-digit divisors should be multiples of 10. Then, to ease the transition to multidigit divisors that are not multiples of 10, have each child build and use a table of multiples similar to those described earlier.

48	48	48	48	48	48	48	48	48
× 1	× 2	× 3	× 4	× 5	× 6	× 7	× 8	× 9
48	96	144	192	240	288	336	384	432

In this instance, division examples with 48 as a divisor are assigned.

```
48) 336   48) 1008   48) 3600   48) 10272
```

Estimating trial quotients is another necessary skill. Several methods have been advocated and taught; none are totally reliable. Children usually discover their own modification. As Green observes, "The *simplest* method is probably the best, since most children will learn by themselves, to improve upon it anyway."[17] Children need to be aware that the first estimate does not always yield the correct number. They need to know how to adjust their first estimate.

```
      5  The quotient is too          4 R41
44) 217  large because 220 is    44) 217
    220  larger than 217. Try 4      176
         as the quotient.             41
```

17. Ibid., p. 238.

A popular method of estimating the quotient is called the *apparent method*. The divisor and dividend are rounded down according to the first digit in the divisor and the first one or two digits in the dividend. If we teach this method, we should begin instruction with examples leading to a correct estimate on the first trial.

Step 1: Estimate the tens.

```
       6
43) 2643      4) 26 is about 6.
  - 2580
      63
```

Step 2: Estimate the ones.

```
      61 R20
43) 2643
  - 2580
      63      4) 6 is about 1.
    - 43
      20
```

Later, introduce division examples that involve adjustments in the quotient.

Step 1: Estimate the tens.

```
       2
57) 1068      5) 10 is about 2. The partial
    1140      quotient is too large.
```

Step 2: Reduce the first estimate.

```
       1
57) 1068
   - 570
     498
```

Step 3: Estimate the ones.

```
      19
57) 1068
   - 570
     498      5) 49 is about 9. The partial
     513      quotient is too large.
```

Step 4: Reduce the first estimate.

```
    18R42
57) 1068
   - 570
     498
     456
      42
```

The apparent method never leads to an underestimate and is simple to use. Divisors in the teens are especially troublesome. Swenson cautions:

> The "teens" numbers might be assumed to be easier divisors than larger two-digit numbers. Actually, estimating the correct quotient for the divisor 14 is much more difficult than estimating the correct quotient for the divisor 74. Rounding off the 14 changes the quotient much more than rounding off 74 does.[18]

The sequence of skills must be well planned and gradual, so that a child does not face too many new ideas or procedures at one time.

Consolidating Activities

We need to help children remember the concepts and refine the skills they learn. Skills need to become habitual and accurate.

In order to vary practice activities, place computation in a less familiar setting. Note the following multiplication wheels.

PROBLEM

Find the product of the number in the center and each number in the inner ring. Write the products in the outer ring.

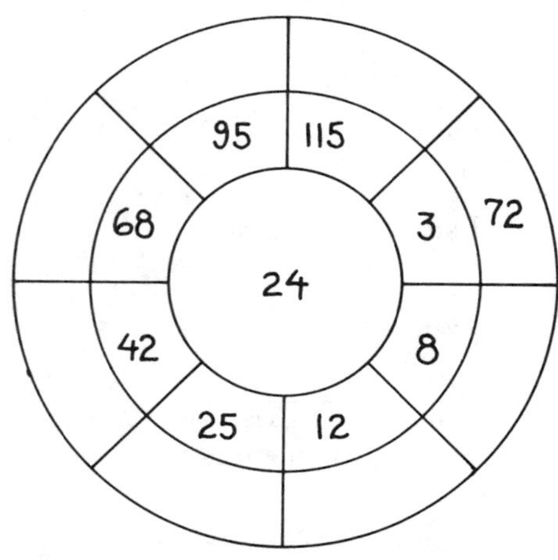

PROBLEM

Find the missing factors by dividing each number in the outer ring by the number in the center. Write the missing factors in the inner ring.

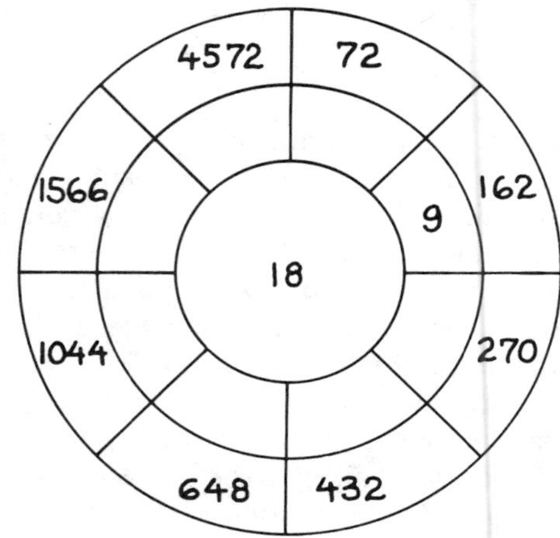

Provide practice for multiplication with self-checking practice tables, as illustrated in Figure 10.2. Also, the multiplication pattern shown in Figure 10.3 can be used as a consolidating activity. The final product is easily verified. By omitting a different set of numbers, this pattern can be made a division activity.

Partially completed multiplication and division problems can provide practice and help develop a child's problem-solving skills. As children become more familiar with the computations, the number of digits provided can be reduced.

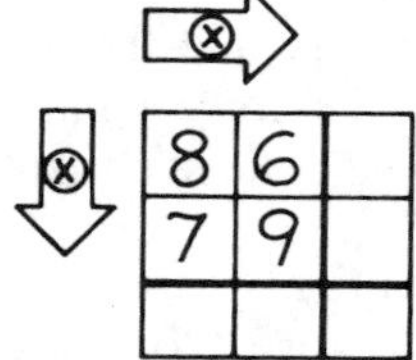

Figure 10.2
A Self-Checking Practice Table

18. Reprinted with permission of Macmillan Publishing Co., Inc. from *Teaching Mathematics to Children*, 2nd ed., by Esther J. Swenson. Copyright © 1973 by Esther J. Swenson.

EXCURSION
How Much Time on Drill and Practice?

Short daily assignments (about 15 minutes) are the best use of time for practice. For long-term retention, assignments around a particular skill should be spread out rather than concentrated within a short time interval. It is better to review a skill at staggered intervals after the initial learning rather than immediately. Research also suggests that practice of just learned mathematical rules, when delayed over several days, is more effective.[19] The amount of practice required varies from child to child.[20]

```
          3 □
1 □ ) □ 8 □
    − 3 6 0
        □ □
      − 2 □
```

```
    2 □
  × □ 5
  1 3 □
  8 □ 0
  □ 4 5
```

Activities should also provide some measure of success and interest to the child. They can involve applications in problem-solving contexts. Other activities can be games between two children or more children at the same level of skill development.

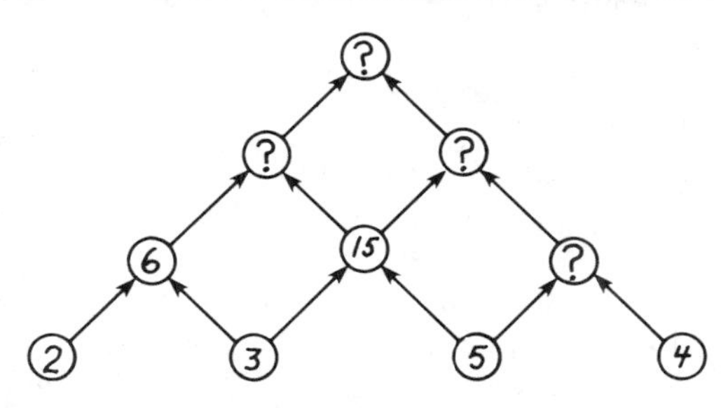

Figure 10.3
A Pattern for Practicing Multiplication

Chinese Checkers Multiplication Game[21]

This game provides practice with estimation and computation.

How to Play:

1. Teams take turn. Pick any two of the numbers.
2. Multiply the numbers picked.
3. Find the answer on the game board. Place your team's mark on it (X or O).
4. The first team to get a path of answers connecting its 2 sides of the game board wins.

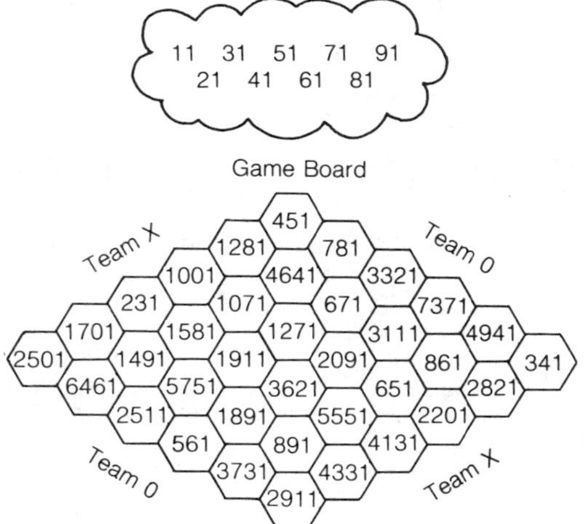

Bean-Bag Toss

To play this game, place two 4′ × 4′ grids on the floor. Label one *divisor* and the other *dividend.* Put appropriate numerals in each 1′ × 1′ square.

19. Stephen Horwitz, "Effects of Amount of Immediate and Delayed Practice on Retention of Mathematical Rules" (Paper presented at the annual meeting of the American Educational Research Association, Washington, D.C., April 1975) ED 120010.

20. Lorraine R. Gay, "Use of Retention Index for Mathematics Instruction," *Journal of Educational Psychology* 63 (October 1972): 466–72.

21. Adapted from Earl Ockenga and Joan Duea, "Ideas," *The Arithmetic Teacher* 27, no. 5 (January 1979): 31.

8	2	5	13
10	6	4	1
12	16	9	14
3	11	15	7

32	28	45	72
68	105	63	142
75	86	93	55
81	125	38	21

How to play:
1. The first player tosses one bean bag onto the divisor square and one onto the dividend square, and performs the indicated division.
2. If the quotient is correct, the remainder is the player's score. If the remainder is zero, the player receives 5 points.
3. Play continues until one person reaches a previously designated winning score.

Bean-bag Toss[22] can be adapted to practice addition, subtraction, or multiplication.

Greatest Product (or Greatest Quotient)
Make 5 cards for each of the ten digits 0–9. Prepare forms similar to Figure 10.4 for multiplication, or Figure 10.5 for division.

How to play:
1. Mix the cards well. Draw one and read the digit.
2. Have each child write the digit in any cell of the table.
3. Continue drawing digits until all cells are filled.

22. L. Carey Bolster et al., *Scott Foresman Mathematics*, Teacher's Edition, Book 5 (Glenview, Ill.: Scott, Foresman, 1980), p. 409.

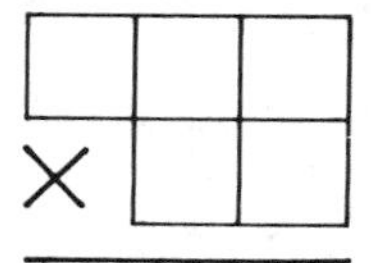

Figure 10.4
Form for Greatest Product

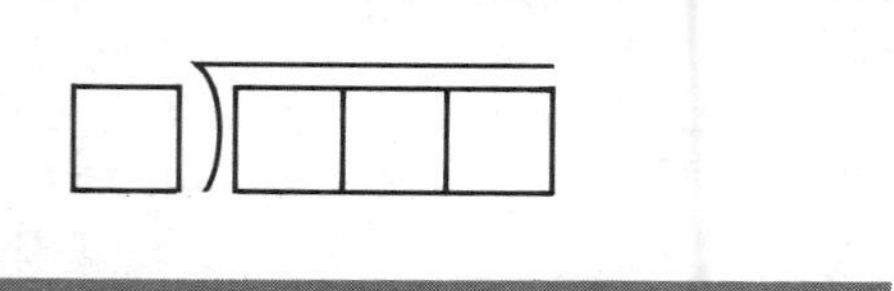

Figure 10.5
Form for Greatest Quotient

4. Children then find their product (or quotient if playing for division). The child with the greatest product (or quotient) wins.

This game can be varied by having the *least* product (or quotient) win.

Laboratory Activities
We may choose to plan laboratory activities to help children consolidate skill with computation. Carefully consider management procedures such as how children will be assigned to tasks, how their work will be monitored, and how individual questions can best be answered when planning these activities. Rising lists five steps to follow in organizing laboratory activities:

1. Establish specific goals in your lessons.
2. Develop clear statements of tasks for your students.
3. Identify and provide all materials needed for each task.
4. Carefully assign students to tasks.
5. Prepare students for the work.[23]

Examples of task cards that can be developed for multiplication and division activities are found in Figures 10.6 and 10.7. In the first activity children construct and play a game. The second activity is designed to help them discover how to develop a schedule. Children may require additional assistance with the latter.

Sets of tasks or work cards as well as laboratory equipment are available commercially. In addition to these, we can construct our own.

23. Gerald R. Rising and Joseph B. Harkin, *The Third "R": Mathematics for Grades K–8* (Belmont, Calif.: Wadsworth, 1978), pp. 181–82.

SPINNER PRODUCTS GAME[24]

Content objective

To find the product of three factors, multiply the first two factors. Then multiply the product by the other factor (based on objective 9.16).

Materials needed

Cardboard and scissors
Paper clips and thumbtacks
Pencil and paper

Procedure

1. Use your scissors to cut out three squares about 3 inches on a side like those shown below.
2. Use paper clips for the pointers. Use the thumbtacks to hold the clips in place.
3. Use these spinners to play the game.

Rules and scoring

1. *First player:* Spin all three pointers. Your score is the product of the three numbers shown.

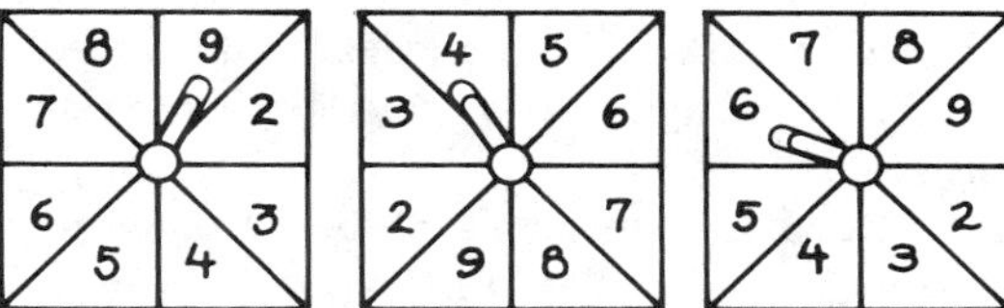

Example: 9 × 4 = 36; 36 × 6 = 216, so the score is 216.

2. *Second player:* Do the same.
3. Continue taking turns. After each turn, add the product of the three numbers shown to your score.
4. After 10 turns each, the player with the highest score wins.
5. You may challenge the other player's score at any time. If the score is wrong, the player loses the score for that turn. If the score is correct, the player receives 10 extra points.

Figure 10.6
Sample Laboratory Activity for Multiplication

PLANNING ASSESSMENT

Effective assessment of understanding and skill with multiplication and division algorithms requires careful planning. Assessment of *skill* with paper-and-pencil procedures is straightforward, and necessitates using ordered lists of skills such as those presented earlier in this chapter.

Assessment of a child's *understanding* of a procedure follows the model presented throughout this text. Begin by focusing on important content objectives. For example, consider assessment procedures we might use for content objective 10.11. What behaviors will help us determine if a child understands this objective? If more than one behavior is observed, we will have greater assurance the child does understand. We might look for the following:

24. Adapted from Bolster, A Spinner Products game.

ARRANGING A SCHEDULE[25]

Content objective

To make a playing schedule, all the possible combinations of one team playing another must be considered.

Materials needed

Paper and pencil

Procedures and Questions

The Cromwell Valley League

Orioles	Clowns	Cubs
Bears	Giants	Red Sox

Each team plays every other team exactly once. Do you know how many games will be played in the league? Think about these questions:

1. How many teams are there?
2. How many games does each team play?
3. How many games will be played in the league? (Be careful not to count each game twice.)
4. If each Cromwell Valley League team plays exactly one game per week, how many weeks does their season last?
   The Northwood League has eight teams, and each team plays every other team exactly once.
5. How many games are played in the Northwood League?
6. Each Northwood team plays exactly one game per week. How many weeks does their season last?
7. The Hillendale League has 7 teams. Each team plays every other team exactly once. Give names to the Hillendale League teams. Make up a schedule for one season, with 3 games each week. Schedule no team for more than one game per week.

Figure 10.7
Laboratory Activity for Multiplication

1. Given a division statement and base 10 blocks, the child represents the dividend, then partitions (separates) the blocks into the number of equivalent subsets named by the divisor.
2. Given a division statement and a place value chart, the child places markers on the board to represent the dividend, then partitions (separates) the markers into the number of equivalent subsets named by the divisor.
3. Given a division statement, the child computes the quotient by beginning at the left of the dividend.

The indicators themselves should suggest ways we can elicit the behaviors. Tasks can be designed for each one listed.

25. Adapted from Gary G. Bitter et al., *McGraw-Hill Mathematics, Grade 6* (New York: Webster Division, McGraw-Hill, 1981), p. 343.

11

Solving Problems and Exploring for Patterns

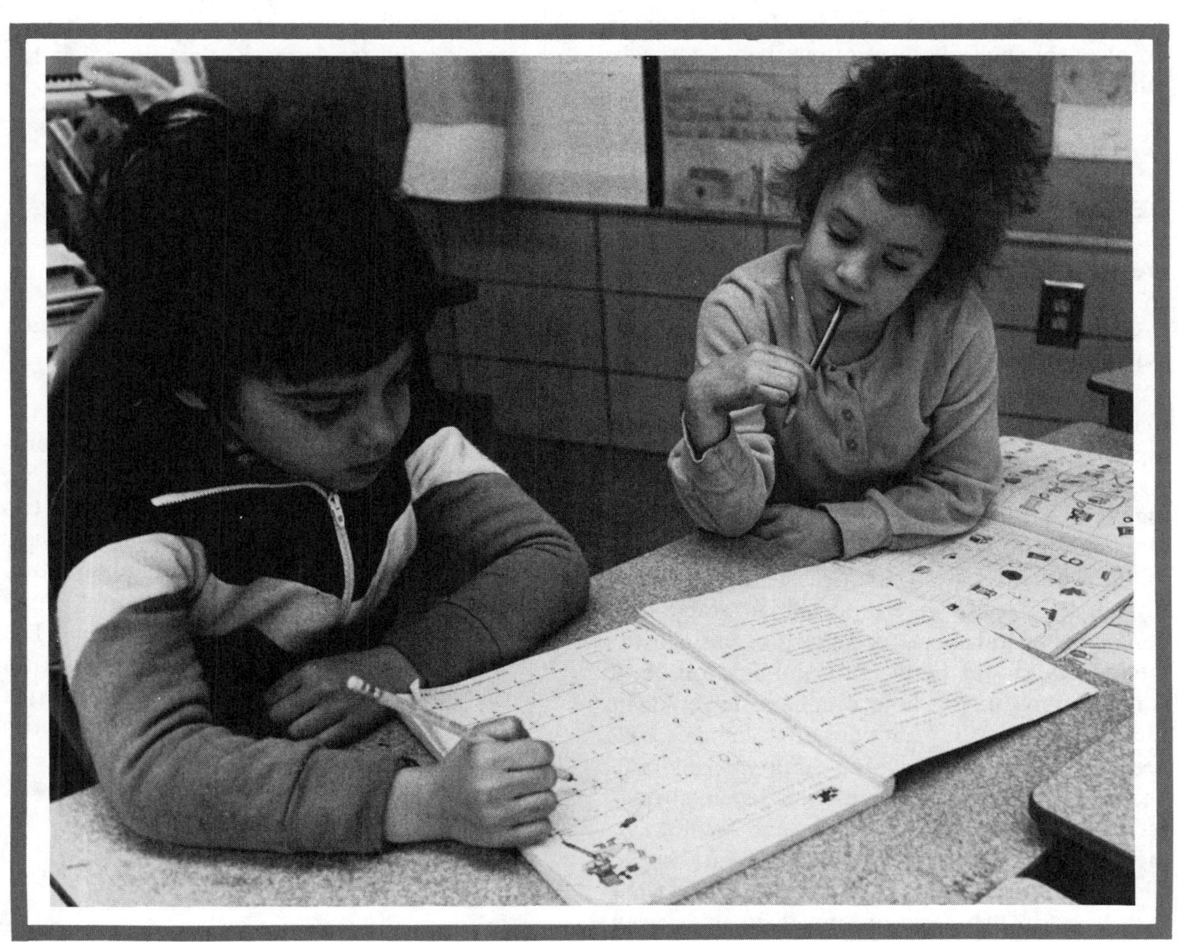

Helping a child become a pattern seeker and problem solver is one of the most important aspects of teaching mathematics. In this chapter problem solving is discussed first, then the closely related activity of looking for patterns.

DEVELOPING PROBLEM SOLVERS

We want to help *each* child become a problem solver. The first recommendation of the National Council of Teachers of Mathematics for the 1980s was that, "Problem solving must be the focus of school mathematics." They suggested that "The mathematics curriculum should be organized around problem solving."[1]

These recommendations reflect concerns highlighted by a 1977–78 survey of the mathematical ability of 9, 13 and 17-year-old students in the United States.[2] Changes in achievement since a 1973 study were reported. Although students appear to have a good grasp of basic arithmetic facts, simple mathematical definitions, and computations with whole numbers, fewer dealt successfully with fractions, decimals and percents. Problem-solving ability generally declined. Children of all ages tested were strongest in their knowledge of facts, and weakest in their ability to apply what they already knew. Clearly, children need to be able to apply what they know in the world around them; they need to be problem solvers.

1. National Council of Teachers of Mathematics, *"An Agenda for Action: Recommendations for School Mathematics of the 1980s"* (Reston, Va.: The Council, 1980), p. 2.
2. National Assessment of Educational Progress, "Mathematical Achievement: Knowledge, Skills, Understanding, Applications," brochure (Denver, Col.: NAEP, undated).

What Is Problem Solving?

Problem solving is viewed in different ways. Varied definitions are offered and, therefore, research on the teaching and learning of it has been somewhat difficult to apply in the classroom.

Polya suggests that,

> To solve a problem is:
>
> - To find a way where no way is known off-hand,
> - To find a way out of a difficulty,
> - To find a way around an obstacle,
> - To attain a desired end, that is not immediately attainable, by appropriate means.[3]

In *Problem Solving in School Mathematics,* Branca views problem solving as a goal, a process, and a basic skill.[4] It is a goal because we want children to become problem solvers. It is also, "The process of applying previously acquired knowledge to new and unfamiliar situations."[5] Finally, it is a basic skill, and not something to be tacked on if there is time. The National Council of Supervisors of Mathematics lists problem solving as the first of ten basic skills that need to be included in school mathematics programs.[6]

It is helpful to consider the child's view. A child regards a situation as a problem if:

3. George Polya, "On Solving Mathematical Problems in High School," in *Problem Solving in School Mathematics,* ed. Stephen Krulik and Robert E. Reys, 1980 Yearbook of the National Council of Teachers of Mathematics (Reston, Va.: The Council, 1980), pp. 1–2.
4. Nicholas A. Branca, "Problem Solving as a Goal, Process, and Basic Skill," ibid., pp. 3–8.
5. National Council of Supervisors of Mathematics, "Position Paper on Basic Skills," *The Arithmetic Teacher* 25, no. 1 (October 1977): 19–22.
6. Ibid.

1. He is called upon to do a task, to take action;
2. The goal is clearly understood; and
3. A readily accessible procedure for reaching that goal is not immediately available—at least some time must be spent in planning.

That which is a problem for one child may not be for another.

Phases of Problem Solving

The process of problem solving is often discussed in terms of four phases listed by Polya in *How to Solve It:*[7]

1. Understanding the problem
2. Devising a plan
3. Carrying out the plan
4. Looking back

To understand a problem, a child must recognize what is unknown, what is known, and the conditions present in the situation. In order to devise a plan for solving the problem, a child must find a connection between given data and the unknown. Polya lists many questions that can help a child devise such a plan. After the plan is carried out, the child should look back and examine the solution; a different way of solving the problem may become apparent.

Research on problem solving defined in such general terms has been of interest to mathematics educators, and involves many variables. Some kind of linkage between what a child already knows and his ability to solve problems must be established. The whole child must also be considered, including physical and emotional states. So far, research on problem solving as a generic activity has limited direct application in the classroom.[8] Some mathematics educators suggest that future research should focus on different *kinds* of problems.

7. George Polya, *How to Solve It,* 2nd ed. (Princeton, N.J.: Princeton University Press, 1973), pp. xvi–xvii.

8. See Marilyn N. Suydam, "Untangling Clues from Research on Problem Solving," in *Problem Solving in School Mathematics,* ed. Krulik and Reys, pp. 34–50. For more recent research, see Marilyn N. Suydam, "Update on Research on Problem Solving: Implications for Classroom Teaching," *The Arithmetic Teacher* 29, no. 6 (February 1982): 56–60.

Types of Problems

Even when problem solving is restricted to mathematics, different types of problems can be identified. Some problems are like puzzles, what is called *recreational mathematics.*[9] Others are verbal problems or word problems that appear in texts. Children also need nontextbook problems.

LeBlanc, Proudfit, and Putt, in their discussion of teaching mathematical problem solving in elementary schools, focus on two types of mathematical problems: *standard textbook* and *process* problems. They note that previously studied operations and algorithms are applied when standard textbook problems are solved, but process problems require other strategies. An algorithm may exist that could solve the process problem, but it is not available to the child. The process of obtaining a solution is stressed instead of the solution itself.[10]

Standard verbal or word problems characteristic of textbooks are discussed and illustrated in the following pages. However, process problems are also beginning to appear. They encourage children to develop problem-solving strategies, providing a context in which children can devise creative solutions. A child can share his plans in working with other children. This way, the child gains confidence in his ability to solve problems.[11]

Verbal Problems

Verbal, word, or story problems can be presented orally or in writing. They should be included along with process problems. We must make sure, however, that word problems are not just disguised computation.

Verbal problems should involve a child in gathering, organizing, and interpreting information so that he can use mathematical symbols to describe real world relationships. The quality and variety of problems needs to be carefully monitored, with practical applications in social situa-

9. For bibliography on mathematical puzzles and recreations, see Sarah F. Mason, "Problem Solving in Mathematics: An Annotated Bibliography," in *Problem Solving in School Mathematics,* ed. Krulik and Reys, pp. 34–50.

10. John F. LeBlanc, Linda Proudfit, and Ian J. Putt, "Teaching Problem Solving in the Elementary School," ibid., p. 105.

11. Ibid.

EXCURSION
Process Problems

How would you solve each of these?

1. Bill was paid $1.85. He received 19 coins in all: quarters, dimes, and nickels. How many of each could he have received? (Is there more than one solution?)
2. Seven people attended a party, and each person shook hands with everyone present. How many handshakes occurred?

What was your plan for solving each of the problems? Did you make a table? A list? A diagram? An equation?

tions. Problems should also include applications to as many fields of study as possible.[12]

Wilson identified three types of meanings associated with the operations of arithmetic: mathematically pure, physical action, and socially significant meanings.[13] *Mathematically pure meanings,* such as, addition is an operation that maps an ordered pair of numbers onto a third number, do *not* help a child decide which operation is appropriate when solving problems. *Physical action meanings,* such as, addition is putting things together, focus on the physical action involved in a problem situation. But physical action definitions are mathematically incorrect. Furthermore, they often mislead children who are trying to choose an operation. A problem situation requiring addition may not describe any physical action at all, or it may actually include taking away. *Socially significant meanings,* such as, addition is an operation used to find a sum when the addends are known, are sometimes called wanted-given definitions. Socially significant meanings, along with appropriate supporting cognitive skills, can help a child choose the appropriate operation. Numbers are seen as addends, sums, factors, or products from the social situation described in the problem.

Categories of Verbal Problems

Verbal problems can be categorized according to application concepts, similar to what Usiskin has suggested for algebra and geometry.[14] Problems would be grouped on the basis of real world applications. For example, *opposite directions* is an application concept for negative numbers, and *rate* is an application concept for division.

Knifong and Burton recommend grouping word problems on the basis of *operator meanings* or *explanations.*[15] For example, intuitive operator meanings listed for addition are counting, sets, linear models, and balance beam. Multiplication includes these, along with repeated addition, arrays, and cross-products of sets. This category scheme encourages the study of word problems while introducing an operation, rather than saving word problems for later applications.

Wilson's types of verbal problems are helpful in introductory work.[16] Each type is defined in terms of what is known and what is to be found, although the types are sometimes further distinguished if physical action is described. These problem types are also useful in breaking down multistep problems to a succession of single steps, simplifying decisions about operations. Wilson's problem types were discussed in Chap-

12. See Sidney Sharron and Robert E. Reys, eds., *Applications in School Mathematics,* 1979 Yearbook of the National Council of Teachers of Mathematics (Reston, Va.: The Council, 1979).
13. John W. Wilson, "What Skills Build Problem-Solving Power," *The Instructor* 76, no. 6 (February 1967): 79–81. Also in *Mathematics Is a Verb,* ed. C. W. Schminke and William R. Arnold (Hinsdale, Ill.: The Dryden Press, 1971), pp. 286–91.

14. Zalman Usiskin, "Applications in Elementary Algebra and Geometry," in *Applications in School Mathematics,* ed. Sharron and Reys, pp. 20–30.
15. J. Dan Knifong and Grace Burton, "Understanding Word Problems: What Does It Mean?" to be published in a forthcoming issue of *The Arithmetic Teacher.*
16. Wilson, "What Skills Build Problem-Solving Power," pp. 79–81.

EXCURSION
Flow Charting

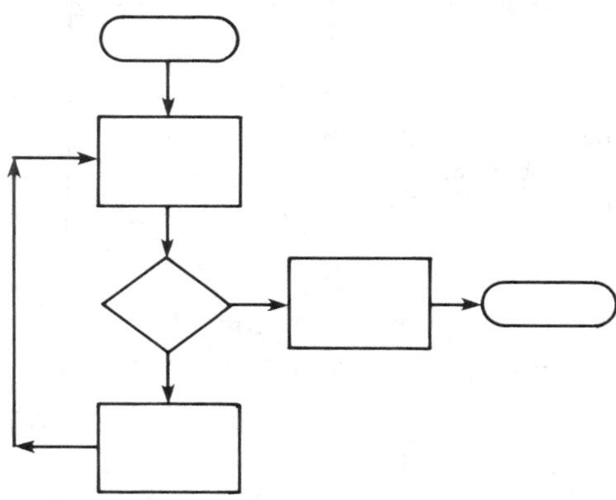

If you want children to learn to break down a process into simpler units and present it in a logical sequence, teach them how to make flow charts. This is an excellent schematizing activity which helps children become better problem solvers.[17]

ters 5 and 9. The problem types for each of the operations are listed and illustrated inside the back cover of this book.

Each child needs to focus on what is known and wanted in terms of the social situation; children should *not* focus on the numbers per se. For example, a child may observe that both the whole or total amount and the number in one part are known. What is wanted is the number in the other part. The child who understands what subtraction is in terms of socially significant meanings will know to subtract regardless of the physical action (if any) and the numbers involved. Children who have difficulty solving word problems frequently do not have a hard time with computation. But such children probably have a limited understanding of the meanings of the operations.

Children need experience in a variety of social situations. You may wonder about the extent to which different types of word problems are present in your program of instruction, in your texts, and in various achievement tests. You may find that the tests you administer include a different type of word problem than those used during instruction.

Processes Used in Solving Verbal Problems

Verbal problem solving involves different processes; a child may be able to do one process and not another. Processes include programming, transforming, and computing.

A child *programs* a word problem by producing an appropriate open number sentence. The number sentence is usually an equation, though other kinds may be appropriate (such as $6 + \square < 15$). An appropriate number sentence presents the relationships in the problem. Each child should write down the number sentence, at least during initial instruction. Different approaches to programming often produce differ-

17. See the following for instruction on making flowcharts: Francis M. Fennell, "Fun with Flowcharting," *School Science and Mathematics* 77, no. 4 (April 1977): 310–14; Sherry P. Hubbard and Robert B. Ashlock, "Using Flowcharts with First Graders," *The Arithmetic Teacher* 24, no. 1 (January 1977): 23–29; Bernard M. Kessler, "A Discovery Approach to the Introduction of Flow–Charting in the Elementary Grades," *The Arithmetic Teacher* 17, no. 3 (March 1970): 220–24; Grayson H. Wheatley, "Mathematical Road Maps: A Teaching Technique," *The Arithmetic Teacher* 23, no. 1 (January 1976): 18–20.

ent number sentences for the same problem. For example:

PROBLEM

> Jerry gave 27 baseball trading cards to Ben, and he now has 68 trading cards left. How many baseball trading cards did Jerry have before he gave some to Ben?

Different approaches to programming can result in the following equations, both of which are appropriate:

A. $\square - 27 = 68$
B. $27 + 68 = \square$

The first equation includes a minus sign, but subtraction is not the operation needed to complete the solution. Therefore, the equation needs to be transformed.

A child *transforms* an open number sentence by substituting an equivalent open number sentence that indicates the operation required to find the unknown. If the unknown in a given equation is isolated to the left or to the right of the equals sign, as in equation B above, no transforming is necessary; that equation already specifies the operation to be used. However, whenever the number sentence obtained by programming does not specify the operation, an equivalent number sentence in which the unknown *is* isolated to the left or to the right of the equals sign must be determined. The operation to be used can then be noted.

These processes can create a dilemma. We must either emphasize an approach to programming that initially produces an equation specifying the operation to be computed, or we must teach children procedures for transforming equations. Thorough instruction in sets of equivalent number sentences is sometimes undertaken so that one number sentence can readily be substituted for another. When this is done, relationships need to be learned so well that a child can easily substitute one equation for another. Sometimes the rules for manipulating equations associated with algebra are taught, but the authors favor the approach to programming that immediately produces a number sentence specifying the operation to be used. Carpenter, Hiebert, and Moser even suggest that during beginning instruction, "it may be that the transformation process is the limiting factor."[18]

When the desired equation is at hand, the unknown number is determined by *computing.* This can be done with a paper-and-pencil algorithm or with a calculator. The solution requires a correct computation.

At first, separate instruction in preparing a number sentence from that in computing; a child may be able to do one but not the other. You need to separate the processes so you can recognize if the child is doing one or both correctly. The child who computes but needs help programming can use a calculator so he can specifically focus on programming. On the other hand, the child who can program but not compute probably needs to develop skill in a more focused context.

Teachers sometimes think difficulty with verbal problems results because the children cannot read and comprehend, or because of errors in computation. Researchers have challenged such beliefs. Knifong and Holtan state that poor reading ability is responsible for less than 10% of the problems missed.[19] Zweng concludes that results from the National Assessment for Education Progress

> indicate rather clearly that deciding which operation to perform (addition, subtraction, and so on) is the major stumbling block to successful problem solving. Neither lack of computational skill nor . . . lack of reading comprehension is a significant factor in the lack of success experienced by many youngsters in problem solving.[20]

We need to focus on helping children decide what operation to use. We can more easily plan instruction that helps children with this

---

18. Thomas P. Carpenter, James Hiebert, and James M. Moser, "Problem Structure and First-Grade Children's Initial Solution Processes for Simple Addition and Subtraction Problems," *Journal for Research in Mathematics Education* 12, no. 1 (January 1981): 38.
19. J. Dan Knifong and Boyd D. Holtan, "A Search for Reading Difficulties Among Erred Word Problems," *Journal for Research in Mathematics Education* 8, no. 3 (May 1977): 229.
20. Marilyn J. Zweng, "The Problem of Solving Story Problems," *The Arithmetic Teacher* 27, no. 1 (September 1979): 2.

decision if we distinguish between two specific approaches to programming: action-sequence programming and wanted-given programming.[21]

Action-sequence programming requires a child to recognize what is going on in the social situation. *Actions* that occur, whether real or imagined, are identified and considered in *sequence.* Quantities are written down in the order encountered, and actions are recorded as operations. Consider these word problems.

PROBLEMS

A. Margaret had 17 dolls in her collection. Friends gave her more dolls for her birthday, and she now has 21 dolls. How many dolls did she receive?

B. Steve had 34 cents, but he spent 25 cents on a snack. How much money did he have left?

C. Six of the apples in a basket are red. The other five apples are yellow. How many apples are in the basket?

Action-sequence programming for problem A involves thinking it through in a manner similar to the following:

She had 17 to start, so write 17.
Then she received some more. That suggests addition, so write +.
Don't know how many she received, so make a box next.
As a result she now has 21, so write "= 21."

$$17 + \square = 21$$

The resulting number sentence will have to be transformed and subtraction used to find the unknown addend, even though the problem includes words like *gave, more,* and *receive.*

Sometimes, as in problem B, the equation resulting from action-sequence programming does not have to be transformed. Problem C illustrates another difficulty frequently encountered when action-sequence programming is used; the word problem contains no reference to physical action at all.

21. Descriptions of both action-sequence and wanted-given programming can be found in John W. Wilson, "The Role of Structure in Verbal Problem Solving," *The Arithmetic Teacher* 14, no. 6 (October 1967): 486–97.

In wanted-given programming, a child must focus on the relationships between what is *wanted* (unknown) and what is *given* (known). The wanted number is examined in terms of the social context and is designated as a sum, an addend, a product, or a factor. Each given number is classified similarly. Each number, known or unknown, is assigned a specific function within the social situation.

Rather than use mathematical terms like addend and sum, younger children may prefer to think of the number for each *part* (or subset) and the total amount. In multiplication and division situations, they identify parts that have the same number; the number of parts is also of interest.

In a wanted-given approach to programming, the operations of arithmetic are viewed as structural relationships between unknowns and knowns. The meanings of operations presented in this text are *descriptions* of structural relationships: for example, subtraction is an operation used to find one addend when the sum and its other addend are known. To program a word problem using a wanted-given approach, a child matches the appropriate operation meaning to the data at hand, then writes a number sentence that includes the operation and shows the relationships described in the problem.

Consider problem A. Programming involves reading the entire problem, then thinking in a manner similar to the following:

21 is the total number or sum.
17 is one of the addends.
The number of dolls she received is the wanted number; it is one of the addends.
Wants to know an addend, and is given a sum and an addend.
Subtraction provides an addend if you know a sum and the other addend, so write, sum minus addend.

$$21 - 17 = \square$$

Wanted-given programming produces the number sentence showing what operation to use when computing the unknown. No transformation is necessary. Problems B and C can be programmed similarily, and the lack of action in problem C creates no difficulty at all.

Instruction in Problem Solving

Three tasks face teachers: choosing and/or creating problems for instruction, planning how to help children understand and analyze problem situations, and planning how to encourage children to verify their solutions.

Choosing and Creating Problems

Each child needs experience in solving problems. They need to be a part of every topic and encountered frequently–even during the in-between times such as just before lunch.

Problems need to be at an appropriate level of difficulty. We must be concerned with vocabulary, length and structure of phrases, complexity of problem solving, as well as size of numbers involved.[22]

The problems also need to be varied. If this is done, each child is more likely to remember processes and use them later on and in different settings. Varied problems also more accurately approximate experiences encountered outside of school, and increase a child's attention span.

One way to add variety is to provide children with experiences in both oral and written problems. Oral problems are especially useful to those who are younger or have difficulty reading; for example, play audiotapes in a learning center accompanied by a sequence of pictures. Children can be helped with reading by writing problems one phrase per line. For example:

Joan bought a book
about postage stamps
for $3.49
and a package
of stamps
for $1.38.
How much did she spend
for the book and stamps?

Many children tend to decide which operation to use by taking note of how many numbers there are in the problem, and whether they are large or small. They apply rules that *they* have invented, such as:

- If the problem has two numbers about the same size, subtract the smaller from the larger.
- If the problem has three numbers, add.

Because children apply rules such as these, they have difficulty solving problems with extraneous data. The problems they encounter outside the classroom often involve extraneous data.

Some problems we prepare should have extraneous data. Occasionally, before giving a list of verbal problems to children, go over it with them deciding which problems have extraneous data and crossing out irrelevant information. Examples include:

PROBLEMS

Nancy is 9 years old. Her father is 24 years older than she is. In 8 years, how old will Nancy be?

Susan and her brother went to the game. Susan's ticket cost $2.00. Her brother's cost $1.50, but he paid for his own ticket. Susan had 75¢ left. How much money did she have before the game?

The untidy problem situations of real life may also lack sufficient data. Some problems have incomplete data: a missing number, or no numbers at all. Examples include:

PROBLEMS

Jane is 51 inches tall. How much taller is she than Mary?

Harry went to the store and bought a newspaper, a 5¢ package of bubble gum, and a 10¢ pencil. How much change did he get from $1.00?

We can also choose problems with no question, working closely with children as they suggest numbers and supply questions. A child may be able to think of more than one question for the situation, or questions for different operations. For example:

---

22. LeBlanc, Proudfit, and Putt, "Teaching Problem Solving in the Elementary School," in *Problem Solving in School Mathematics*, ed. Krulik and Reys, pp. 106–107.

PROBLEMS

Ann weighed herself and her cat together. Then she weighed herself.

Jack's brother spent $3.75 for lunches this week.

In choosing and creating problems, we sometimes oversimplify real world situations. Consider the following:

PROBLEM

To make a rug that measures 10 feet by 13 feet, how many square yards of carpeting will George need to buy?

The fact that carpeting is actually purchased in fixed widths of 9, 12, and 15 feet needs to be considered. The problem might better be written:

PROBLEM

George is going to buy carpeting to make a rug that measures 10 feet by 13 feet. When he went to the store he learned that carpeting comes in rolls that are either 9, 12, or 15 feet wide. Which width will George need to buy? How many square yards of carpeting will George need to make his rug?

We can also include some problems involving familiar names and situations. Emphasize everyday experiences and interesting themes.

Some use of the absurd is warranted, just for fun. Occasionally, throw in a silly problem. Consider this:

PROBLEM

If Bob weighs 40 kilograms standing on one foot, how much does he weigh standing on two feet?

Sometimes the structure of a silly problem can be studied to determine what operation to use.

PROBLEM

Tim is building a whoople. He needs 6 more glugs to finish. If glugs cost 35¢ each, how much will it cost Tim to buy what he needs?

Also, have children identify what makes the problems silly.

PROBLEMS

Nancy said she needed 9 inches of ribbon. Gail told her that she needed only $\frac{1}{4}$ of a yard of ribbon. Who was right?

John bought some radishes at 15¢ a bunch. How much did he pay for them?

Children also enjoy the variety of trying to solve word problems found in old texts. The following is from a nineteenth century arithmetic book.

PROBLEM

A stone mason worked $23\frac{1}{3}$ days, and after paying $\frac{3}{7}$ of his earnings for board and other expenses, had $\$53\frac{1}{3}$ left. What did he receive a day?

Children tend to rely on words and phrases such as the following to determine the appropriate operation. These are often called *key words.*

Left	In all	Put into groups
Spent	Put with	Separate
From	Have together	Gone

Key words are not really cues. So, in order to help each child focus on information in the problem and on relationships between numbers, use *distorted cues*. A distorted cue is a word used in a problem that is solved by an operation *different* from the commonly associated operation. Find the distorted cues in this problem.

PROBLEM

After Christmas there were 27 red balls left and there were 18 blue balls left. How many balls remained?

Children sometimes need to create their own word problems. They can write problems suggested by graphs or charts that we supply; they can also write problems for specific open number sentences. Also, we can clip cartoons from newspapers, deleting any printed words and

EXCURSION
Calculators in Schools

The calculator is gradually being recognized as an instructional tool and incorporated into programs at all levels. Children must be taught how and when to use it; they need to become aware of its capabilities and its limitations. They will probably be using calculators for the rest of their lives; many already have their own.[23]

To understand one limitation, multiply each of the following with a calculator:

75 × 1000	7968 × 100
4678 × 100	532 × 10000

It would be faster to do these examples in your head.

Some people fear that if children are given calculators in school they will not be motivated to master basic facts and algorithms, and that it will become a substitute for learning computational skills. Research generally runs contrary to such fears.

Calculators can be helpful when teaching problem solving. Problems are more realistic if they are not limited to the child's scope of computational abilities at a given time. Calculators also allow investigation of a variety of interesting questions. Consider the following:

PROBLEM

Robert is offered a job that pays 1¢ the first day, 2¢ the second day, 4¢ the third day, 8¢ the fourth day, and so on. Each day the salary is double the salary paid the previous day. If Robert accepts this job for just 30 days, how much will he earn during the 30 days?

For ideas on using calculators when teaching problem solving, see Janet Morris, *How to Develop Problem Solving Using a Calculator* (Reston, Va: National Council of Teachers of Mathematics, 1981).

writing one or more operation signs in the corner. Have each child write a problem using a cartoon.

On occasion, have children create word problems while working as a group. Individually or group written problems can be exchanged among children.

Understanding and Analyzing Problems

Practice results in increased ability to solve problems, yet there are things that can help children improve even more.

A list of general *guidelines* follows.[24] Keep them in mind in planning instruction and talking to children.

When *planning* instruction remember to:

1. Allow plenty of time for children to solve problems. For more complex problems, allow more than one day; give children time to reflect.

2. Provide for a variety of problems to be solved by the same technique, and for a variety of techniques to be used in solving one problem. Do

23. See Marilyn N. Suydam, "The Use of Calculators in Precollege Education: A State-of-the-Art Review," Report of the Calculator Information Center, 1200 Chambers Rd., Columbus, Ohio (May 1979).

24. Adapted from Suydam, "Untangling Clues from Research on Problem Solving," *Problem Solving in School Mathematics*, ed. Krulik and Reys, pp. 34–50.

not create the impression there is a different technique for each problem or type of problem.

Also, as you *talk* to a child while he is solving a problem, remember to:

3. Emphasize the process to be used in solving the problem, and why it is appropriate.

4. Emphasize general rather than specific statements or questions. For example, say, "Have you used all of the information?" and, "What do you already know that might help you solve this problem?" rather than, "Remember the rules on page 62? They are very important."

5. Emphasize the structural characteristics of a problem situation; do not focus all of a child's attention on details.

6. Frequently ask, "Can it be solved another way?"

Children can learn specific *techniques* to help them understand and analyze problem situations. The following techniques can be useful, regardless of the approach to programming. Each child should gain experience with several, so he can determine which techniques are especially helpful.

1. Restate the problem in several different ways. Or, have the child reread the problem and substitute nouns for pronouns.

2. Identify assumptions. A study by Kennedy, Eliot, and Krulee suggests that this kind of thinking is helpful to children.[25] For example, if one driver is going 60 kilometers per hour and the other is going 80 kilometers per hour, we can assume that one will cover a greater distance within a given amount of time. We can also assume that if the cars are identical, the one going faster will run out of gas more quickly. Or, if a mixture with two ingredients is described and it is stated that one ingredient accounts for 35% of the mixture, we can assume the other ingredient accounts for 65% of the mixture.

25. George Kennedy, John Eliot, and Gilbert Krulee, "Error Patterns in Problem-Solving Formulations," *Psychology in the Schools* 7, no. 1 (January 1969): 93–99.

3. Make a drawing. A freehand drawing, possibly a diagram or a map, will often clarify relationships. Note how a drawing for the following shows the problem solver that the number of pieces is not as important as the number of cuts.

Problem

A log is cut into five pieces in 12 seconds. At the same rate, how long would it take to cut the same log into six pieces?

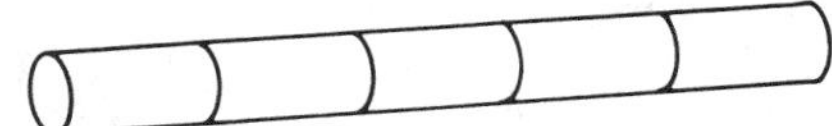

4. Make an analogy. For many children, it is helpful to restate the problem with very small numbers. The same logical structure is maintained, but the relationships may become self-evident.

5. Consider other problems. Recall problems that have already been solved. Were any of them similar to the new problem? Was the same information needed?

6. Study mathematical vocabulary. A child may be unfamiliar with units of measurement or may have difficulty picturing geometric shapes.

7. Act out the problem. For some children, it helps to clarify the situation by dramatizing the story.

8. Use trial and error. Davis and McKillip believe systematic trial and error is a good problem-solving procedure, especially when calculators are available.[26] Consider this problem:

Problem

Mr. Jones has 36 meters of fence, and wants to enclose a rectangular garden. What are the measurements of the rectangular garden with the largest possible area?

26. Edward J. Davis and William D. McKillip, "Improving Story-Problem Solving in Elementary School Mathematics," in *Problem Solving in School Mathematics,* ed. Krulik and Reys, pp. 89–91.

In a problem such as this, alternatives may be tried and the results recorded. Frequently, a pattern will be observed in the data.

Instruction in problem solving should include group activities. The quality of thought can often be improved if we ensure that each child can present his ideas within a climate of mutual respect.

The following group activities may help each child become a better problem solver.

1. Discuss every other problem on a page with a group of children before they solve them independently.

2. Identify unseen problems. Give a sheet of two-step problems to a group and have them identify unseen one-step problems.

3. Estimate and analyze. After children have estimated answers to problems individually, initiate a group discussion on how different results were obtained. Children can give their views on why one solution might be correct while others are likely to be wrong. At first, you may want to provide answers in a multiple-choice format.

4. Occasionally analyze incorrect solutions. Give children problems that have been solved incorrectly. Have the children determine what went wrong and why the answer is incorrect.

5. When possible, have children measure directly. Problem solving can take place in the gym or outdoors or during camping programs. Problems like the following require students to measure.

PROBLEMS

Can you roll a ball across the floor as fast as it is legal for a car to drive past the school?

How fast is the average car going when it passes the school?

How far is it from the flagpole to the big sycamore tree across the brook?

Verifying Solutions

There is no set procedure for making sure that a solution is correct. We need to continuously encourage a child to check solutions for reasonableness, and confirm them in a variety of ways.

EXCURSION
How Can We Use Calculators?

The Instructional Affairs Committee of the National Council of Teachers of Mathematics lists the following applications for calculators in schools. The calculator can be used:

- To encourage students to be inquisitive and creative as they experiment with mathematical ideas
- To assist the individual to become a wiser consumer
- To reinforce the learning of the basic number facts and properties in addition, subtraction, multiplication, and division
- To develop the understanding of computational algorithms by repeated operations
- To serve as a flexible answer key to verify the results of computation
- As a resource tool that promotes student independence in problem solving
- To solve problems that previously have been too time-consuming or impractical to be done with paper and pencil
- To formulate generalizations from patterns of numbers that are displayed
- To decrease the time needed to solve difficult computations[27]

---

27. National Council of Teachers of Mathematics, "A Position Statement on Calculators in the Classroom" (Reston, Va.: The Council, September 1978).

Estimation, for example, has an important role in finding a solution and can be a way of indicating the reasonableness of a result. An alternate solution by a different procedure is another excellent form of verification. Some of the techniques suggested to help understand and analyze a problem can also be used for verifications, such as a drawing or an analogy.

A method of verification is often unique to the type of problem solved. If a solution was suggested by a pattern observed in data generated in a few examples, a few more examples might indicate if the solution is still appropriate. When the problem originates from a real world situation, the solution can sometimes be reapplied to that situation for verification. The child should make sure a solution is reasonable and, if possible, correct.

EXPLORING FOR PATTERNS

Closely related to problem solving is the process of exploring for patterns. Children who are good problem solvers are alert for these, looking for patterns in sets of numbers and in geometric settings. Many of the concepts children apply while studying the topics in this book are learned by exploring for patterns in mathematics.

Mathematics *is* the study of patterns.[28] Sawyer defines it as the classification and study of all possible patterns.[29] When we teach children to be aware of patterns and to look for numerical and geometric relationships, we actually provide them with additional insights into the structure of mathematics.

The concept of a pattern is so broad that it is difficult to define. For instructional purposes, think of a pattern as a recurring configuration or relationship.

Planning the Content

Patterns reflected in the following content objectives in this chapter are only samples, but they suggest important discoveries children can make.

Number Patterns

Many patterns result from manipulating numbers. Certain sets are important in elementary schools.

11.1
A *natural number* is any number in the set {1, 2, 3, 4, 5, 6, 7, . . .}. The set of natural numbers is also called the set of *counting numbers.*

11.2
A *whole number* is any number in the set {0, 1, 2, 3, 4, 5, . . .}.

11.3
The whole numbers that end in 0, 2, 4, 6, or 8 are called *even numbers.* The set of even numbers can be represented as {0, 2, 4, 6, 8, . . .}.

11.4
Any whole number that is not an even number is an *odd number.* The set of odd numbers can be represented as {1, 3, 5, 7, 9, 11, . . .}.

11.5
The set of whole numbers with their opposites are called the set of *integers.* They can be represented as {. . . , ⁻3, ⁻2, ⁻1, 0, 1, 2, 3, . . .}.

11.6
The set of *rational numbers* contains the set of integers and the set of *fractional numbers.*

The relationships between the special sets described in the content objectives can be shown as follows:

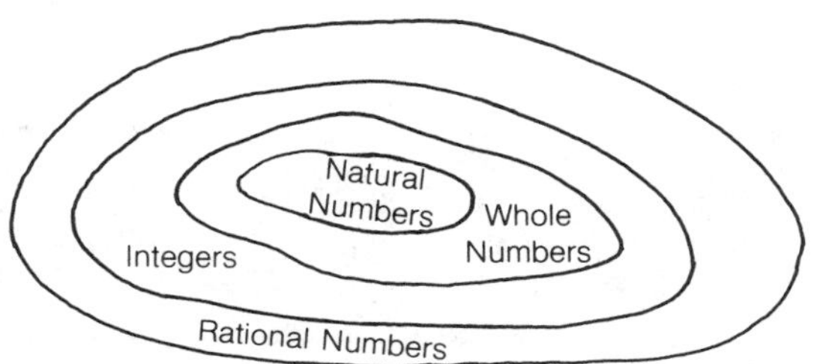

Among the most interesting of the natural numbers are those that can be represented geometrically.

28. William H. Glenn, et. al., *Number Patterns* (St. Louis, Mo.: Webster Publishing, 1961), p. 2.
29. W. W. Sawyer, *Prelude to Mathematics* (Baltimore: Penguin Books, 1955), p. 12.

11.7
A *triangular number* is any number in the set {1, 3, 6, 10, 15, 21, . . .}.

Triangular numbers can be represented in this pattern.

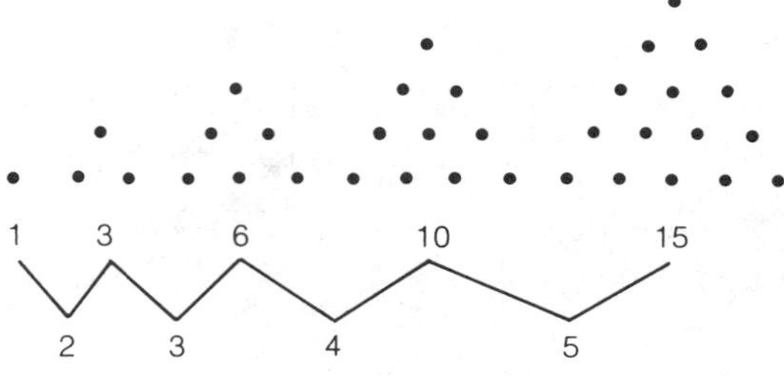

11.8
A *square number,* or *perfect square,* is any number in the set {1, 4, 9, 16, 25, 36, . . .}.

Square numbers can be represented as follows.

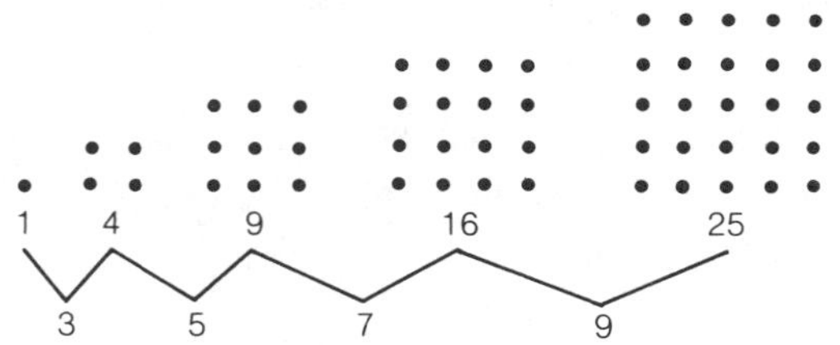

11.9
A *pentagon number* is any number in the set {1, 5, 12, 22, 35, . . .}.

A geometric representation for these numbers is illustrated.

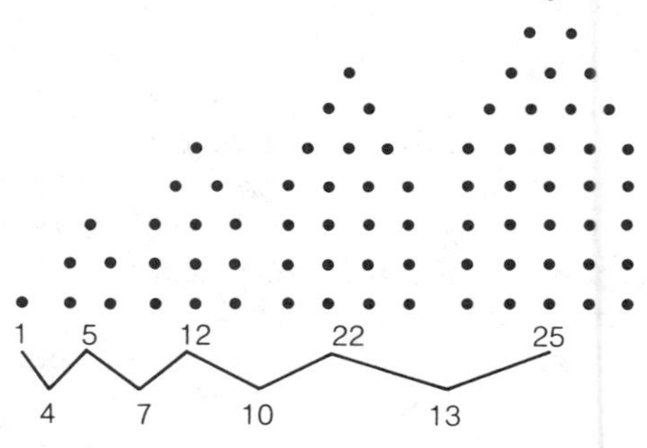

11.10
If two or more numbers are multiplied, each number is a *factor* of the product.

11.11
A *multiple* is the product of two or more numbers.

The only whole number factors of 24 are 1, 2, 3, 4, 6, 8, 12, and 24, as in the following illustration. Similarly, 24 is a multiple of each of these.

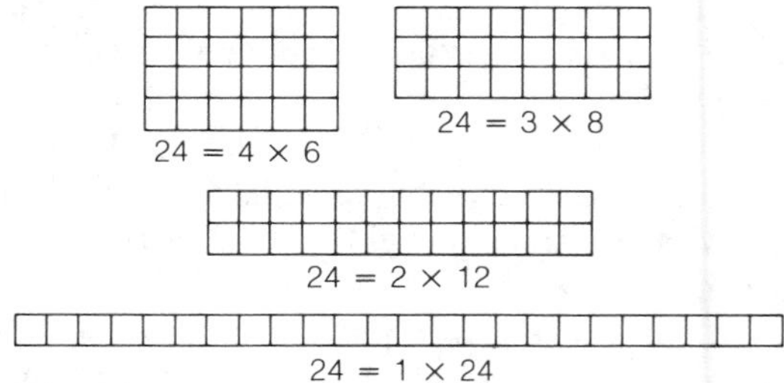

Three classes of natural numbers are often identified:

1. The number 1
2. The numbers that have exactly two different factors—the number itself and 1.
3. The numbers that have more than two different factors

EXCURSION
Cubic Numbers

Children build up squares to make larger squares. They can also investigate the growth of cubes by building up smaller cubes to make larger cubes.

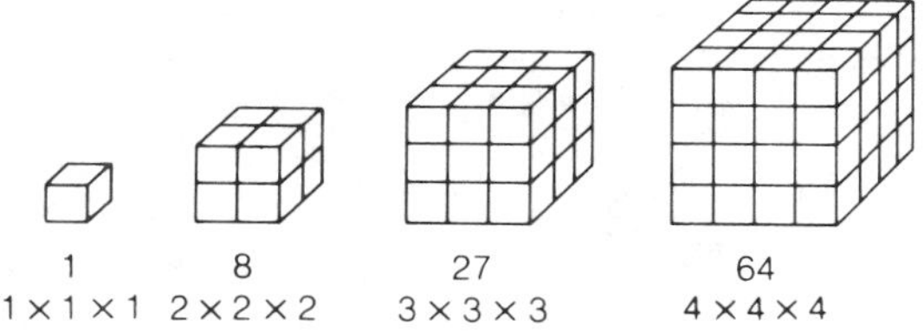

What is the sixth number in this set?

> **11.12**
> A whole number that has exactly two different factors is called a *prime number.*

The only factors of a prime number are 1 and the number itself. Prime numbers include 2, 3, 5, 7, 11, 13, ...

> **11.13**
> A whole number greater than 1 that is not a prime number is called a *composite* number.

The first few composite numbers are 4, 6, 8, 9, 12, 14, 15, ... The natural number 1 is a special set, neither prime nor composite. Thus, every natural number is a member of one of the three classes specified.

The number 4 is a factor of 36, but 4 is not a prime factor because it is not a prime number. When a composite number is expressed as a product of factors that are all prime, the factors are called the *prime factors* of the composite number.

> **11.14**
> Every composite number can be expressed as a product of prime numbers in a unique way. This is called the *Fundamental Theorem of Arithmetic.*

This name for a whole number, called the *prime factorization*, is unique in the sense that, apart from the order of the factors, one and only one set of primes can be used. For example, the prime factorization for 36 is $2 \times 2 \times 3 \times 3$.

> **11.15**
> Two numbers are *relatively prime* to each other if they have no factors in common except 1.

All prime and some composite numbers are relatively prime to each other. For example, 8 and 9 are relatively prime to each other because they have no factors in common except 1.

Numbers that are not prime to each other can have one or more of the same factors. Knowing the greatest of the common factors of two or more numbers can be helpful, for example, in renaming fractions to lowest terms.

EXCURSION
Goldbach's Conjecture

Every even number greater than 2 can be expressed as the sum of two prime numbers. For example,

$4 = 2 + 2$	$12 = 5 + 7$	$20 = 3 + 17$
$6 = 3 + 3$	$14 = 7 + 7$	$22 = 11 + 11$
$8 = 3 + 5$	$16 = 5 + 11$	$24 = 11 + 13$
$10 = 3 + 7$	$18 = 7 + 11$	$26 = 13 + 13$

This discovery was made in 1742 by the Russian mathematician, Christian Goldbach. An exception has never been found; however, no one has been able to make a formal proof. Therefore, we speak of it as Goldbach's conjecture.[30]

11.16
The *Greatest Common Factor* (G.C.F.) of two numbers is the greatest of the factors common to both numbers.

Prime factorizations for 48 and 84 are:

$$48 = 2 \times 2 \times 2 \times 2 \times 3$$
$$84 = 2 \times 2 \times 3 \times 7$$

Prime factors common to both numbers are 2, 2, and 3. The number $2 \times 2 \times 3$ is common to both; therefore, 12 is the Greatest Common Factor.

30. C. W. Schminke, Norbert Maertens, and William Arnold, *Teaching the Child Mathematics*, 2nd ed. (New York: Holt, Rinehart and Winston, 1978), p. 313.

When a number is a multiple of more than one number, it is called their *common multiple.*

11.17
The *Least Common Multiple* (L.C.M.) of two or more numbers is the smallest number evenly divisible by each.

A child who can find the Least Common Multiple can use the *least* of *common denominators* for several fractions, thereby avoiding renaming. Use of prime factorizations is the most efficient way to do this. For example, to find the L.C.M. for 15 and 18, determine the prime factorization for each.

$$15 = 3 \times 5 \qquad 18 = 2 \times 3 \times 3$$

EXCURSION
Using the G.C.F.

If we want to simplify the fraction $\frac{48}{84}$ we can divide both the numerator and the denominator by the G.C.F.; that is, by 12.

$$\frac{48}{84} = \frac{48 \div 12}{84 \div 12} = \frac{4}{7}$$

What is the relationship between this procedure and the fact that 1 is the multiplicative identify?

EXCURSION
Using the L.C.M.

How would you use the L.C.M. for 15 and 18 to add the fractions $\frac{7}{15} + \frac{5}{18}$?

$$\frac{7}{15} \times \frac{?}{?} = \frac{?}{90}$$
$$+$$
$$\frac{5}{18} \times \frac{?}{?} = \frac{?}{90}$$

The L.C.M. of the whole numbers is used as the Least Common Denominator (L.C.D.) for the two fractions.

How is the fact that 1 is the multiplicative identity applied during the computation?

The L.C.M. is the product of all factors occurring in one or both of the factorizations. A factor appearing in both factorizations is used only once.

$$2 \times 3 \times 3 \times 5 = 90$$

Thus, 90 is the L.C.M. of 15 and 18.

A divisibility test determines if a given number is evenly divisible by another number. Such tests are useful whenever a child needs to determine factors. Divisibility tests for 2, 3, 4, 5, 6, and 9 are often taught at the elementary school level.

11.18
If the last digit in a whole number is even (0, 2, 4, 6, or 8), the number is divisible by 2.

11.19
If the sum of the digits of a whole number is divisible by 3, the original number is divisible by 3.

11.20
If the last two digits (tens and ones) of a whole number are divisible by 4, the original number is divisible by 4.

11.21
If the last digit of a whole number is either 0 or 5, the number is divisible by 5.

11.22
If the last digit in a whole number is even and the sum of the digits is divisible by 3, the original number is divisible by 6.

11.23
If the sum of the digits of a whole number is divisible by 9, the original number is divisible by 9.

Geometric Patterns

In the process of looking for patterns in geometric settings, children can be introduced to two forms of mathematical reasoning.

One method used by mathematicians to discover new relationships is to perform experiments, collect data, and then examine the data for patterns. This is called the *experimental method,* and involves inductive reasoning.

11.24
Reasoning in which a general conclusion is made after considering specific examples is called *inductive reasoning.*

Consider a circular cake sliced so that the cuts do not pass through the same point more than twice (see page 254). How many pieces of cake do you get when 1 cut is made? What is the largest number you can get when 2 cuts are made? 3 cuts? 4 cuts? A chart similar to the one below is helpful in predicting the number of pieces.

Number of Cuts	Largest Number of Pieces	Increase in Number of Pieces
0	1	
1	2	1
2	4	2
3	7	3
4	11	4
5	?	?

Since it is impossible to test all possible cuts within a circular region, we cannot be sure that our conclusion will apply to every situation. Thus the inductive method tells us an answer that is only *probably* true.

We can arrive at mathematical conclusions using a different method. For example, if the polygons illustrated in Figure 11.1 are divided into triangles, we can assume that a polygon can be divided into $(N-2)$ triangles, 2 less triangles than the number of sides in the polygon.

We can also assume that the sum of the measures of the interior angles of a triangle is 180°, and that the sum of the measures of the interior angles in a polygon will be the sum of the measures of all the angles in the triangles. Therefore, the sum of the measures of the angles of any polygon is $(N-2) \times 180°$, when N is the number of sides in the polygon.

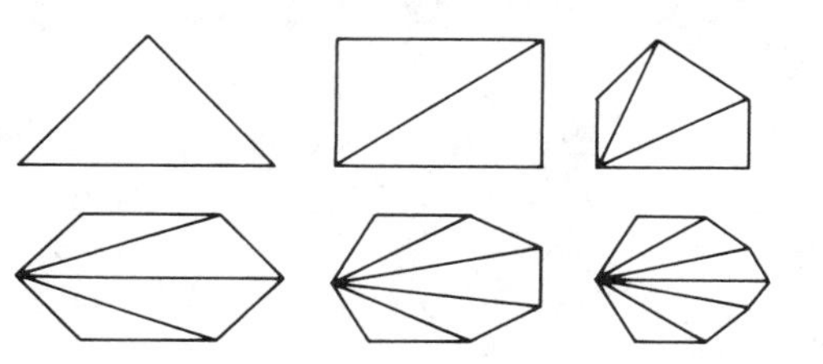

Figure 11.1
Polygons Divided Into Triangles

We started with several ideas we assumed to be true, reasoning with them to reach a conclusion.

> **11.25**
> Reasoning in which a specific conclusion is made from other ideas or assumptions is called *deductive reasoning.*

The truth of our conclusions, of course, depends upon the truth of our original assumptions.

A *tessellation* is a repetitive pattern similar to a mosaic which uses one or more geometric shapes. As in a mosaic or tiling pattern, the shapes fit together with neither holes nor gaps. They are placed in a repeated pattern so that, once begun, a flat surface can be covered by simply repeating the pattern.

Perhaps the most common tessellations are the brick patterns in a wall, a checkerboard, a honeycomb, and floor tiles. Figure 11.2 contains examples of *regular tessellations;* these are constructed from polygons all of a single shape and size.

Ordering the Content

The study of patterns is an ongoing process. Children can begin looking for them in many early abstracting and schematizing activities, stating the rule associated with the concept or algorithm. Searching for patterns is a way of introducing many new ideas.

The order of presentation will vary; however, for a given idea, make sure each child is ready for it, and that it relates to other ideas presently being considered. For example, the proce-

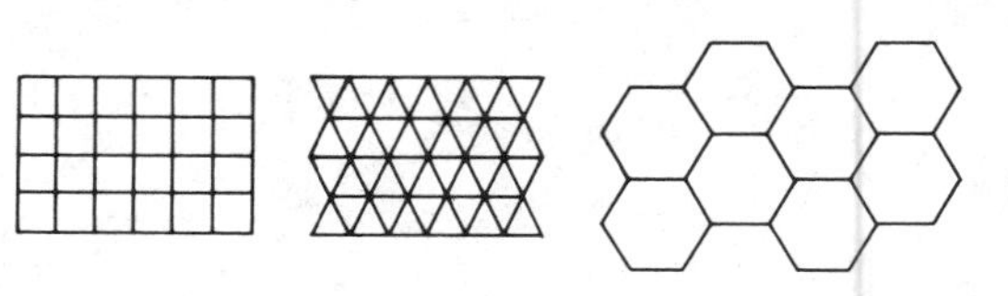

Figure 11.2
Examples of Regular Tessellations

dure for finding the L.C.M. is appropriately presented when it can be applied to addition and subtraction of fractions. Other topics, such as those related to geometry, can be used for enrichment.

Planning Instruction

The study of mathematics must include more than computation. As Jerman and Beardslee note, "The essence of mathematics lies in problem solving, pattern investigation, the study of structures that underlie axiomatic systems, and modeling."[31] Many of the concepts learned by working with patterns are also applicable to computation: the L.C.M., the G.C.F., and the divisibility rules are important when working with fractions. Some pattern-seeking activities provide opportunities to consolidate skills.

Exploring patterns in mathematics is closely associated with the process of *guided discovery*. As a teaching strategy, guided discovery requires much sensitivity. Rising and Harkin note:

> The teacher must be adroit at setting the stage, asking exactly the right questions at the right time, encouraging students to think, identifying what students mean when they respond, and rewarding creativity.[32]

We need to include many activities that enable children to discover patterns on their own.

Number Patterns

When children learn names and meanings of numbers, initiating activities can be planned for *odd and even numbers*. Then, as children develop numeration skills, these concepts can be extended to larger numbers.

Children can be asked to arrange collections of objects in rows of two. Ask: "How many objects are there? Can they be arranged in twos? Which collections cannot be arranged in twos?" (The pattern boards from the Stern materials are especially useful.) As the list of numbers for various sets increases in length, the 0–2–4–6–8 and 1–3–5–7–9 patterns of endings become apparent. Then ask other questions, "Is 24 an odd or even number?[33]

Many opportunities to observe patterns occur as the *basic facts* are taught. The relationship between addition and subtraction and commutativity for addition can be seen in families of facts. Zero as the additive identity, the zero property and the identity element for multiplication, the commutative properties, and the relationship between multiplication and division can be noted. Planning instruction for these objectives was discussed in Chapters 6 and 9.

But other patterns can be found in a table of products. (See Figure 11.3.) These include:

- Each row contains a set of multiples of the factor at the beginning of the row.
- Each column contains a set of multiples of the factor at the top of the column.
- When both factors are odd numbers, the product is odd.
- When both factors are even numbers, the product is even.
- When one factor is an even number and one factor is odd, the product is even.
- The main diagonal (upper left to lower right) consists of square numbers.
- The diagonals on either side of the main diagonal contain the same set of numbers. (See Figure 11.4.)

With skillful questions we can draw a child's attention to the relationships between numbers on the main and adjacent diagonals. For example, the ratios of the first number in each pair to the second are $\frac{1}{2}, \frac{2}{3}, \frac{3}{4}, \frac{4}{5} \ldots$.[34]

A *prime number* can be defined as a whole number that has exactly two different factors: 1 and the number itself. Children quickly grasp this idea when they are asked to cut as many rectan-

31. Max E. Jerman and Edward C. Beardslee, *Elementary Mathematics Methods* (New York: McGraw-Hill, 1978), p. 200.
32. Gerald R. Rising and Joseph B. Harkin, *The Third "R"—Mathematics Teaching for Grades K–8* (Belmont, Calif.: Wadsworth 1978), pp. 201–202.
33. George Green Jr., *Elementary School Mathematics Activities and Materials* (Lexington, Mass.: D.C. Heath, 1974), p. 257.
34. See Frances B. Cacha, "Exploring the Multiplication Table and Beyond," *The Arithmetic Teacher* 26, no. 3 (November 1978): 46–48.

×	0	1	2	3	4	5	6	7	8	9
0	0	0	0	0	0	0	0	0	0	0
1	0	1	2	3	4	5	6	7	8	9
2	0	2	4	6	8	10	12	14	16	18
3	0	3	6	9	12	15	18	21	24	27
4	0	4	8	12	16	20	24	28	32	36
5	0	5	10	15	20	25	30	35	40	45
6	0	6	12	18	24	30	36	42	48	54
7	0	7	14	21	28	35	42	49	56	63
8	0	8	16	24	32	40	48	56	64	72
9	0	9	18	27	36	45	54	63	72	81

Figure 11.3
Table of Products

gular arrays as they can for each counting number from graph paper. Figure 11.5 shows the results for numbers 1–6, and clearly illustrates that 1, 4, and 6 are not prime numbers. Each child should also be asked to make a table of the factors. The worksheet illustrated in Figure 11.6 can be used to record observations as both primes and composites are identified.

×	1	2	3	4	5	6	7	8	9
1	1	2							
2	2	4	6						
3		6	9	12					
4			12	16	20				
5				20	25	30			
6					30	36	42		
7						42	49	56	
8							56	64	72
9								72	81

Figure 11.4
Main Diagonal of a Table of Products

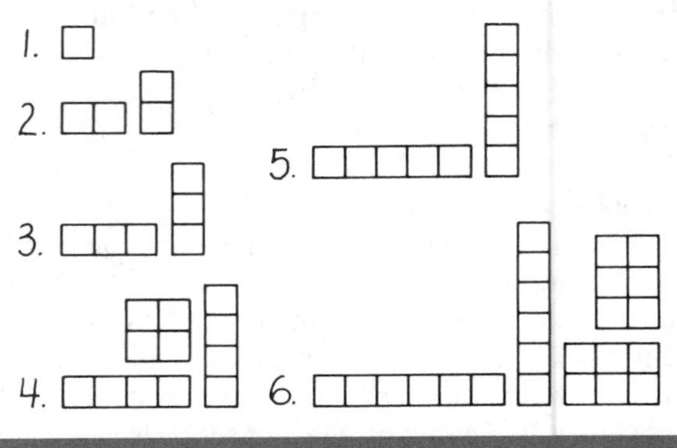

Figure 11.5
Rectangular Arrays for Discovering Primes

Children need a procedure for finding the prime factorization of a composite number. One particularly appealing method is to make a *factor tree*. Children soon realize that they always get the same prime factors no matter how they proceed and that there is only one prime factorization for each number.

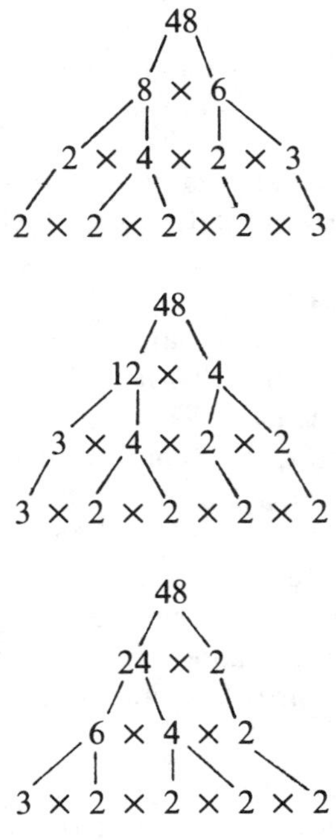

Geometric Patterns

Only a few geometric patterns are described here, but others can be found in references such as *The*

Name Cathy

Number	Factors	Prime or Composite?
2	1, 2	P
3	1, 3	P
4	1, 2, 4	C
5	1, 5	P
6	1, 2, 3, 6	C
7	1, 7	P
8		
9		
10		
11		
12		
13		
14		
15		
16		
17		
18		
19		
20		

Figure 11.6
Worksheet for Primes and Composites

Arithmetic Teacher. One area for combining the exploration of patterns with the study of geometry is the study of *formulas*.

Each child must learn formulas in working with perimeter, area, and volume. Do not tell children the formulas; rather, attempt to have each child discover them through careful guidance.

For example, in teaching the formula for area of a triangular region, focus initially on the relationship between the surface areas of a triangle and a rectangle. Begin with the special case of the right triangle, as it is most closely related to the rectangle. Draw around a cardboard or plastic model of a right triangle, then rotate the triangle about the midpoint of its longest side so that the new and the old positions form a rectangle (Figure 11.7). Children can easily see that the area of the triangle is half that of the rectangle, and the length and width of the rectangle can be determined by measuring the two shorter sides of the triangle.

Move on to the general rule for area of any triangle. Any triangle, whatever its shape, can be made into two right triangles with a common side; then children can focus on related rectangles for the two right triangles. To show this, draw a triangle with unequal sides and circumscribe the triangle with a rectangle, as shown in Figure 11.8. Cut out the rectangular region, and fold the longer side upon itself so that the fold passes through a vertex of the triangle.

A child can easily see that each of the two rectangles contains two right triangles. In Figure 11.8, the original triangle has a side (base) of 12 units which is now partitioned by an altitude into two parts, 4 and 8 units long. The altitude is 7 units in length. Since each of the two triangles is

EXCURSION
Calculator Mathematics

Calculators can be used to find patterns. Predict the next number in this series:

$$\begin{aligned} 1 \times 9 + 2 &= 11 \\ 12 \times 9 + 3 &= 111 \\ 123 \times 9 + 4 &= 1111 \\ 1234 \times 9 + 5 &= 11111 \\ 12345 \times 9 + 6 &= \ ? \end{aligned}$$

Verify your prediction by using a calculator.

Can you detect the following patterns? Write down the missing numbers, then check your prediction by using a calculator.

9×6	$= 54$	7×7	$= 49$
99×66	$= 6534$	67×67	$= 4489$
999×666	$= 665334$	667×667	$= 444889$
9999×6666	$=\ ?$	?	$= 44448889$

Interesting number patterns occur when factors are renamed as decimals. Use your calculator to change each of the following fractions to a decimal. Remember that each decimal repeats infinitely, but the calculator shows only a few digits. As soon as you see the pattern, predict the next decimal equivalent before using the calculator.

$\frac{1}{11} = .09090909\ldots$ $\quad \frac{2}{11} =$ ________ $\quad \frac{3}{11} =$ ________

$\frac{4}{11} =$ ________ $\quad \frac{5}{11} =$ ________ $\quad \frac{6}{11} =$ ________

Use the same procedure to find the pattern for the following:

$\frac{1}{9} =$ ________ $\quad \frac{1}{99} =$ ________ $\quad \frac{1}{999} =$ ________

a right triangle, the area of the original triangle can be found as follows:

$$\text{Area of A} = \frac{1}{2} \times 4 \times 7 = 14 \text{ Square Units}$$

$$\text{Area of B} = \frac{1}{2} \times 8 \times 7 = 28 \text{ Square Units}$$

$$\text{Area of Whole Triangle} = 42 \text{ Square Units}$$

Some children will observe that computation can be shortened by first adding the two parts of the base and then multiplying their sum (12) by 7 and by $\frac{1}{2}$, giving 42 square units as the area. Point out that by doing this they are applying the distributive property.

The study of *tessellations* provides an excellent way to combine art and mathematics in geometry and spatial problem solving. A collection of cardboard tiles made from basic geometric shapes can be used to introduce children to area and perimeter measurement, and to symmetry and pattern searching. They begin to compre-

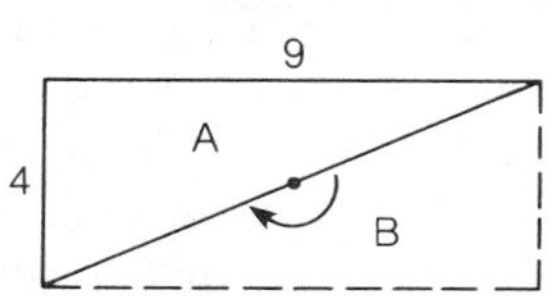

Figure 11.7
Right Triangle Rotated to Form a Rectangle

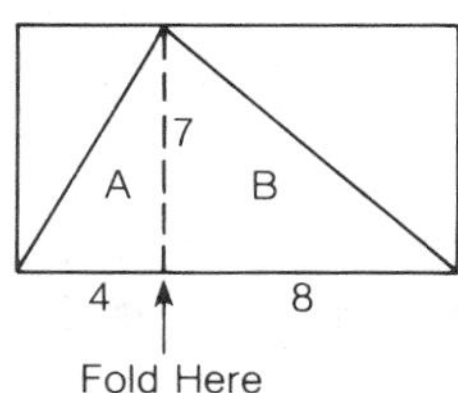

Figure 11.8
Paper Folded to Develop Formula for Area of a Triangle

hend the notion of a geometric pattern by covering surfaces with a repeated number of tiles. Color can be incorporated later.

Planning Assessment

Mathematical patterns should be incorporated as a part of daily instruction, at all levels and with all types of content. Schematizing activities typically involve pattern-related tasks.

When the search for patterns is rooted in daily study, they can be assessed during ongoing instruction. Consider the objective:

> **13.18**
> Multiplying a decimal by 10 or a power of 10 has the effect of moving the digits in the numeral the same number of places to the left of the decimal point as there are zeros in the power of 10.

EXCURSION
Sieve of Eratosthenes

Eratosthenes, a Greek mathematician who lived in Alexandria about 240 B.C., developed a sieve to find prime numbers from a list of natural numbers.

1	2	3	4	5	6	7	8	9	10
11	12	13	14	15	16	17	18	19	20
21	22	23	24	25	26	27	28	29	30
31	32	33	34	35	36	37	38	39	40
41	42	43	44	45	46	47	48	49	50
51	52	53	54	55	56	57	58	59	60
61	62	63	64	65	66	67	68	69	70
71	72	73	74	75	76	77	78	79	80
81	82	83	84	85	86	87	88	89	90
91	92	93	94	95	96	97	98	99	100

First, circle 2, the first prime. Then cross out all of the multiples of 2 because they are not primes, such as, 4, 6, 8, 10, 12, ... Next, circle the next number which is 3 and cross out its multiples such as, 6, 9, 12, 15, 18, ... This process is continued with succeeding numbers not previously crossed out.

Use of the sieve is not as laborious as it looks. A number less than 100 that is not a prime must have a factor less than 10, because $10 \times 10 = 100$. Thus, sieving out multiples of the primes through 7 will identify all primes from 2–100.

Can you extend the process to find all the primes from 2–200? (You only have to sieve out primes less than 15, because $15 \times 15 = 225$.)

One strategy for approaching the objective is to present a series of multiplication examples:

$$\begin{array}{r} 2.54 \\ \times\ 10 \\ \hline 25.4 \end{array} \quad \begin{array}{r} 2.54 \\ \times\ 100 \\ \hline 254 \end{array} \quad \begin{array}{r} 2.54 \\ \times\ 1000 \\ \hline 2540 \end{array} \quad \begin{array}{r} 2.54 \\ \times\ 10{,}000 \\ \hline 25{,}400 \end{array} \quad \begin{array}{r} 2.54 \\ \times\ 100{,}000 \\ \hline 254{,}000 \end{array}$$

Questions such as these may draw a child's attention to the pattern.

1. How is each problem alike?
2. How is each problem different?
3. When you multiply by 10, how far do the digits move to the left of the decimal point; how far when you multiply by 100?
4. How is the power of 10 in the factor related to the moving of digits in the product?

Children with experience searching for patterns should be able to state a general rule for determining the product.

A child's understanding of numerical and geometric patterns can often be assessed with paper and pencil. The following are a few sample tasks.

PROBLEMS

Find the pattern and fill in the missing numbers.

6, 12, 18, 24, _____, _____, _____
1, 4, 9, 16, _____, _____, _____
39, 33, 27, 21, _____, _____, _____
1, 3, 6, 10, _____, _____, _____
4, 6, 8, 9, 10, _____, _____, _____
$\frac{1}{3}, \frac{7}{12}, \frac{5}{6}, 1\frac{1}{12}$, _____, _____, _____
$10, 5, 2\frac{1}{2}, 1\frac{1}{4}$, _____, _____, _____
1, 8, 27, 64, _____, _____, _____

Draw the missing picture for each of the following. In every case, does A change to B like C changes to D?

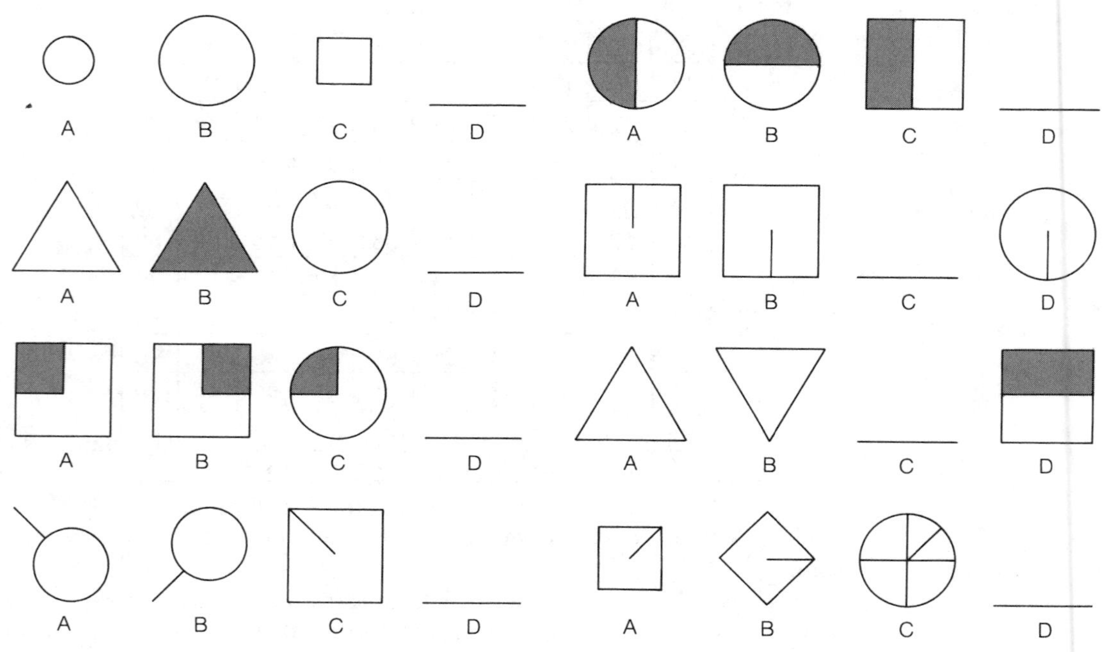

EXCURSION
The Fibonacci Sequence

The Fibonacci sequence is a series of numbers such that each number after the first two is the sum of the two preceding it. These sequences are named after Leonardo di Pisa (1170–1250), a medieval mathematician nicknamed Fibonacci.

Generating a Fibonacci sequence is interesting addition practice. For example, the numbers 0, 1, 2, 3, 5, 8, 13, 21, 34, . . . form a Fibonacci series. Any two numbers can be used to start a sequence. Five and 2, for instance, give the series 5, 2, 7, 9, 16, 25, 41, 66, 107, 173, ...

The sum of any ten Fibonacci numbers can be found quickly, because it is 11 times the seventh number. Thus, the sum of the sequence 5 through 173 is 451. Try this pattern with another series of ten Fibonacci numbers.

Teaching Rational Number Concepts and Operations

12

Introducing Rational Numbers

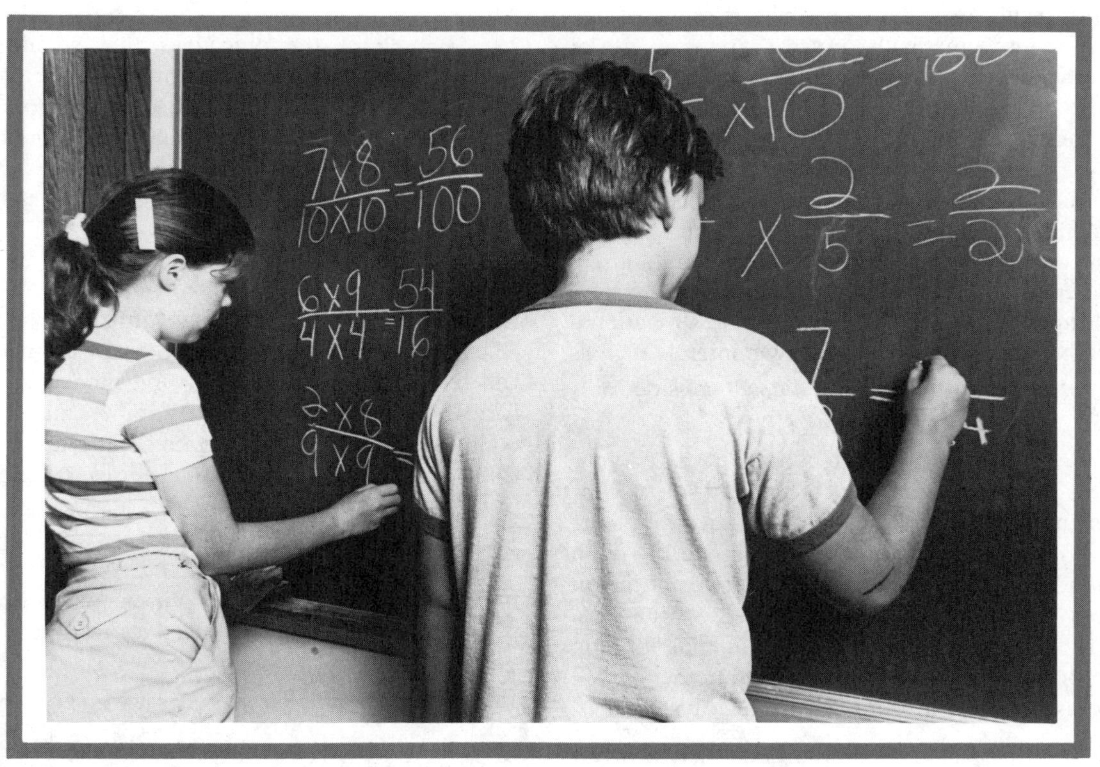

Three boys wish to equally share two candy bars. Children often solve this problem by breaking each bar into three parts of the same size; each child then takes two parts. Expression of the answer mathematically, however, requires use of numbers other than whole numbers.

If machinists' rulers were graduated only in inches, or merchants used scales limited to pounds, they would find their measuring instruments unsatisfactory. For greater precision, the machinist and merchant need instruments partitioned into smaller parts such as halves, quarters and eighths. Such needs gave rise to a new system of numbers called *rational numbers.* Rational numbers may be represented by numerals such as $\frac{3}{4}$, $\frac{^-5}{8}$, and $\frac{7}{4}$. In this chapter we will consider only non-negative rational numbers.

Fractions have often been hard for children. Historically, the development of a convenient notational system for rational numbers was slow, and algorithms for computing with fractions were cumbersome. A given fraction often has different uses in different situations. Further, fractions involve two numerals, one on top of the other. In addition, each number can be expressed in an infinite variety of ways; that is, $\frac{1}{2} = \frac{2}{4}, \frac{3}{6}, \frac{4}{8}, \frac{5}{10}, \frac{6}{12} \ldots$ Admonitions such as invert and multiply can be confusing when given without any real meaning attached.

The child's difficulty with fractions usually results from insufficient exploration and discovery experiences in the early grades. Often the child only partially grasps what a fraction is, and has limited opportunities to schematize the many important relationships.

PLANNING THE CONTENT

Children need to understand rational number concepts before moving on to operations and algorithms. In order to prevent difficulties with rational numbers, the teacher must first slowly and carefully develop specific underlying concepts.

Some of the objectives that should be chosen to introduce rational numbers are listed below. Content objectives are designed for our use as teachers and are not usually the words we use with children or expect them to use. As you read, think of appropriate student activities that would facilitate understanding.

Meaning and Uses of Rational Numbers

The non-negative rational numbers are defined as the set of numbers that may be named in the form a/b where a is a whole number and b is a counting number. Non-negative rationals have three principal interpretations or uses: fractional part, division, and ratio.

Fractional Parts

For rational numbers used to indicate fractional parts, important content objectives include:

12.1
The fraction a/b may be interpreted as a of b equal size parts. This is the fraction or partition interpretation.

12.2
A fraction may be used to express the number for part of a whole thing, or part of a set of things.

12.3
A fraction may be used to express the number for more than one whole.

12.4
The *denominator,* indicated by the numeral below the fraction bar, gives the name to the fraction. It tells how many parts of the same size there are in the whole.

12.5
The *numerator,* indicated by the number above the fraction bar, tells how many of the particular fractional parts are indicated.

12.6
Partitioning a region or a line segment, or a solid into congruent subregions, line segments, or solids is a model for the fractional part(s) interpretation of a non-negative rational.

The need for rational numbers to describe physical situations is obvious: a pie is subdivided into parts, or the edge of a piece of material does not correspond to an exact number of units used for measuring. Fractional units are required.

Symbols such as $\frac{2}{3}$ and $\frac{8}{5}$ express rational numbers, which are sometimes called fractional numbers. The word *fraction* is often used to refer to both a rational number and its numeral.

An analogy to the concept of partition division becomes clear when using a fraction for equivalent subsets. Consider the following example for the fraction $\frac{2}{3}$.

PROBLEM

Mary cuts a cake into 3 pieces of equal size. She keeps 2 of the pieces for herself. What fraction of the cake does she keep?

When a fraction is used in this way, the numeral written above the horizontal bar, the numerator, may be thought of as the *numberer.* It tells how many parts are being considered. The numeral below the bar is called the denominator, and it is the *namer* of the fractional parts.

2	Numerator	Numberer	How many
3	Denominator	Namer	What size

Rational numbers are used to designate fractional parts in three distinct situations, and children need to be familiar with each.

1. Subdividing a region into equal-sized parts. Fractions may indicate the size of each of the parts as compared to the whole unit.

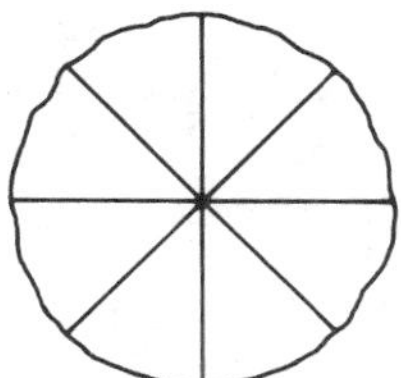

If a pizza is cut into 8 pieces of the same size, each piece is 1/8 of the pizza.

2. Subdividing a line segment into equal-sized segments. Fractions can indicate the length of each part as compared to the length of the original segment.

If an inch on a ruler is divided into 4 segments each segment is of equal length, 1/4 of an inch.

3. Subdividing a set into equal-sized subsets. Fractions can indicate one or more equivalent subsets.

If a set of 6 children is divided into 3 sets of the same size, each subset is 1/3 of the original set.

Division

Rational numbers are also used to indicate division.

12.7
A fraction may be interpreted as an indicated division: $a/b = a \div b$.

12.8
A rational number is determined by dividing a whole number (the numerator) by a counting number (the denominator); therefore, zero cannot be a denominator.

12.9
The numerator indicates a product; the denominator a known factor.

12.10
Partitioning a set of discrete elements into equivalent disjoint subsets is a model for the division interpretation of a non-negative rational number.

Rational numbers may be used to indicate the division of one number by another. This may result from a partitioning situation.

If 5 apples are to be divided equally among 2 persons, the quotient may be expressed as 5/2. The indicated division is 5/2, or 2-1/2.

It may also result from a measurement situation.

If 5 quarts of water are poured into containers holding 2 quarts each, the number of containers may be expressed as a fraction. The indicated division is 5/2, or 2-1/2.

Rational numbers make the division of any pair of whole numbers possible, except where the divisor or known factor is zero. For example, the solution to $2 \div 3 = N$ can be expressed as the fraction $\frac{2}{3}$. The numerator of the fraction names the product, and the denominator names the known factor.

Ratio

A third interpretation of non-negative rational numbers is that of *ratio* or *rate pair*.

12.11
A fraction may be interpreted as a *ratio* or *rate pair*: *a* to *b*.

12.12
The size (length, weight, etc.) of two things may be compared by using a ratio.

12.13
The ratio of one number to another may be shown by a fraction.

12.14
The numerator indicates the number in one set. The denominator indicates the number in a second set being compared by the ratio of the first set to the second set.

12.15
Comparing sets (many-to-many correspondence) is a model for the ratio or rate pair interpretation of a non-negative rational.

A relationship between two numbers may be shown by stating a comparison of one number to the other. When a fraction is used to express this relationship, it is called a ratio. The most common ways to express a ratio are 1:2 and $\frac{1}{2}$ (read as one to two).

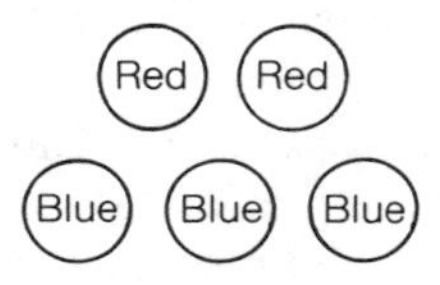

If you compare the number of red marbles to the number of blue marbles, the ratio is 2 to 3 and may be expressed as 2/3 or 2:3. Comparison of the number of red marbles to the total number of marbles results in the ratio of 2 to 5.

The idea of ratio, or rate pair, can be extended to include relationships between such everyday things as miles walked and minutes of time, number of candy bars and cost, number of books and children, and miles driven and gallons of gasoline used. For example, to express the relationship 2 candy bars for 35¢, write $\frac{2}{35¢}$ or simply $\frac{2}{35}$.

Two important types are *probability* and *percent* ratios. The study of probability has become increasingly important, and will help children understand many statements about everday affairs, such as weather predictions and odds on the World Series. For instance, probability provides the foundation upon which insurance rates are based.

PROBLEM

A drawer contains 16 red socks and 8 blue socks. If you randomly select a sock from the drawer, what is the probability you will select a blue sock?

$$\text{Probability Ratio} = \frac{\text{Successful Ways}}{\text{All Possible Ways}} = \frac{8}{24} = \frac{1}{3}$$

The probability you will select a blue sock drawn at random is $\frac{1}{3}$.

A *percent* ratio expresses the ratio between any whole or fractional number and 100. The word percent means per hundred, and the symbol % expresses a denominator of 100. When 35 is used as a percent, it expresses the ratio 35 of 100, and is written as 35%. In elementary school the study of percent is usually restricted to percents of from 1 through 100.

Special Subsets of Rational Numbers

Objectives related to important subsets in the set of non-negative rational numbers include:

12.16
Two fractions that name the same non-negative rational are called *equivalent fractions.*

12.17
A set of all equivalent fractions is called an *equivalence class.*

12.18
The set of whole numbers is a subset of the non-negative rationals.

12.19
Two fractions $\frac{a}{b}$ and $\frac{c}{d}$ are equivalent if and only if the product of the numerator of the first and the denominator of the second, equals the product of the denominator of the first and the numerator of the second; that is, $\frac{a}{b} = \frac{a}{b}$ only if $ad = bc$.

12.20
Two fractions $\frac{a}{b}$ and $\frac{c}{d}$ are equivalent if there is a nonzero number n such that $\frac{a}{b} \times \frac{n}{n} = \frac{c}{d}$.

Any whole number can also be expressed as a fraction. For example, 3 may be thought of as $\frac{3}{1}, \frac{6}{2}, \frac{9}{3}, \frac{12}{4}\ldots$ Each whole number can be expressed in an infinite number of ways, thus generating an equivalence class for each.

To see if two fractions are equivalent, compare their cross products. If the cross products are equal, the fractions are equivalent.

$\frac{2}{3} \stackrel{?}{=} \frac{12}{18}$	$\frac{3}{4} \stackrel{?}{=} \frac{15}{18}$
$\frac{2}{3} \stackrel{?}{=} \frac{12}{18}$	$\frac{3}{4} \stackrel{?}{=} \frac{15}{18}$
$2 \times 18 \stackrel{?}{=} 3 \times 12$	$3 \times 18 \stackrel{?}{=} 4 \times 15$
$36 = 36$	$54 \neq 60$
Equivalent	Not Equivalent

Equivalency of two fractions can also be determined by applying the multiplicative identity. For example, in determining if $\frac{3}{4}$ and $\frac{12}{16}$ are equivalent, find some name for one such that $\frac{3}{4} \times 1 = \frac{12}{16}$. We find that:

$$\frac{3}{4} \times \frac{4}{4} = \frac{12}{16}$$

To emphasize the fact that we are multiplying by 1 to change the name of the number $\frac{3}{4}$, draw a large 1 around $\frac{4}{4}$.

Relations Among Rational Numbers

Often children have difficulty when first comparing two rational numbers. A common mistake is thinking that $\frac{1}{3}$ is larger than $\frac{1}{2}$. Emphasizing the following objectives should be helpful.

12.21
As fractional parts get smaller, the denominator of the fraction gets larger.

12.22
Four of the fourths (or two halves, three thirds, seven sevenths, and so on) make a whole.

12.23
Two fractional parts identified with the same fraction are the same size only when they are parts of the same whole object, or of some other whole object of identical size. Parts of wholes identified with the same fraction are *not* the same size if they are parts of different sized wholes.

12.24
If two fractions have the same numerators, the fraction with the smaller denominator names the greater number.

12.25
If two fractions have the same denominators, the fraction with the greater numerator names the greater number.

Determining the greater of two rational numbers is easiest if they have either the same denominator or numerator.

$\frac{3}{4} > \frac{1}{4}$ because 3 is greater than 1.

$\frac{3}{5} > \frac{3}{7}$ because 5 is less than 7.

If the denominators of two fractions being compared are not equal, the fractions can be renamed so they both have the same denominator.

Which is greater, $\frac{5}{8}$ or $\frac{2}{3}$?

$$\frac{5}{8} = \frac{5 \times 3}{8 \times 3} = \frac{15}{24}$$

$$\frac{2}{3} = \frac{2 \times 8}{3 \times 8} = \frac{16}{24}$$

$\frac{2}{3} > \frac{5}{8}$ because $\frac{16}{24}$ is greater than $\frac{15}{24}$.

Denseness of Rational Numbers

In the set of whole numbers, the next larger whole number can always be named: if n is any whole number, the next larger is $n + 1$. But the next larger number cannot be named in the set of rational numbers: there is always a point on a number line that is midway between the points for any two numbers.

Consider $\frac{1}{4}$ and $\frac{1}{2}$. In order to find the number midway between them, you can find their average.

$$\left(\frac{1}{4} + \frac{1}{2}\right) \div 2 = \frac{3}{4} \times \frac{1}{2} = \frac{3}{8}$$

This property of rational numbers is described as *denseness.* We can say that the set of rational numbers is *dense.*

12.26
The number of rational numbers between any two rational numbers is limitless.

12.27
If a and b are non-negative rationals and $a < b$, there is an infinite set of rationals such as $\frac{a + b}{2}, \frac{a + b}{3}, \frac{a + b}{4}, \ldots$ betweeen a and b.

Notational System

A notational system is used to record rational numbers and the vocabulary associated with different kinds of fractions.

12.28
A *rational number* is an idea; the symbol for this number idea is called a numeral or a fraction.

12.29
The term *fraction* is commonly used to refer to either the symbol or to the rational number.

12.30
A fraction has three parts: a bar, a numeral above it, and another numeral below it.

12.31
A number named by using both whole number and fraction names is a *mixed numeral* or *mixed form.* These expressions are other names for the *sum* of two numbers, for example, $2\frac{1}{2} = 2 + \frac{1}{2}$.

12.32
If the numerator of a fraction is greater than the denominator, the rational number can be renamed as a whole number or in mixed form.

12.33
A fraction for a rational number is in *simplest form* when it shows a numerator and a denominator that have no common factor greater than 1.

12.34
A fraction can be renamed to *higher* or *lower terms* by multiplying both the numerator and denominator by the same number, specifically, $\frac{n}{n}$

Table 12.1 provides a quick reference to the different kinds of fractions that can be used.

Fractions such as $\frac{7}{2}$ and $\frac{6}{5}$ are often called *improper fractions* or, more accurately, *fractions greater than one.*

Mixed forms combine a whole number and a fraction, naming a rational number.

$$2\tfrac{1}{2} = 2 + \tfrac{1}{2}$$
$$= (1 + 1) + \tfrac{1}{2}$$
$$= (\tfrac{2}{2} + \tfrac{2}{2}) + \tfrac{1}{2}$$
$$= \tfrac{4}{2} + \tfrac{1}{2}$$
$$= \tfrac{5}{2}$$

To write a fraction such as $\frac{7}{3}$ as a mixed numeral, think of the fraction as an indicated division. Thus:

$$\tfrac{7}{3} = 7 \div 3 = 2\tfrac{1}{3}$$

When the quotient for $7 \div 3$ is expressed as $2\frac{1}{3}$, the $\frac{1}{3}$ is not a remainder.

Mechanical rules or short cuts, such as the rule for changing a mixed numeral to a fraction greater than 1 (denominator times the whole number, plus the numerator; write it over the denominator) should be introduced *only* after a child has seen a more meaningful mathematical explanation.

A fraction is *renamed,* rather than changed or reduced. The manner of expressing the number, and not the value or size, is altered.

A fraction can be renamed to higher or lower terms by multiplying both numerator and denominator by the same number.

Higher Terms	Lower Terms
$\frac{2}{3} = \frac{2 \times 4}{3 \times 4} = \frac{8}{12}$	$\frac{15}{24} = \frac{15 \times \frac{1}{3}}{24 \times \frac{1}{3}} = \frac{5}{8}$
$\frac{2}{3}$ and $\frac{8}{12}$ are equivalent.	$\frac{15}{24}$ and $\frac{5}{8}$ are equivalent.

Table 12.1
Distinguishing Features of Fractions

Kinds of Fractions	*Special Features*	*Examples*
Unit Fractions	Numerator = 1	$\frac{1}{4}$, $\frac{1}{3}$
Proper Fractions (Fractions less than 1)	Numerator < Denominator	$\frac{3}{4}$, $\frac{1}{2}$
Improper Fractions (Fractions greater than 1)	Numerator > Denominator	$\frac{4}{3}$, $\frac{7}{5}$
Fractions Equal to 1	Numerator = Denominator	$\frac{3}{3}$, $\frac{4}{4}$
Mixed Numerals	Whole Number Numeral Plus Fraction	$1\frac{1}{2}$, $2\frac{2}{3}$
Fractions for Whole Numbers	Fraction Simplifies to a Whole Number	$\frac{6}{2}$, $\frac{12}{3}$
Like Fractions	Same Denominators	$\frac{1}{4}$, $\frac{3}{4}$
Unlike Fractions	Different Denominators	$\frac{1}{4}$, $\frac{2}{3}$

Changing to lower terms can be taught by dividing both numerator and denominator by the same number, in this case, 3.

$$\frac{15}{24} = \frac{15 \div 3}{24 \div 3} = \frac{5}{8}$$

Relating Fractions to Decimals

12.35
In a place value numeration system, each digit within a numeral is assigned a place or position.

12.36
Rational numbers can be expressed with *decimals.* Decimals extend the Hindu-Arabic numeration system to places with values less than 1.

12.37
The digits in a decimal are arranged horizontally with respect to a reference point called the decimal point.

12.38
In a decimal, each place has a value 10 times as great as the place to its right.

12.39
The first three places to the right of the decimal point are tenths, hundredths, and thousandths.

12.40
A decimal has an implied numerator and denominator. The name of the place of the last digit to the right in a decimal indicates the denominator. The numeral itself names the numerator.

12.41
When two non-negative decimals have the same denominator, the decimal with the greater numerator is the larger of the two.

12.42
Fractions and mixed numerals may be renamed as decimals.

The expression *decimal* can describe any base 10 numeral; however, whole numbers are usually not referred to as decimals. The term is generally used for a numeral with digits written to the right of the decimal point. Numerals with a whole number and a decimal are called mixed decimals.

To read a decimal, first look to the place where the last digit is written. This tells you the denominator of the fraction. The numeral itself tells what the numerator is. For example:

- .07 is read *seven hundredths.*
  hundredths
- .265 is read *six hundred twenty-five thousandths.*
  thousandths

The whole number portion of a mixed decimal is read first, followed by the decimal. For example,

- 2.75 is read *two and seventy-five hundredths.*

The word *and* should only be used in reading numerals to indicate the separation of a whole and a fractional number (fraction or decimal).

Two decimals expressed in the same power of 10 can be compared by examining their numerators: for example, .42 is greater than .36. To compare two decimals with *unlike* denominators (different number of decimal places), rename them so that both have the same denominator.

- Which is greater, .3 or .28?
- Rename .3 as hundredths → .30.
- Since .30 is greater than .28, .3 is greater than .28.

Renaming Fractions, Decimals, and Percents

12.43
Fractions and decimals can be renamed as *percents.*

12.44
Every rational number can be expressed either as a terminating or repeating decimal.

12.45
A *terminating decimal* can be written with a finite number of digits, such as, .25 or .69324.

12.46
A *repeating decimal* cannot be written with a finite number of digits. Some digits or series of digits repeat infinitely, such as, $.333\overline{3}$ or $.2727\overline{27}$.

In renaming a fraction as a decimal, the denominator must be renamed as a power of 10. For some fractions, the easiest method is to find an equivalent fraction by using the property of 1.

$$\frac{3}{5} = \frac{3 \times 2}{5 \times 2} = \frac{6}{10} = .6$$

$$\frac{3}{8} = \frac{3 \times 125}{8 \times 125} = \frac{375}{1000} = .375$$

Recognizing the needed name for 1 that will yield a power of 10 is not always easy. A procedure may be used that relies upon the interpretation of a fraction as an indicated division. To rename the fraction as a decimal, divide the numerator by the denominator. The quotient is the decimal.

$$\frac{3}{4} = 4\overline{)3.00}^{\,.75} \quad \frac{3}{4} = .75 \qquad \frac{5}{8} = 8\overline{)5.000}^{\,.625} \quad \frac{5}{8} = .625$$

When the division comes out even (that is, no additional digit in the quotient except zero) the resulting decimal is called a *terminating decimal.* When the division does not come out even, the decimal is called a *repeating decimal.*

$\frac{2}{3} = 3\overline{)2.000}$, quotient $.666$ or $.666\frac{2}{3}$

```
  .666
3)2.000
  1 8
    20
    18
     20
     18
      2
```

$\frac{1}{7} = 7\overline{)1.000000}$, quotient $.142857$ or $.142857\frac{1}{7}$

```
  .142857
7)1.000000
   7
   30
   28
    20
    14
     60
     56
      40
      35
       50
       49
        1
```

Because the last remainder is the same number as the original dividend, the pattern will repeat itself. Such a quotient cannot be completed except to incorporate the remainder in the quotient. A bar is often used to accomplish this.

$$\frac{2}{3} = .6666\ldots = .\overline{6}$$

$$\frac{1}{7} = .142857\ldots = .\overline{142857}$$

The bar indicates that these digits will continue to repeat themselves infinitely.

To express a fraction as a percent, rename the fraction as a two-place decimal; that is, as hundredths. Focus on the number of *hundredths* in order to translate the decimal to a percent. For example:

- .05 is 5 hundredths, so .05 = 5%
- 1.25 is 125 hundredths, so 1.25 = 125%
- .075 is 7.5 hundredths, so .075 = 7.5% or $7\frac{1}{2}$%

To translate from a percent to a decimal, the number of hundredths is again the key. For example,

- 9% means 9 per 100 or $\frac{9}{100}$;
  therefore, $\frac{9}{100}$ = 9 hundredths = .09.
- 125% means 125 per 100 or $\frac{125}{100}$;
  therefore, $\frac{125}{100}$ = 125 hundredths = 1.25.
- 7.5% means 7.5 per 100 or $\frac{7.5}{100}$ or $\frac{75}{1000}$;
  therefore $\frac{75}{1000}$ = 75 thousandths = .075.

Exponential and Scientific Notation

12.47
Numbers written with a large number of digits can be written in a shorter form by using exponential notation.

12.48
Numbers represented in *exponential notation* contain a number called the base, and a superscript (raised numeral) called the exponent.

EXCURSION
Irrational Numbers

Some decimals do not repeat *or* terminate. Numbers named by such decimals are called *irrational* numbers. An example of an irrational number is the $\sqrt{2}$. The set of rational numbers and the set of irrational numbers are both infinite; together they form the set of real numbers.

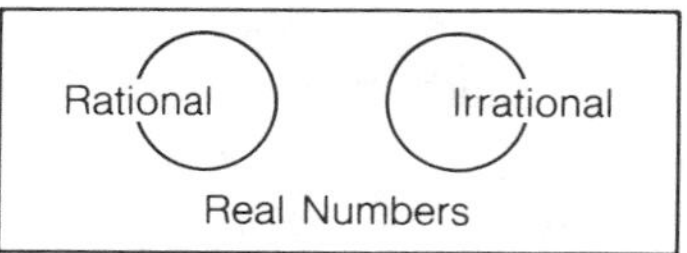

12.49
An *exponent* tells how many times the base is taken as a factor.

12.50
Numbers expressed in *scientific notation* are written as the product of two factors: one is a number between 1 and 10, and the other is a power of 10 expressed in exponential notation.

Numbers such as 425 can be expressed in different ways with expanded notation.

$$\begin{aligned}425 &= 400 + 20 + 5\\ &= (4 \times 100) + (2 \times 10) + (5 \times 1)\\ &= (4 \times 10 \times 10) + (2 \times 10) + (5 \times 1)\\ &= (4 \times 10^2) + (2 \times 10^1) + (5 \times 10^0)\end{aligned}$$

The latter incorporates exponential notation. The product 10×10 is written 10^2. The exponent, the raised numeral, tells how many times the base 10 is used as a factor. Other ways to write 425 using exponential notation include:

$$\begin{aligned}425 &= 42.5 \times 10 \text{ or } 42.5 \times 10^1\\ &= 4.25 \times 100 \text{ or } 4.25 \times 10^2\\ &= .425 \times 1000 \text{ or } .425 \times 10^3\end{aligned}$$

When 425 is written as 4.25×10^2, the number is said to be expressed in *scientific notation*: one factor is between 1 and 10, and the other is a power of 10 expressed with an exponent. Here are other examples of scientific notation:

Number	Expanded Notation	Scientific Notation
42	4.2×10	4.2×10^1
425	4.25×100	4.25×10^2
4250	4.25×1000	4.25×10^3
42,500	$4.25 \times 10{,}000$	4.25×10^4
425,000	$4.25 \times 100{,}000$	4.25×10^5
4,250,000	$4.25 \times 1{,}000{,}000$	4.25×10^6

Often, a large number is rounded off before it is written in scientific notation. For example, the distance from the earth to the sun is approximately 149,668,000 kilometers. If we round 149,668,000 to the nearest million, the result is 150,000,000.

$$150{,}000{,}000 = 1.5 \times 100{,}000{,}000 \text{ or } 1.5 \times 10^8$$

Scientific notation, as the name suggests, is an efficient way for mathematians and scientists to write and work with large numbers.

PLANNING INSTRUCTION

Meaning and Uses of Rational Numbers

Before selecting specific objectives for a child, find out what prerequisite skills and concepts the child possesses. Most children enter school with some understanding of rational numbers. Statements like, "I want the bigger half" and, "That's not fair; her half is bigger than mine" are common. The interest children have in fractional parts can be used to introduce fractions and clarify any misconceptions.

At first, initiating and abstracting activities should focus on the ways fractions can designate parts.

Subdividing a Region into Equal-size Parts

Pies, cakes, and rectangular regions are often pictured in elementary texts. Wooden pattern blocks are also useful for developing fraction concepts. For example: a regular hexagon can be used to show $\frac{1}{6}, \frac{1}{2}, \frac{1}{3}$, and $\frac{2}{3}$.

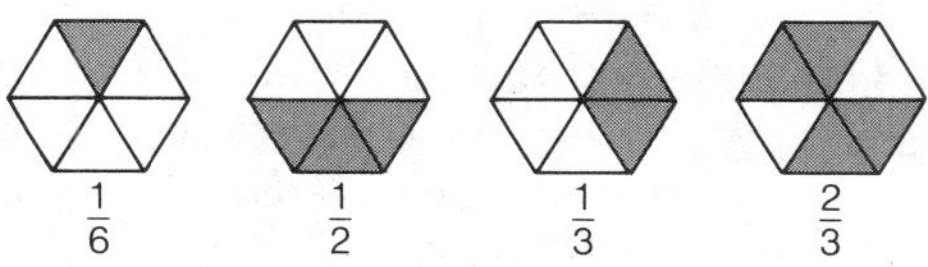

Because many children erroneously believe that half means two parts, and third means three parts, children must also see fractional parts that *do not* suggest $\frac{1}{2}, \frac{1}{3}$, etc.

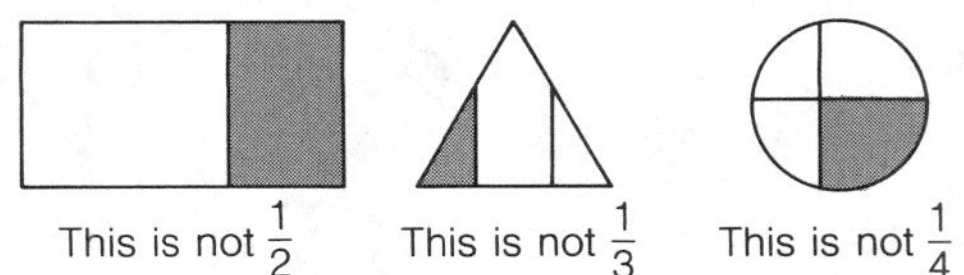

This is not $\frac{1}{2}$ This is not $\frac{1}{3}$ This is not $\frac{1}{4}$

Children should fold, color, and cut figures into equal-size parts. Such activities should help them discover that the size of a half or a third depends on the size of the whole.

A similar representation can be made with Cuisenaire rods. One rod is used to show a unit, and shorter rods of *one color* are placed on top (or at the side) of the unit rod. If the train of shorter rods is the same length as the longer, the *number* of shorter rods is the denominator illustrated. The number of shorter rods actually present at a given time is the numerator.

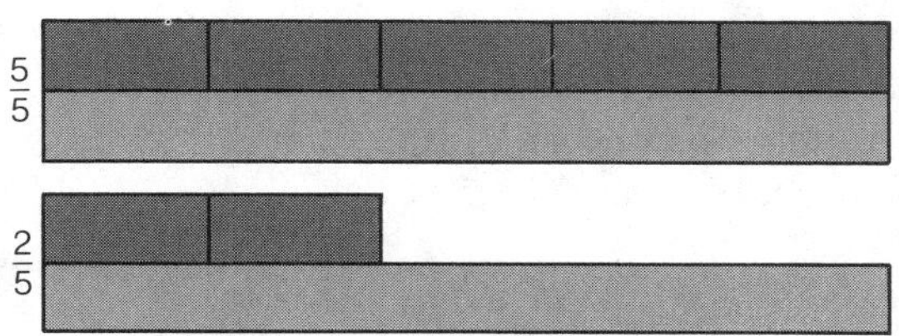

Children should also see that equal-sized parts of the same whole come in different shapes. Have children partition a region into fourths in different ways, then into eighths in as many ways as possible. Such activities help children transfer knowledge of a fraction to less familiar situations; they can be creative as they design partitioned regions.

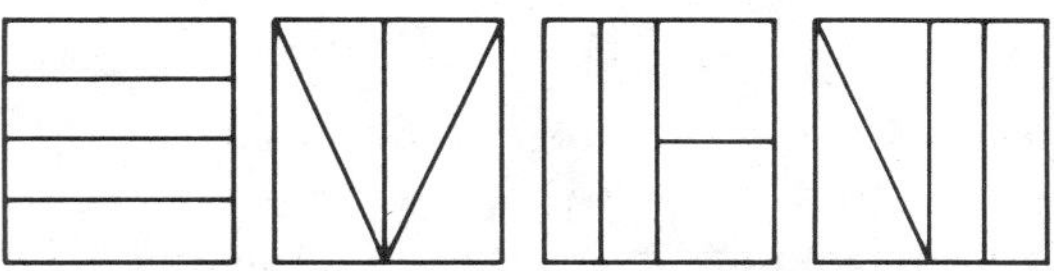

Be sure you use unit regions that vary in shape. For instance, if you use too many circular regions, you may lead children to believe that a fraction is something shaped like a piece of pie.

Subdividing a Set into Subsets of Equal Size

Sets can be made from marbles, apples, dolls, and toy cars. Ask students to identify fractional parts of each set.

What fractional part of this set of toy cars is gray?

2 of the 5, or 2/5 of the cars, are gray.

Also, refer to children in classroom situations: 3 of the 8 children are finished; therefore, $\frac{3}{8}$ are finished.

Egg cartons can be used to suggest a set. Children can place objects in the sections to show different fractions.

Subdividing a Line Segment into Segments of Equal Length

A child's experiences with the number line may be extended to include fractions.

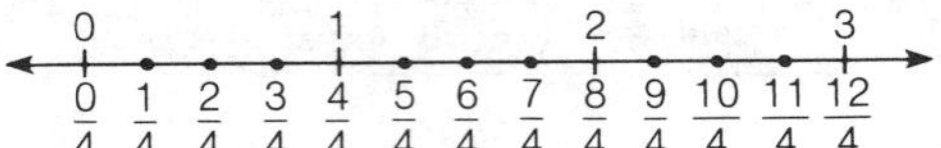

In this number line, each unit segment is divided into 4 segments of equal length. If each fourth is labeled, a child can observe that $\frac{4}{4}$ is another name for 1, and $\frac{8}{4}$ is another name for 2. The number line can also help children see that a fraction such as $\frac{5}{4}$ can be expressed in mixed form ($1\frac{1}{4}$). The number line model is more abstract than models involving regions or sets, but can be used to illustrate a greater variety of fractions and concepts.

When children are ready for a consolidating activity, ask them to identify points on a number line. Label each point with a letter.

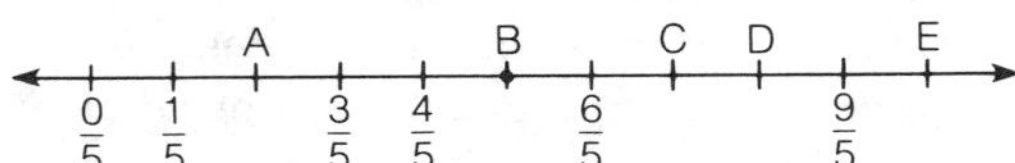

Ratio

Children often use ratios in everyday situations; for instance, in trading and collecting baseball cards. They buy candy in quantities of 3 for a certain price, and listen to their parents talk about gas mileage per gallon.

Because ratios are applied automatically in everyday life without use of the word itself, classroom study of ratios may seem more complicated than necessary.[1] To establish a clearer understanding of the ratio concept, use commonplace things in initiating/abstracting activities. Show children a set of 12 marbles, 4 white and 8 black, and ask them to compare the number of white marbles to the number of black.

A child's response may be that there are 4 more black marbles than white; the child understands that subtraction can be used to compare sets. A more pointed question, such as "How many black marbles are there for every white marble?" may elicit a response related to ratio. Pairing the marbles 1 to 2 will help make the relationship clear. A ratio is an *ordered* pair of numbers, so be sure children view the materials from the correct side. Otherwise, they may see a different ratio.

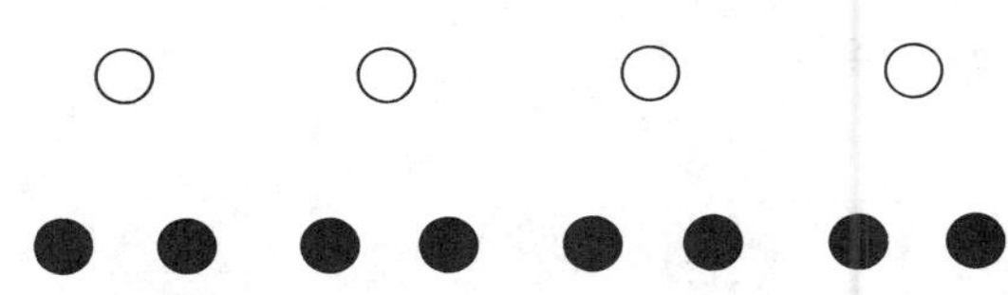

1. John Marks, et al., *Teaching Elementary School Mathematics for Understanding,* 4th ed. (New York: McGraw-Hill, 1975), p. 204.

EXCURSION Name the points shown on each line.

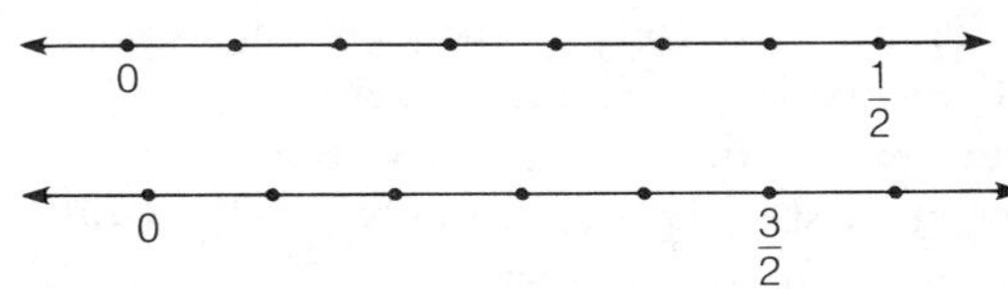

This is an excellent transferring activity for some children.

Cuisenaire rods and centimeter cubes can also be used as models for the concept of ratio.

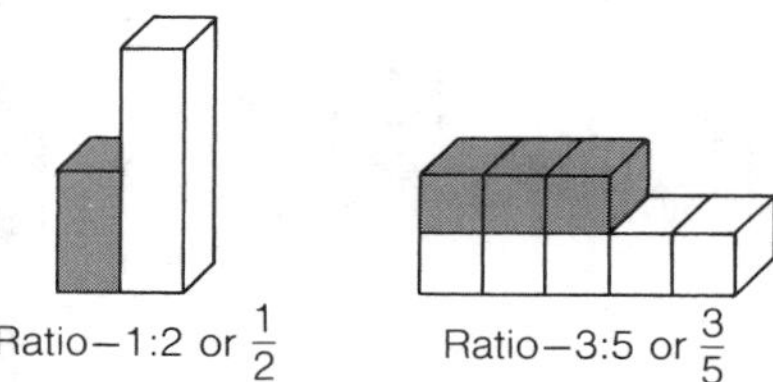

Children should have many experiences with ratios. For instance, compare the number of boys in the classroom to the number of girls, centimeters in a meter, Saturdays in a week, inches in a foot, pennies in a nickel, and items correct on a test to the number of test items.

Then introduce notation for recording a ratio. A mathematician prefers to write a ratio in the form 2:3 rather than $\frac{2}{3}$. The distinction between a ratio and a fraction is subtle and is usually not made in the elementary school. Since both forms of expressing a ratio are commonly used, children should be introduced to both.

Price relationships may also be expressed as a ratio. For example, 3 cans of corn for 95¢ may be expressed as the ratio $\frac{3}{95¢}$. Such situations are common and should be included in the study of ratios.

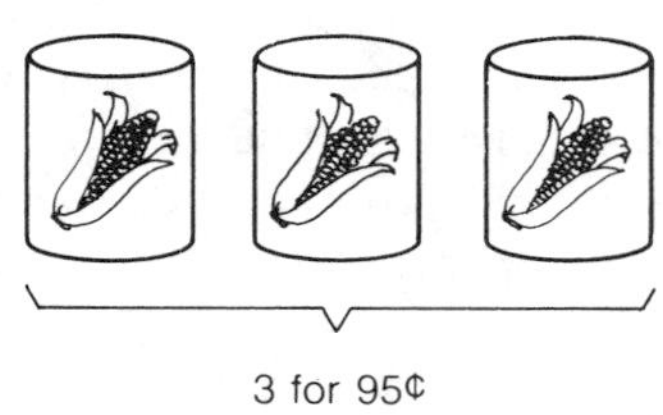

Numbers Greater Than 1

Fractions equal to and greater than 1 should be illustrated with regions and sets. Several examples follow. The key to understanding these fractions is the concept of a reference unit. What is the reference unit for each fraction illustrated?

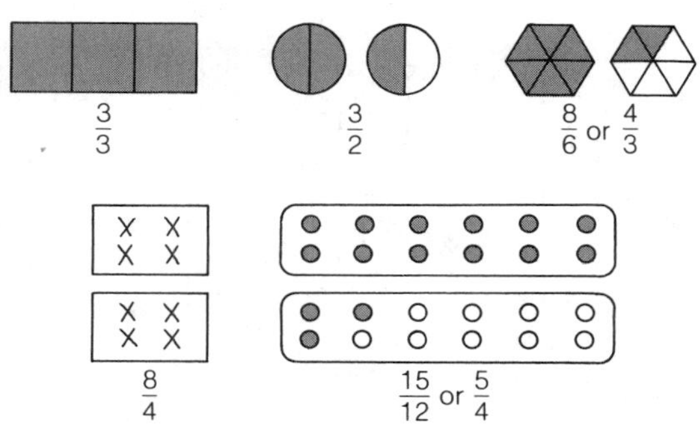

Relations Among Rational Numbers: Fractions

Children have many informal experiences with equivalent fractions: half of a pound of butter may consist of 2 sticks or 2 quarters, and a half dozen eggs is 6 of the 12 eggs in the box.

Given an opportunity to experiment with fractional parts, children soon discover that more than one fraction can be used to identify a particular part of a set or region. For instance, parts of circular regions or a fraction board can be used to illustrate that $\frac{1}{2}$, $\frac{2}{4}$, and $\frac{4}{8}$ name the same part of a region.

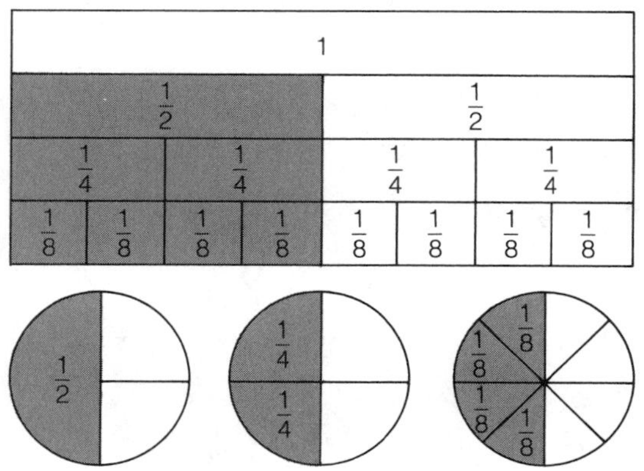

A number line can be used to identify families of fractions (see page 278, top). The child can see that $\frac{1}{2}$, $\frac{2}{4}$, $\frac{3}{6}$ and $\frac{6}{12}$ all name the same point. Note that one may be named many ways ($\frac{1}{1}$, $\frac{2}{2}$, $\frac{3}{3}$, $\frac{4}{4}$, $\frac{6}{6}$, . . .). Fractional numbers greater than 1, such as $\frac{3}{2}$, also have other names.

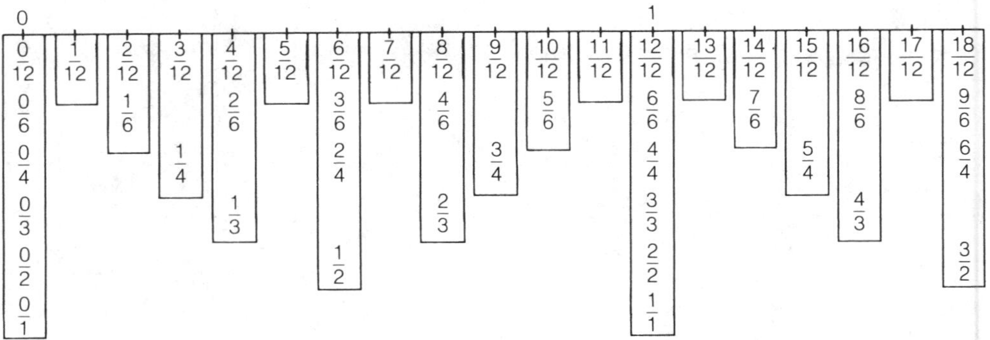

The child who has acquired the rudiments of the equivalence idea is ready to compare fractional numbers. Have students compare two fractions by using parts for the same region and by referring to points on a number line. Later, introduce children to more abstract procedures.

Usually, the first method presented involves cross products. Later children rename the fractions so they have the same denominators.

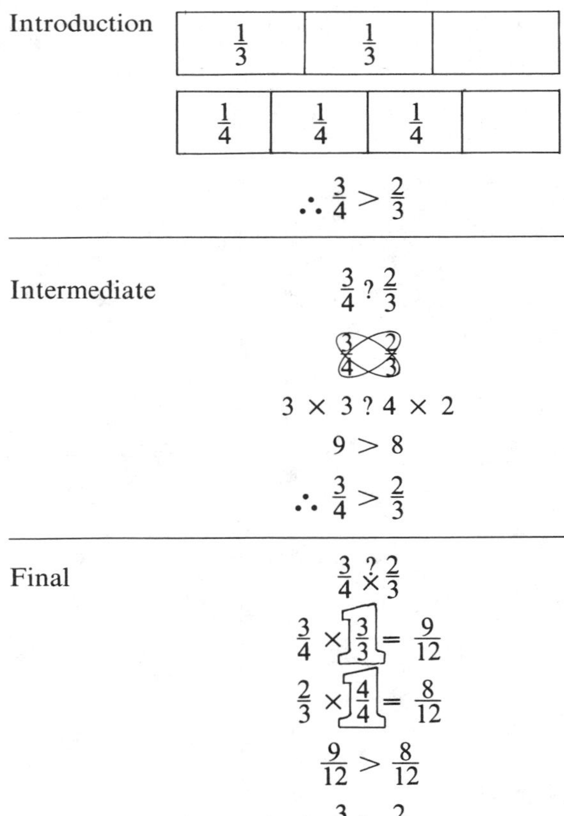

Renaming Rationals: Fractions and Mixed Numerals

A chart listing many equivalent fractions for each of several different numbers will suggest a method for generating equivalent fractions. A child can be asked to consider the set of fractions equivalent to $\frac{1}{4}$ (that is, $\frac{1}{4}, \frac{2}{8}, \frac{3}{12}, \frac{4}{16}, \ldots$). Guidance as follows may be appropriate:

- Compare the first fraction, $\frac{1}{4}$, to the second fraction, $\frac{2}{8}$.
- Notice that the numerator of the second fraction is 2 times the numerator of the first fraction.
- How do the denominators compare?
- How do the numbers in the third fraction compare to the numbers in the first? (Etc.)

A child can be led to see that multiplying the numerator and denominator of a fraction by the same number produces another fraction equivalent to the original. A child can remember this application of the multiplicative property of 1 by expressing the operation this way:

$$\frac{1}{4} = \frac{1 \times 3}{4 \times 3} = \frac{3}{12}$$

If we provide extensive activities with equivalent fractions, the task of *simplifying* fractions becomes an application and extension of concepts learned when raising fractions to higher terms. For example, to write the fraction $\frac{12}{18}$ in lowest terms, a child needs only to find the "family name" for the equivalence class for which $\frac{12}{18}$ is a member. To determine the simplified fraction, the child finds the integer that is the greatest factor of both the numerator and denominator. For some children, finding the greatest common factor (G.C.F.) may be more time consuming than simplifying the fraction in stages. Three ways of simplifying the fraction $\frac{12}{18}$ are illustrated. Procedure 1 uses the G.C.F.

Procedure 1

$$\frac{12}{18} = \frac{12 \div 6}{18 \div 6} = \frac{2}{3}$$

Procedure 2

$$\frac{12}{18} = \frac{12 \div 3}{18 \div 3} = \frac{4}{6}$$

$$\frac{4}{6} = \frac{4 \div 2}{6 \div 2} = \frac{2}{3}$$

Procedure 3

$$\frac{12}{18} = \frac{12 \div 2}{18 \div 2} = \frac{6}{9}$$

$$\frac{6}{9} = \frac{6 \div 3}{9 \div 3} = \frac{2}{3}$$

Children should see all three procedures and, with time and experience, should be encouraged to use the method that is most efficient for them.

Many of the things that have been said about equivalent fractions and renaming fractions are illustrated in the activity plans in Figures 12.1 (pages 279–80) and 12.2.

The procedure for renaming a mixed numeral so it is a fraction greater than 1 should be taught with models such as circular regions to illustrate the following reasoning.

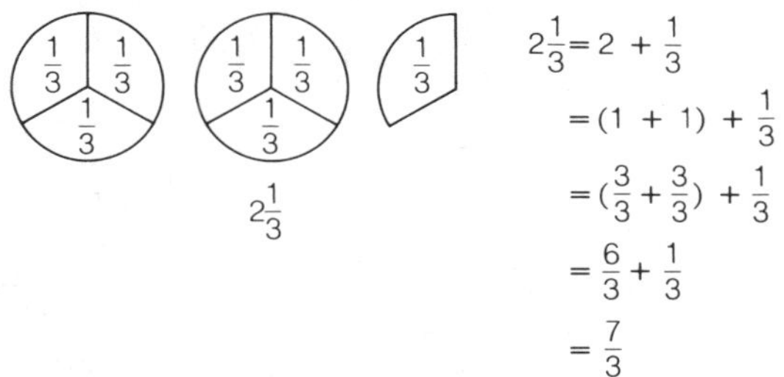

$2\frac{1}{3}$
$2\frac{1}{3} = 2 + \frac{1}{3}$
$= (1 + 1) + \frac{1}{3}$
$= (\frac{3}{3} + \frac{3}{3}) + \frac{1}{3}$
$= \frac{6}{3} + \frac{1}{3}$
$= \frac{7}{3}$

Reverse the process to rename a fraction such as $\frac{7}{3}$ as a mixed numeral. After children understand this process, have them divide to find the mixed numeral.

$$\frac{7}{3} \rightarrow 3\overline{)7} \text{ (quotient 2; } 7 - 6 = 1\text{)} \rightarrow 2\frac{1}{3}$$

ACTIVITY PLAN

Equivalent Fractions — ***2-D/Rep and 3-D/Rep → Symbolic*** — ***Abstracting/Schematizing***

Content objectives

12.16

Materials/Exemplars

Candy bars
Circular regions partitioned into fractional parts
Shaded regions showing fractional parts
Number line

Behavioral indicators

The child:

1. Compares two fractions by comparing their models as part of a region or line segment.
2. Applies the term equivalent fractions to any two fractions that name the same fractional number.
3. Constructs a set of equivalent fractions for a specified fraction.

Procedure

1. Take two candy bars of the same size. Cut one into halves and the other into quarters. Hold up half of one candy bar and ask each child to write a fraction to name this part. Do the same for the two quarters of the candy bar.

2. Say, "Which is greater, $\frac{1}{2}$ or $\frac{2}{4}$?"
3. Write $\frac{1}{2} = \frac{2}{4}$ and say, "Two fractions that name the same fractional part are called 'equivalent fractions.' The fractions $\frac{1}{2}$ and $\frac{2}{4}$ are equivalent fractions."
4. Show two circular regions of the same size partioned into thirds and sixths. Point to one of the thirds. Ask a child to find how many sixths it will take to cover $\frac{1}{3}$ of the other region. Say. "Do $\frac{2}{6}$ and $\frac{1}{3}$ name the same fractional part?" Write the statement $\frac{1}{3} = \frac{2}{6}$.

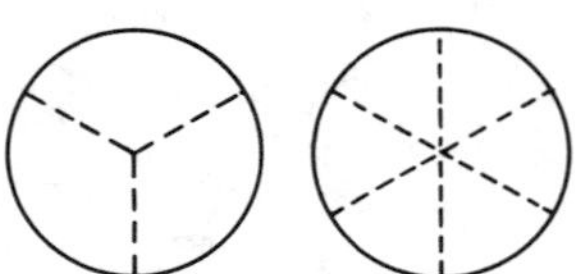

5. Show two regions of the same size partitioned into fourths and eighths with $\frac{3}{4}$ of each shaded. Say, "How many parts are shaded in the first figure? How many in the second?"

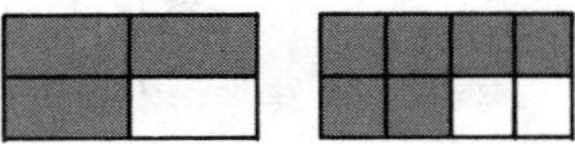

6. Write the fractions $\frac{3}{4}$ and $\frac{6}{8}$. Ask, "Are these fractions equivalent? How can you show that they are equivalent?"
7. Show two number lines partitioned into fifths and tenths. Show $\frac{2}{5}$ by marking the point on the first line. Say, "Can you find an equivalent fraction for $\frac{2}{5}$ on the second line?" After a child does this, write $\frac{2}{5} = \frac{4}{10}$. Ask children to show that $\frac{3}{5}$ and $\frac{6}{10}$ are equivalent.

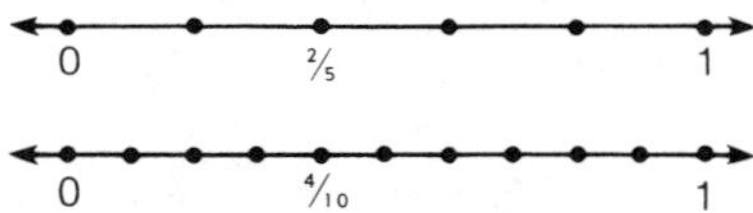

8. Provide each child with two rectangular regions, partitioned into fourths and eighths. Say, "Show by shading the two regions that $\frac{1}{4}$ and $\frac{2}{8}$ are equivalent fractions."

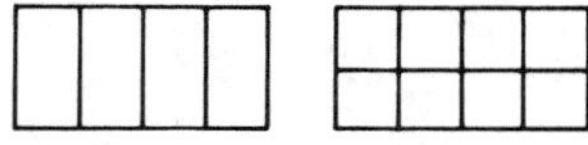

9. Repeat the procedure using regions for $\frac{2}{3}$ and $\frac{4}{6}$; and for $\frac{1}{5}$ and $\frac{2}{10}$. Do the same using number lines for $\frac{1}{4}$ and $\frac{2}{8}$; and for $\frac{4}{5}$ and $\frac{8}{10}$.
10. Ask, "What do you call two fractions that name the same fractional numbers? Can you find two more fractions equivalent to $\frac{3}{4}$?"

Figure 12.1
Sample Activity Plan for Equivalent Fractions

ACTIVITY PLAN

Equivalent Fractions | *2-D/Rep → Symbolic* | *Abstracting/Schematizing*

Content objective

12.20

Materials/Exemplars

1. Shaded regions showing fractional parts
2. Chalkboard and chalk

Behavioral indicators

The child:

1. States that any fraction multiplied by 1 is equal to itself.
2. Renames a fraction by multiplying by 1 in the form of *n/n*.
3. Constructs a set of equivalent fractions for a specified fraction.

Procedure

1. Show each child two shaded regions of the same size, one partitioned into quarters and the other into eighths. Say, "Write a fraction to show the shaded part of each region."
2. Say, "Show by comparing the regions that $\frac{3}{4}$ and $\frac{6}{8}$ are equivalent fractions. Show by using cross products that $\frac{3}{4}$ and $\frac{6}{8}$ are equivalent fractions."
3. Ask, "Look at the fraction $\frac{6}{8}$. How do its numerator and denominator compare with the numerator and denominator of $\frac{3}{4}$?" (Both are twice as large.)
4. Write $\quad \times \quad = \frac{6}{8}$ and say, "If you mulitply the fraction $\frac{3}{4}$ by $\frac{2}{2}$, the new fraction is $\frac{6}{8}$. What is another name for $\frac{2}{2}$?"
5. Say, "What fraction is produced if you multiply $\frac{3}{4}$ by $\frac{3}{3}$?" Write $\frac{3}{4} \times \boxed{\frac{3}{3}} = \frac{9}{12}$. Show each child a rectangular region the same size as the other two, partitioned into twelfths. Ask, "Is the new fraction, $\frac{9}{12}$, equivalent to the other two fractions?"
6. Ask each child to show that $\frac{3}{4}$ and $\frac{9}{12}$ are equivalent by comparing regions and cross products.
7. Write the fraction $\frac{1}{4}$ on the board. Ask each child to multiply $\frac{1}{4}$ by $\frac{2}{2}$. Say, "Any fraction multiplied by 1 is equal to itself." Emphasize this by using the large 1 to show that $\frac{2}{2} = 1$.
8. Write $\frac{1}{2} = \frac{2}{4} =$ ____ $=$ ____ $=$ ____ on the board. Ask, "Can you find 3 more equivalent fractions for $\frac{1}{2}$?"
9. Ask each child to complete these sentences: Any fraction multiplied by ____ is equal to itself. To make a fraction equivalent to $\frac{2}{3}$ you would multiply by ________.

Figure 12.2
Sample Activity Plan for Procedure to Rename Fractions

Consolidating activities for renaming fractions and mixed numerals can include games and other stimulating practice situations.

Rummy

Ten or more sets of four cards are prepared with different names for the same fraction on each of the four cards. (For example: $\frac{1}{3}, \frac{4}{12}, \frac{3}{9}, \frac{6}{18}$.) To play, each student must match corresponding fractional names.

Old Maid

The game Old Maid can be adapted to drill with fraction and decimal equivalents by constructing pairs of cards naming equivalent decimals and fractions (for example: $\frac{3}{5} = .60$). One monster card is inserted in the deck. Children play the game by matching pairs. Later the game can be adjusted to include equivalent percents.

Geofraction

Geofraction cards may be constructed by selecting a fractional sequence (for example, $\frac{1}{12}, \frac{1}{6}, \frac{1}{4}, \frac{1}{3}, \frac{5}{12}, \frac{1}{2}, \frac{7}{12}, \frac{2}{3}, \frac{3}{4}, \frac{5}{6}, \frac{11}{12}, \frac{12}{12}$).[2] Write these fractions inside geometric shapes.

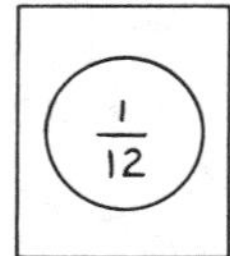

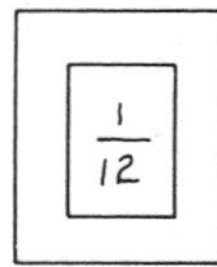

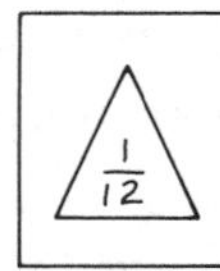

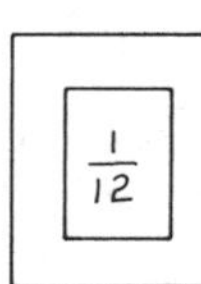

The fractions replace the normal number and face cards found on playing cards. Using a deck containing 12 circles, 12 squares, 12 triangular and 12 rectangular cards, you can have students play most popular card games. By creating different sets of geofractions cards, a child can learn to order fractional numbers and recognize equivalent fractions.

Bingo

Fraction Bingo is played like the well-known game.[3] Playing cards consist of 3 rows of fractions with 3 fractions in a row; the center space is a free space. Various names are used for the more frequently used fractions, such as $\frac{2}{4}$ and $\frac{3}{6}$ for $\frac{1}{2}$. Fractions such as $\frac{5}{10}$ and $\frac{25}{100}$ that can be renamed as decimals may also be used. The caller picks a card containing a fraction from a pile and any player who has on his card some equivalent form of the fraction covers it with a disc. Three in a row wins.

Fraction Bingo		
$\frac{1}{3}$	$\frac{2}{14}$	$\frac{5}{10}$
$\frac{9}{15}$	Free Space	$\frac{6}{9}$
$\frac{25}{100}$	$\frac{2}{10}$	$\frac{10}{12}$

This game may be adjusted in complexity for more able children by increasing the number of rows and columns, or by having a mix of decimals, fractions, and percents.

Probability

Certain basic concepts of probability provide another context for using rational numbers. Topics related to probability are intriguing and motivating because the answer is speculative and cannot be precisely predicted. Children enjoy guessing and then testing their guesses against data they collect.[4]

Typical initating activities involve flipping coins, rolling dice or drawing cards from a deck. Experiments can be introduced in class and repeated at home. After frequent repetitions of an experiment in which each result is recorded, a child will learn that the outcome of a particular trial is uncertain. Children realize that the probability of a coin landing heads-up is $\frac{1}{2}$; the

2. Adapted from R. T. Heimer and C. R. Trueblood, *Strategies for Teaching Children Mathematics* (Reading, Mass.: Addison-Wesley, 1977), pp. 245–46. Copyright © 1977. Reprinted with permisssion.

3. Nancy Cook, "Fraction Bingo," *The Arithmetic Teacher* 17, no. 3 (March 1970): 237–39.

4. George Green, *Elementary School Mathematics: Activities and Materials* (Lexington, Mass.: D.C. Heath, 1974), p. 440.

probability of drawing a spade from a deck of 52 cards is $\frac{1}{4}$, and the probability of rolling a five with a dice is $\frac{1}{6}$. Most children grasp these probabilities intuitively.

Later, ask them to predict what will happen if *two* coins are tossed at the same time. Most will predict three outcomes: one head and one tail, two heads, and two tails; and that each outcome will occur about one-third of the time.

When children actually flip two coins, they will discover that the outcomes are significantly different from their predictions. To help them figure out why, the experiment should be done with two different coins, such as a penny and a nickel. Children are able to see that coins can land *four* ways—the one head and one tail occurring either when the nickel lands heads-up and the penny lands tails, or when the nickel lands tails and the penny lands heads. Ask children to complete this chart.

	Outcomes		
Event	Penny	Nickel	Probability
1	H	H	$\frac{1}{4}$
2	H	T	
3	T	H	
4	T	T	

For a transferring activity, ask children to analyze what will happen if they flip three coins. After making the prediction, the children should flip three different coins numerous times, and compare the outcomes to their predictions. Other questions for investigation follow.

PROBLEMS

Given a box containing 5 red marbles, 3 green marbles, and 2 blue marbles, what is the probability of drawing a red marble at random?

Given the spinner illustrated, what is the probability of having the arrow stop on red? on yellow? or green?

Given a pair of dice, what is the probability of rolling a sum of 6, a sum of 7, a sum of 3?

Children should also predict the theoretical probability of an event based on the experimental probability. Figure 12.3 describes a plan to introduce this concept.

Proportion

The concept of ratio can be further extended by introducing proportion. Problems similar to the following may be used to initiate the concept. If students are given candy bars and play money, they can quickly discover a solution to the problem.

PROBLEM

Candy bars are 2 for 25¢. What is the cost of 6 bars?

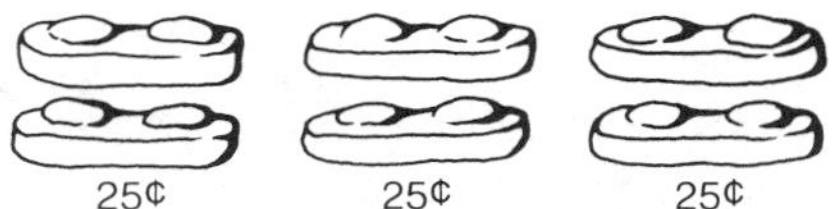

Candy bars are 2 for 25¢. What is the cost of 6 bars?

The solution is achieved by matching a quarter with 2 candy bars. This is repeated until the child has 6 bars; counting the money indicates a total of 75¢ is required.

Decimals

In the past, fractions and the associated operations were introduced before decimals. Today, decimals are likely to be introduced earlier due to increased use of the metric system and hand-held calculators. Therefore, children must thoroughly understand numeration concepts for whole numbers.

Decimals are typically introduced as an extension of the place value system used for whole numbers. Decimals should be related to their fraction equivalents if fractions have been introduced; for example, $.3 = \frac{3}{10}$. Emphasize the symmetry of the extended place value system, gradually developing a chart similar to that shown on page 284 (top).

An effective aid for initial work with decimals is a set of base 10 blocks. Redefine the values associated with the blocks, possibly

1	1,000,000	$\frac{10000000}{1}$	1 million
1	100,000	$\frac{1000000}{1}$	1 hundred thousand
1	10,000	$\frac{100000}{1}$	1 ten thousand
1	1,000	$\frac{10000}{1}$	1 thousand
1	100	$\frac{100}{1}$	1 hundred
1	10	$\frac{10}{1}$	1 ten
1	1	$\frac{1}{1}$	1 one
1	0.1	$\frac{1}{10}$	1 tenth
1	0.01	$\frac{1}{100}$	1 hundredth
1	0.001	$\frac{1}{10000}$	1 thousandth
1	0.0001	$\frac{1}{100000}$	1 ten thousandth
1	0.00001	$\frac{1}{1000000}$	1 hundredth thousandth
1	0.000001	$\frac{1}{10000000}$	1 millionth

regarding a flat as one unit, a long as a tenth of the unit, and the individual "units" as hundredths of the new unit.

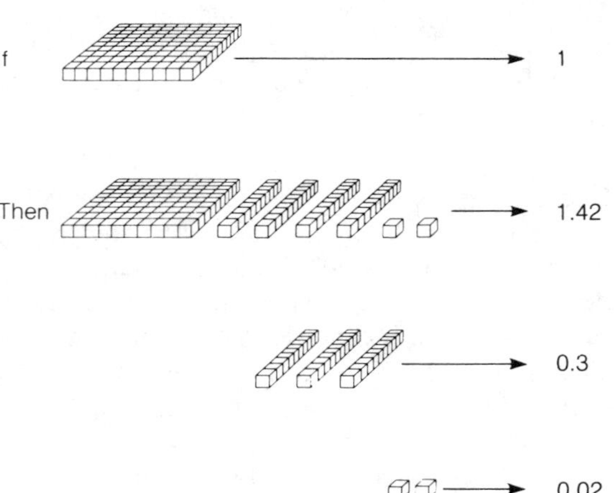

An abacus can also be used with decimals. The value of each rod must be redefined and a decimal point inserted as a marker. As with whole numbers, abacus activities are more abstract than those using base blocks. Place value charts can also be adapted (see top of next column).

Number lines involving tenths and metric containers are also helpful in understanding decimals.

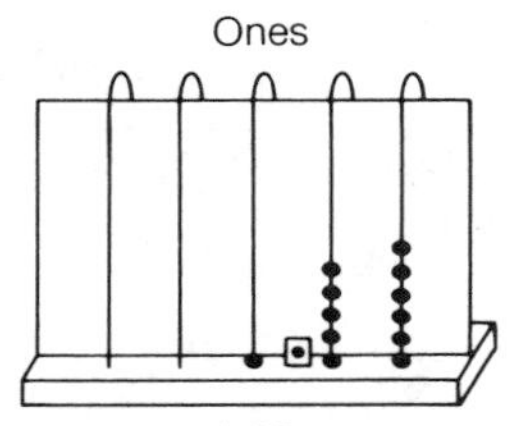

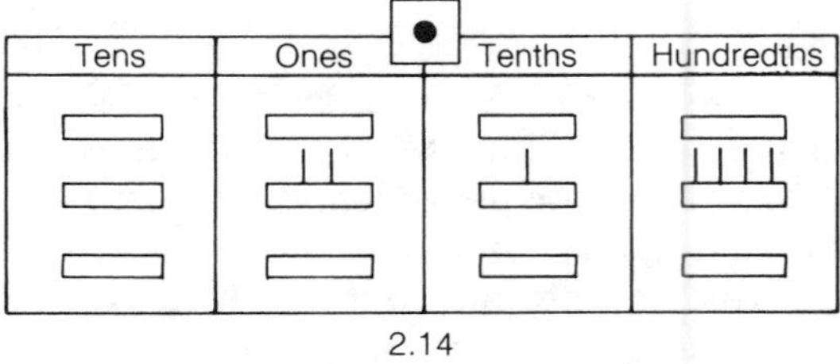

In teaching children to read decimals, place emphasis on the fact that *and* is used when reading both mixed decimal and mixed fraction numerals. Thus, 2.75 is read as two *and* seventy-five hundredths. Emphasis on the clear pronunciation of *th* is necessary from the very beginning.

Children's early experiences with decimals are often related to money. Because of this, initiating activities involving money are sometimes used. But even when a child has learned to read and write symbols for dollars and cents, the child may still not understand cents as hundredths of a dollar, or dimes as tenths of a dollar. Mastery of

ACTIVITY PLAN

Probability | *3-D Rep → Symbolic* | *Initiating/Abstracting*

Content objective

Predictions can be made as a result of observations made during an experiment.

Exemplars/Materials

12 marbles (9 red and 3 black)
A bag
Paper and pencil

Behavioral indicators

Given a bag containing 12 marbles, the child:

1. Predicts the ratio of red marbles to the total number of marbles in the bag.
2. Predicts the ratio of black marbles to the total number of marbles in the bag.

Procedure

1. Give each child a bag containing 12 marbles (9 red and 3 black).
2. Say, "You have a bag containing 12 marbles. Some are red and some are black. *At no time may you look in the bag*. Draw marbles, one at a time, from your bag, record the color, replace the marble, shake the bag and draw another marble. Repeat this procedure 24 times."
3. As needed, help individuals to record their results, making certain that they do not peek into the bag.
4. After 24 selections, ask "Based on the data you have recorded, how many of the 12 marbles in your bag are red? How many are black?"

Figure 12.3
Sample Activity Plan for Probability

this difficult relationship may be postponed until the child has had more experiences with decimals.

Relations Among Rational Numbers: Decimals

It is much easier for children to compare the size of two decimals than to compare two fractions. Have children compare two different regions of a grid containing one hundred squares, then the decimals for the regions. This should lead to the generalization that to order decimals with the same denominator (same number of decimal places), only the numerators need to be examined (CO 12.41). This rule is similar to the rule for ordering fractions (CO 12.25).

If the denominators of two decimals are unequal (different number of decimal places), the student need only apply the rule for comparing two fractions; that is, rename the given decimals so they have the same denominator. For example:

- Which is greater, .6, .62, or .593?
- Rename .6 as thousandths →.600
- Rename .62 as thousandths→.620
- 593 thousandths ——————→.593

Thus, the order from smallest to largest is .593, .6, and .62. A number line can be used to verify the relationships involved.

Renaming Rationals: Decimals, Fractions, and Percents

In some situations, decimals are more useful than fractions for comparing two rational numbers or for performing certain calculations. Skill in changing from one to the other form is needed. For example, for renaming tenths, show a child a strip of 10 stamps and ask:

- How many stamps are on this strip?
- After I tear off 5 stamps, how can I represent the remaining portion?

- I can write 5 tenths as .5. Is there another way I can write 5 tenths?
- Is there a simpler way to write $\frac{5}{10}$?
- Prove that your answer is correct by showing that 5 stamps are one-half of a strip of 10.

A 10 × 10 square of graph paper can be used to show tenths and hundredths, and to illustrate the renaming of such commonly used fractions as .25 and .75.

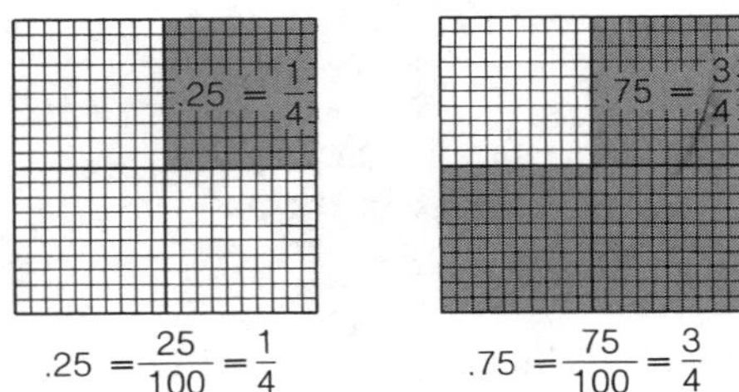

Because only powers of 10 can serve as denominators of decimals, renaming a decimal as a fraction is simple; find an equivalent fraction with a power of 10 as the denominator. Students who have learned how to generate an equivalent fraction with a specified denominator already know how to do this.

$$\frac{2}{5} = \frac{2 \times 2}{5 \times 2} = \frac{4}{10}$$

Some fractions cannot be renamed with denominators of 10 and are renamed as hundredths or thousandths.

$$\frac{7}{20} = \frac{7 \times 5}{20 \times 5} = \frac{35}{100} = .35$$

$$\frac{9}{250} = \frac{9 \times 4}{250 \times 4} = \frac{36}{1000} = .036$$

But for some denominators, no such whole number multiplier exists.

As soon as children learn to divide with decimals, a more general method for all fractions should be introduced. Because a fraction indicates division, the numerator can be divided by the denominator. The result is a decimal that names the same number as the fraction. The numerator (dividend) must be written in a form suitable for division. Thus, 5 is written as 5.000.

$$\frac{5}{8} \rightarrow 8\overline{)\,5.000} \qquad \begin{array}{r} .625 \\ 4\,8 \\ \hline 20 \\ 16 \\ \hline 40 \\ 40 \\ \hline \end{array} \qquad \frac{5}{8} = .625$$

Children soon generalize that any fraction can be renamed as a decimal by dividing the numerator by the denominator, annexing zeros after the decimal point if needed. Begin with terminating decimal fractions. The more difficult repeating fractions should be saved until the students feel comfortable with the renaming procedure.

A percent is a special kind of fraction or ratio with a denominator of 100. In fact, percent means *per hundred.* The fraction $\frac{25}{100}$, the ratio 25 per 100, and 25% all express the same quantity. This relationship may be shown using tags on a hundreds board. Also, strings of 100 beads or spools can illustrate the percent concept.

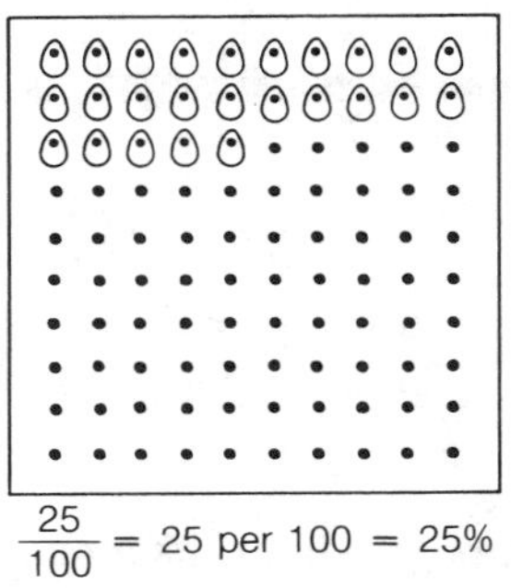

Many of the activities used to develop an understanding of ratio, fractions, and decimals can also be used when teaching percent; graphic representation, for example. On duplicated 10 × 10 squares, students may color various sections, 30 for example, to show 30 out of 100, or 30%. Ask students to show that 7 out of every 10, $\frac{7}{10}$, is the same as 70 out of 100 or 70%. Rectangular regions similar to those for comparing fractions can also be used for percent.

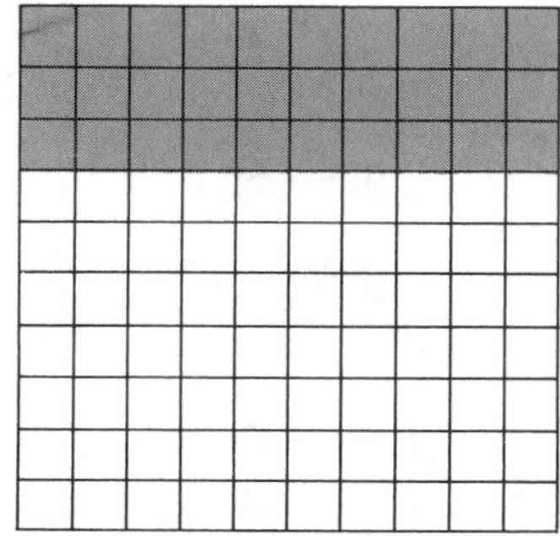

100%					
50%			50%		
33⅓%		33⅓%		33⅓%	
25%	25%		25%	25%	
20%	20%	20%	20%	20%	

Because certain fraction, decimal, and percent equivalents are used so frequently, they should be stressed during abstracting and schematizing activities. Games and other consolidating activities should be used to help each child memorize the important equivalencies. Table 12.2 can be used as a reference.

PLANNING ASSESSMENT

To assess which content objectives have been learned, we need to identify related behavioral indicators that will provide criteria against which we can evaluate each child's performance. A child who can illustrate a concept in many ways has a superior level of understanding.

First, determine the content objective (or cluster of objectives) to be assessed. For example:

> **12.2**
> A fraction may be used to express the number for part of a whole thing, or part of a set of things.

With the objective in mind, prepare behavioral indicators that involve different translations across exemplar classes.

Consider the following behaviors that involve translating from two-dimensional to symbolic representations.

Table 12.2
Fraction/Decimal/Percent Equivalents

Fraction	Decimal	Percent	Fraction	Decimal	Percent
$\frac{1}{20}$	.05	5%	$\frac{1}{2}$	.50	50%
$\frac{1}{10}$	.10	10%	$\frac{3}{5}$	.60	60%
$\frac{1}{8}$	$.12\frac{1}{2}$	$12\frac{1}{2}\%$	$\frac{5}{8}$	$.62\frac{1}{2}$	$62\frac{1}{2}\%$
$\frac{1}{6}$	$.16\frac{2}{3}$	$16\frac{2}{3}\%$	$\frac{2}{3}$	$.66\frac{2}{3}$	$66\frac{2}{3}\%$
$\frac{1}{5}$	.20	20%	$\frac{7}{10}$	.70	70%
$\frac{1}{4}$	.25	25%	$\frac{3}{4}$	.75	75%
$\frac{3}{10}$	.30	30%	$\frac{4}{5}$	.80	80%
$\frac{1}{3}$	$.33\frac{1}{3}$	$33\frac{1}{3}\%$	$\frac{5}{6}$	$.83\frac{1}{3}$	$83\frac{1}{3}\%$
$\frac{3}{8}$	$.37\frac{1}{2}$	$37\frac{1}{2}\%$	$\frac{7}{8}$	$.87\frac{1}{3}$	$87\frac{1}{3}\%$
$\frac{2}{5}$	.40	40%	$\frac{9}{10}$	.90	90%

- Given a pictorial representation of a fractional number, state and write the appropriate fraction.
- Given a number line representation of a fractional number, state and write the appropriate fraction.

Tasks that could be used to elicit these behaviors include:

1. Show the child three shaded regions.

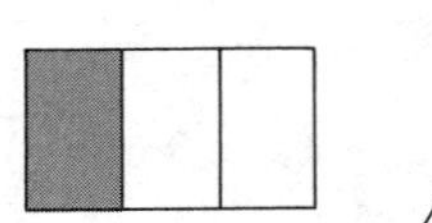
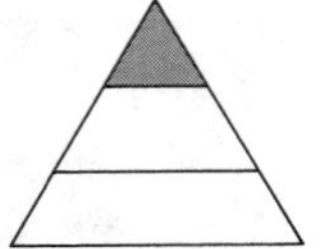
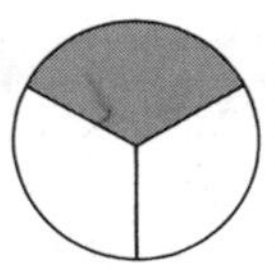

Ask, "Which regions show $\frac{1}{3}$?"

2. Show the child a number line.

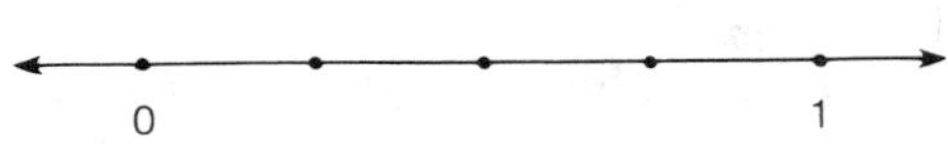

Say, "Put your finger on the point on the line which names $\frac{3}{4}$."

3. Show the child a picture of a set of triangles.

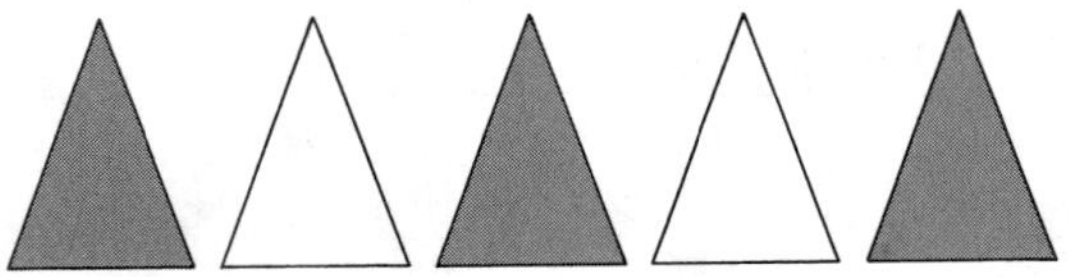

Ask "What part of the set is shaded?"

4. Show the three shaded regions.

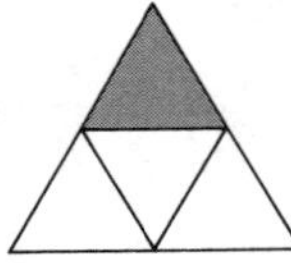
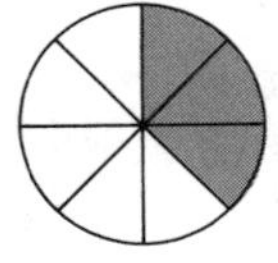
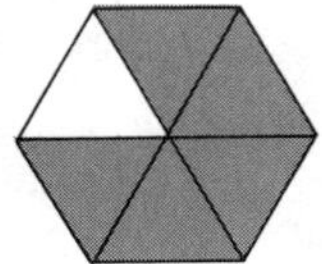

Say, "Write a fraction to show the shaded part of each region."

5. Show the child a number line.

Say, "Write a fraction to name point A on the number line."

Each child should be able to reverse this procedure; translate from a symbolic to a two-dimensional representation. Consider this behavior:

- The child constructs a pictorial representation of a fraction by drawing a ring around a part of a set or shading a region.

Tasks suggested by this behavior include:

1. Show the child a region partitioned into eighths.

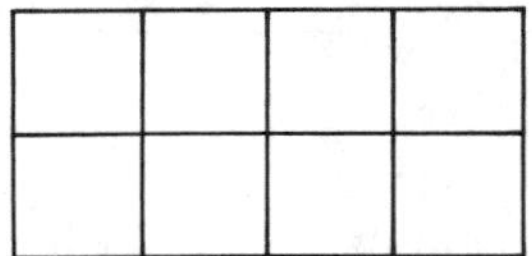

Say, "Shade part of this region to show $\frac{3}{4}$."

2. Show the child a picture of 6 oranges.

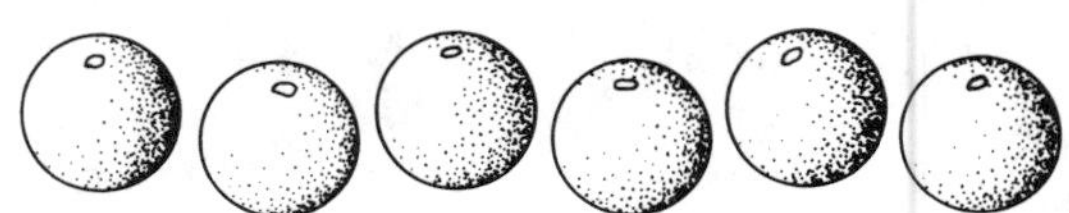

Say, "Draw a ring around the part of the set that shows $\frac{2}{3}$."

The following behavior involves a translation from symbolic to three-dimensional representations.

- The child partitions a set of discrete elements into equivalent disjoint subsets as a model for the fractional part.

To assess the behavior:

1. Place a set of 12 crayons on the desk.

Say, "Place these crayons into sets. Make each set contain exactly $\frac{1}{3}$ of the crayons."

2. Repeat, and have the child construct sets containing $\frac{1}{4}$ and $\frac{1}{6}$ of the crayons.

EXCURSION

Construct an assessment task for this behavior:

Given a set of discrete elements partitioned into equivalent disjoint subsets, write a fraction for the part represented by each subset.

13

Developing Computational Procedures with Decimals and Percents

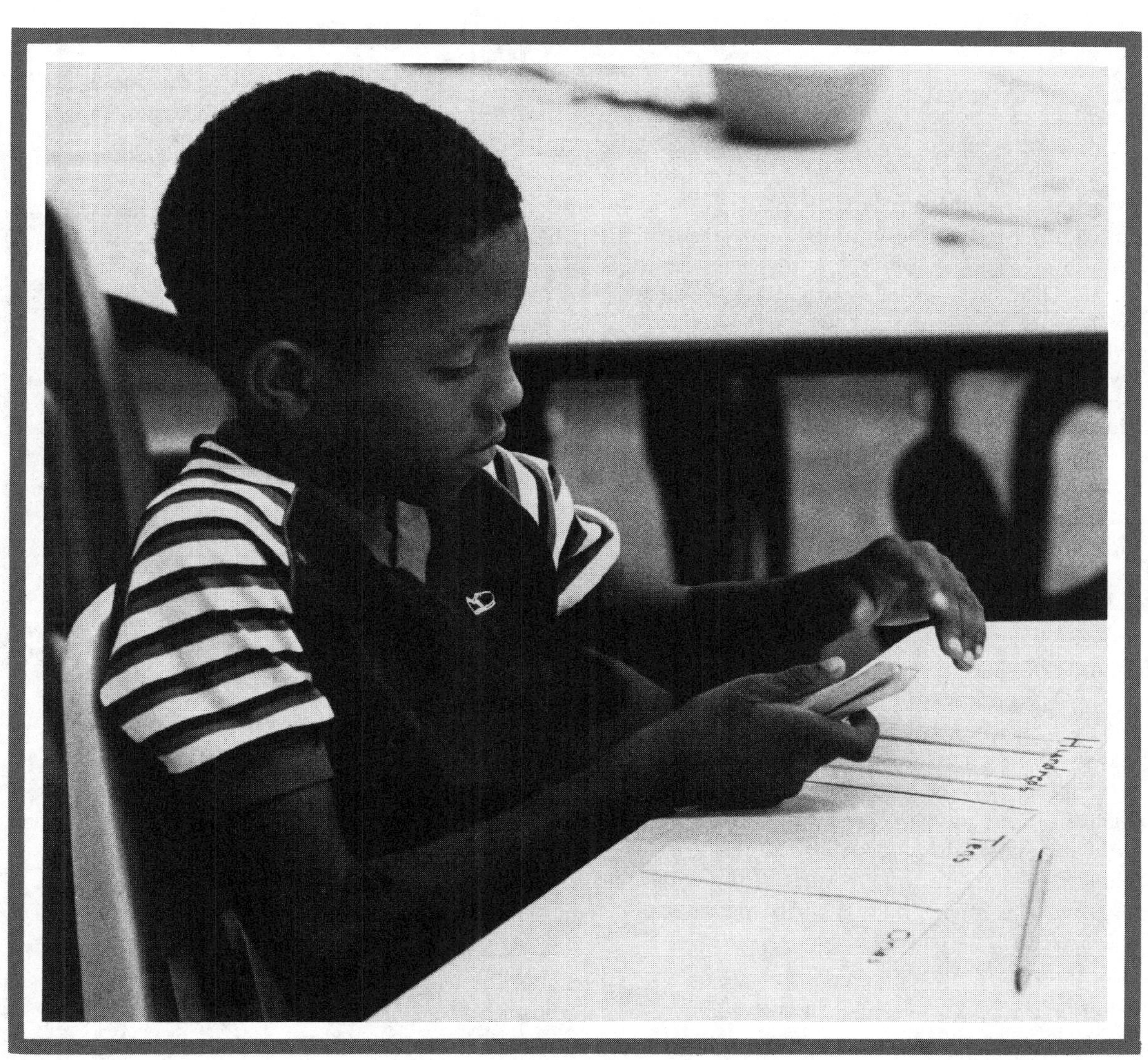

Symbols are used to record the results of operations on numbers; we do not perform the operations on symbols. Therefore, the term *adding decimals* is technically inaccurate; the precise expression is, *adding numbers expressed as decimals.* Throughout this chapter, more commonly accepted and less cumbersome expressions such as *adding and subtracting decimals* are used, but we should be aware of this subtle distinction as we teach children.

PLANNING THE CONTENT

Even if we follow a textbook or a curriculum guide, we must decide what objectives to teach each child. Which content is he ready to learn? Does he have the prerequisite knowledge and skills? Teaching time is limited and the objectives chosen must be the most important for each child.

The concepts and procedures used in computing with whole numbers are required for operations on decimals, including the basic facts, regrouping, and the addition, subtraction, multiplication and division algorithms. Computing with decimals can be done as though they were whole numbers, until the results of the computation must be interpreted, then the decimal point is placed in the answer.

Addition and Subtraction with Decimals

The following objectives relate to properties of the operations on rational numbers, algorithms for the operations, and ragged decimals.

Properties

Some of the properties that apply for addition and subtraction of whole numbers also apply to these operations on rational numbers.

13.1
Closure—The sum of any two rational numbers is always a rational number.

13.2
Commutativity—The order or sequence in which two rational numbers are added has no effect on their sum.

13.3
Associativity—The way in which three rational numbers are grouped together in addition has no effect upon their sum.

13.4
Identity—Zero added to or subtracted from any rational number yields a sum or difference equal to the original number.

Algorithms

13.5
Decimals are added and subtracted much in the same manner as whole numbers.

13.6
When decimals are added or subtracted, decimal points are aligned vertically.

EXCURSION
Measuring with Precision[1]

Machinists use micrometers to measure distances to the nearest hundredth of a millimeter. The object to be measured is placed between the jaws of the micrometer and the trimble is tightened until the object fits snugly. You can measure the distance by taking the reading from the barrel and head.

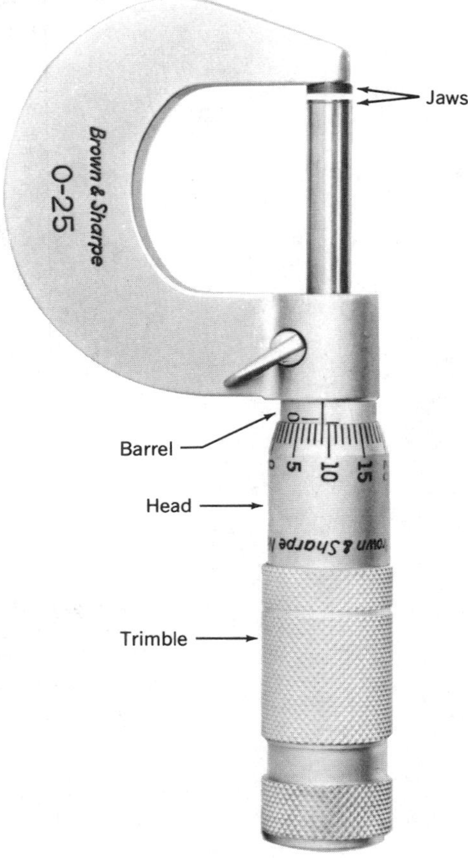

The reading on the barrel is to the nearest 0.5 mm.

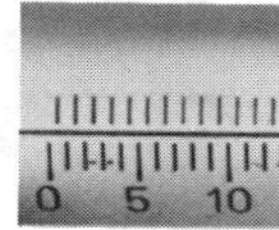

The reading on the head is to the nearest 0.01 mm.

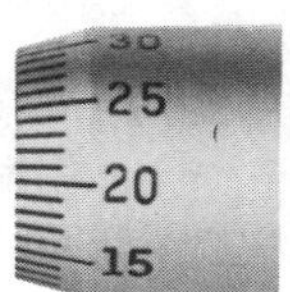

1. Adapted from Jack Price et al., *Mathematics for Everyday Life* (Columbus, Ohio: Charles E. Merrill, 1978), pp. 66–67.

Can you see how the measurements are obtained?

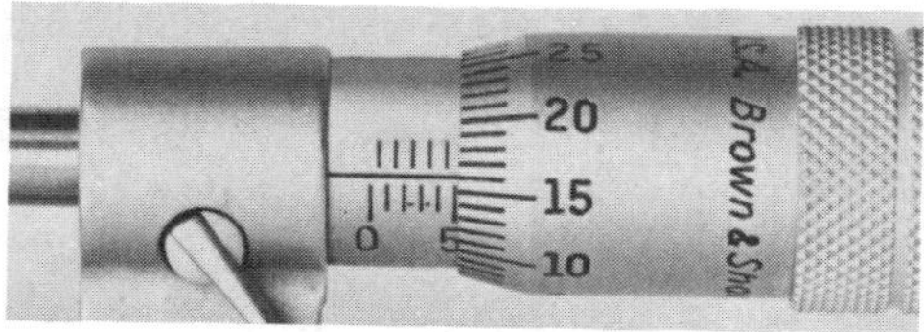

Barrel reading 5.00 mm
Head reading .16 mm
5.16 mm

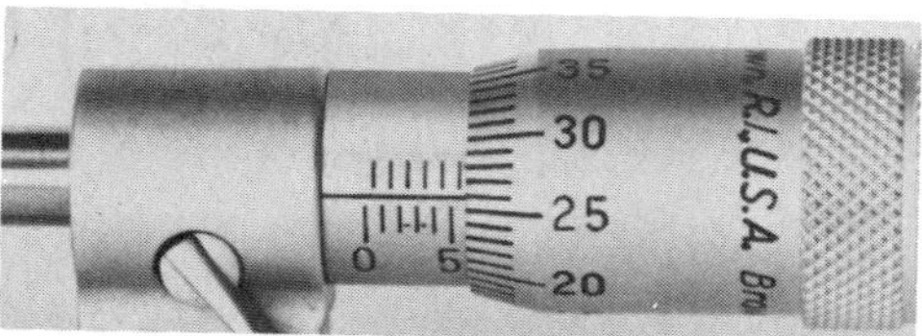

Barrel reading 5.50 mm
Head reading .26 mm
5.76 mm

Find the measurement of each.

a.

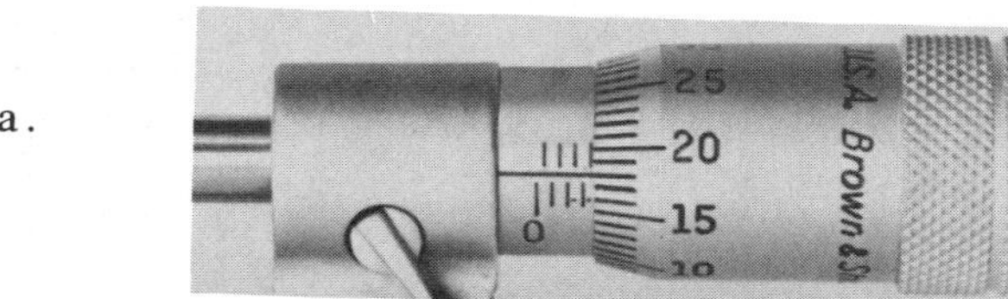

b.

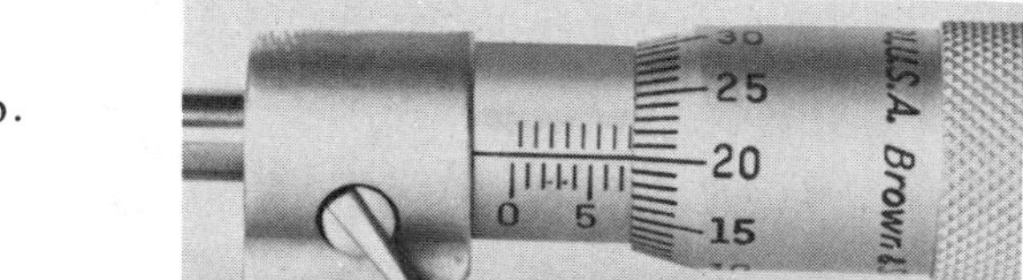

13.7
Tenths are added to tenths, with the resulting sum in tenths; hundredths are added to hundredths, with the sum in hundredths; and so on.

13.8
Tenths are subtracted from tenths, with the resulting difference in tenths; hundredths are subtracted from hundredths, with the difference in hundredths; and so on.

Digits should be placed under each other according to their respective place values for computational efficiency.

$$\begin{array}{r} 4.13 \\ 2.32 \\ +\ \underline{1.12} \\ 7.57 \end{array} \qquad \begin{array}{r} 4.25 \\ +\ \underline{2.98} \\ 7.23 \end{array} \qquad \begin{array}{r} 7.5 \\ -\ \underline{2.3} \\ 5.2 \end{array} \qquad \begin{array}{r} 6.1 \\ -\ \underline{3.8} \\ 2.3 \end{array}$$

Ragged Decimals

Ragged decimals are those in which the number of digits to the right of the decimal point varies, for example, .75 + .1 + .386, or 6.2 − 3.75. Generally, the need to add or subtract fractions or decimals arises from situations in which measurements are typically expressed in the same units of measure. Problems involving ragged decimals are often contrived and not related to everyday situations. Children are usually taught to annex zeros as required.

Many elementary mathematics teachers instruct children in addition and subtraction of ragged decimals because such examples are included in certain standardized tests.

Multiplication with Decimals

Properties

Again, some of the properties for multiplication of whole numbers also apply to multiplication with rational numbers.

> **13.9**
> *Closure*—The product of any two rational numbers is always a rational number.
>
> **13.10**
> *Commutativity*—The order or sequence in which two rational numbers are multiplied has no effect on their product.
>
> **13.11**
> *Associativity*—The way in which three rational numbers are grouped together in multiplication has no effect upon their product.
>
> **13.12**
> *Identity*—The product or quotient of any rational number and the number 1 is equal to the given number.

> **13.13**
> *Distributivity*—Rational numbers under the operation of multiplication are distributive with respect to addition.

The mechanics for computing products when numbers are expressed as decimals are not particularly complex. But, as with algorithms for whole numbers, making the procedure seem reasonable requires careful instruction. Unless children make sense of what they are doing they will operate by rote, which often results in difficulties with retention.[2]

Algorithms

> **13.14**
> The product of a whole number and a decimal in tenths is expressed in tenths; the product of two decimals in tenths is expressed in hundredths; the product of a decimal in tenths and a decimal in hundredths is expressed thousandths; and so on.
>
> **13.15**
> The algorithm for multiplying decimals may be considered in two phases.
> a. Ignore the decimal points and compute as though the factors were whole numbers.
> b. Locate the decimal point in the product with as many digits to the right of the point as there are to the right of both decimal points in the factors.

If one of the factors (the multiplier) is a whole number, the relationship between multiplication and addition can be shown. For example:

$$4 \times 7.28 = 29.12 \qquad \begin{array}{r} 7.28 \\ 7.28 \\ 7.28 \\ +\ 7.28 \\ \hline 29.12 \end{array}$$

2. Francis Mueller, *Arithmetic, Its Structure and Concepts*, 2nd ed. (Englewood Cliffs, N.J.: Prentice Hall, 1964), p. 270.

If, however, both factors are decimals, do not try to show that an addend appears a fractional number of times. Examples such as the following suggest a rule for locating the decimal point in a product.

A.
```
     .2←1 place
   × 4←0 place
     .8←1 place
```

C.
```
     .8←1 place
   × .9←1 place
    .72←2 places
```

B.
```
     .62←2 places
   × .04←2 places
   .0248←4 places
```

D.
```
    .035←3 places
   × .14←2 places
  .00490←5 places
```

In examples B and D, the number of digits in the product is less than required. Therefore, the required number of zeros is annexed in front of the digits.

Division with Decimals

13.16
The algorithm for division with decimals may be considered in two phases.
a. Locate the decimal point in the quotient.
b. Ignore the decimal point and divide as though the dividend and divisor were whole numbers.

13.17
If both dividend and divisor are multiplied by the same nonzero number, the quotient does not change.

If the divisor is a whole number (other than zero), the division example can be interpreted as either measurement or partitive division.

PROBLEMS

If $1.20 is to be divided equally among four children, how much will each receive? (partitive division)

John earns $2 an hour. How many hours did he work if he earned $8.50? (measurement division)

However, if the divisor is a decimal, the example may be interpreted only as a measurement situation.

PROBLEM

How many pieces of steel .2 meter long may be cut from a longer piece 1.2 meters long?

The division algorithm that we use for decimals will probably be the same as for whole numbers. The new element is the placement of a decimal point in the quotient. In this chapter, the authors use the distributive algorithm.

If the divisor is a whole number, place the decimal point in the quotient directly above the decimal point in the divisor, dividing as with whole numbers.

When the divisor is a decimal, the process becomes more difficult. A widely used method for locating the decimal point in the quotient involves substituting an equivalent division statement for the given one. The two divisions, the original and the substitute, yield the same quotient. For a more difficult division, substitute a simpler one by multiplying both the divisor and dividend by some power of 10.

Original Division		Substituted Division
1. .4) 4.32 → .4) 4.32 (Multiply by 10)		10.8 over 4) 43.2
2. .43) 77.4 → .43) 77.40 (Multiply by 100)		180. over 43) 7740. (Annex Zero)

When children understand the use of a substitute division, they can write carets (∧) directly on the original division to locate its decimal points. Writing both the original division and the substitute is unnecessary; one set of numerals is sufficient.[3]

.28∧) 6.44∧   .025∧).0875∧

3. Mueller, *Arithmetic*, p. 279.

EXCURSION
How Far Away Is the Storm?[4]

Thunder and lightning occur at the same time. Light travels almost instantly, and sound travels about .206 miles per second. To find how far away a storm is, follow these steps.

1. When you see the flash, count the number of seconds it takes before you hear the thunder.
2. Multiply the number of seconds the thunder took to reach you by .206 in order to find the distance in miles.

For example, if it takes 5 seconds for thunder to reach you after you see the lightning, how far away is the storm?

Distance = .206 × Seconds It Takes Thunder to Reach You
= .206 × 5
= 1.030 Miles

The storm is about a mile away.

Products and Quotients Involving Powers of 10
The systematic effect of multiplying by a power of 10 can be extended to include multiplying and dividing decimals by powers of 10.

> **13.18**
> *Multiplying* a decimal by 10 or a power of 10 has the effect of moving the digits in the numeral the same number of places to the left of the decimal point as there are zeros in the power of 10.

1000 × 6.38 = 6380 (Note: Zero has to be annexed)
3 zeros 3 places

> **13.19**
> *Dividing* a decimal by 10 or a power of 10 has the effect of moving the digits in the numeral the same number of places to the right of the decimal point as there are zeros in the power of 10.

48.2 ÷ 1000 = .0482 (Note: Zero has to be annexed)
3 places 3 zeros

Percent Equations
Of all the ratios, the percent is the most frequently encountered. However, it is often the most difficult for students to interpret.

> **13.20**
> Every basic percent problem involves one of three situations, determined by the unknown term in the basic relationship.
> a. Finding a part or percent of a number. For example,
>
> 25% of 20 is ______.
>
> b. Finding what part or percent one number is of another. For example,
>
> ______% of 15 is 5.
>
> c. Finding a number, when a certain part or percent of that number is known. For example,
>
> 20% of ______ is 6.

4. Adapted from Sadie C. Braggs, *General Mathematics Skills and Applications* (Morristown, N.J.: Silver Burdett, 1977), p. 115.

EXCURSION
Fuel Performance[5]

To check your car's fuel performance:

1. Fill your tank completely and record the odometer reading.
2. Drive a reasonable distance, until your tank is about $\frac{1}{4}$ full.
3. Refill the tank, record the amount of gas purchased and record the odometer reading.
4. Find the difference between the odometer readings to measure the distance driven.
5. Divide the distance driven by the amount of gasoline purchased to find the fuel performance.

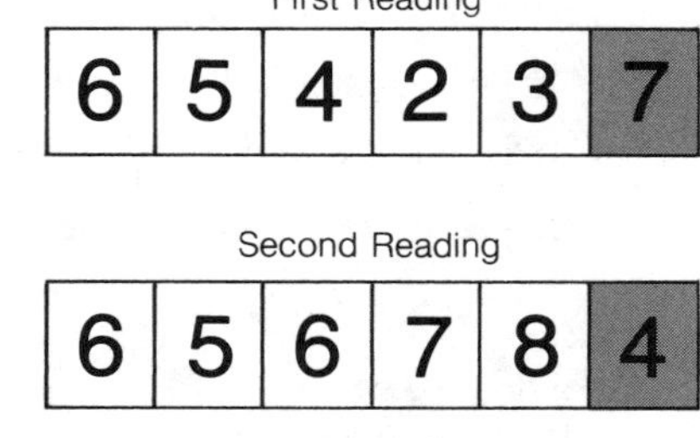

65,678.4	Second Odometer Reading
−65,423.7	First Odometer Reading
254.7	

$$\text{Miles per Gallon} = \frac{\text{Miles Driven}}{\text{Number of Gallons Used}} = \frac{254.7}{10.5}$$

The fuel economy is 24.3 miles per gallon.

Ordering the Content

We do not have to present operations with decimals in a fixed order. In some textbook series, operations with fractions precede those with decimals. In others, decimals are presented first. The merits of both presentations can be argued; the authors treat decimals first.

Work with whole numbers can be extended to include decimals, starting with addition and subtraction. Multiplication with decimals should be taught before division. The order in which we sequence the content for division is particularly important. Begin with division of a decimal by a whole number. Follow this with division of a decimal by a decimal, then present division of a whole number by a decimal.

PLANNING INSTRUCTION

Before teaching algorithms that involve decimals, review prerequisite concepts and skills. Plan diagnostic activities to make certain that children:

5. Adapted from Braggs, *General Mathematics*, p. 137.

- Understand the meanings of addition and subtraction
- Can add and subtract with whole numbers, and explain the use of place values and renaming
- Can name and rename decimals

Initiating, Abstracting, and Schematizing Activities

Whole number procedures can be consolidated in extending addition and subtraction to decimals: the algorithms are similar.

Addition and Subtraction

The most common difficulty with adding and subtracting decimals is the proper placement of digits and the decimal point. Children solve problems that involve money long before they are formally introduced to decimals, and such problems help them abstract and schematize the important idea of lining up decimal points before adding or subtracting.

Most children are also familiar with the use of decimals in situations involving distance. Such experiences can be incorporated into initiating activities for addition and subtraction of decimals.

PROBLEMS

On the way to school Joe stopped to pick up Scott. How far does Joe have to walk to school?

Fred lives how much farther from school than Scott?

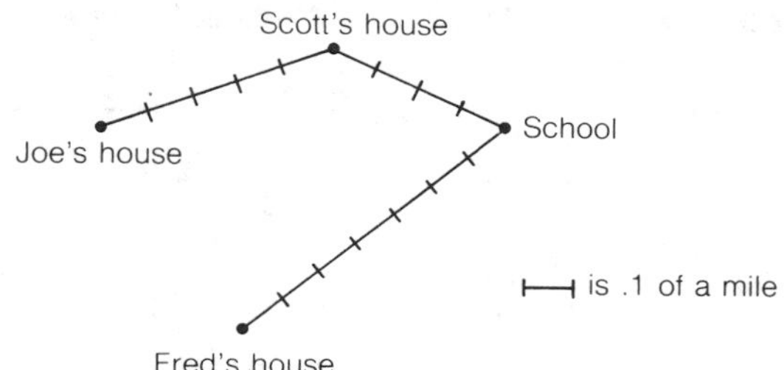

Each child can be led to record:

5 tenths		.5
+4 tenths		+.4
9 tenths	or	.9

and

7 tenths		.7
−4 tenths		−.4
3 tenths	or	.3

A pocket chart and an abacus are both effective devices to use when relating these operations with decimals to the whole number algorithms. For example, have each child use a pocket chart to find the sum of 2.7 and .5. Discuss the pattern involved. (See Figure 13.1.)

We can also use a pocket chart to develop rules for subtracting. For 2.1 − .8, see Figure 13.2.

Step 1: Represent 2.1. There are not enough tenths, so

Step 2: Rename 1 as 10 tenths.

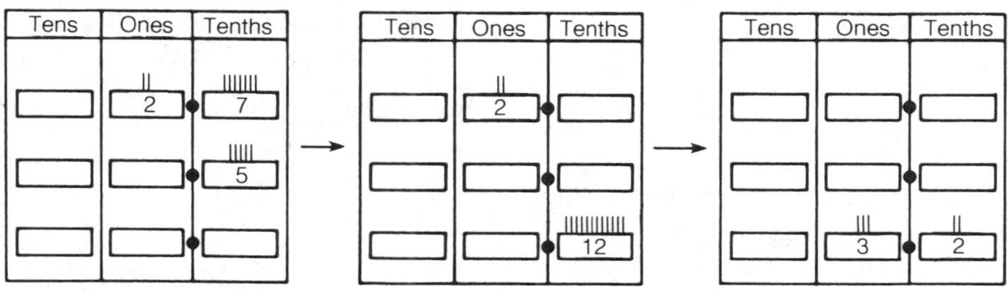

FIGURE 13.1
Pocket Chart for Adding Decimals

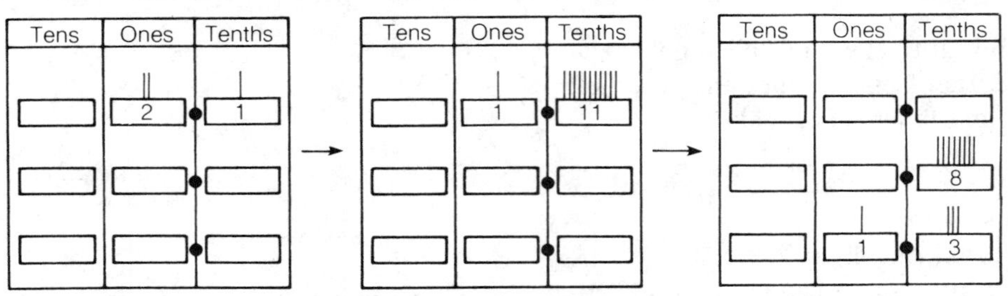

FIGURE 13.2
Pocket Chart for Subtracting Decimals

Step 3: Subtract 8 tenths.

When mathematics programs introduce operations with fractions before decimals, we can relate addition and subtraction of decimals to prior experiences with fractions. For example:

$.5 + .3 = ?$	$.28 + .4 = ?$
$.5 \rightarrow \frac{5}{10}$	$.28 \rightarrow \frac{28}{100} \rightarrow \frac{28}{100}$
$+.3 \rightarrow \frac{3}{10}$	$+.4 \rightarrow \frac{4}{10} \rightarrow \frac{40}{100}$
$? \quad \frac{8}{10} = .8$	$? \quad \frac{68}{100} = .68$
$.9 - .3 = ?$	$.7 - .42 = ?$
$.9 \rightarrow \frac{9}{10}$	$.7 \rightarrow \frac{7}{10} \rightarrow \frac{70}{100}$
$.3 \rightarrow \frac{3}{10}$	$-.42 \rightarrow \frac{42}{100} \rightarrow \frac{42}{100}$
$? \quad \frac{6}{10} = .6$	$? \quad \frac{28}{100} = .28$

Multiplication

The inverse relationship between multiplication and division should be stressed when teaching algorithms for decimals, although multiplication is introduced prior to division. Furthermore, make sure each child has the following *before* introducing the multiplication algorithm:

- Understanding of the meaning of multiplication
- Competence in multiplying whole numbers
- Competence in using decimal numeration

If the child grasps these, only one new idea—that of the placement of the decimal point in the product—needs to be stressed.

An excellent initiating activity for multiplication with decimals involves base 10 blocks. Flats (which represent hundreds in work with whole numbers) must be defined as equal to 1, longs (which represent tens) as $\frac{1}{10}$, and units (ones) as $\frac{1}{100}$. Figure 13.3 shows how a child might display the product 3×3.6. The approach can be extended for examples like 2.3×3.6, as in Figure 13.4. In both cases, a child regroups (trades) as necessary and counts to determine the product.

The use of such arrays helps a child schematize the algorithm for multiplication. The following record can be made from the array in Figure 13.4:

3.6	
× 2.3	
.18	.3 × .6
.90	.3 × 3
1.20	2 × .6
6.00	2 × 3
8.28	2.3 × 3.6

EXCURSION
A Maze

Can you go from the outside to the center? Add together every number that you cross. Find a path with a sum of 17.39. Find a sum of 20.

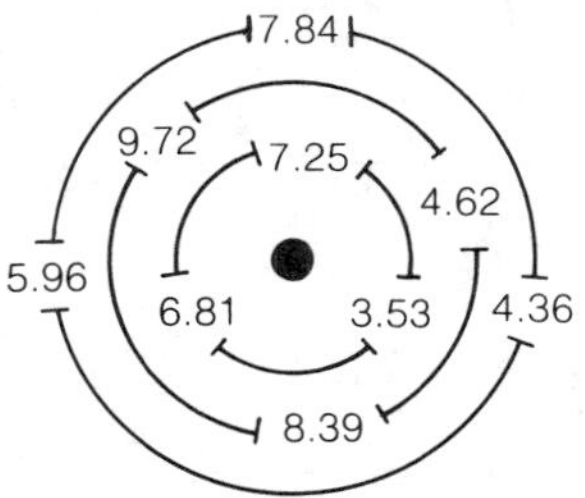

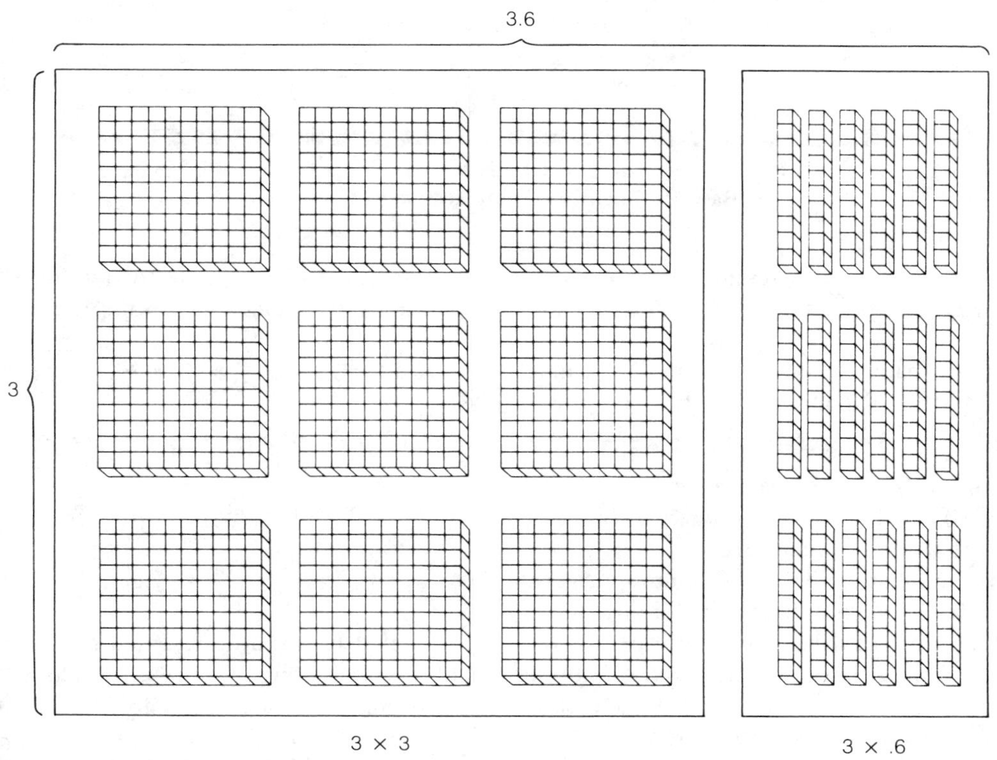

FIGURE 13.3
Base 10 Blocks for Multiplication of Decimals

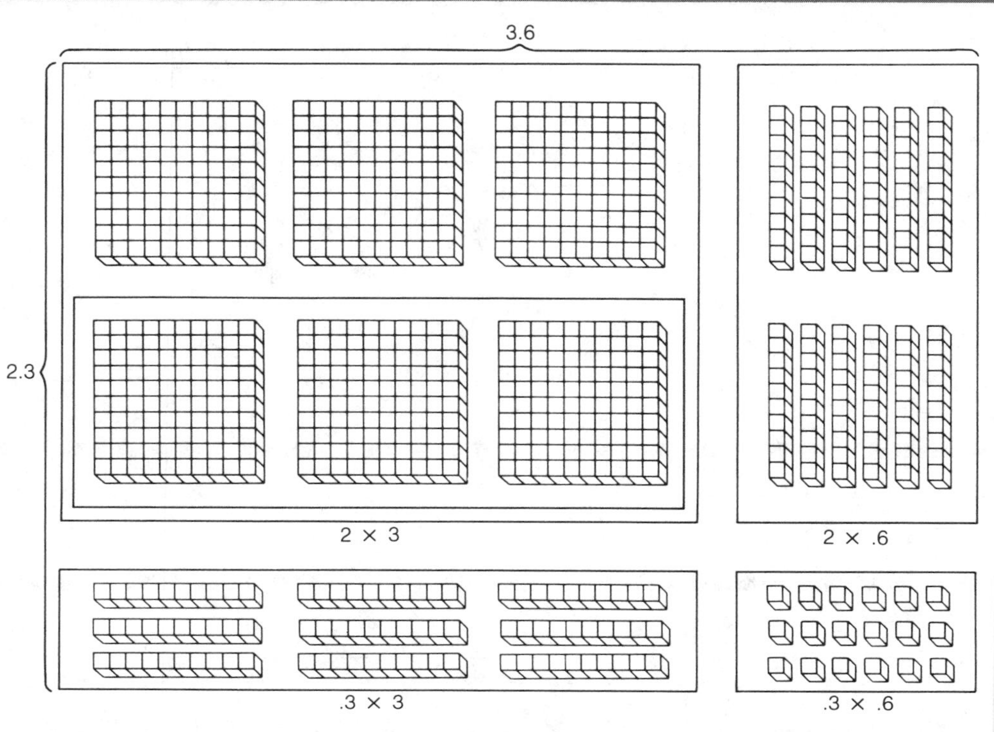

FIGURE 13.4
Base 10 Blocks for Multiplication of Decimals

Because of the physical representation, the placement of decimal points appears logical to the child.

Graph paper is another exemplar; a unit region is partitioned into 10 or 100 congruent parts. In Figure 13.5, each small part is $\frac{1}{100}$ of the unit region in the representation for 3 × 3.6. When the shaded regions are counted, a child can easily see that the product is 10.8.

An understanding of the multiplication algorithm for whole numbers is based on numeration ideas and the distributive property. We need to make sure children understand that these relationships also hold for rational numbers. Experiences with arrays made from base 10 blocks or graph paper as previously described will help children understand this application of distributivity.

If operations with fractions precede work with decimals, help children discover the rule for placing the decimal point by relating the two operations. Consider these examples:

$$3 \times 3.6 = 3 \times 3\tfrac{6}{10} = 3 \times \frac{36}{10} = \frac{108}{10} = 10.8$$

$$.5 \times 3.2 = \frac{5}{10} \times 3\tfrac{2}{10} = \frac{5}{10} \times \frac{32}{10} = \frac{160}{100} = 1.60$$

$$3.2 \times 2.4 = 3\tfrac{2}{10} \times 2\tfrac{4}{10} = \frac{32}{10} \times \frac{24}{10}$$

$$= \frac{768}{100} = 7.68$$

If decimals are presented before fractions and decimal computation is developed as an extension of the algorithm for whole numbers, introduce the rule for placing the decimal point by having children round both factors to the nearest whole number and estimate a reasonable product. Begin activities by having each child select the product from among several possible answers.

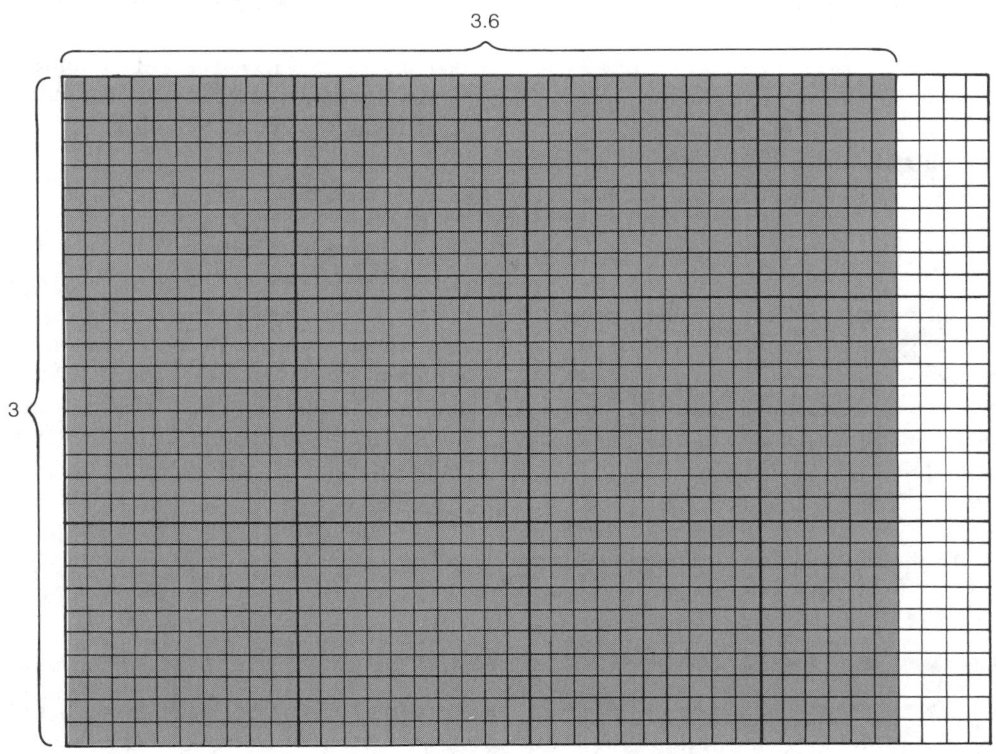

FIGURE 13.5
Graph Paper for Multiplication of Decimals

2.8 × 4 = ?	112, 11.2, 1.12, .112
3.5 × 6.2 = ?	2170, 217.0, 21.70, 2.170

The substitution of 3 × 4 in the first example and 4 × 6 in the second leads to the conclusion that 11.2 and 21.7 are reasonable products.

However, we want children to eventually be able to establish upper and lower bounds for products. This involves two approximations, one less than and the other greater than the desired product. For example, when asked to find the product of 2.7 × 4.6, a child may think: 3 × 5 = 15 and 2 × 4 = 8, so the product is between 8

EXCURSION

Select the best estimate for each. Then check with your calculator.

1. 3.7 × .8	(a) 4	(b) 40	(c) 400
2. 27.6 × 9.8	(a) 28	(b) 280	(c) 2800
3. 12.75 × 7.38	(a) 13	(b) 130	(c) 1300

and 15. Estimating skills help children compute with confidence, enabling them to check for reasonable answers.

Division

Essential understandings and skills for division with decimals include:

- An understanding of the meaning of the operation of division
- Competence in division with whole numbers
- Competence in rounding numbers and making estimates
- Competence in renaming decimal numbers, that is, renaming tenths as hundredths, ones as tenths, and so on

Children who do not have these prerequisites should learn them before they are taught division with decimals. Most children benefit from a brief review of these essential concepts and skills.

Division is usually introduced with problems that have natural number divisors. Select dividends and divisors carefully to avoid introducing too many ideas at once. For example:

PROBLEM

If 3 girls wish to share $2.25 equally, how much will each receive?

The dividend in this problem can be represented with play money (see Figure 13.6). The partitioning into 3 equivalent parts is shown in a manner similar to whole number division; that is, the 2 is exchanged for 20 tenths, then the 22 tenths are partitioned. The 1 tenth that is left is exchanged for 10 hundredths. The 15 hundredths are then partitioned.

EXCURSION
The Cost of Electricity

The cost of electricity is based on the number of kilowatt hours used. A kilowatt is 1000 watts, and a kilowatt hour is the amount of electricity used by a 1000-watt appliance in 1 hour. To find the number of kilowatt hours used by an appliance, use this formula: kw = (Watts ÷ 1000) × No. of Hours.

How much does it cost to run a 210-watt television for 60 hours a month if the cost of electricity is $.0525 per kilowatt hour?

$$\begin{aligned} kw &= (\text{Watts} \div 1000) \times \text{No. of Hours} \\ &= (210 \div 1000) \times 60 \\ &= .210 \times 60 \\ &= 12.6 \text{ kilowatt hours} \end{aligned}$$

To find the cost, multiply the number of kilowatt hours by the cost per kilowatt hour.

$$12.6 \times .0525 = \$.66150 \approx \$.66$$

The cost is about $.66.

Find the cost to operate:

1. A 100-watt bulb for 500 hours.
2. A 5750-watt clothes dryer for 20 hours.

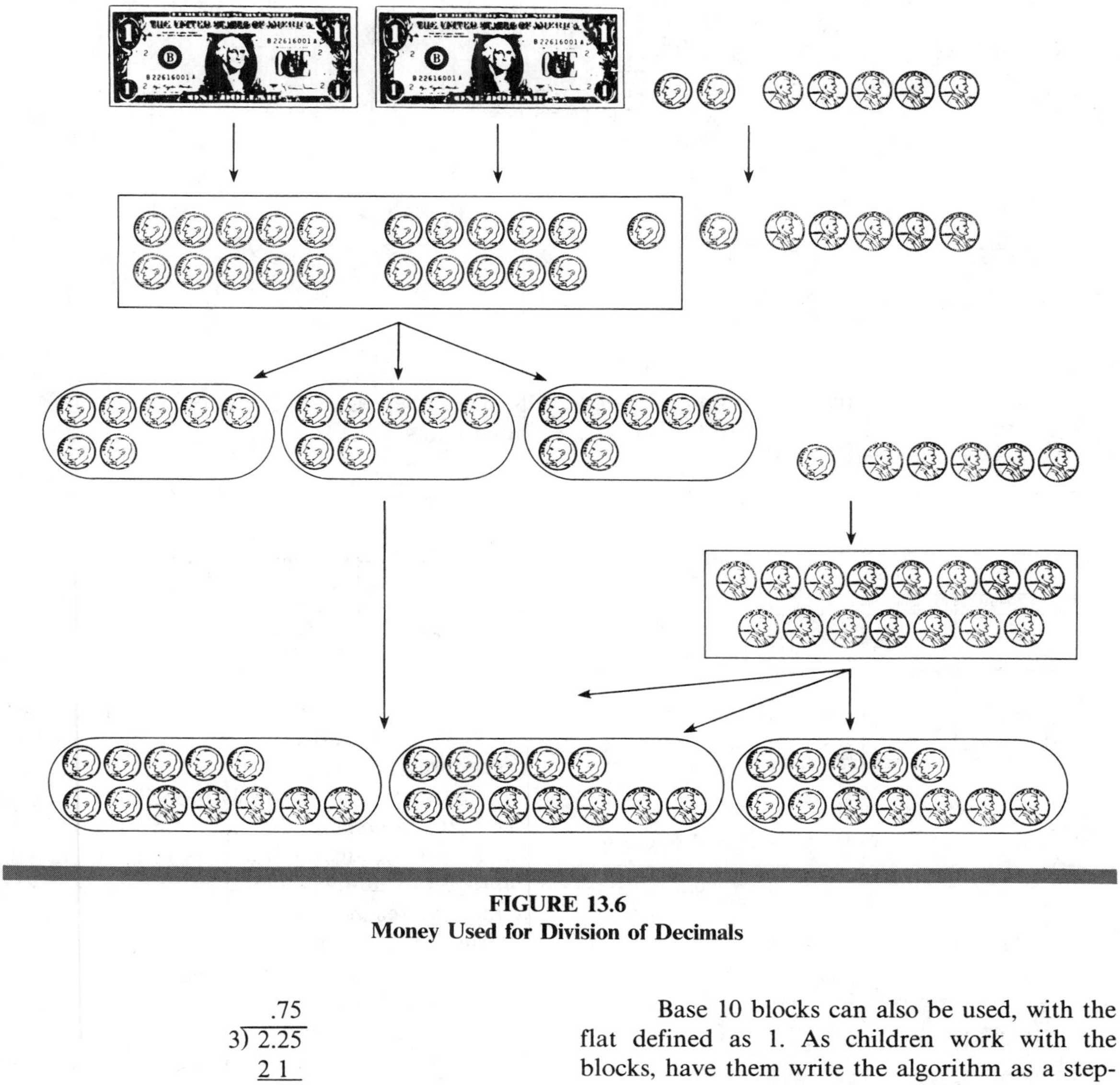

FIGURE 13.6
Money Used for Division of Decimals

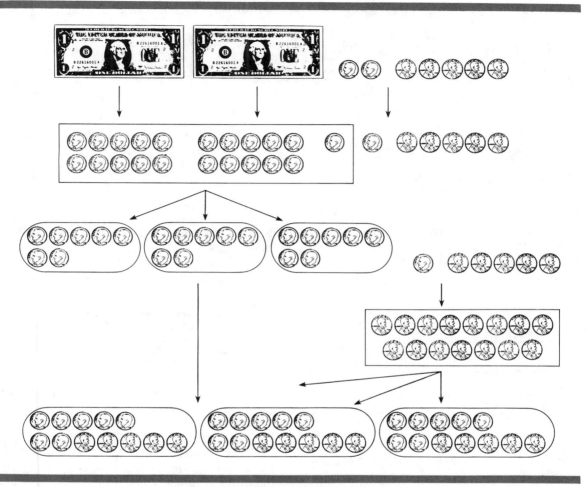

```
   .75
3) 2.25
   2 1
     15
     15
```

Base 10 blocks can also be used, with the flat defined as 1. As children work with the blocks, have them write the algorithm as a step-by-step record of what they did with the blocks.

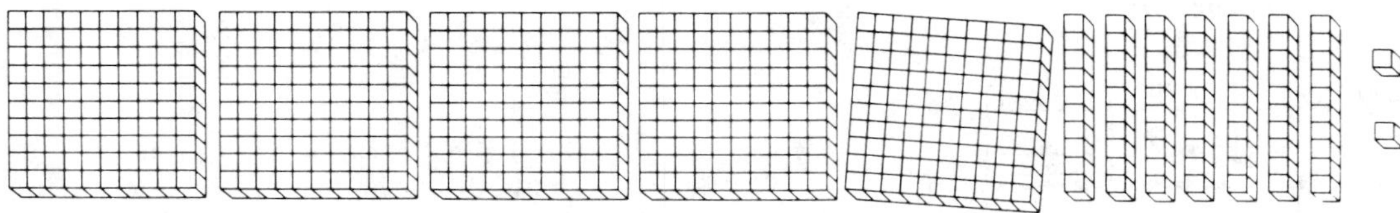

4) 5.72

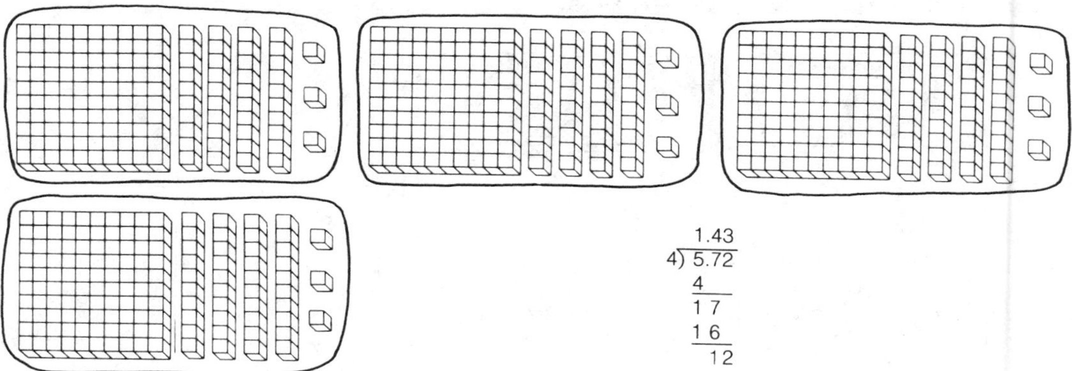

```
    1.43
4) 5.72
   4
   1 7
   1 6
     12
```

A more abstract representation of division with decimals, but a closer model of the process, can be shown with a pocket chart.

Step 1: Represent 3.24 ÷ 2.

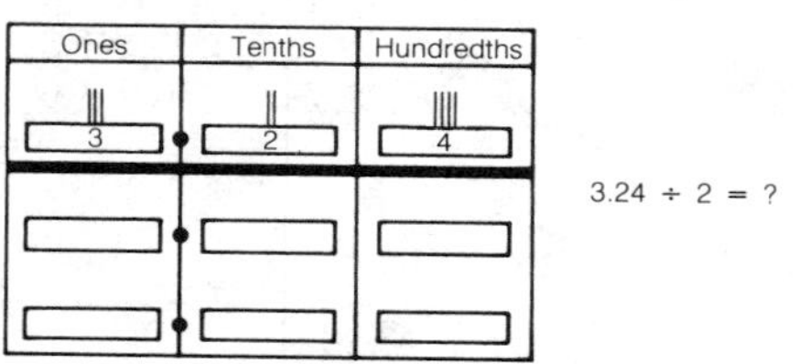

3.24 ÷ 2 = ?

Step 2: Separate 2 ones into two sets. 1 one is left over.

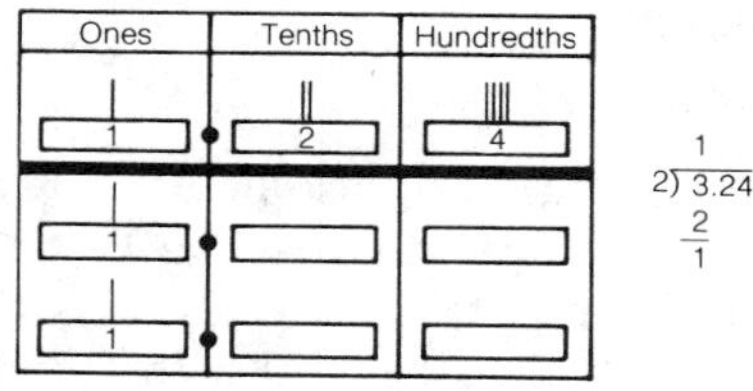


```
   1
2) 3.24
   2
   1
```

Step 3: Exchange 1 one for 10 tenths.

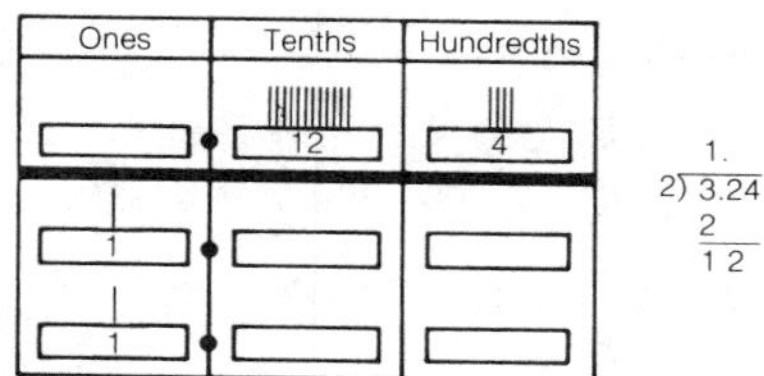


```
   1.
2) 3.24
   2
   1 2
```

Step 4: Separate 12 tenths and then 4 hundredths.

Ones	Tenths	Hundredths
1	6	2
1	6	2

```
   1.62
2) 3.24
   2
   1 2
   1 2
      4
      4
```

As a child works through problems of this type, he begins to abstract and schematize the procedure.

In order to divide with decimals, children need to understand that annexing zeros to the right of a decimal point does not change the value of the number. Figure 13.7 (pages 307–8) is a plan to teach this concept when introducing division and decimals.

Eventually lead children to recall a previously learned generalization: a quotient is unaffected if both dividend and divisor are multiplied by the same nonzero number. This idea also applies to examples like 3 ÷ .2 = ? Multiplying $\frac{3}{2}$ by 1 does not change the value of the number.

$$\frac{3}{.2} = \frac{3}{.2} \times \boxed{\frac{10}{10}} = \frac{30}{2} = 15$$

The quotient 3 ÷ .2 can be renamed to the equivalent 30 ÷ 2 and solved by a whole number procedure.

$$.2\overline{)3} \rightarrow .2\overline{)3.0} \rightarrow 2\overline{)30}^{\,15}$$

ACTIVITY PLAN

Annexing Zeroes | *3-D→Symbolic* | *Initiating/Abstract*

Content objectives

Annexing zeroes to a decimal does not change the value of the number.

Exemplars/materials

Base 10 blocks
Chalk and chalkboard
Paper and pencil

Behavioral indicators

The child:

1. Affixes a zero to a decimal when required in order to divide.
2. States that annexing a zero to a decimal does not change the value of the number.

Procedures

1. Using base 10 blocks, place one flat and 26 longs on a table. Tell the child that the flat is a kilogram of candy and each long represents a tenth of a kilogram of candy. Place the flat aside.

2. Ask, "How many pieces do I have?" After they are counted, write, "26 pieces, each with a mass of a tenth of a kilogram."

3. Ask, "What is another way to write 26 tenths using a whole number and a decimal?" Write "2.6" on the chalkboard. Show that 20 tenths = 2 ones by comparing 20 longs to 2 flats.

4. Ask, "I want to divide these (26 tenths) equally among 4 persons. How much will each get?" Write $4\overline{)\,2.6}$ on the chalkboard.

5. Say, "Use the blocks to show each person's share." Ask, "How many tenths does each get? How many are left over?"

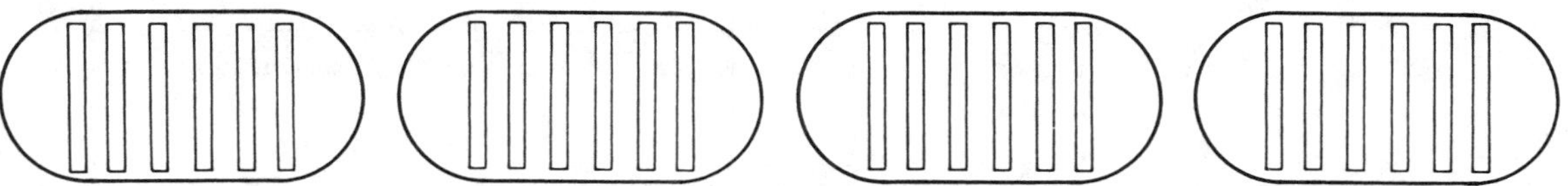

6. Record the division.

$$\begin{array}{r} .6 \\ 4\overline{)\,2.6} \\ \underline{2\,4} \\ 2 \end{array}$$

Say, "There are 2 tenths left. That isn't enough to distribute 1 to each person. Can we exchange the 2 tenths? If we exchange them for hundredths, how many would we have?"

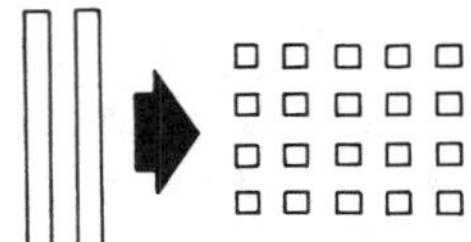

7. Ask, "How many of these will each person get if these are distributed equally?"

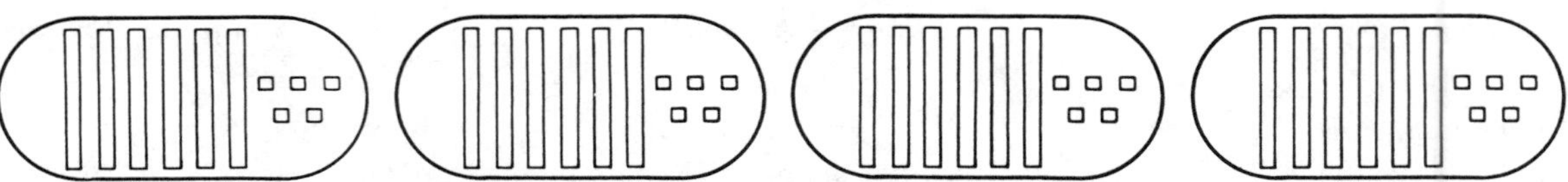

8. Record the division.

```
      .65
.4) 2.60
    2 4
      20
      20
       0
```

Say, "Notice that we place a zero after the 6. This is the way we show that tenths have been exchanged for hundreds. This enables you to complete the division."

9. Say, "Now, with the blocks, show how to complete this example: 4.2 ÷ 5 = ? Make a record of what you do."

10. Check to see that the child affixes a zero as required. Ask, "When you made this zero, did you change the value of the dividend?"

FIGURE 13.7
Activity Plan for Annexing Zeros

Have children rewrite each division problem, then introduce a shortcut for locating the decimal point in the quotient. The caret method should be plausible.

```
   15.
.2)3.0
```

While helping children with correct placement of decimal points, extend their skill in estimating. For example, when dividing 6.4 ÷ 2, children need to think, 6.4 is between 6 and 7, so when I divide by 2 a reasonable answer would be near 3. Children should recognize that quotients such as 32 or .32 are not reasonable.

EXCURSION
Unit Pricing

Which has the lower unit price?

An 18 oz. container for $.93

```
    $.0517 ≈ $.052 per oz.
18) $.9300
```

A 28 oz. container for $1.35

```
    $ .0482 ≈ $.048 per oz.
28)$1.3500
```

The 28 oz. container has a lower *unit price*.

Find the lower unit price.

1. 7 oz. for $.61 *or* 10 oz. for $.85
2. 340 g for $1.07 *or* 567 g for $1.72

The same skill can be applied to quotients of more difficult problems. For example, to locate the decimal point in the quotient of 2.34 ÷ .3, a child's thought process might be, I want to know how many groups of 3 tenths are contained in 2.34. There are about 3 groups of 3 tenths in one, so there are about 6 groups in 2. There is one additional group in .34. Therefore, the quotient should be a little more than 7. The quotient is 7.8.

```
     78        78?  7.8?  .78?
.3)2.34
   2 1
     24
     24
```

Percentage

Few new mathematical ideas are involved in solving percent equations. Previously learned concepts are applied in a different context; therefore, make sure that children:

- Understand the meaning of percent
- Can rename percents as decimals and as fractions
- Can rename fractions and decimals as percents
- Can write and solve proportions if the ratio method is to be taught

Initiating activities for percent equations should involve tasks in which children use familiar things and what they already know to work out a solution.

PROBLEM

What is 20% of $10?

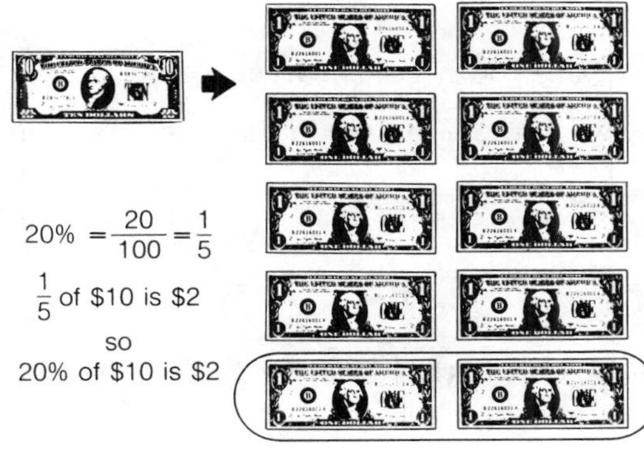

PROBLEM

Three out of 12 eggs are broken. What percent of the eggs are broken?

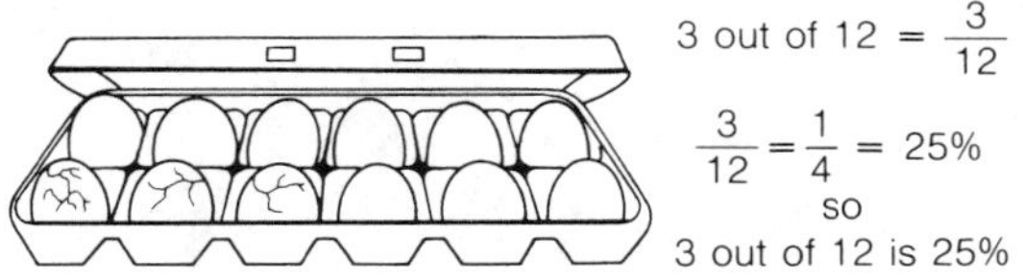

Beyond this stage, consider the types of problems used and the method taught to solve them. Three types of problems occur frequently; first with decimals and fractions, then later with percent.

1. Finding a part or percent of a number
2. Finding what part or percent one number is of another
3. Finding the number, when a certain part or percent of that number is known

The only difference among these three types is the particular term needed. Therefore, all three types can be approached with one of three possible methods.

The *ratio method* has recently become the most widely used way to solve percent problems. Many children find it easier to understand.

The ratio method uses a proportion to answer questions involving percent. Two numbers are compared in a ratio. In a proportion, the equality of two ratios is expressed. By the time children study percent problems, they usually know that cross multiplying terms of a proportion yields equal products. For example, the fact that $\frac{3}{8} = \frac{6}{16}$ implies that $3 \times 16 = 8 \times 6$.

Three examples of the ratio method follow.

PROBLEM

Scott had 3 out of 4 questions on a test correct. What percent did he have correct?

Think: Three out of 4 is $\frac{3}{4}$, and n compared to 100 is the same ratio as $\frac{3}{4}$.

EXCURSION
Percentage in Sports

This table shows the standings for one year of the Eastern Division of the National Conference of the NFL. Each team in the division is ranked according to games won. The percent is usually written as a three-place decimal.

St. Louis' Pct. $= \dfrac{\text{Games Won}}{\text{Games Played}}$

$$\frac{n}{100} = \frac{11}{14}$$

$$14n = 1100$$

$$n = .786$$

Team	Games Won	Games Lost	Pct.
St. Louis	11	3	.786
Dallas	10	4	.714
Washington	9	5	?
New York	7	7	?
Philadelphia	?	?	.357

Can you find the missing parts of the table? Remember that all teams play the same number of games in a season.

Write: To show that the ratios are equivalent, write $\frac{n}{100} = \frac{3}{4}$.

Solve: To find the percent, use the idea that the cross products are equal in a proportion.

$$\frac{n}{100} = \frac{3}{4}$$

$$4 \times n = 100 \times 3$$

$$4n = 300$$

$$n = 75$$

Scott had 75% of the questions correct.

PROBLEM

Christa saved 25% of her monthly allowance. If she receives \$20 a month, how much does she save?

Think: 25% means 25 out of 100, or $\frac{25}{100}$. Christa saves part of the \$20, $\frac{n}{20}$, at the same ratio.

Write: To show that the ratios are equivalent, write $\frac{n}{20} = \frac{25}{100}$.

Solve: To find how much she saves, use the idea that in a proportion, the cross products are equal.

$$\frac{n}{20} = \frac{25}{100}$$

$$100 \times n = 500$$

$$n = 5$$

Christa saved \$5 a month.

PROBLEM

Carol has \$6. This is 30% of what she will need to buy a new dress. How much does the new dress cost?

Think: 30% means 30 out of 100, or $\frac{30}{100}$. Six is what part of the original amount? That ratio is $\frac{6}{n}$.

Write: To show that the ratios are equivalent, write $\frac{30}{100} = \frac{6}{n}$.

Solve: Use cross products to solve the proportion.

$$\frac{30}{100} = \frac{6}{n}$$
$$30 \times n = 100 \times 6$$
$$30n = 600$$
$$n = 20$$

The dress costs \$20.00.

In developing the insight required to set up a proportion expressing the relationship in a situation, carefully select experiences and questions that gradually increase in difficulty. Initially selected problems should offer no computational difficulties or hidden meanings. More difficult problems can be introduced as each child experiences some success.

To solve a percent problem by the *factor factor product method,* express the problem as a simple percent statement in the form of percent of a number equals a number or ___% × ___ = ___. A child is asked to relate his understanding of the factor, factor, product relationship to the situation. Each number is identified as a factor or the product.

The factor, factor, product method is more widely accepted in the secondary school because of its similarity to the procedure for solving algebraic word problems.

The *formula method,* which requires that the students memorize three artificial categories of percent problems, is rarely taught today. A basic equation, $p = br$, is used. P stands for percentage (part), b for base (total amount), and r for rate (percent).

To find the percent, the equation is solved for r and written $r = \frac{p}{b}$. To find the base, the equation is written in terms of b: $b = \frac{p}{r}$. The child had to choose the appropriate formula for a given problem. For example:

Type 1: 25% of 60 = _?_.
Type 2: _?_% of 60 = 15.
Type 3: 25% of _?_ = 15.

Consolidating Activities

Children need to continually refine new skills with decimals and percent. Routine drill and practice, although essential, may become boring and cause a loss of interest. It is, therefore, important that consolidating activities be designed to motivate. The activities that follow have proven to be effective.

EXCURSION
Percentage in Engineering

Engineers use percentage in many problems. A highway engineer measures the steepness of a hill by a percent called the *grade* of the road.

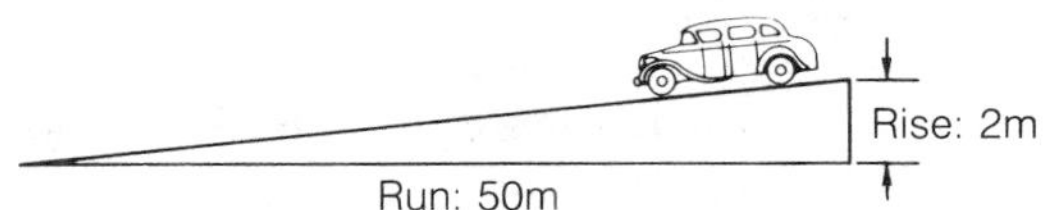

A road that rises 2 for every 50 meters measured along level ground (run) has a grade of 4%.

$$\text{Road grade} = \frac{\text{rise}}{\text{run}} \qquad \frac{1}{100} = \frac{2}{50}$$
$$50n = 200$$
$$n = 4 \text{ or } 4\%$$

The run portion of a road is 1200 meters long and rises 72 meters. What is the percent of grade?

EXCURSION
Discounts

How much does an eight-track deck cost that lists for $95.00 but is offered at a 25% discount?

$95.00	List price	$95.00	List price
× .25	Rate of discount	− 23.75	Discount
$23.75	Discount	$71.25	Net price

The net price (discount price) is $71.25.

Find the net price of each.

1. A $25.00 heater with 20% discount
2. 100 rose bushes (individually priced at $6.75 each) with a quantity discount of $33\frac{1}{3}$%.

Magic Squares
If the sum of the numbers in each row, column, and diagonal is the same, then the square is a magic square. See if this is a magic square.

8.2	1.9	6.4
3.7	5.5	7.3
4.6	9.1	2.8

Riddles
Activities such as the following provide highly motivating practice.[6]

To find the hidden message on the rock at the top of the next page,

1. do the exercises.
2. locate the answers on the rock (they run across, left to right).

6. Walter E. Rucker et al., *Heath Mathematics, Teacher's Edition—Level 6* (Lexington, Mass.: D.C. Heath, 1979), p. 224–b.

EXCURSION

Magic squares can also be used for practice with subtraction. Complete this magic square.

4.1	?	7
1.85	2.75	?
?	4.55	1.4

New magic squares may be constructed by multiplying or dividing each number in a given magic square by the same number.

WHERE WAS KING SOLOMON'S TEMPLE?

O 9	~~T 1~~	~~T .9~~	~~E 5~~	N 3	I 6	T 7	I .0	S 5	T 7	O 2	N .3
H 1	O 1	W .8	T 1	H 2	E .8	S 3	E .6	C 5	A .4	N 9	I .5
D 1	E 8	N 5	S 0	E .6	O 6	F .2	T 6	T .9	H 9	I .8	S 3
H 9	E 6	A 1	R 8	T .5	R 4	E 4	A .2	D 7	F 2	I .9	N 1

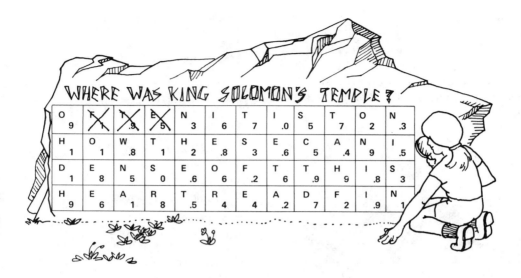

3. cross out each box that contains part of an answer. (There will be 18 boxes left.)

The letters in the remaining boxes give the message.

$\begin{array}{r} .65 \\ \times\ 3 \\ \hline 1.95 \end{array}$	$\begin{array}{r} 7.45 \\ \times\ 9 \\ \hline \end{array}$	$4\overline{)9.2}$	$\begin{array}{r} .302 \\ \times\ 6 \\ \hline \end{array}$	$6\overline{)3.90}$
$.06\overline{).0294}$	$\begin{array}{r} 9.2 \\ \times\ 5.5 \\ \hline \end{array}$	$.8\overline{)5.52}$	$\begin{array}{r} 3.04 \\ \times\ 6.1 \\ \hline \end{array}$	$.17\overline{).4947}$

Tic-Tac-Toe

One version of Multiplication Tic-Tac-Toe was described earlier. The following is another version.[7]

Two players play this game using decimals. One player writes with a black pencil; the other uses a red pencil. The object of the game is similar to any form of tic-tac-toe, namely, to get three correct sums in a column, a row, or a decimal. When a player succeeds, a point is scored. Another point is scored if the player correctly adds the three numbers in a row. Children should be encouraged to make their own tic-tac-toe games.

+	7.2	.65	4
2.6	9.8		
4.7		1.12	
3.62	10.82	4.27	4.02

Bean Bag Toss

Another activity that can be used to reinforce operations with decimals is bean bag toss.[8] On a bean bag, tape a decimal expressed in tenths, hundredths, etc. Make the playing region from a large sheet of paper that can be taped to the floor. Children stand back and toss the bean bag at the playing region. One point is earned if the child correctly multiplies the decimal on the bean bag times the number on the square where it lands.

.6	.02	3	.12
.09	.32	.9	4
1.5	.7	.08	5
2	1.02	.06	2.1

7. Adapted from Rucker et al., *Heath Mathematics*, p. 187.

8. Adapted from Rucker et al., *Heath Mathematics*, p. 205.

Cross Number Puzzles

Cross number puzzles are completed in the same manner as crossword puzzles. Only one digit can be written in a space, and it must fit both horizontally and vertically. Many different skills may be reviewed along with those that are consolidated.[9]

ACROSS

1. $\frac{5}{8} = \frac{\quad}{1000}$
3. 80% of _____ = 56
4. $8 \div \frac{1}{5}$ = _____
5. 471 ÷ 19 = 24 r _____
6. _____% of 72 = 7.2
8. _____ × 3 = 192
9. The L.C.M. of 9, 2, and 12 is _____
11. The product of 2.4 and 7.5 is _____
12. 24 = _____% of 32
14. $24 \times 2\frac{1}{8}$ = _____
17. The average of 35, 22, and 9 is _____
19. 50% of 82 = _____
21. .588 ÷ .007 = _____
22. _____% of 75 = 60
24. $\frac{3}{4} = \frac{\quad}{28}$
25. 45 is _____ percent of 30.
27. $3\frac{3}{8} + 6\frac{7}{24} + 4\frac{1}{3}$ = _____
28. Fifteen hundredths = _____%
29. 13% of 5000 = _____
30. $\frac{13}{25} = \frac{\quad}{100}$
31. $83\frac{1}{3}\%$ of 864 = _____
33. Named as a decimal, $2\frac{1}{2}\%$ is _____
35. 400% of 72 = _____
37. Named as a percent, $2\frac{1}{2}$ is _____

DOWN

1. $\frac{3}{5}$ names _____ percent.
2. 486 ÷ .9 = _____
3. (2 × 15) ÷ (9 × 5) = _____
5. _____ is to 35 as 2 is to 5.
6. At $1.20 per pound, _____ ounces of candy can be bought for $.90.
7. Simplify. $1\frac{1}{2} \times 6 \div \frac{1}{7}$ = _____
8. To the nearest whole number, the quotient for 40.94 ÷ .06 is _____
10. The reciprocal of $\frac{1}{64}$ is _____
13. $\frac{5}{6}$ × _____ is 45.
15. 5 × 3 × 3 × 4 = _____
16. $8\frac{1}{4} \div \frac{11}{12}$ = _____
18. 5321 − 3064 = _____
20. $2.9\overline{)333.5}$ = _____
22. $37\frac{1}{2}\%$ of 224 = _____
23. $1\frac{4}{5} \times 3\frac{1}{3}$ = _____
26. 548.7 rounded to the nearest hundred is _____
27. $\frac{3}{4}$ pound = _____ ounces
29. 2024 ÷ 1746 + 896 + 1416 = _____
30. 20% of _____ = 11
32. $8\frac{1}{4} \times \frac{5}{6} \times 3\frac{1}{5}$ = _____
34. 8 yards 2 feet = _____ feet
36. _____% of 20 = 17

9. From LEARNING TO COMPUTE 2, Third Edition, by Wilmer L. Jones, p. 114. Copyright © 1973 by Harcourt Brace Jovanovich, Inc. Reprinted by permission of the publisher.

Transferring Activities

Transferring activities involve problem solving, and provide new situations for previously learned knowledge and skills. Problem-solving procedures can be extended and reviewed, for instance, through the study of decimals and percent.

Begin with simple one-step verbal problems involving addends and a sum, or factors and a product. Provide different types of verbal problems and focus on the analysis of the problem. Have each child identify the addends, sum, factors, or product.

Consumer problems can easily be brought into the classroom. A drawing can make the problem more realistic. Consider these examples.[10]

4. Socks normally sell for $1.00 a pair. How much is saved by buying 4 pairs at the sale price?
   a. $.84 b. $3.16 c. $4.00 d. not given

5. Scott and his father each agreed to pay half of the sale price of a pair of walking shorts. How much did each pay?
   a. $2.85 b. $4.50 c. $2.50 d. $2.25

To work problems 1–5, look at the picture to find the price of each item. Do not include sales tax in your computation.

1. How much would you pay for 4 beach towels at the sale price?
   a. $7.60 b. $6.60 c. $9.60 d. not given
2. How much would 4 knit shirts cost?
   a. $4.95 b. $9.90 c. $19.80 d. $8.90
3. How much do you save by buying walking shorts now?
   a. $5.75 b. $.50 c. $1.25 d. not given

Analyze tests on problem solving before administering them. Do children have experience with the different types of problems? The previous example is taken from a proficiency test developed by one of the authors for the Baltimore City Public Schools. Similar kinds of items may also be found on commercial standardized tests.

PLANNING ASSESSMENT

The following illustrates certain types of difficulty children may have when adding decimals. Be alert for each when you administer assessment tasks.

10. Adapted from Wilmer Jones, *Mathematics Proficiency Test, Level Q* (Baltimore, Md.: Baltimore City Public Schools, 1977), p. 6.

1.

$$\begin{array}{r} .9 \\ .3 \\ +\ .6 \\ \hline 1.8 \end{array}$$

The sum is greater than 10 tenths. The main element of difficulty is involved in placing the decimal point properly.

2.

$$\begin{array}{r} 6.75 \\ +\ 2.48 \\ \hline 9.23 \end{array}$$

This problem involves renaming in two places. The difficulty is in renaming.

3. 46.2 + 9.5 + 2.75 = 58.45 This problem contains ragged decimals. A child must copy the numbers, place them correctly into columns, and locate the decimal point in the sum.

A similar analysis can be made for other operations with decimals. For example, in division, examine a child's ability to compute different types of examples.

1.

$$3\overline{)9.3} = 3.1$$

This problem causes very little difficulty other than placing the decimal point in the quotient.

2.

$$5\overline{).25} = .05$$

The new element is the insertion of zero to fill an empty place in the quotient.

3. 3 ÷ 6 = ?

$$6\overline{)3.0} = .5$$

In this problem, a zero was annexed in the dividend so that the division could be completed.

4.

$$.3\overline{)2.82} = 9.4$$

This problem involves division of a decimal by a decimal. Before dividing, a child must place the decimal point correctly in the quotient.

5.

$$2.7\overline{).081} = .03$$

Besides dividing a decimal by a decimal and placing the point in the quotient, a child must insert a zero to fill the empty space.

These examples are paper-and-pencil assessment tasks. To elicit more information about a child's understanding of decimals and operations on decimals, plan tasks involving two and three-dimensional exemplars. Many of the instructional tasks presented earlier in this chapter can also be used for assessment.

14

Developing Computational Procedures with Fractions

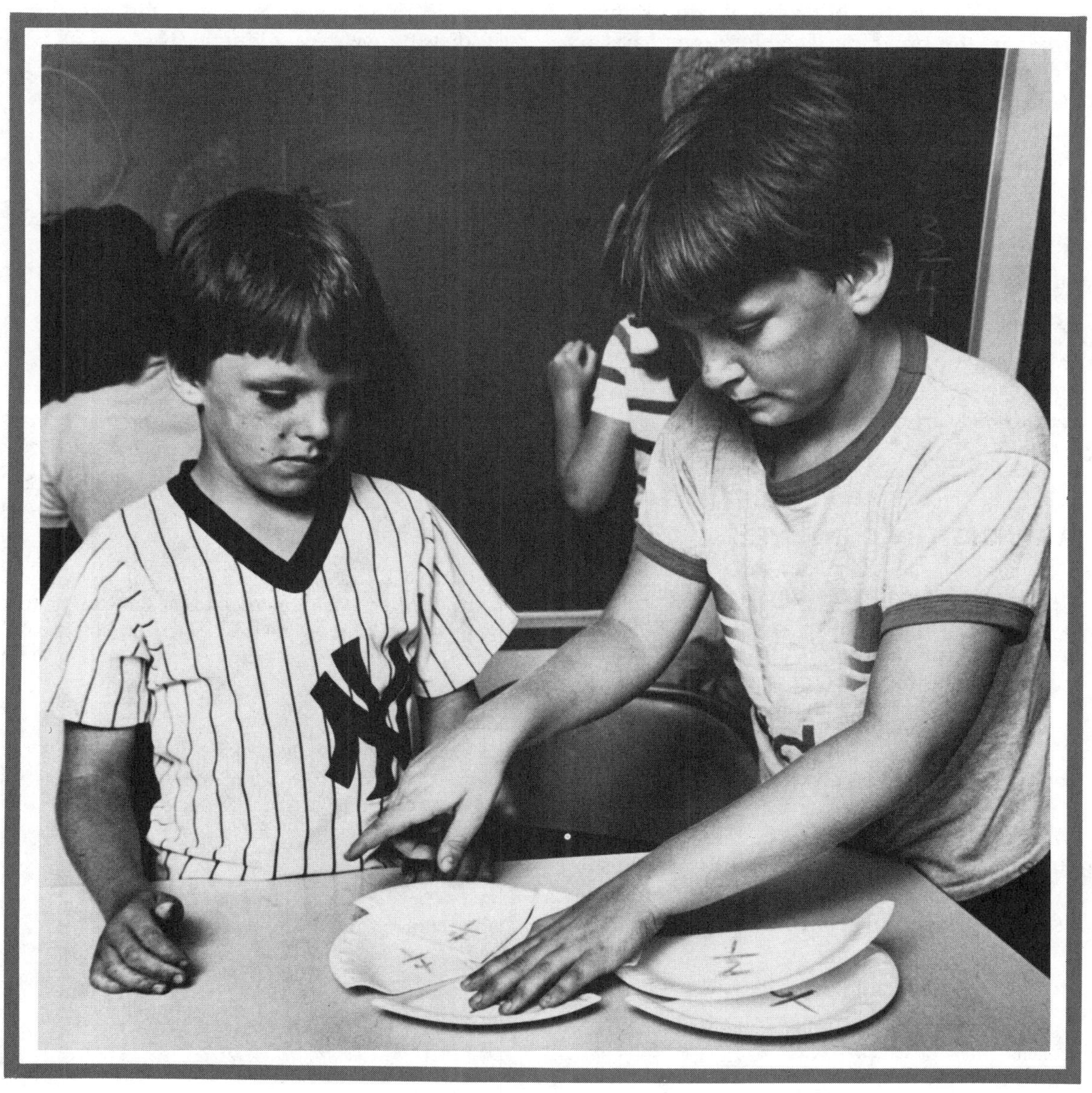

Fractions were invented before decimals, and they have received greater emphasis in elementary school programs. With increasing use of hand-held calculators and the metric system, work with decimals will probably be introduced earlier in the program. Therefore, emphasis on decimals should be increased, with less time spent on algorithms for operations with fractions. Children do not need to master the most difficult examples.

Despite the increased emphasis on decimals, many fraction concepts and terms, such as *halves, thirds* and *fourths,* are here to stay. It is unlikely that .5 will altogether replace $\frac{1}{2}$, or that .333 will replace $\frac{1}{3}$ as a common symbol. Fractions with denominators of 5 or less are also widely used. Others with 60ths and 360ths are used to measure time and rotation.[1]

PLANNING THE CONTENT

There are different algorithms for each of the operations with fractions. As you read about the more conventional procedures in this chapter, anticipate possible teaching activities.

Addition and Subtraction with Fractions

Properties

Although the set of whole numbers is only a subset of the set of rational numbers, the fundamental properties and structures developed for whole numbers apply to all rational numbers. Children find that addition and multiplication remain commutative and associative. Zero and 1 continue to be the identity elements for addition and multiplication, respectively; and multiplication is still distributive over addition. These ideas were expressed in content objectives 13.1 through 13.4, and in 13.9 through 13.13.

Algorithms

When explaining the standard algorithms, fractions are usually interpreted in terms of parts of a whole, or parts of a set.

Some of the important content objectives related to addition and subtraction of fractions include:

14.1
To add or subtract rational numbers expressed as fractions, the fractions must have the same denominators; they must be *like* fractions.

14.2
When adding like fractions, add the numerators to get the numerator of the sum. The denominator of the sum is the same as the denominator in each fraction.

14.3
When subtracting like fractions, subtract the numerators to get the numerator of the difference. The denominator of the difference is the same as the denominator in each fraction.

14.4
A multiple of a rational number is a product of that number and a whole number.

14.5
The Least Common Denominator (L.C.D.) for two or more fractions is the Least Common Multiple of their denominators.

1. Don Firl, "The Role of Fractions in the Mathematics Curriculum," *Mathematics Flyer* 9, no. 2 (Minneapolis: Minnesota State Dept. of Education, Winter 1975): 5.

EXCURSION
Baseball Fractions

When adding two fractions, many children make the error of adding the denominators as well as the numerators. When fractions are used to show a *ratio*, this procedure is sometimes correct. For example, in a double header, Amy went 2 for 4 (2 hits out of 4 times at bat) and 3 for 5 (3 hits out of 5 times at bat). To find the total number of hits per times at bat for the two games, add the fractions as follows:

$$\frac{2}{4} + \frac{3}{5} = \frac{5}{9}$$

Thus, Amy had 5 hits out of 9 times at bat. When else may fractions be added in this way?

14.6
To add (or subtract) when denominators are different:
a. Find the L.C.D.
b. Write equivalent fractions using the L.C.D.
c. Add (or subtract).
d. Simplify the sum (or difference).

Children usually learn addition and subtraction with like denominators first, then addition and subtraction with *un*like denominators. The process for adding or subtracting fractions with like denominators is similar to the procedure for whole numbers. The following two algorithms are often taught together.

$$\frac{3}{8} + \frac{2}{8} = \frac{3 + 2}{8} = \frac{5}{8}$$

$$\frac{5}{8} - \frac{2}{8} = \frac{5 - 2}{8} = \frac{3}{8}$$

A preliminary procedure is needed before adding and subtracting unlike fractions. These fractions must be renamed with common denominators, preferably the L.C.D.

To find the sum of two numbers, for example $\frac{1}{4} + \frac{2}{3}$, a child can search for a common denominator among the sets of equivalent fractions for each addend.

$$\frac{1}{4}: \left\{\frac{1}{4}, \frac{2}{8}, \frac{3}{12}, \frac{4}{16}, \frac{5}{20}, \frac{6}{24}, \ldots\right\}$$

$$\frac{2}{3}: \left\{\frac{2}{3}, \frac{4}{6}, \frac{6}{9}, \frac{8}{12}, \frac{10}{15}, \frac{12}{18}, \frac{14}{21}, \frac{16}{24}, \ldots\right\}$$

A fraction from each set with the same denominator is selected. The equivalent fractions are substituted for the originals for computation.

$$\begin{array}{rl} \frac{1}{4} = & \frac{3}{12} \\ + \frac{2}{3} = & \frac{8}{12} \\ \hline & \frac{11}{12} \end{array} \qquad \text{or} \qquad \begin{array}{l} \frac{1}{4} + \frac{2}{3} \\ = \frac{3}{12} + \frac{8}{12} \\ = \frac{3 + 8}{12} \\ = \frac{11}{12} \end{array}$$

The procedure for subtracting unlike fractions is the same. Consider $\frac{3}{4} - \frac{1}{6} = \square$.

$$\frac{3}{4}: \{\frac{3}{4}, \frac{6}{8}, \boxed{\frac{9}{12}}, \frac{12}{16}, \frac{15}{20}, \frac{18}{24}, \ldots\}$$

$$\frac{1}{6}: \{\frac{1}{6}, \boxed{\frac{2}{12}}, \frac{3}{18}, \frac{4}{24}, \frac{5}{30}, \ldots\}$$

$$\begin{array}{rcl} \frac{3}{4} & = & \frac{9}{12} \\ -\frac{1}{6} & = & \frac{2}{12} \\ \hline & & \frac{7}{12} \end{array} \qquad \text{or} \qquad \begin{aligned} &\frac{3}{4} - \frac{1}{6} \\ &= \frac{9}{12} - \frac{2}{12} \\ &= \frac{9-2}{12} \\ &= \frac{11}{12} \end{aligned}$$

Like denominators other than the L.C.D. could be selected, but fractions with larger denominators result, making computation more difficult.

In order to become proficient, however, children must learn a more systematic procedure for determining Least Common Denominators. In one procedure, the prime factorizations of denominators are used to find the L.C.D.

For example, to add $\frac{7}{12}$ and $\frac{2}{15}$, first find the prime factorization for each denominator.

$$\frac{7}{12} = \frac{7}{2 \times 2 \times 3}$$

$$\frac{2}{15} = \frac{5}{3 \times 5}$$

$$\{2, 2, 3\} \cup \{3, 5\} = \{2, 2, 3, 5\}$$

Next, determine the *union* of the sets of factors incorporated in the prime factorizations. The set of factors in the union is used to make a prime factorization; that number (2 × 2 × 3 × 5, or 60 in this example) is the Least Common Multiple (L.C.M.) of the denominators. The L.C.M. of the denominators is also the L.C.D. for the two fractions. To rename each fraction as 60ths, multiply by the appropriate fraction for 1.

$$\begin{array}{rcl} \frac{7}{12} & = \frac{7}{12} \times \boxed{\frac{5}{5}} = & \frac{35}{60} \\ + \frac{2}{15} & = \frac{2}{15} \times \boxed{\frac{4}{4}} = & \frac{8}{60} \\ \hline & & \frac{43}{60} \end{array}$$

Addition and Subtraction with Mixed Numerals

A mixed numeral can be expressed as the sum of a whole number and a fraction. The shortest procedure involves adding or subtracting the fraction and whole number parts separately.

14.7
To add with mixed numerals:
a. Find the least common denominator.
b. Write equivalent fractions for the fraction part.
c. Add the fraction part.
d. Add the whole number part.
e. Simplify the sum.

14.8
To subtract a number expressed as a mixed numeral from a whole number:
a. Rename the whole number as a mixed numeral.
b. Subtract the fraction part.
c. Subtract the whole number part.
d. Simplify the difference.

14.9
To subtract one number expressed as a mixed numeral from another:
a. Find the least common denominator.
b. Write equivalent fractions using the L.C.D.
c. If necessary, rename the larger mixed numeral.
d. Subtract the fraction part.
e. Subtract the whole number part.
f. Simplify the difference.

In the following examples, the addition implied by a mixed numeral is clearly shown.

$$\begin{array}{rcl} 4\frac{3}{8} & = & 4 + \frac{3}{8} \\ + 1\frac{1}{8} & = & 1 + \frac{1}{8} \\ \hline & & 5 + \frac{4}{8} = 5 + \frac{1}{2} \text{ or } 5\frac{1}{2} \end{array}$$

$$\begin{array}{rcl} 7\frac{7}{12} & = & 7 + \frac{7}{12} \\ - 5\frac{3}{12} & = & 5 + \frac{3}{12} \\ \hline & & 2 + \frac{4}{12} = 2 + \frac{1}{3} \text{ or } 2\frac{1}{3} \end{array}$$

EXCURSION
Egyptian Fractions

The ancient Egyptians wrote fractions as sums of unit fractions. (A unit fraction has a numerator of 1.) For example, $\frac{7}{8}$ was $\frac{1}{2} + \frac{1}{4} + \frac{1}{8}$. See if you can write $\frac{7}{16}$ as the sum of 3 unit fractions.

Problems similar to these appeared in the Rhind papyrus, which dates about 1650 B.C.

When the fractional parts are unlike, they must be renamed as equivalent fractions before the computation can be completed.

$$\begin{array}{r} 2\frac{2}{3} = 2\frac{2}{3} \times \frac{4}{4} = 2\frac{8}{12} \\ +\ 3\frac{1}{4} = 3\frac{1}{4} \times \frac{3}{3} = 3\frac{3}{12} \\ \hline 5\frac{11}{12} \end{array} \qquad \begin{array}{r} 4\frac{1}{2} = 4\frac{1}{2} \times \frac{3}{3} = 4\frac{3}{6} \\ -\ 1\frac{1}{3} = 1\frac{1}{3} \times \frac{2}{2} = 1\frac{2}{6} \\ \hline 3\frac{1}{6} \end{array}$$

In subtracting with mixed numerals, a child should remember that numbers have many different names. Consider $3\frac{1}{3} - 1\frac{2}{3} = \square$. The fractions already have common denominators, but the numerators cannot be subtracted. Therefore, $3\frac{1}{3}$ must be renamed.

$$\begin{aligned} 3\tfrac{1}{3} &= (2 + 1) + \tfrac{1}{3} \\ &= (2 + \tfrac{3}{3}) + \tfrac{1}{3} \\ &= 2 + (\tfrac{3}{3} + \tfrac{1}{3}) \\ &= 2 + \tfrac{4}{3} \text{ or } 2\tfrac{4}{3} \end{aligned}$$

By renaming 3 as 2 + 1 and expressing 1 as $\frac{3}{3}$, the number $3\frac{1}{3}$ is rewritten as $2\frac{4}{3}$. Subtraction is then possible.

$$\begin{array}{r} 3\frac{1}{3} + 2\frac{4}{3} \\ -\ 1\frac{2}{3} = 1\frac{2}{3} \\ \hline 1\frac{2}{3} \end{array}$$

The same situation exists if the minuend is a whole number. To subtract $3\frac{5}{8}$ from 6, find a way to write 6 as a mixed number.

$$6 = 5 + 1 = 5 + \frac{8}{8} \text{ or } 5\frac{8}{8}$$

Subtraction is then possible.

$$\begin{array}{r} 6 = 5\frac{8}{8} \\ -\ 3\frac{5}{8} = 3\frac{5}{8} \\ \hline 2\frac{3}{8} \end{array}$$

Each child needs to understand that whenever the fractional part of the minuend is less than the fractional part of the subtrahend, the minuend must be renamed.

Multiplication with Fractions and Mixed Numerals

Models for multiplication

Fractional computations for multiplication and division can be easily taught. However, teaching what the operations *mean* involves ways of modeling these operations as well as procedures for computing.

14.10
Repeated addition of equal fraction addends is a model for multiplication of a fraction by a whole number.

Repeated addition is useful only when the fraction is multiplied by a whole number.

> **14.11**
> An array is a model for multiplication of a fraction by a fraction or a whole number by a fraction.

The array concept can often be extended to show multiplication with fractions. When the first factor is a fraction and the second a whole number or a fraction, arrays can interpret the situation, as illustrated in Figure 14.1. Arrays can also clarify why multiplication is used in what appear to be division situations.[2]

For $\frac{1}{4} \times 12 = 3$, 12 discs are arranged in an array with four columns because 4 is the denominator. Three of the discs are encircled, $\frac{1}{4}$ of the total. To show $\frac{2}{3} \times \frac{3}{4}$, a unit region is subdivided into a 3 by 4 array. The denominators indicate the subdivisions required. Three-fourths of the array is shown by shading 3 of the 4 columns, with $\frac{2}{3}$ indicated by shading 2 of the 3 rows. The product of $\frac{2}{3} \times \frac{3}{4} = \frac{6}{12}$. The intersection of 6 sections is shaded both ways. The $\frac{6}{12}$ is 6 of the 12 sections of the original unit.

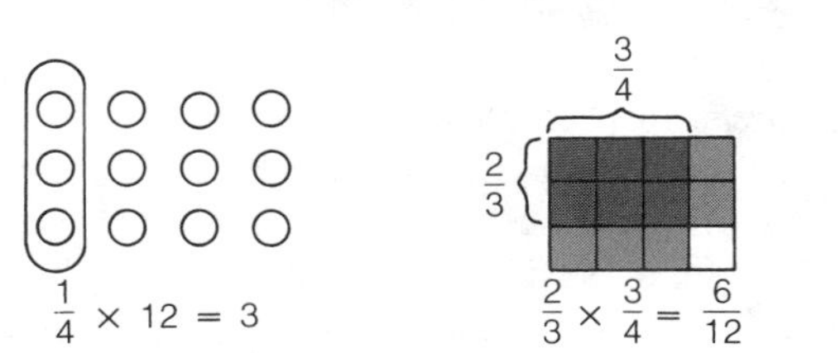

FIGURE 14.1
Arrays Showing Multiplication with Fractions

> **14.12**
> The ratio concept of multiplication is a model for multiplication when one or both of the factors are fractions.

The ratio concept of multiplication can be used to interpret any multiplication situation involving fractions. Think of multiplication as finding the product when the ratio of the product to the second factor is the same as the ratio of the first factor to 1. This concept represents a more abstract way of interpreting multiplication and is not useful when first introducing multiplication with fractions. However, older children may find it helpful.

$4 \times 6 = \square$	$\frac{1}{4} \times 12 = \square$
The product 24 has the same ratio to 6 as 4 has to 1.	The product 3 has the same ratio to 12 as $\frac{1}{4}$ has to 1.
$\frac{24}{6} = \frac{4}{1}$	$\frac{3}{12} = \frac{\frac{1}{4}}{1}$ or $\frac{3}{12} = \frac{1}{4}$
$24 \times \frac{1}{2} = \square$	$\frac{1}{2} \times \frac{3}{4} = \square$
The product 12 has the same ratio to $\frac{1}{2}$ as 24 has to 1.	The product $\frac{3}{8}$ has the same ratio to $\frac{3}{4}$ as $\frac{1}{2}$ has to 1.
$\frac{12}{\frac{1}{2}} = \frac{24}{1}$ or $\frac{24}{1} = \frac{24}{1}$	$\frac{\frac{3}{8}}{\frac{3}{4}} = \frac{\frac{1}{2}}{1}$ or $\frac{1}{2} = \frac{1}{2}$

Properties

In addition to the properties of multiplication with rational numbers listed in content objectives 13.9 through 13.13, observe one additional property:

2. Leonard M. Kennedy, *Guiding Children to Mathematical Discovery*, 3rd ed. (Belmont, Calif.: Wadsworth, 1980), pp. 327–28.

> **14.13**
> *Inverse*—Every positive rational number has a reciprocal or multiplicative inverse. The product of a positive rational number and its multiplicative inverse is 1, the identity for multiplication.

For example, the inverse of $\frac{2}{3}$ is $\frac{3}{2}$, because $\frac{2}{3} \times \frac{3}{2} = 1$. This property is applied in one of the algorithms for dividing with fractions.

Algorithms

Content objectives closely related to multiplication with fractions and with mixed numerals include:

> **14.14**
> To multiply with fractions, multiply the numerators to obtain the numerator of the product, and multiply the denominators to find the denominator of the product.
>
> **14.15**
> To simplify a fraction before multiplying, divide both the numerator and the denominator of the fraction by the same nonzero number.
>
> **14.16**
> Whenever a factor is expressed as a mixed numeral or a whole number, rename the factor as a fraction to multiply.

Fractions are multiplied using a simple rule: numerator times numerator, and denominator times denominator. For example:

$$\frac{2}{3} \times \frac{7}{8} = \frac{2 \times 7}{3 \times 8} = \frac{14}{24} \text{ or } \frac{7}{12}$$

A process called *cancelling* eliminates the need to change products to lower terms, and is an application of the identity property. For the example $\frac{5}{8} \times \frac{12}{25}$, divide 5 and 25 by 5, and divide 12 and 8 by 4.

$$\frac{\cancel{5}^{1}}{\cancel{8}_{2}} \times \frac{\cancel{12}^{3}}{\cancel{25}_{5}} = \frac{1 \times 3}{2 \times 5} = \frac{3}{10}$$

If one of the factors is a whole number, the whole number can be shown as a fraction with a denominator of 1. For example:

$$4 \times \frac{3}{8} = \frac{4}{1} \times \frac{3}{8} \rightarrow \frac{\cancel{4}^{1}}{1} \times \frac{3}{\cancel{8}_{2}} = \frac{3}{2} = 1\frac{1}{2}$$

When teachers introduce examples like $\frac{2}{3} \times 12$, they sometimes tell children that *times means of*, which can lead to misunderstandings. For example, 4 of 5 may be interpreted correctly as a fraction rather than as 4×5. But if a child takes $\frac{2}{3}$ of a set of 12 things, and also multiplies $\frac{2}{3} \times 12$, the result will be 8. When a child has made several such observations, she can conclude that a *fraction of a number* and a *fraction times a number* have the same result.[3]

> **14.17**
> A fraction *of* a number and a fraction *times* a number have the same result.

One method for computing with mixed numerals is to rename them as fractions. For example:

$$\frac{5}{8} \times 1\frac{1}{3} = \frac{5}{8} \times \frac{4}{3} \rightarrow \frac{5}{\cancel{8}_{2}} \times \frac{\cancel{4}^{1}}{3} = \frac{5}{6}$$

$$1\frac{1}{4} \times 2\frac{2}{3} = \frac{5}{4} \times \frac{8}{3} \rightarrow \frac{5}{\cancel{4}_{1}} \times \frac{\cancel{8}^{2}}{3} = \frac{10}{3} = 3\frac{1}{3}$$

However, if one factor is a whole number, a procedure applying the distributive property may be preferred. For example:

$$24 \times 2\frac{1}{2} = \begin{array}{rl} 24 & \\ \times\ 2\frac{1}{2} & \\ \hline 12 & \leftarrow (\frac{1}{2} \times 24) \\ 48 & \leftarrow (2 \times 24) \\ \hline 60 & \leftarrow (2\frac{1}{2} \times 24) \end{array}$$

3. John L. Marks et al., *Teaching Elementary School Mathematics for Understanding,* 4th ed. (New York: McGraw-Hill, 1975), p. 176.

EXCURSION
Adjusting a Recipe[4]

A recipe usually serves a specified number of people. When you need to increase or decrease the number of servings, you can multiply to adjust each ingredient.

Peanut Butter Crunch
(Serves 12)

$\frac{2}{3}$ Cup Water
$\frac{1}{2}$ Tsp. Salt
$1\frac{1}{2}$ Cups Sugar
$1\frac{3}{4}$ Cups Molasses
6 qt. Popped Corn
$\frac{7}{8}$ Cup Peanut Butter
$\frac{1}{4}$ Tsp. Baking Soda

1. How many cups of sugar are needed if the recipe is to be adjusted to serve 36 persons?

$$3 \times 1\frac{1}{2} = \frac{3}{1} \times \frac{3}{2} = \frac{9}{2} = 4\frac{1}{2} \text{ cups}$$

2. How many teaspoons of salt are needed if the recipe is adjusted to serve 30 persons?
3. How many cups of molasses are needed if the recipe is adjusted to serve 6 persons?

Division with Fractions and Mixed Numerals

14.18
The partition of a universal set into equivalent disjoint subsets is a model for division with fractions.

14.19
Repeated subtraction is a model for division with fractions.

Models for Division

Division is the least used operation with both whole numbers and fractions. In fact, many adults still do not understand the procedures for division with fractions.[5]

Division with fractions should be related to division with whole numbers. Both partitive and measurement situations can be extended to include rational numbers.

Whenever a whole number is divided by a fraction, for example, $6 \div \frac{1}{2}$, ask a measurement question, such as, how many halves are in 6 wholes? A partitive interpretation does not produce a meaningful question.

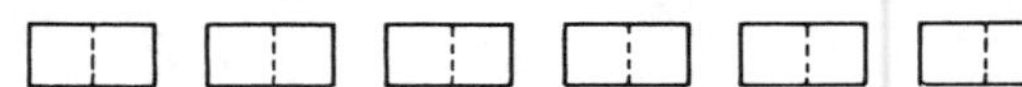

4. Adapted from Wilmer L. Jones, *Mathematics for Today, Level Green* (New York: Sadlier, 1979), p. 78.
5. Esther J. Swenson, *Teaching Mathematics to Children*, 2nd ed. (New York: Macmillan, 1973), p. 339.

$$6 \div \frac{1}{2} = \frac{6}{1} \times \frac{2}{1} = 12$$

However, if the dividend is a fraction and the divisor a whole number, a different interpretation is required.

PROBLEM

If $\frac{1}{2}$ of a pizza is to be divided equally among 4 persons, what is the size of the piece each person will get?

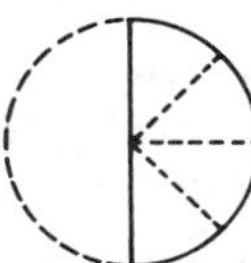

$$\frac{1}{2} \div 4 = \frac{1}{2} \times \frac{1}{4} = \frac{1}{8} \text{ of the original pizza}$$

In this situation only the partitive interpretation can be given. It makes no sense to ask,How many whole pieces of size 4 are in $\frac{1}{2}$?

In another type of division problem, both numbers are expressed as fractions. The simplest illustration is the situation in which the divisor is *less* than the dividend.

PROBLEM

How many $\frac{1}{8}$ yd. pieces of copper can be cut from a $\frac{3}{4}$ yd. piece?

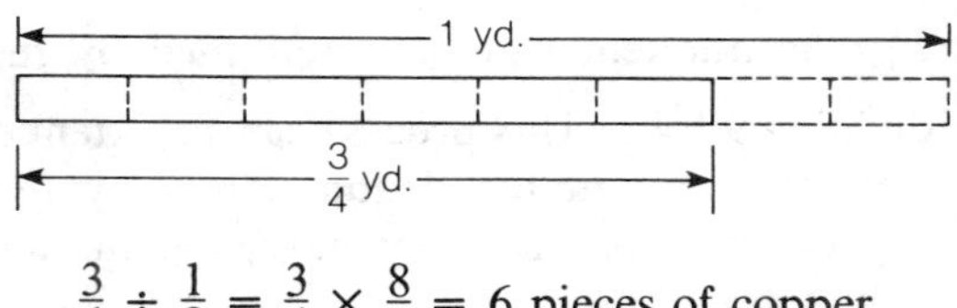

$$\frac{3}{4} \div \frac{1}{8} = \frac{3}{4} \times \frac{8}{1} = 6 \text{ pieces of copper}$$

Whenever the divisor is *greater* than the dividend, the quotient is always a fraction and is only meaningful when the ratio interpretation is applied.

PROBLEM

A recipe calls for $\frac{3}{4}$ of a pound of butter. Carol only has $\frac{1}{2}$ pound. What portion of the recipe can she make?

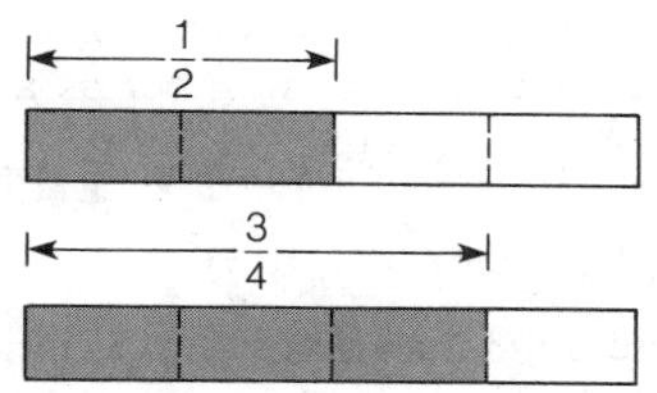

$$\frac{1}{2} \div \frac{3}{4} = \frac{1}{2} \times \frac{4}{3} = \frac{2}{3} \text{ of the recipe}$$

Algorithms

Two procedures can be used to divide with fractions. The method called *invert and multiply* is frequently taught in schools. Another, the *common-denominator method,* is introduced for enrichment. The content objectives that follow relate to the invert and multiply method.

> **14.20**
> If a nonzero number is expressed as a fraction, interchange the numerator and denominator to find the reciprocal.
>
> **14.21**
> The quotient of a given number divided by 1 is the given number.
>
> **14.22**
> When the numerator and the denominator of a fraction are multiplied by the same nonzero number, the value of the fraction remains the same.
>
> **14.23**
> To divide one fraction by another, multiply the dividend by the reciprocal of the divisor.

The concept of a reciprocal is important when developing the division algorithm. The reciprocal of a number is defined in various ways, such as: two numbers are reciprocals of one another if their product is 1; or the reciprocal of a number is 1 divided by that number. For example, $\frac{1}{3}$ and 3 are reciprocals and $\frac{3}{4}$ and $\frac{4}{3}$ are reciprocals.

A reciprocal is closely associated with the inverse relationship between multiplication and division; sometimes it is called the *multiplicative inverse.* Multiplying a number by its reciprocal undoes the effect of that number as a multiplier, changing it to 1. For example, $\frac{2}{3}$ (original factor) $\times$ $\frac{3}{2}$ (the reciprocal of $\frac{2}{3}$) $= 1$.

Teaching the invert and multiply rule (often called the inversion algorithm) by rote causes difficulty. Students become confused about inverting the first or second fraction; some of this carries over into later studies in algebra. Children need to know *why* inverting and multiplying produces a quotient.

Each child should understand that the inversion algorithm is a shortcut based on several mathematical principles, the steps for which are omitted when the invert and multiply procedure is used. Consider these steps for finding $\frac{3}{4} \div \frac{3}{8}$.

Step 1: Write the division statement as a complex fraction.

$$\frac{3}{4} \div \frac{3}{8} = \frac{\frac{3}{4}}{\frac{3}{8}}$$

Step 2: In order to have a denominator of 1, multiply $\frac{3}{8}$ by its reciprocal, $\frac{8}{3}$.

$$\frac{\frac{3}{4}}{\frac{3}{8} \times \frac{8}{3}}$$

Step 3: Also multiply the numerator by $\frac{8}{3}$. This multiplies the original fraction by 1.

$$\frac{\frac{3}{4} \times \frac{8}{3}}{\frac{3}{8} \times \frac{8}{3}}$$

Step 4: Now that the denominator of the fraction is 1, it can be eliminated (content objective 14.21).

$$\frac{\frac{3}{4} \times \frac{8}{3}}{\frac{3}{8} \times \frac{8}{3}} = \frac{\frac{3}{4} \times \frac{8}{3}}{1}$$

Step 5: Thus the original division statement is replaced by an equivalent multiplication statement.

$$\frac{3}{4} \div \frac{3}{8} = \frac{3}{4} \times \frac{8}{3}$$

Step 6: Multiply, and the result is the quotient for the original example.

$$\frac{\overset{1}{\cancel{3}}}{\underset{1}{\cancel{4}}} \times \frac{\overset{2}{\cancel{8}}}{\underset{1}{\cancel{3}}} = 2$$

Lead each child to the generalization that, to divide with fractions, multiply the dividend times the *reciprocal* of the divisor, rather than the common expression, to divide with fractions, invert and multiply.[6] The latter rule is mathematically meaningless.

The common denominator procedure is easier to explain to children than the inversion method. For example, $2\frac{1}{4} \div \frac{1}{4}$ can be represented as follows:

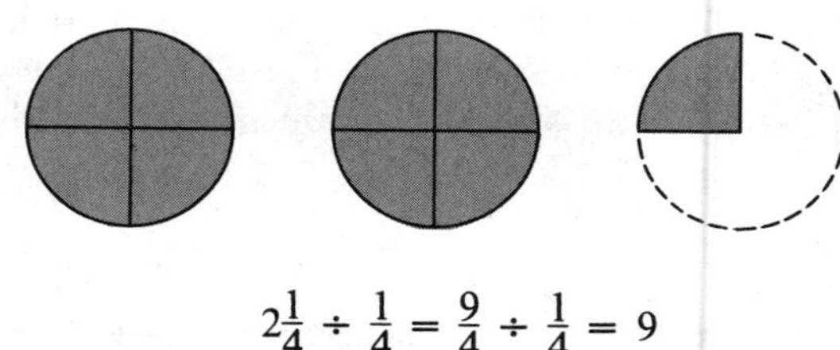

$$2\frac{1}{4} \div \frac{1}{4} = \frac{9}{4} \div \frac{1}{4} = 9$$

It is easy to see that nine-fourths are contained in $2\frac{1}{4}$; the quotient can be determined by merely dividing 9 by 1. This process can be extended to fractions with unlike denominators by renaming the fractions so they have a common denominator.

$$\frac{7}{8} \div \frac{1}{3} = \left(\frac{7}{8} \times \frac{3}{3}\right) \div \left(\frac{1}{3} \times \frac{8}{8}\right)$$

$$= \frac{21}{24} \div \frac{8}{24} = 21 \div 8 = 2\frac{5}{8}$$

6. Kennedy, *Guiding Children*, p. 344.

Like multiplication, division with a mixed numeral involves only one additional step, that of renaming the number as a fraction. Therefore, computation with mixed numerals is often taught along with fractions.

Ordering the Content

Before introducing algorithms, make sure each child has prerequisite skills, and understands the following: concepts related to the four basic operations with whole numbers, the meanings and models for fractions, and the basic properties. The child should also be able to rename a fraction to higher and lower terms.

Traditionally, multiplication with fractions was not taught until after addition and subtraction of fractions with unlike denominators. However, this may involve teaching the more difficult concept first. As Firl notes, "It is a more 'natural' skill to multiply numerators and to multiply denominators than it is to find a common denominator, to replace fractions with equivalent fractions, and then to write the sum of the numerators over the common denominator."[7] Furthermore, in renaming a fraction to higher terms (as is required when adding or subtracting with unlike fractions), you must multiply by 1 in the form of a fraction $\frac{n}{n}$.

Different operations with fractions may be presented in various orders, depending upon the preference of the teacher or the school curriculum guide and textbook. One logical sequence is as follows:

- Addition and subtraction with like fractions
- Multiplication
- Addition and subtraction with unlike fractions
- Division

If a child has difficulty with mathematics, postpone division with fractions until junior high school.

7. Firl, *Role of Fractions*, p. 7.

PLANNING INSTRUCTION

Because children have already developed some skill with whole number algorithms, you may be tempted to give them rules for operating with fractions, illustrate how the rules work, and then provide them with consolidating activities. However, when algorithms for fractions are taught by rules alone, children typically have little or no understanding of what or why they are doing something. Make sure activities are planned so that mathematical *ideas* are exemplified.

Initiating, Abstracting and Schematizing Activities

Addition and Subtraction

Activities for computation with fractions should involve familiar situations. For example:

PROBLEM

> Scott and his sister each had $\frac{1}{4}$ of a pizza. What part of the pizza did they have together?

Geometric shapes can be used to picture the addition situation involved in this problem.

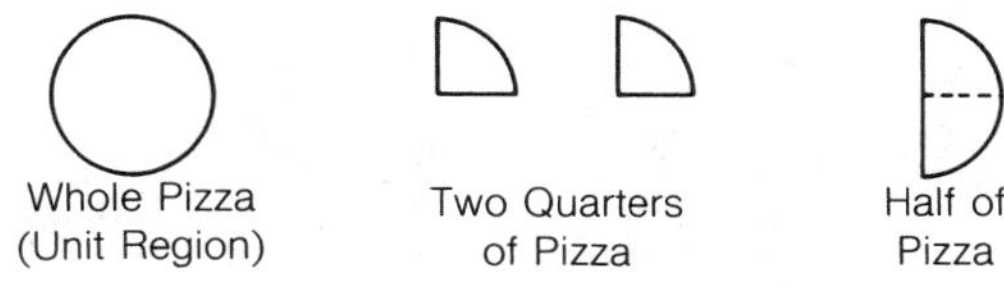

Strips of cardboard and the number line may also be used to illustrate addition of like fractions.

$\frac{3}{8} + \frac{2}{8} = \frac{5}{8}$ $\frac{2}{5} + \frac{1}{5} = \frac{3}{5}$

a. $\frac{5}{8}$; $\frac{3}{8}$; $\frac{2}{8}$; 0, $\frac{1}{8}$, $\frac{2}{8}$, $\frac{3}{8}$, $\frac{4}{8}$, $\frac{5}{8}$, $\frac{6}{8}$, $\frac{9}{10}$, $\frac{8}{8}$

b. $\frac{2}{5}$ $\frac{1}{5}$

If any of these exemplifications are to make sense to children, show the unit (region or line segment) along with the fractional parts. Children must be able to see the relationship between each part and a unit.

Gradually, each child can be led to observe that the sum is found by adding the numerators of the fractions. At first have children write the entire procedure.

$$\frac{2}{5} + \frac{1}{5} = \frac{2 + 1}{5} = \frac{3}{5}$$

$$\frac{3}{8} + \frac{2}{8} = \frac{3 + 2}{8} = \frac{5}{8}$$

Because the algorithm for subtracting with like fractions is similar to the algorithm for adding, both procedures can be introduced at the same time.

When introducing the subtraction procedure, illustrate different subtraction situations.

1. Subtraction is used to determine the fractional part remaining after a part is removed.

PROBLEM

The plate contains $\frac{5}{8}$ of a cake. After Jake eats a piece that is $\frac{2}{8}$ of the original cake, how much cake will be left?

$$\frac{5}{8} - \frac{2}{8} = \frac{3}{8}$$

2. Subtraction is used to determine the difference in size of two fractional parts.

PROBLEM

The distance from Janet's house to school is $\frac{7}{8}$ of a mile. Christa lives $\frac{5}{8}$ of a mile from school. How much farther is Janet's house from school than Christa's?

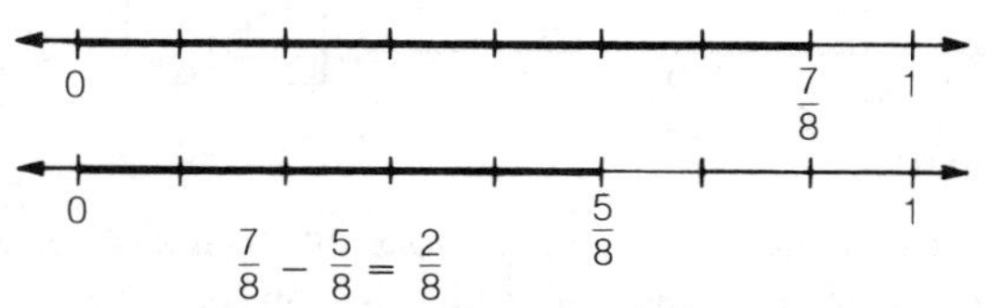

3. Subtraction is used to determine how large a fractional part must be joined with a second part of known size, to yield a third part of known size.

PROBLEM

Carol must walk $\frac{7}{8}$ of a mile to work. After she has walked $\frac{3}{8}$ of a mile, how much farther does she have to go?

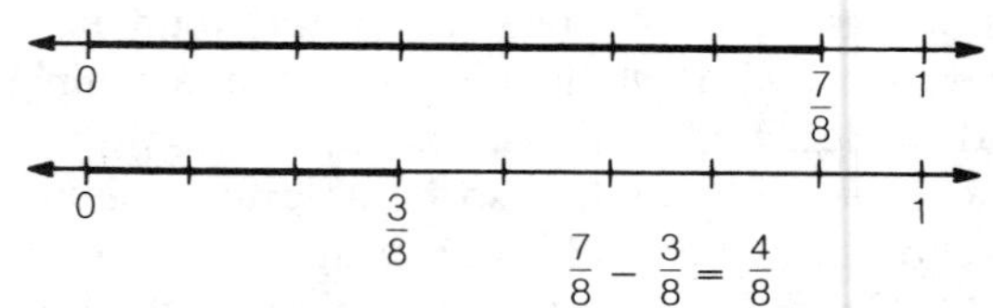

Writing examples in the form shown in Figure 14.2 emphasizes the fact that the numerator of a fraction tells how many. This helps each child see that the numerators are added, and the denominators tell what is added, discouraging children from adding both.

During early work with addition and subtraction, children can record their computations horizontally and vertically. Examples in which answers need to be simplified can be introduced gradually.

$$\begin{array}{r} \frac{3}{8} \\ -\ \frac{1}{8} \\ \hline \frac{2}{8} \end{array} \rightarrow \ \frac{2}{8} = \frac{1}{4} \qquad \begin{array}{r} \frac{3}{5} \\ +\ \frac{2}{5} \\ \hline \frac{5}{5} \end{array} \rightarrow \ \frac{5}{5} = 1 \qquad \begin{array}{r} \frac{3}{4} \\ +\ \frac{2}{4} \\ \hline \frac{5}{4} \end{array} \rightarrow \ \frac{5}{4} = 1\frac{1}{4} \qquad \begin{array}{r} \frac{7}{8} \\ +\ \frac{7}{8} \\ \hline \frac{14}{8} \end{array} \rightarrow \ \frac{14}{8} = 1\frac{6}{8} = 1\frac{3}{4}$$

Initially, most children encounter some difficulty with fractions having unlike denomina-

$$\begin{array}{r} \frac{3}{8} \\ +\ \frac{2}{8} \\ \hline \frac{5}{8} \end{array} \quad \begin{array}{l} \longrightarrow \text{3 eighths} \\ \longrightarrow \text{2 eighths} \\ \longrightarrow \text{5 eighths} \end{array}$$

FIGURE 14.2
Introductory Form for Addition

tors; a change in denominators implies a change in units used to state quantities. This can be partially overcome by using examples with money. For instance, to find the total value of 8 nickels and 3 quarters, the value of the quarters may be expressed as nickels.

$$\begin{array}{rl} 8 \text{ nickels} = & 8 \text{ nickels} \\ \underline{+\ 3 \text{ quarters}} = & \underline{15 \text{ nickels}} \\ & 23 \text{ nickels} \end{array}$$

The difficulty children have with the renaming steps depends upon denominators. There are three different situations.

1. One of the denominators is a common denominator.

$$\begin{array}{r} \frac{3}{4} \\ \underline{+\frac{1}{2}} \end{array} \qquad \begin{array}{r} \frac{3}{4} \\ \underline{-\frac{3}{8}} \end{array}$$

2. The denominators are relatively prime. Their product is the Least Common Denominator.

$$\begin{array}{r} \frac{2}{3} \\ \underline{+\frac{1}{2}} \end{array} \qquad \begin{array}{r} \frac{3}{4} \\ \underline{-\frac{2}{3}} \end{array}$$

3. The denominators are not relatively prime; some other number is the Least Common Denominator.

$$\begin{array}{r} \frac{3}{4} \\ \underline{+\frac{5}{6}} \end{array} \qquad \begin{array}{r} \frac{5}{6} \\ \underline{-\frac{2}{9}} \end{array}$$

Allow children to experiment when beginning instruction in addition and subtraction with unlike denominators. For example, to find $\frac{1}{4} + \frac{1}{2}$, ask the child to improvise a solution using fractional parts of circular regions. She will soon see that $\frac{1}{2}$ can be shown with $\frac{2}{4}$, and the sum is $\frac{3}{4}$, as in Figure 14.3.

Sums for fractions such as $\frac{1}{2}$ and $\frac{1}{3}$ can be found by comparing rectangular regions, as shown in Figure 14.4

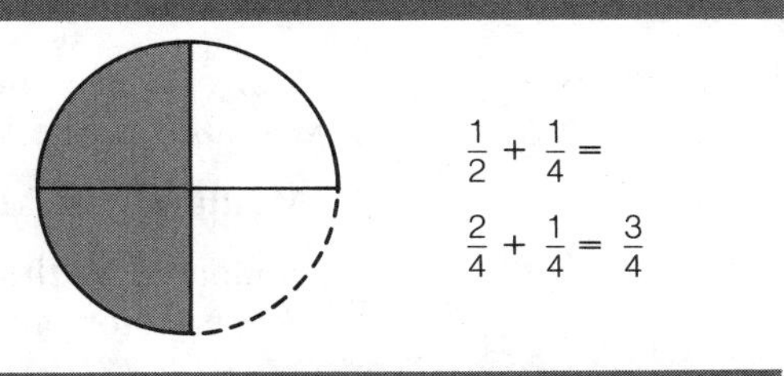

FIGURE 14.3
Addition with Parts of Circular Regions

Before more formal procedures are taught, introduce each child to common multiples of whole numbers and to the concepts of common denominators and Least Common Denominator (L.C.D.). If prime factorizations are used to find the L.C.D., children should learn about prime numbers, factoring a number, and the *prime factorization* for a number.

Children are able to guess the L.C.D. in the easier examples; they often determine it by inspection. However, each child needs to learn a procedure that is useful for working with all types of denominators, for later mathematics study.[8]

To add and subtract with unlike denominators, a child must also understand equivalent fractions. Provide each child several opportunities to act out the algorithm with Cuisinaire rods or fractional parts of unit regions before introducing two-dimensional exemplars. To show the addition of $\frac{3}{4}$ and $\frac{1}{2}$, a child can use fractional parts (see page 330).

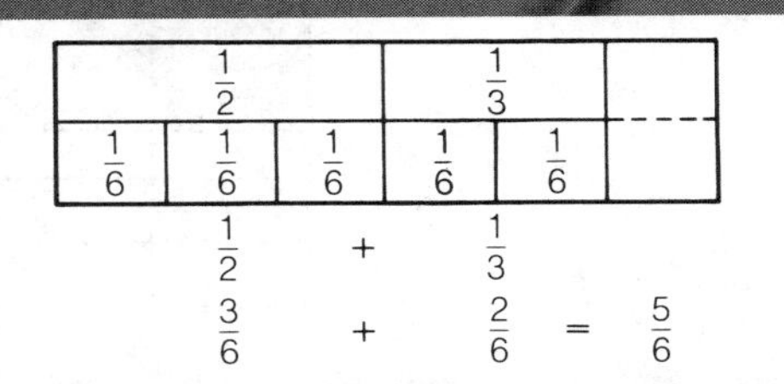

FIGURE 14.4
Comparing Rectangular Regions

8. Donald D. Paige et al., *Elementary Mathematical Methods* (New York: John Wiley, 1978), pp. 91–92.

EXCURSION
Looking Ahead

Many of the mathematical ideas learned by younger children are quite essential to the study of algebra. For example, finding the Least Common Multiple (L.C.M.) of 6, 9, and 15 in order to add $\frac{5}{6}$, $\frac{1}{9}$, and $\frac{4}{15}$ can be compared to the solution of a similar problem in algebra. Note the use of prime factorizations.

$$\frac{5}{6} + \frac{1}{9} + \frac{4}{15}$$

$$6 = 2 \times 3 \qquad 9 = 3 \times 3 \qquad 15 = 3 \times 5$$

$$\text{LCM} = 2 \times 3 \times 3 \times 5 = 90$$

$$\frac{5}{6} \times \frac{15}{15} = \frac{75}{90}$$

$$\frac{1}{9} \times \frac{10}{10} = \frac{10}{90}$$

$$+\ \frac{4}{15} \times \frac{6}{6} = \frac{24}{90}$$

$$\frac{109}{90} = 1\frac{19}{90}$$

$$\frac{5}{x^2 - 1} + \frac{3}{x^2 + 2x + 1}$$

$$x^2 - 1 = (x + 1)(x - 1) \qquad x^2 + 2x + 1 = (x + 1)(x + 1)$$

$$\text{LCM} = (x + 1)(x - 1)(x + 1)$$
$$= x^3 + x^2 - x - 1$$

$$\frac{5}{x^2 - 1} \times \frac{x + 1}{x + 1} = \frac{5x + 5}{x^3 + x^2 - x - 1}$$

$$\frac{3}{x^2 + 2x + 1} \times \frac{x - 1}{x - 1} = \frac{3x - 3}{x^3 + x^2 - x - 1}$$

$$\frac{8x + 2}{x^3 + x^2 - x - 1}$$

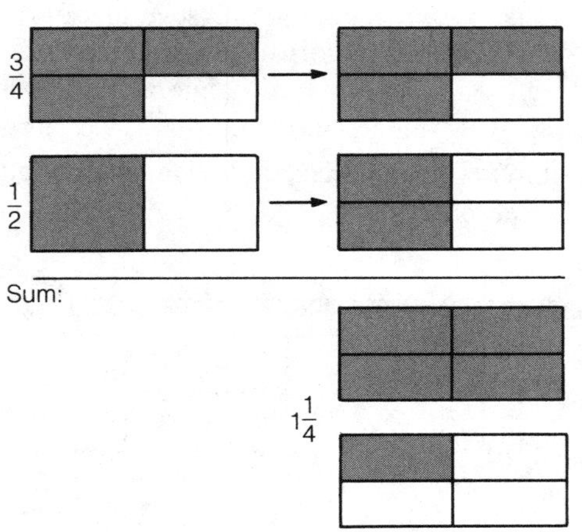

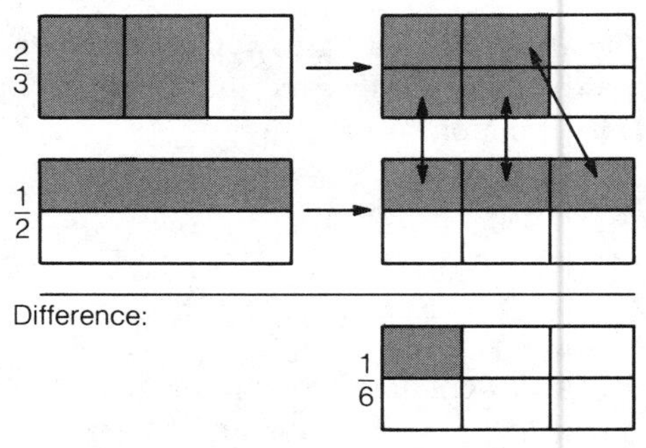

To subtract or find a difference, a similar procedure can be followed, illustrated for $\frac{2}{3}$ and $\frac{1}{2}$. The solution (see above right) uses a comparison model for subtraction. The procedure can be varied to show subtraction by removing a subset if desired.

Addition with mixed numerals does not usually require teaching new skills; however, subtraction may be more difficult. Children should be alert for whole numbers that need to be renamed. This occurs when the minuend is a whole number, and when the minuend is a mixed numeral in which the fractional part is less than the fractional part of the subtrahend.

$$\begin{array}{r} 2 \\ -\ \frac{1}{4} \\ \hline \end{array} \qquad \begin{array}{r} 3 \\ -\ 1\frac{3}{5} \\ \hline \end{array} \qquad \begin{array}{r} 2\frac{1}{4} \\ -\ \frac{3}{4} \\ \hline \end{array} \qquad \begin{array}{r} 4\frac{1}{3} \\ -\ 2\frac{5}{6} \\ \hline \end{array}$$

EXCURSION
Using Flow Charts

Some teachers have their students describe addition and subtraction procedures with fractions by making a flow chart.

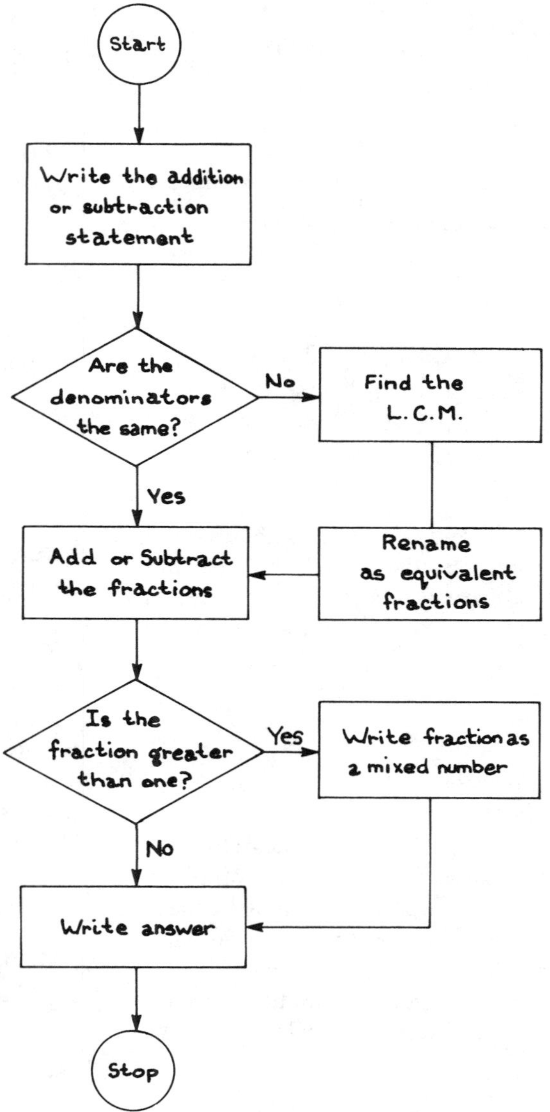

Make a flow chart to describe the procedure for subtracting $2\frac{1}{2} - 1\frac{3}{4}$.

Circular regions can help exemplify the procedure. For instance, in subtracting $2\frac{1}{4} - \frac{3}{4}$, illustrate the minuend as follows.

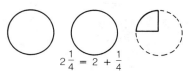

$2\frac{1}{4} = 2 + \frac{1}{4}$

To subtract $\frac{3}{4}$, a 1 must be renamed as $\frac{4}{4}$.

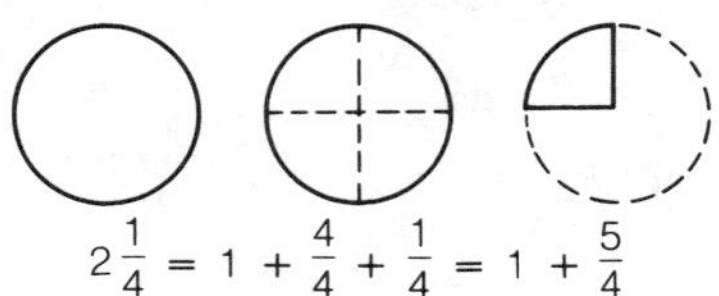

When the fractional part for the known addend $\left(\frac{3}{4}\right)$ is set aside, the unknown addend or remainder is seen to be $1\frac{2}{4}$, or $1\frac{1}{2}$.

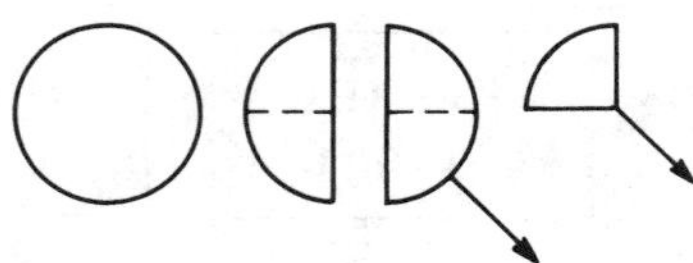

The procedure makes sense to children who work through it with manipulatives and make a paper-and-pencil record. They are apt to avoid the common error illustrated in Figure 14.5.

Multiplication with Fractions and Mixed Numerals

Several ways of interpreting multiplication with fractions were noted earlier. Despite limitations of the repeated addition and array models, these interpretations are most useful when initially explaining multiplication with fractions.

Multiplication involving fractions occurs in three different arrangements: whole number times a fraction, fraction times a whole number, and fraction times a fraction. Each case is treated separately in interpreting multiplication statements, but children should learn a generalized procedure for computing all three.

Initiating activities should begin with realistic problems.

$$2\frac{1}{4} = 1\frac{11}{4}$$
$$-\frac{3}{4} = \frac{3}{4}$$

FIGURE 14.5
A Common Error

PROBLEM

Christa takes swimming lessons for $\frac{3}{4}$ of an hour, 6 days a week. How many hours of lessons does she have each week?

For this problem of a *whole number times a fraction*, make a time line out of strips of paper of uniform size (each represents $\frac{1}{4}$ of an hour).

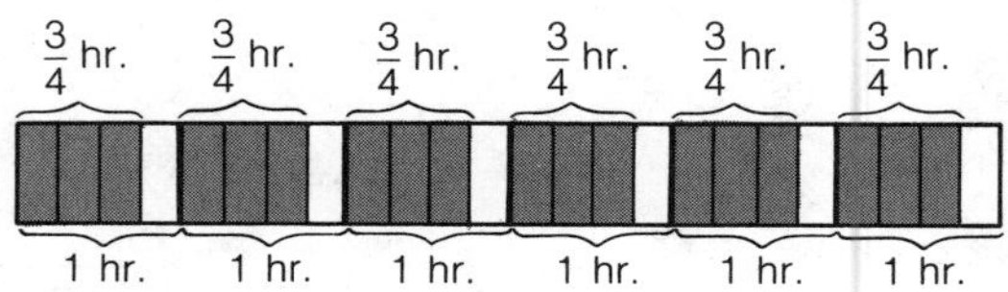

Each set of 4 strips represents an hour. For each hour that Christa takes lessons, 3 of the strips are shaded. The number of hours practiced can be determined either by counting the sets of 4 shaded strips, or by rearranging the strips in the following fashion.

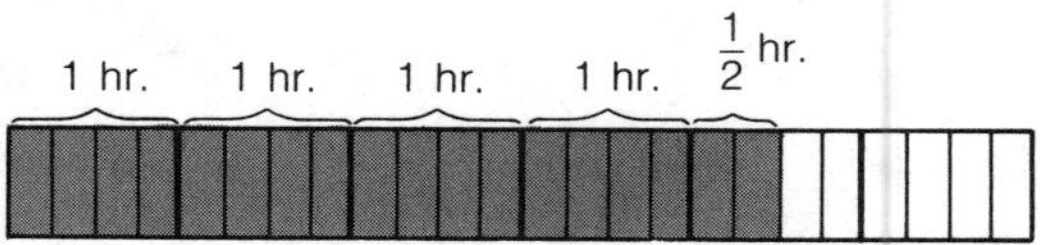

It can be seen that $6 \times \frac{3}{4} = 4\frac{1}{2}$. Christa takes $4\frac{1}{2}$ hours of swimming lessons each week.

Circular regions can be used similarly. Problems using representative materials help the child conclude that the whole number times the number of parts (numerator) gives the total number of parts.

Some teachers invoke the commutative property, to introduce the case of a *fraction times a whole number*. Because $4 \times \frac{2}{3} = 2\frac{2}{3}$ we know that $\frac{2}{3} \times 4 = 2\frac{2}{3}$. However, this is inadequate treatment: in certain respects, the two situations are quite different.[9]

The multiplication of a whole number times a fraction suggests a unit for measuring, while the multiplication of a fraction times a whole number involves partitioning.

PROBLEM

Carol used $\frac{2}{3}$ dozen eggs to make cookies. How many eggs did she use?

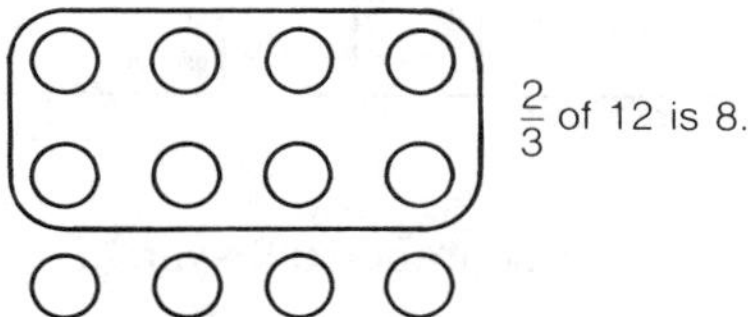

Partitioning is involved because we must find the size of one of the equal subsets in which the total set is divided. In this case, $\frac{1}{3}$, or one row, has 4 eggs; therefore two rows, or $\frac{2}{3}$, has 8 eggs.

How do we help each child understand that $\frac{2}{3}$ of 12 is the same as $\frac{2}{3} \times 12$? As indicated earlier, a child can be led to observe that a fraction *of* a number and a fraction *times* a number *have the same result*. Avoid the erroneous statement *of means times*.

In cases involving a *fraction times a fraction*, extend ideas the child already knows.

PROBLEM

Scott saw $\frac{2}{3}$ of a cake left on the plate, and proceeded to eat $\frac{1}{2}$ of it. What part of the original cake did he eat?

For the child who knows that $\frac{1}{2}$ of 6 and $\frac{1}{2} \times 6$ are equivalent, it follows that $\frac{1}{2}$ of $\frac{2}{3}$ is the same as $\frac{1}{2} \times \frac{2}{3}$. A simple way to illustrate this situation is to partition a rectangular region; a rectangular piece of paper can be folded into thirds to illustrate $\frac{2}{3}$.

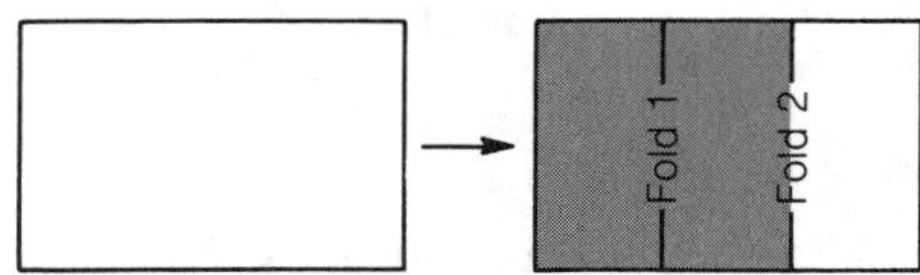

Shade $\frac{2}{3}$ of the cake to show the portion remaining on the plate. To show that $\frac{1}{2}$ of the remaining $\frac{2}{3}$ was eaten, fold the paper in half in the other direction, thus forming sixths of the original cake.

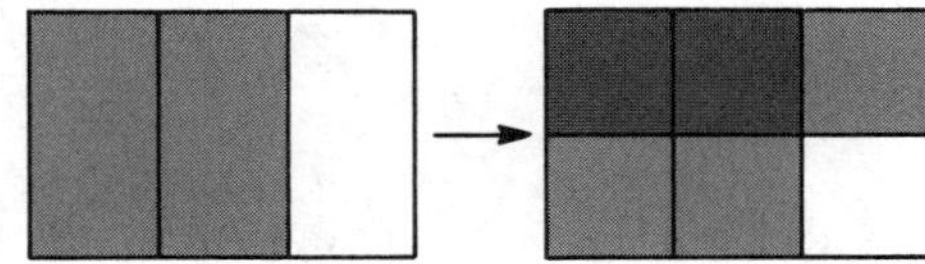

By counting, a child can determine that $\frac{1}{2}$ of $\frac{2}{3}$ is $\frac{2}{6}$ (or $\frac{1}{3}$) of the original cake.

During repeated experiences, have each child record the examples and the products. Eventually children will observe that, in every case, the product (before it is changed to simpler terms) has a numerator that is a product of the numerators in the factors, and it has a denominator that is the product of the denominators in the factors. This pattern is in fact the algorithm for multiplication with fractions. Once the rule has been stated, each child should have schematizing activities in which she can verify the rule with other examples using manipulatives. We then need to increase the difficulty of computations.

The shortcut of *cancelling* can be introduced at this point. Some children may initially be reluctant to alter an already learned procedure, but if they see that cancelling makes a solution easier, as in Figure 14.6, they will be encouraged to use it.

9. Swenson, *Teaching Mathematics*, p. 327.

$$\frac{18}{25} \times \frac{15}{24} \to \frac{\overset{3}{\cancel{18}}}{\underset{5}{\cancel{25}}} \times \frac{\overset{3}{\cancel{15}}}{\underset{4}{\cancel{24}}} = \frac{9}{20}$$

FIGURE 14.6
Cancelling

Each child should realize there is more than one way to compute the product. Examples for $\frac{3}{4} \times \frac{8}{12}$ include:

Method 1 $\quad \frac{3}{4} \times \frac{8}{12} = \frac{24}{48} = \frac{1}{2}$

Method 2 $\quad \frac{\overset{1}{\cancel{3}}}{\underset{\underset{1}{\cancel{2}}}{\cancel{4}}} \times \frac{\overset{\overset{1}{\cancel{2}}}{\cancel{4}}}{\underset{\underset{2}{\cancel{4}}}{\cancel{12}}} = \frac{24}{48} = \frac{1}{2}$

Method 3 $\quad \frac{\overset{1}{\cancel{3}}}{\underset{1}{\cancel{4}}} \times \frac{\overset{\overset{1}{\cancel{2}}}{\cancel{8}}}{\underset{\underset{2}{\cancel{4}}}{\cancel{12}}} = \frac{24}{48} = \frac{1}{2}$

Problems involving three or more factors should also be introduced. Application of the associative and commutative properties may simplify the computation.

Multiplication with mixed numerals requires no new skills; a child simply changes both numbers to fractions and multiplies, as shown in Figure 14.7.

$$2\frac{1}{2} \times 1\frac{3}{5} = \frac{5}{2} \times \frac{8}{5} \to \frac{\overset{1}{\cancel{5}}}{\underset{1}{\cancel{2}}} \times \frac{\overset{4}{\cancel{8}}}{\underset{1}{\cancel{5}}} = 4$$

FIGURE 14.7
Multiplication with Mixed Numerals

Division with Fractions and Mixed Numerals

Introduce division of fractions with a practical problem that can be represented easily.

PROBLEM

Bill makes wooden toys as a hobby. He uses a piece of wood that is $\frac{1}{2}$-yard long for each. How many pieces can he cut from a 3-yard-long piece of wood?

3 yd.

$\frac{1}{2}$ yd.	$\frac{1}{2}$ yd.	$\frac{1}{2}$ yd.	$\frac{1}{2}$ yd.	$\frac{1}{2}$ yd.	$\frac{1}{2}$ yd.

By counting the number of $\frac{1}{2}$-yard pieces in this situation, a child can see that $3 \div \frac{1}{2} = 6$.

Children also need experiences with partitive division problems. Parts of a circular region show the solution to this problem.

PROBLEM

Mary has $\frac{3}{4}$ of a cake and cuts it into 6 equal pieces. What part of the whole cake will each piece be?

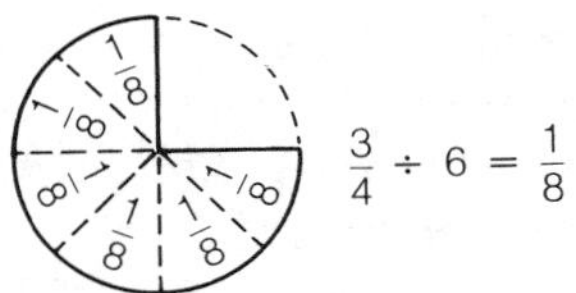

Make sure each child understands that the answer to the problem ($\frac{1}{8}$) indicates a part of the *whole* cake.

Before the division algorithm is presented, the concept of a reciprocal should be introduced. The typical approach involves unit fractions and whole numbers.

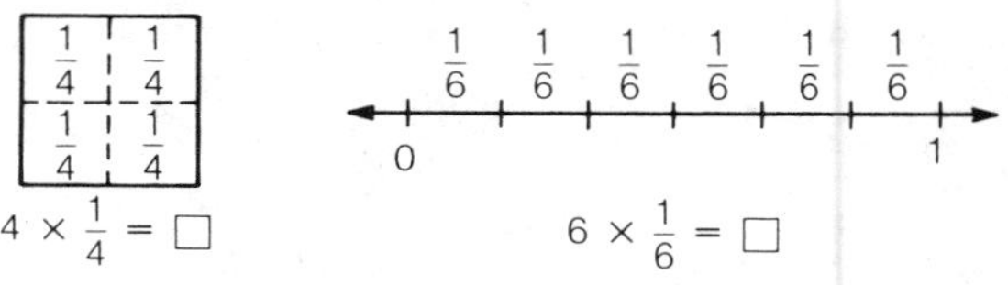

Follow the previous illustrations with number statements:

$$\frac{1}{3} \times \frac{3}{1} = \square$$

$$\frac{2}{3} \times \frac{3}{2} = \square$$

$$\frac{1}{6} \times \square = 1$$

$$\frac{3}{4} \times \square = 1$$

Children need to eventually realize that two numbers are reciprocals if their product is 1.

In partitive division situations, the divisor tells the number of parts created. The idea that something is divided into $\frac{2}{3}$ parts is awkward and useless. Therefore, carefully choose those situations for presenting the algorithm.

A whole number divisor can be used. One interpretation of $\frac{1}{4} \div 2$ is $\frac{1}{4}$ separated into two parts of equal size. A child can illustrate this by replacing a region whose measure is $\frac{1}{4}$ with two regions with a measure of $\frac{1}{8}$ each. In the same manner, $\frac{3}{4} \div 6$ can be shown by replacing a region that measures $\frac{3}{4}$ with six regions of $\frac{1}{8}$ measure each.

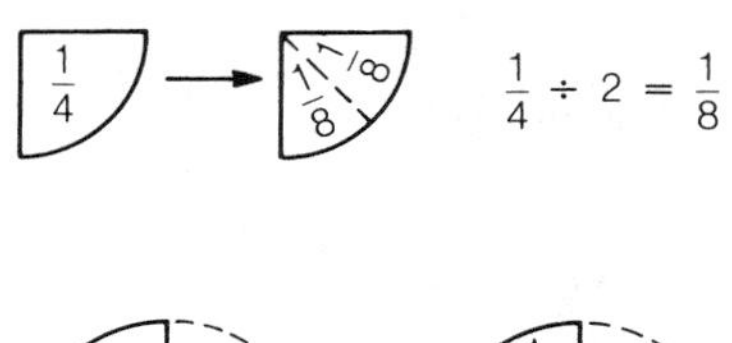

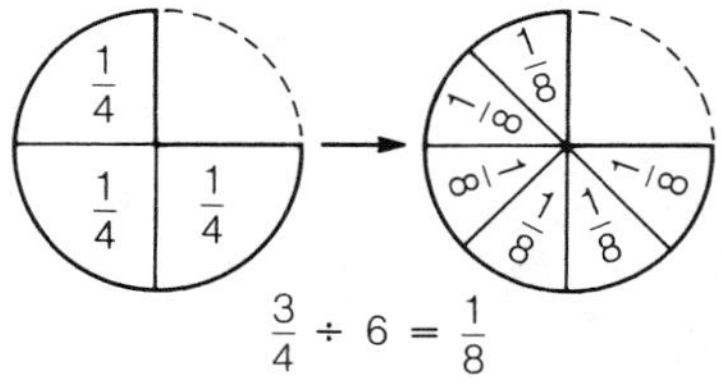

As children examine these and other partitive situations with whole number divisors, lead them to see that $\frac{1}{4} \div 2$ and $\frac{1}{4} \times \frac{1}{2}$ yield the same answer. Likewise, $\frac{3}{4} \div 6 = \frac{3}{4} \times \frac{1}{6}$ and $\frac{3}{4} \div \frac{6}{1} = \frac{3}{4} \times \frac{1}{6}$. Use guiding questions to lead each child to the generalization that to divide by any number, multiply by the reciprocal of the number.

Division by a fraction usually occurs in measurement situations such as these:

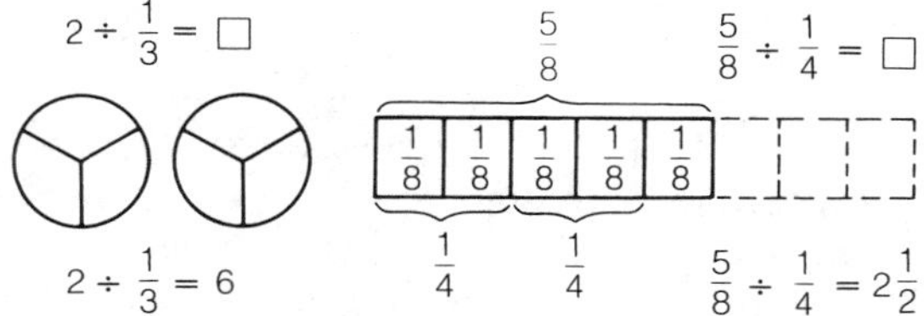

Children readily see that the solutions for such problems make sense. Have them determine if the reciprocal rule also applies when the divisor is a fraction, and conclude that the reciprocal rule can be used whether the divisor is a whole number or a fraction.

According to Piaget, most children enter the formal operational stage at about age eleven, and begin to reason more abstractly. Rather than develop the division algorithm through concrete materials, a more abstract approach may be used. Children can be introduced to the invert and multiply algorithm by applying mathematical ideas as follows:

Steps	*Mathematical Ideas*
1. $\frac{3}{4} \div \frac{1}{8} = \dfrac{\frac{3}{4}}{\frac{1}{8}}$	A fraction can be used to indicate division: $a \div b = \frac{a}{b}$.
2. $\dfrac{\frac{3}{4} \times \frac{8}{1}}{\frac{1}{8} \times \frac{8}{1}}$	If both numerator and denominator are multiplied by the same number, the value of the fraction is unchanged: $\frac{a}{b} = \frac{a \times c}{b \times c}$
3. $\dfrac{\frac{3}{4} \times \frac{8}{1}}{1}$	A number and its reciprocal yield a product of 1: $a \times \frac{1}{a} = 1$

4. $\frac{3}{4} \times \frac{8}{1}$ — Division by 1 always yields a quotient that is the same as the dividend: $a \div 1 = a$.

5. $\frac{3}{\not{4}_{1}} \times \frac{\not{8}^{2}}{1} = 6$ — To multiply two fractions, multiply numerator and denominator times denominator:

$$\frac{a}{b} \times \frac{c}{d} = \frac{a \times c}{b \times d}$$

Children who are guided through this process should work through several examples before stating the generalization that, to divide fractions, multiply the dividend by the reciprocal of the divisor. By developing the division algorithm in this fashion, children are introduced to mathematical reasoning.

Children need to understand the process involved in the division algorithm. Regardless of whether it is introduced through three- or two-dimensional representations or through a more abstract procedure, children should apply their knowledge to problem-solving situations. In contrast, rote learning often results in forgetting or in rules that are confusing.

Another algorithm for dividing with fractions is the common denominator method. Because this procedure is similar to the process used for adding fractions, some teachers initially introduce the common denominator algorithm.

The aids used with the invert and multiply algorithm can be utilized. Rectangular regions are effective to illustrate $\frac{5}{8} \div \frac{1}{8}$, for example.

$\frac{5}{8} \div \frac{1}{8} = \square$ $\frac{5}{8}$ How many $\frac{1}{8}$s are in $\frac{5}{8}$?

$\frac{1}{8}$	$\frac{1}{8}$	$\frac{1}{8}$	$\frac{1}{8}$	$\frac{1}{8}$			

$\frac{5}{8} \div \frac{1}{8} = 5$

EXCURSION
Building Stairs

Carpenters frequently use fractions and mixed numerals. When they build stairs, for example, the rise plus the run should be about $17\frac{1}{2}$ inches for each step. If the rise for a set of steps is $7\frac{3}{8}$ inches, what should the run be?

Solution:

$$\text{Rise} + \text{Run} = 17\tfrac{1}{2}$$
$$7\tfrac{3}{8} + \text{Run} = 17\tfrac{1}{2}$$
$$\text{Run} = 17\tfrac{1}{2} - 7\tfrac{3}{8}$$
$$\text{Run} = 17\tfrac{4}{8} - 7\tfrac{3}{8}$$
$$\text{Run} = 10\tfrac{1}{8} \text{ inches}$$

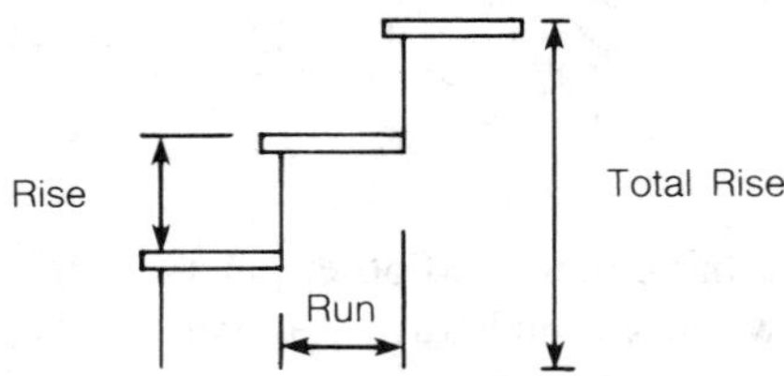

A flight of stairs is to have a total rise of $8\frac{3}{4}$ feet. How many steps will be needed if each rise is to be $7\frac{1}{2}$ inches?

Solution : No. of steps $= \dfrac{\text{Total Rise}}{\text{Rise}}$

$$= \frac{105}{7\frac{1}{2}}$$

$8\frac{3}{4} \times 12$ inches $= 105$ inches

$$= 105 \div \frac{15}{2}$$

$$= \frac{\overset{7}{\cancel{105}}}{1} \times \frac{2}{\underset{5}{\cancel{15}}}$$

No. of steps = 14

1. If the rise for a set of steps is $7\frac{5}{8}$ inches, what should be the length of a run?

2. A flight of steps is to have a total rise of $10\frac{5}{6}$ feet. Sixteen steps are to be used. What should be the length of each rise?

EXCURSION
Representing Division with Paper Folding[10]

Paper folding can be a very effective method to represent division with fractions. To show $\frac{3}{4} \div \frac{1}{8}$, first fold a strip to picture $\frac{3}{4}$.

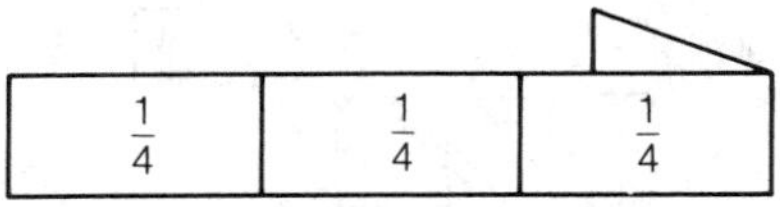

$\frac{1}{4}$	$\frac{1}{4}$	$\frac{1}{4}$

Fold the unused portion behind. Fold another strip into eighths.

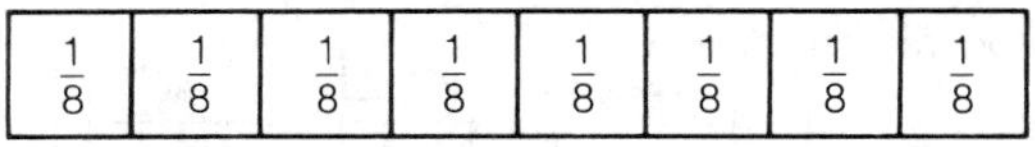

$\frac{1}{8}$	$\frac{1}{8}$	$\frac{1}{8}$	$\frac{1}{8}$	$\frac{1}{8}$	$\frac{1}{8}$	$\frac{1}{8}$	$\frac{1}{8}$

Compare the eighths with the $\frac{3}{4}$. How many eighths are in $\frac{3}{4}$?

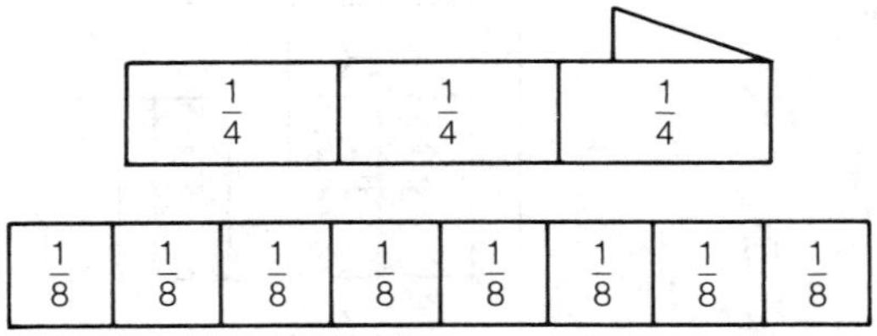

$\frac{1}{4}$	$\frac{1}{4}$	$\frac{1}{4}$

$\frac{1}{8}$	$\frac{1}{8}$	$\frac{1}{8}$	$\frac{1}{8}$	$\frac{1}{8}$	$\frac{1}{8}$	$\frac{1}{8}$	$\frac{1}{8}$

There are six one-eighths in $\frac{3}{4}$, therefore $\frac{3}{4} \div \frac{1}{8} = 6$. Can you represent $\frac{5}{6} \div \frac{1}{3}$ by this paper-folding procedure?

10. Wayne R. Scott, "Fractions Taught by Folding Paper Strips," *The Arithmetic Teacher* 58, no. 5 (January 1981): 20.

Rename the fractions as like fractions when dividing unlike denominators.

PROBLEM

How many pieces $\frac{1}{3}$ in size are contained in a larger piece $\frac{5}{6}$ in size? Children can easily see that there are two of the one-thirds plus $\frac{1}{2}$ of another.

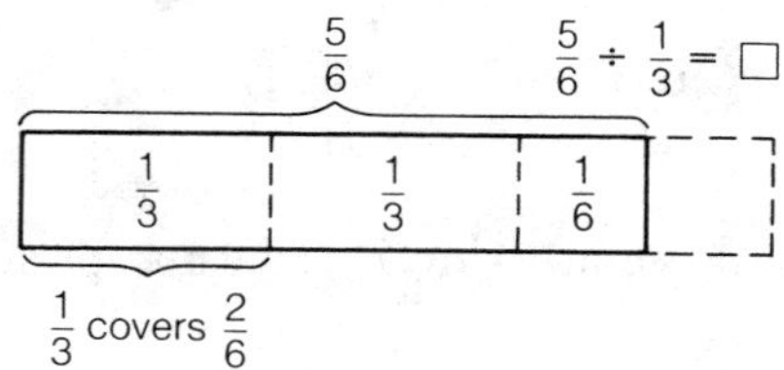

$$\frac{5}{6} \div \frac{1}{3} = \frac{5}{6} \div \left(\frac{1}{3} \times \frac{2}{2}\right) = \frac{5}{6} \div \frac{2}{6}$$

$$= \frac{5 \div 2}{6 \div 6} = \frac{2\frac{1}{2}}{1} = 2\frac{1}{2}$$

Physical models cannot be used to show division of a fraction by a whole number when introducing the common denominator method; the application setting must be partitive. However, the algorithm itself is useful when the divisor is a whole number.

$$\frac{3}{4} \div 4 = \frac{3}{4} \div \left(\frac{4}{1} \times \frac{4}{4}\right) = \frac{3}{4} \div \frac{16}{4}$$

$$= \frac{3 \div 16}{4 \div 4} = \frac{3}{16}$$

Consolidating Activities

Many consolidating activities should be games.

Tic-Tac-Toe Fraction Addition[11]

In this game practice is provided for translating from two-dimensional exemplars to symbols. Before beginning the game, make a transparency of the information in Figure 14.8.

11. Earl Ockenga and Joan Duea, "Ideas," *The Arithmetic Teacher* 25, no. 4 (January 1978): 28–32.

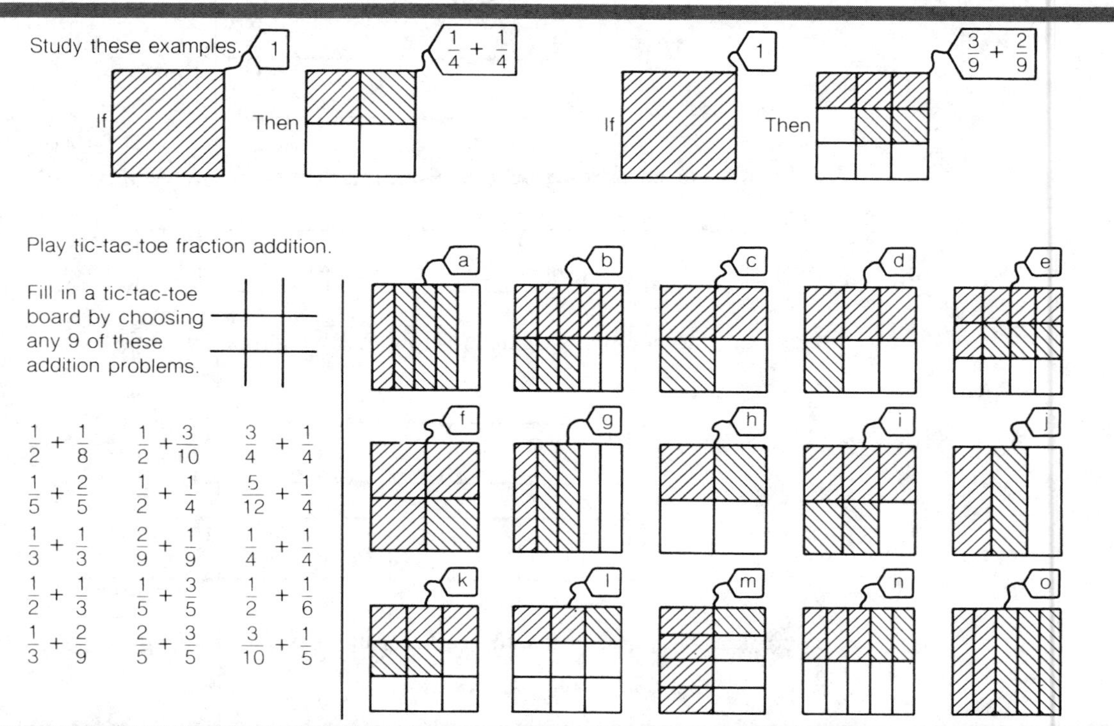

FIGURE 14.8
Tic-Tac-Toe Fraction Addition

1. Place the transparency on the overhead projector. Before you turn on the projector, cover up each of the fifteen lettered unit squares. Plastic discs or pennies work well.

2. Provide each child with a tic-tac-toe board, or let the children draw their own. Have them write within the nine empty spaces on the board any nine of the addition phrases shown on the transparency.

3. Begin the game by uncovering one of the unit squares. Each player must then decide what addition phrase is shown by the shaded parts of the uncovered square. If the corresponding addition phrase is on the player's tic-tac-toe board, she writes the tag letter beside it.

4. Play continues similarly as other unit squares are uncovered.

5. The winner is the first player to letter three spaces in a row, or to letter all four corners.

Tic-Tac-Toe Fraction Multiplication[12]

This game is played in the same manner as Tic-Tac-Toe Fraction Addition. Use the information from Figure 14.9 to make your transparency.

Fraction Bingo[13]

Make several Bingo cards similar to Figure 14.10; use the same numbers, but vary their location on the card. Make 3 × 5 problem cards, each with one of the examples listed below. Give each child a Bingo card, shuffle the problem cards, then draw at random one problem card at a time. Read the example to the class, omitting the answer. Each child computes the sum or difference, and places a marker on her card where the

13. Adapted from *SRA Mathematics: Learning System Text,* Level 5, by M. Vere DeVault et al., p. 163. Copyright © 1978, 1974, Science Research Associates Inc. Reprinted by permission of the publisher.

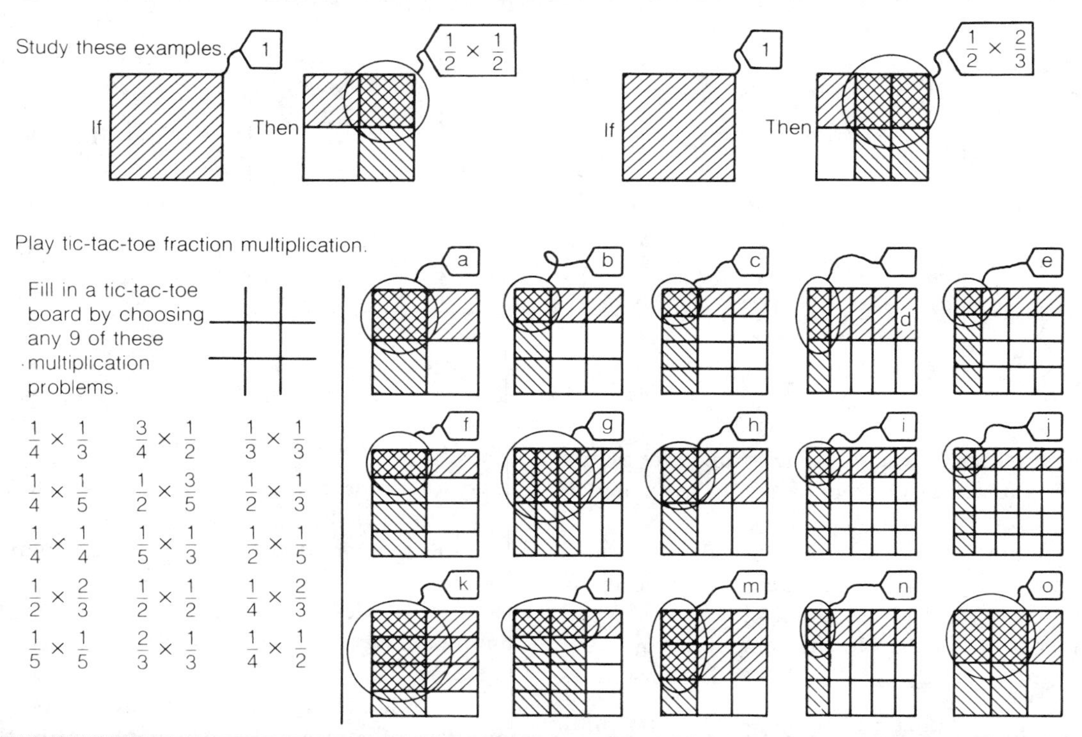

FIGURE 14.9
Tic-Tac-Toe Fraction Multiplication

B	I	N	G	O
$3\frac{1}{2}$	$\frac{5}{8}$	$\frac{21}{25}$	$\frac{3}{20}$	$\frac{3}{4}$
$\frac{5}{18}$	$\frac{1}{3}$	$\frac{7}{8}$	$\frac{3}{10}$	$\frac{3}{8}$
$1\frac{1}{5}$	$\frac{2}{3}$	$\frac{7}{10}$	$2\frac{1}{2}$	$\frac{11}{12}$
$\frac{1}{9}$	$\frac{7}{9}$	$\frac{11}{12}$	$1\frac{8}{9}$	$\frac{5}{6}$
$\frac{3}{100}$	$\frac{1}{2}$	$1\frac{2}{3}$	$\frac{11}{14}$	$\frac{3}{10}$

FIGURE 14.10
Fraction Bingo

answer is recorded. The winner is the first person to put a marker on five numbers (answers) in a row, column, or diagonal, or the first to put a marker on all four corners.

$\frac{2}{3} + \frac{3}{10} = \frac{7}{10}$	$\frac{7}{8} - \frac{1}{2} = \frac{3}{8}$	$\frac{7}{7} - \frac{1}{2} = \frac{3}{8}$
$1\frac{2}{3} + \frac{2}{9} = 1\frac{8}{9}$	$\frac{1}{5} + \frac{2}{5} + \frac{3}{5} = 1\frac{1}{5}$	$1\frac{1}{5} - \frac{13}{15} = \frac{1}{3}$
$\frac{3}{4} - \frac{1}{4} = \frac{1}{2}$	$\frac{7}{10} + \frac{14}{100} = \frac{21}{25}$	$\frac{43}{100} - \frac{4}{10} = \frac{3}{10}$
$\frac{4}{9} - \frac{1}{3} = \frac{1}{9}$	$\frac{5}{6} + \frac{1}{12} = \frac{11}{12}$	$\frac{1}{3} + \frac{1}{6} + \frac{2}{6} = \frac{5}{6}$
$\frac{5}{6} + 1\frac{2}{3} = 2\frac{1}{2}$	$2\frac{1}{2} - \frac{5}{6} = 1\frac{2}{3}$	$\frac{7}{10} - \frac{2}{5} = \frac{3}{10}$
$\frac{1}{3} + \frac{5}{12} = \frac{3}{4}$	$\frac{3}{4} - \frac{1}{12} = \frac{2}{3}$	$\frac{4}{7} - \frac{3}{14} = \frac{11}{14}$
$\frac{3}{8} + \frac{1}{4} = \frac{5}{8}$	$3 + \frac{1}{2} = 3\frac{1}{2}$	$\frac{1}{9} + \frac{2}{3} = \frac{7}{9}$
$1\frac{1}{3} - \frac{5}{12} = \frac{11}{12}$	$\frac{11}{18} - \frac{1}{3} = \frac{5}{18}$	$\frac{13}{20} - \frac{1}{2} = \frac{3}{20}$

This game can be altered to accommodate multiplication and division by changing the examples and answers on the Bingo cards.

Games that encourage speed in calculation provide good practice. Select a fraction or mixed numeral and ask each child to add the number to itself, then to the sum, continuing for a specified number of additions. For example, start with $\frac{3}{4}$, adding $\frac{3}{4}$ four times. The first child to complete the addition wins.

This game can be varied by selecting a number and asking each child to subtract a given fraction from it a specified number of times. For example, subtract $\frac{5}{8}$ five times from 12.

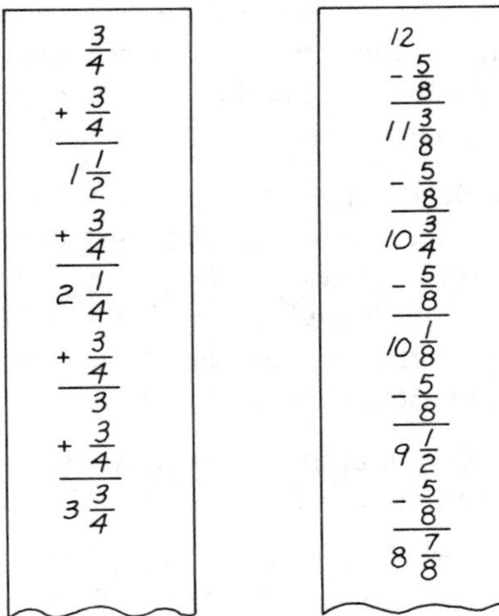

Magic squares can also provide practice. They can be used for fractions in all four operations.

4	$\frac{3}{4}$	$\frac{1}{2}$	$3\frac{1}{4}$
$1\frac{1}{4}$	$2\frac{1}{2}$	$2\frac{3}{4}$	2
$2\frac{1}{4}$	$1\frac{1}{2}$	$1\frac{3}{4}$	3
1	$3\frac{3}{4}$	$3\frac{1}{2}$	$\frac{1}{4}$

Find the sum for each row, each column, and along each diagonal.

Use a magic square such as the one illustrated and ask children to add a fraction (possibly $\frac{5}{8}$) to each number. Determine if the new array of numbers is a magic square. The same procedure can be repeated by asking each child to subtract, multiply or divide each number in the original by a specific fraction. In every case, the newly developed array should result in another magic square, thus the solution is self-checking.

Fraction Bars is a commercially prepared program of games, activities, manipulative materials, workbooks, and tests for teaching fractions.

The games and activities are sequenced to provide a gradual transition from whole numbers to fractions. The total or partial program can be purchased from many distributors of mathematical materials.

Transferring Activities

Transferring activities provide new situations for previously learned knowledge and skills. When designing activities that involve verbal problems, include those that require translation from one of the following to the others: a number sentence, a verbal problem, and a two- or three-dimensional representation. The worksheets in Figure 14.11 provide an example. A copy of the form is available in the student handbook that accompanies this text.

Transferring activities vary greatly and depend on other topics being studied and the age and interests of children. Listed below is a potpourri of activities that may be used.

PROBLEMS

The area of this triangular region is $3\frac{1}{3}$ square yards. If the base of the triangle is 4 yards, what is the height?

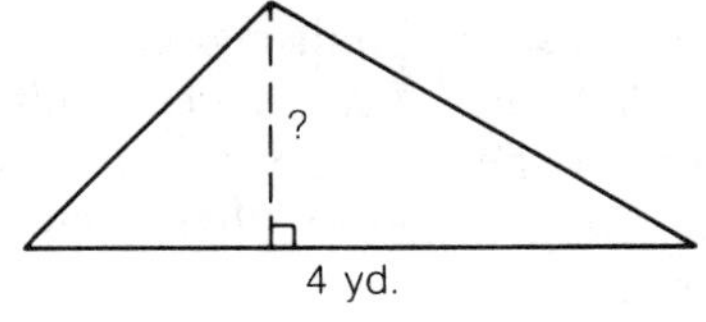

How many cups of flour are needed to make $2\frac{1}{2}$ dozen biscuits?

Biscuits
(makes 1 dozen)

$1\frac{1}{4}$ cups flour

$1\frac{1}{3}$ tsp. baking power

$\frac{1}{2}$ tsp. sugar

2 tbsp. butter

6 tbsp. milk

$\frac{1}{2}$ tsp. salt

The following is Fred's time card for last week. How many hours did he work? At $4.00 an hour, how much did he earn?

Time Card

Monday	8
Tuesday	$7\frac{1}{2}$
Wednesday	$6\frac{3}{4}$
Thursday	$8\frac{1}{2}$
Friday	9
Total Hours	_____

Name_______
Number Sentence:
$\frac{1}{2} + \frac{3}{4} = 1\frac{1}{4}$
Problem:________
Picture:

Name_______
Number Sentence:
Problem: Mary's allowance is $ 2.00. If she saves 1/4 of her allowance, how much does she save?
Picture:

Name_______
Number Sentence:
Problem:________
Picture:

FIGURE 14.11
Worksheets for Translations

PROBLEMS

A man who weighs 180 pounds can eat approximately $\frac{1}{50}$ of his body weight daily. How much can he eat daily?

Africa provides $\frac{1}{8}$ of the world's petroleum. One-half of African petroleum comes from Libya. What fraction of the world's petroleum comes from Libya?

Find the length of A and B in this drawing.

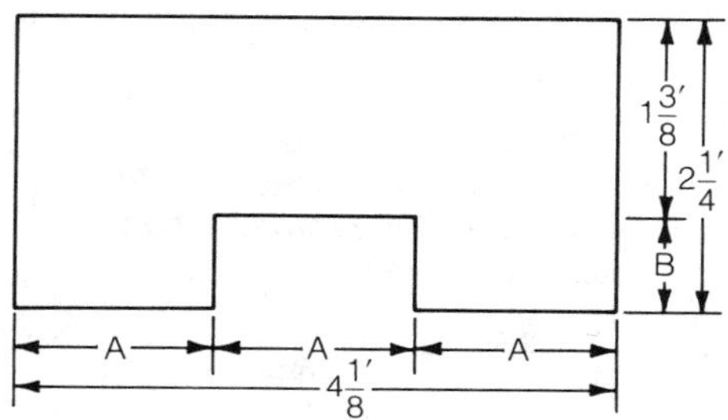

PLANNING ASSESSMENT

Each child should compute with speed and accuracy. However, skill with paper-and-pencil procedures is just one area of assessment. We also need to determine how well a child understands a computational procedure.

What could a child do to indicate she understands multiplication with fractions and the concepts involved? Note the following behaviors, then think of others.

1. Given a unit region with overlapping shaded parts for factors that are fractions, the child identifies and names the two factors and the product.
2. Given a multiplication number sentence containing two fractions and several unit regions with shaded overlapping parts, the child matches the number sentence with the appropriate shaded region.
3. Given a multiplication number sentence with two fractions as factors, the child constructs a unit region with shaded overlapping parts to represent the multiplication.
4. Given a unit region with shaded overlapping parts, the child writes the matching multiplication sentence.

The last two behaviors constitute an example of going both ways between a number sentence and a related model. If a child can do only one of the two translations, then her understanding is probably still incomplete.

How do we elicit these behaviors? Tasks might be prepared similar to these:

1. Show a child this shaded region.

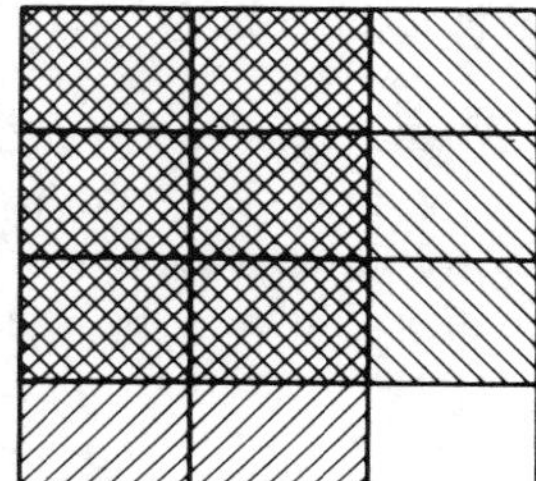

   Say, "What are the factors? What is the product?"
2. Give a child several unit regions with overlapping shaded parts. Then write a multiplication sentence on the chalkboard and say, "Show me the shaded region that goes with the number sentence."
3. Give each child a piece of graph paper. Then write a multiplication sentence on the chalkboard and say, "Use the graph paper to draw a shaded region for this number sentence."
4. Draw a shaded region on the chalkboard, asking each child to write a number sentence for it.

You may want to give children experience with the multiple-choice format.

1. What should replace the □ in the number sentence?

$$\frac{3}{8} + \frac{2}{8} = \frac{3+2}{\square}$$

a. 0 b. 5 c. 8 d. 16

2. What should replace the □ in the number sentence?

$$\frac{5}{8} + \frac{1}{2} = \frac{5 + \square}{8}$$

a. 1 b. 2 c. 3 d. 4

3. To rename $\frac{3}{4}$ as eighths:

a. Multiply by 2 b. Multiply by $\frac{2}{2}$

c. Add 4 d. Add $\frac{4}{4}$

4. The Least Common Denominator for $\frac{5}{8} + \frac{1}{6}$ is:

a. 16 b. 24 c. 48 d. 36

5. What is the sum? $\frac{2}{3} + \frac{3}{4}$

a. $\frac{5}{7}$ b. $\frac{7}{12}$ c. $1\frac{1}{2}$ d. $1\frac{5}{12}$

IV

Teaching Geometry and Measurement

15

Nonmetric Geometry

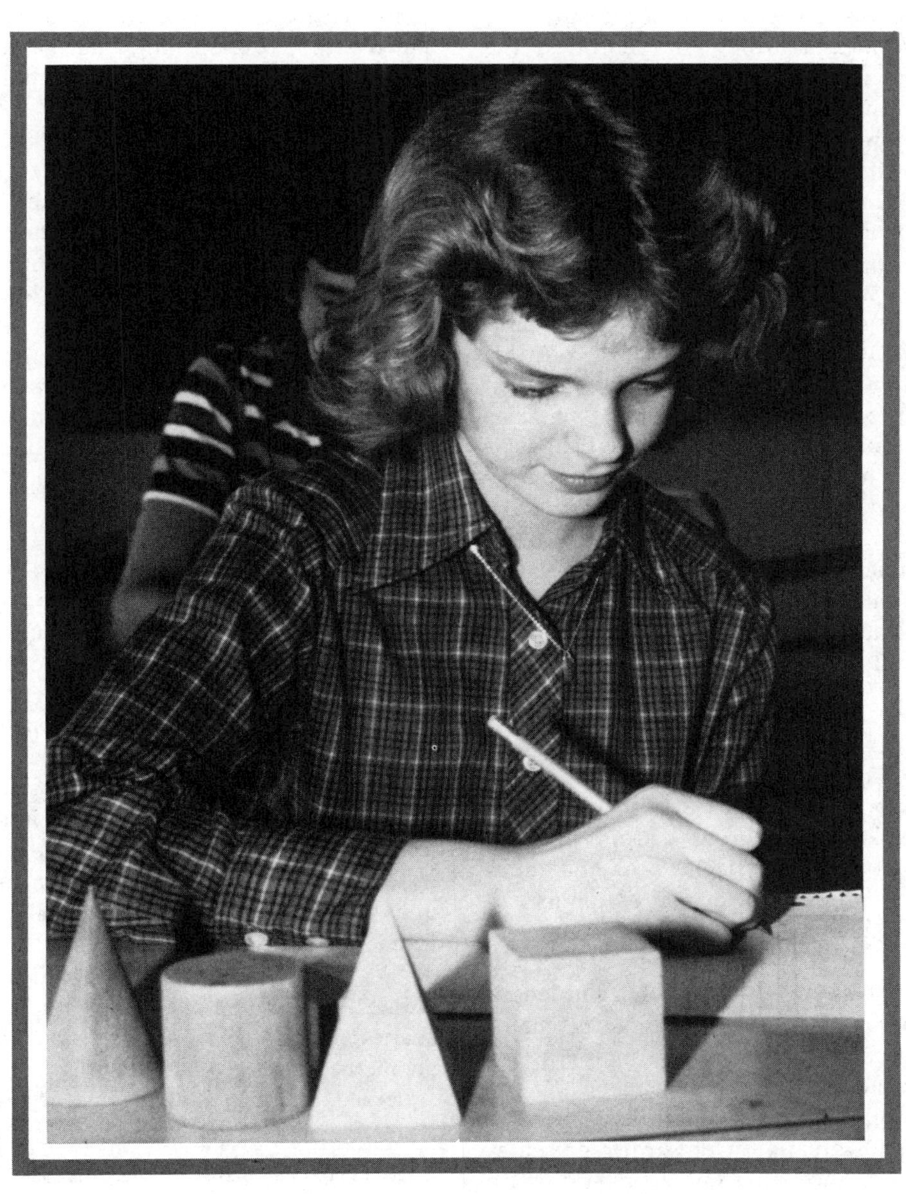

The term *geometry* is derived from the Greek words *geos* for earth and *metros* for measure, and means *earth measurement*. For centuries, the study of geometry has been considered important because of its immediate application to everyday problems. However, geometry was not emphasized in elementary schools because it was such a precise, deductive science. Little geometry beyond properties of familiar polygons was taught in elementary schools through the 1970s.

The 1980s have ushered in a new interest in the teaching of geometry at the elementary school level. Interest has grown because of more insight into how children learn, an increased awareness of the many mathematical patterns and interrelationships found in our physical environment, and more clarity about how geometric and numerical concepts interrelate. With the advent of devices that facilitate computation, such as the microcomputer and hand-held calculators, more emphasis can be placed on geometry as an important component of the elementary school curriculum.

PLANNING THE CONTENT

Mathematical knowledge grows out of a person's experiences with the physical world and takes various forms, differing in degree of sophistication as an individual's experiences change. A child begins at a very early age to notice and compare attributes such as size, color and length of objects. Much can be learned about geometry and geometrical relationships by simply observing attributes and patterns.

Elementary school geometry is highly visual and depends greatly on physical representation. A person notices that one geometric figure has straight sides while another has curved sides, another figure has three sides while another has four, etc. This type of geometry is called *informal geometry* because such relationships can be determined without establishing a unit of measure and measuring. The authors refer to this informal, nonmeasuring study of geometry as *nonmetric geometry*. Although this chapter is entitled Nonmetric Geometry and Chapter 17 is Measurement and Metric Geometry, the ideas contained in both are closely interrelated and constitute the essentials for a thorough study of elementary school geometry.

What geometric content should be included in an elementary mathematics program? One can respond in terms of the discipline of mathematics. Mathematical structures exist that can guide the teacher who is selecting geometric content. As each idea is considered in relation to others, selected ideas can be arranged into a logical pattern that indicates one idea is prerequisite to another.

Another way to decide what content to include, and possibly how to sequence it, is to consider the research and writings of Jean Piaget. In his books *The Child's Conception of Space*[1] and *The Child's Conception of Geometry*[2], Piaget argues that a child's conceptual development of space and geometry should heavily influence the sequence of geometric ideas found in the curriculum. Piaget postulates that a child's concept of space proceeds through developmental stages, and that children acquire topological before Euclidean concepts.

Topology is the branch of geometry that studies properties of figures having nonrigid shapes. It is often referred to as rubber sheet geometry because plane figures can be pulled, squeezed, and moved about in every way, but not

1. Jean Piaget and Barbel Inhelder, *The Child's Conception of Space* (New York: Humanities Press, 1964).
2. Jean Piaget, *The Child's Conception of Geometry* (New York: Basic Books, 1960).

torn. As a result, a shape can undergo many changes. Piaget states that for a child before age three, shape is not perceived as rigid, but is constantly changing; even faces appear to change constantly.

Children perceive attributes of objects and evaluate what they see by using mathematical relations, especially the spatial relations of proximity, separation, order, and enclosure. A child uses the first and most elementary relation, that of *proximity*, when an attempt is made to determine the nearness or farness of an object as it compares to some point of reference. *Separation* involves both the ability to distinguish between objects and the ability to distinguish between an object and its parts. A child who is not yet able to employ this relation would find it difficult to distinguish between figures such as rectangles and squares. *Order* refers to the sequence of events, and is used to answer questions like: Which bead was placed on the table first? Which occurred first, the click of the light switch or the bulb beginning to glow? The fourth relation, *enclosure*, allows a child to consider concepts such as the inside or outside of a curve, and the concept of betweenness.

According to Piaget, each child initially considers space on the basis of these topological concepts. Around the age of eight, the child is ready to consider Euclidean concepts related to rigid shapes, distance, straight lines and angles.[3] Thus, geometry at the preschool level and in the primary grades should emphasize topological ideas and should stress familiarity with objects from the three-dimensional world in which the child lives. Only after a knowledge of the three-dimensional world has been acquired should emphasis be placed on two-dimensional figures.

As a result of Piaget's position, many topological ideas have been included in elementary school geometry programs. These ideas provide variety and are intrinsically motivating. However, the limited data available from experimental research does not support such a strong position in regard to ordering the content. Based on his own research and a thorough analysis of Piaget's work, Martin[4] concluded that some Euclidean concepts are developed before the topological concepts of order, equivalence and continuity. Further, Brumbaugh[5] found that preschoolers could perceive solid shapes, correctly matching them with pictures.

There is little research to suggest that basing the curriculum on Piaget's work will result in a different level of knowledge of geometry in children. Sequencing and selecting the content is still contingent upon each teacher's judgment. A list of content objectives in this chapter details some basic nonmetric concepts that should normally be explored in grades K through 6.

Undefined Terms

The study of geometry begins with basic content. As in number work, where the idea of a number is an abstraction for which no definition is offered, we begin with a set of undefined terms that are the building blocks of geometry.

> **15.1**
> In geometry, *point, line* and *plane* are undefined.

These terms designate abstractions, yet we try to find ways to represent them just as we represent a number with a symbol. The notions of a point, a line, and a plane can be represented physically within the environment. The tip of a child's pencil, a small dot on the chalkboard, and the corner of a book are representations of a point. The edge of a book, a taut string, and a beam of light are representations of a line. Physical representations of a plane include the top of a dining table, a wall in a room, and a classroom floor. By convention, points are designated by capital letters, such as *point A*. Since a line can be

3. See Richard Copeland, *How Children Learn Mathematics*, 3rd ed. (New York: Macmillan, 1979).

4. J. Larry Martin, "An Analysis of Some of Piaget's Topological Tasks from a Mathematical Point of View," *Journal for Research in Mathematics Education* 7, no. 1 (January 1976): 8–25.

5. Douglas K. Brumbaugh, "Isolation of Factors that Influence the Ability of Young Children to Associate a Solid with a Representative of that Solid," *The Arithmetic Teacher* 18, no. 1 (January 1971): 49–52.

EXCURSION

What Is the Role of Geometry in the Elementary School?

Many opinions have been expressed on the role of geometry. In the mid-60s, Lamb suggested that teaching geometry in the primary grades was not worth the effort given because of the many subjects that need to be covered in most curriculums.[6] Goldmark asserted that no practical application could be found for the geometry taught in the primary grades.[7]

Today, mathematics educators argue that a study of geometry in elementary schools should go beyond a study of geometric forms, and has its greatest value in providing a means for children to precisely describe their environment.

Children are asked to observe, think, and generalize. Geometry includes numerous concepts and skills that emerge from the physical environment and progress to more advanced mathematics. Development of many sophisticated ideas later in high school and college is based upon conceptualizations formed during childhood on the playground or with toys.

Geometry allows us to present challenging and interesting problems to children even before they are capable of working with numbers. Because of its strong emphasis on observation, description and generalization, it can help us integrate mathematics, social studies, language arts and science.

drawn through any two distinct points, the two points can be used to determine a line. A line drawn through points A and B is referred to as line AB, and is represented with the symbol $\overleftrightarrow{AB}$. Three points not all on the same line determine a plane. A plane is represented as a drawing of a rectangular region as shown in Figure 15.1.

Geometry in the Plane

Lines, Segments, Rays, and Angles

Points identify a particular position in the plane. However, in a mathematical sense, points have no width, depth, or length, and must therefore be imagined. Similarly, much imagination is needed to conceptualize a line in a plane. A line has length—it goes on infinitely—but it has no width or depth. Certain subsets of lines are important.

> **15.2**
> A *line segment* is a portion of a line consisting of two points together with all points in between.

> **15.3**
> A *ray* is a half-line that includes its end point—a closed half-line.

Figure 15.2 identifies different portions of lines. Figure 15.2 (2) is an example of a line segment; it is denoted segment AB or $\overline{AB}$ because points A and B are the endpoints. Figure 15.2 (3) indicates a *closed half-line*, or a *ray*, written as ray

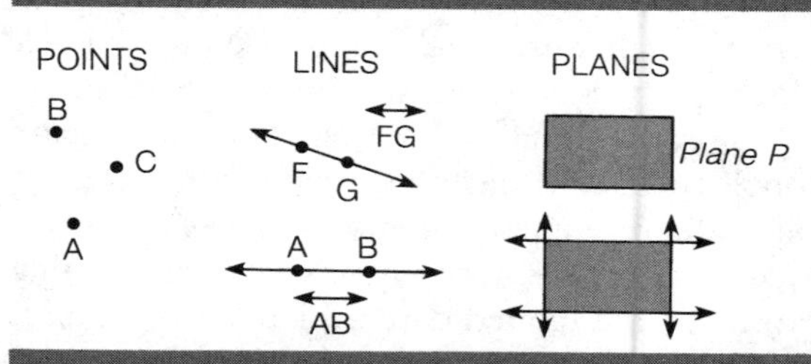

FIGURE 15.1
Representations of Basic Ideas

6. P. M. Lamb, "Geometry for Third and Fourth Graders," *The Arithmetic Teacher* 10, no. 4 (April 1963): 193–94.

7. B. Goldmark, "Geometry in the Primary Grades," *The Arithmetic Teacher* 10, no. 4 (April 1963): 191–92.

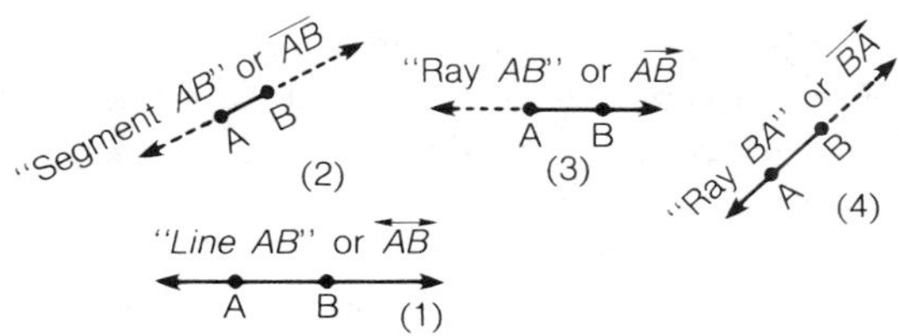

FIGURE 15.2
Symbolism for Line, Segment and Ray

AB or $\overrightarrow{AB}$. Notice that $\overrightarrow{AB}$ is a different subset of points in the line than $\overline{BA}$ (Figure 15.2 [4]).

Rays and line segments are abstractions, as they are subsets of the set of points in a line. Yet everyday situations allow us to imagine such ideas: the markings on a football field suggest line segments; representations of a ray include a one-way sign and light beaming from a flashlight.

We can also conceptualize an angle.

> **15.4**
> The union of two noncolinear rays (not in the same line) having a common endpoint is called an angle.

In some treatments of geometry, collinear rays are allowed in the definition of an angle, resulting in a straight angle $\overleftrightarrow{A \quad B \quad C}$ (Angle ABC is ray BA ∪ ray BC). However this is often referred to simply as a *straight line.*[8]

Angles are named in a variety of ways. Point B in the following angle is the common endpoint, and is called the *vertex*. The angle can be named by referring to the vertex (∠ B) or by listing the vertex as the middle letter (∠ CBA).

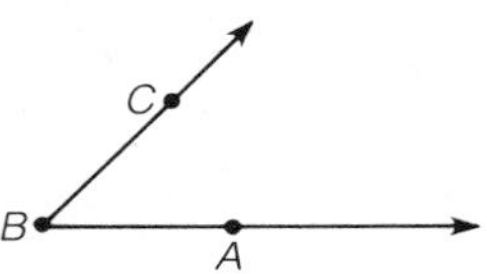

8. In Chapter 17, the measure of a straight line (straight angle) is given as 180 degrees (180°). This is typical metric treatment.

Curves

In geometry, much time is given to drawing plane figures and identifying their distinguishing characteristics. A figure that can be drawn from one point to another without lifting the pencil from the surface is called a *plane path*. A plane path can be conceived as tracing a set of points in the plane.

> **15.5**
> A *curve* in a plane is a set of points that can be traced without lifting the pencil from the paper.
>
> **15.6**
> A curve is a *closed curve* if the tracing begins and ends at the same point.
>
> **15.7**
> A curve is a *simple closed curve* if it is a closed curve, and the curve does not cross or retrace itself.

Curves include paths that are straight as well as paths that are not.

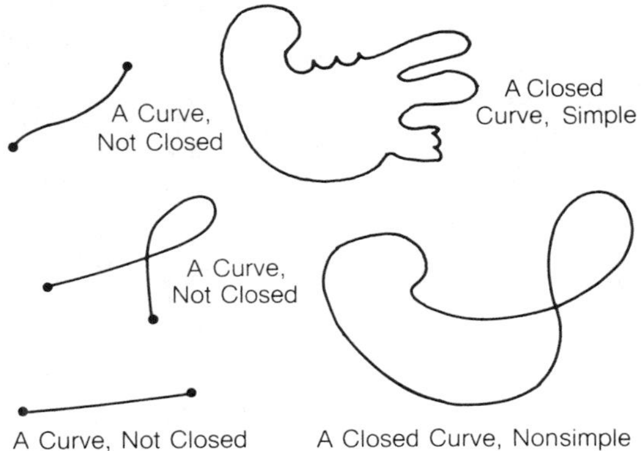

Simple closed curves tend to separate the plane into clearly distinguishable regions. Of particular interest are the interior and exterior of a simple closed curve: points in the interior are *inside* the curve, while points in the exterior are *outside* the curve. This is illustrated in Figure 15.3.

One immediate application of interior and exterior is found in objective 15.8.

EXCURSION
Ideas From Topology: Networks

A *network* is a collection of simple open curves and the points these curves have in common or *vertices*. Some vertices are endpoints of only one curve while others are the endpoint of two or more curves. A network can be illustrated with a drawing such as the following:

Networks are a popular topic of study for topologists. Some of the basic ideas include:

1. The number of lines connected to a vertex is called the degree of the vertex. A vertex of degree 1 would look like this:

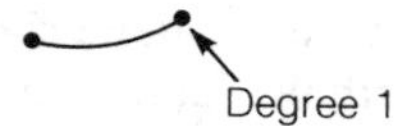

A vertex of degree 5 would be pictured this way:

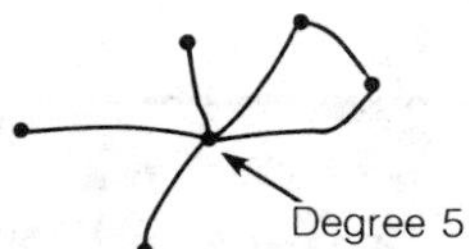

2. If the degree of a vertex is an even number the vertex is *even*. If the degree of a vertex is odd, the vertex is *odd*.

Look closely at the networks in Figure 15.4. Inspect each network and fill in the table using your findings.

Network	*Total number of vertices?*	*Number of even vertices?*	*Number of odd vertices?*	*Can the network be traced without retracing a curve?*
a				
b				
c				
d				
e				
f				
g				
h				

What relationships do you see? Can you write some questions that can be answered by looking at the table? For instance, can a network always be traced:

- If it has an even number of vertices?
- If it has an odd number of vertices?
- If it has more even than odd vertices?
- If it has more than two odd vertices?

In 1735, the famous Swiss mathematician Leonhard Euler (pronounced "oiler") announced that he had a solution to the Bridges of Konigsberg problem.

In the town of Konigsberg in Prussia there is an island called Kneiphof, with the two branches of the river Pregel flowing around it. Seven bridges cross the two branches as shown. Can a person take a walk in which he crosses each bridge once, but not more than once?

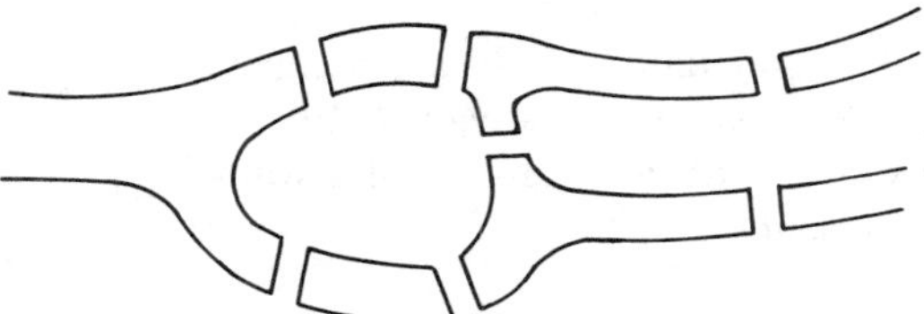

Can you use your information about networks to venture an answer to this question? Euler's answer is available in books on topology, or in Chapter 9 of *Exploring Elementary Mathematics* by Julian Weissglass.[9]

15.8
A *circle* is a simple closed curve having a point in its interior that is equidistant from each point on the curve.

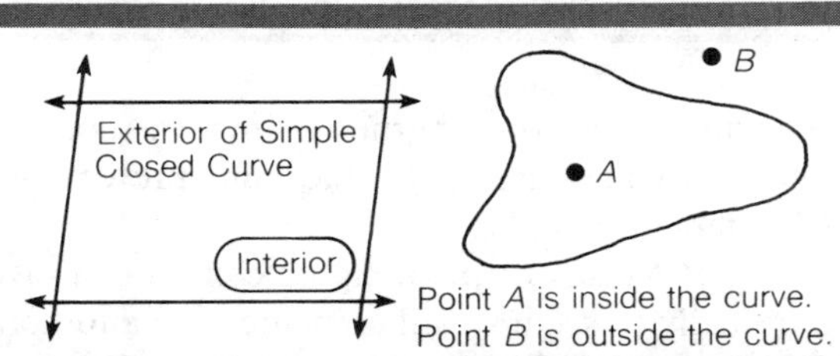

FIGURE 15.3
Interior and Exterior of Simple Closed Curves

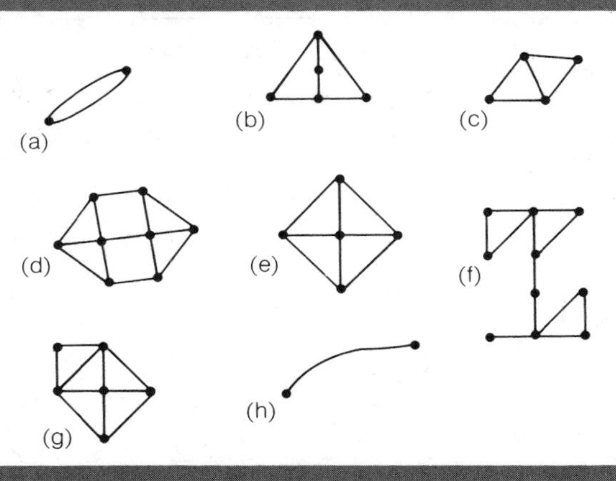

FIGURE 15.4
Networks

9. Julian Weissglass, *Exploring Elementary Mathematics* (San Francisco: W. H. Freeman, 1979).

Polygons

A study of polygons and their characteristics has been important in elementary school geometry programs.

15.9
A *polygon* is a simple closed curve consisting of the union of line segments. Each line segment is called a *side* of the polygon.

15.10
Polygons are named as follows: 3 sides—triangle, 4 sides—quadrilateral, 5 sides—pentagon, 6 sides—hexagon, 7 sides—heptagon, 8 sides—octagon, 9 sides—nonagon, 10 sides—decagon.

Common plane figures such as triangles, squares, and rectangles are all polygons. Figure 15.5 includes examples of polygons usually studied in elementary schools.

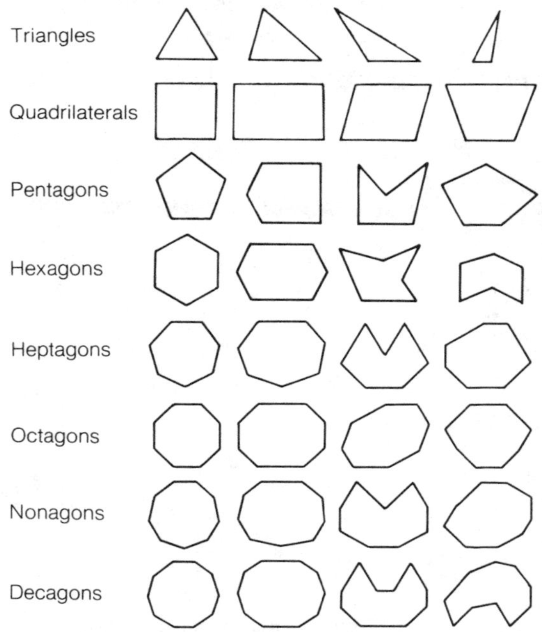

Triangles

Quadrilaterals

Pentagons

Hexagons

Heptagons

Octagons

Nonagons

Decagons

FIGURE 15.5
Polygons

Many different polygons can be drawn for each category and some of them require special consideration.

15.11
A polygon having all sides equal in length is called *equilateral.*

15.12
A polygon having all angles equal is called *equiangular.*

15.13
Polygons that are both equiangular and equilateral are *regular* polygons.

In Figure 15.5, the first polygon illustrated in each category is a regular polygon.

Congruence

Identification of the likenesses and differences among geometric figures is enhanced as children determine if two figures can somehow coincide.

15.14
Any two plane figures that are exact replicas of each other are said to be *congruent.*

15.15
A congruence is a description of how figures can be made to coincide—a correspondence that tells which points (line segments, angles) on one figure correspond to which points (line segments, angles) on another figure.

15.16
Two figures are congruent if they have the same shape and size.

A formal study of congruence involves concepts related to measurement which are presented in Chapters 16 and 17.

Because one plane figure cannot be moved physically to see if it will coincide with another, a different method of comparison is needed. Usually, one figure is traced and the traced figure moved to determine congruence. The movement is equivalent to *transforming* or *moving* the figure within a plane so that properties such as shape and size are preserved.

Symmetry

Some plane figures seem to possess either line or point symmetry. Can you imagine a line passing through this figure, dividing it into two parts that fit exactly on top of each other as indicated?

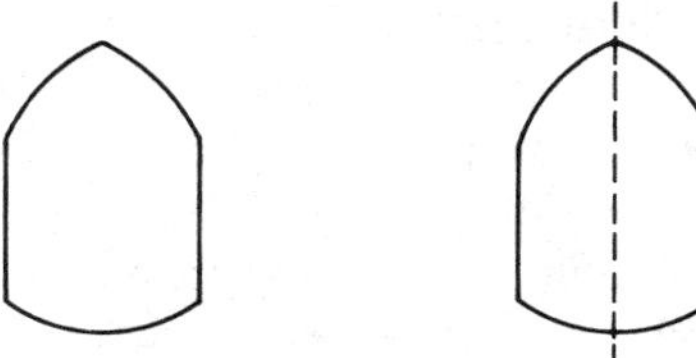

In such situations the line is called a *line of symmetry*. Plane figures that have a line of symmetry appear to be balanced. One part appears to be the mirror image of the other. A line of symmetry has been identified in the first two figures below. Is there a line of symmetry in the others? You may find more than one such line for a figure, or no such line at all.

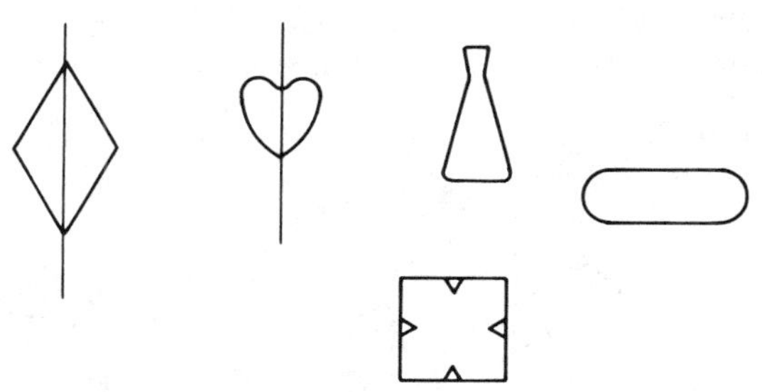

Some plane figures possess *point symmetry*, with symmetry around a point. Consider this figure "S".

It can be rotated around point p. When point p is kept constant, possibly by placing a pin on it, the figure S will fit on itself. A regular triangle also possesses point symmetry, as in the following.

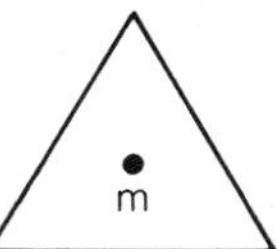

See if the figures below have point symmetry. Can you find the turn points?

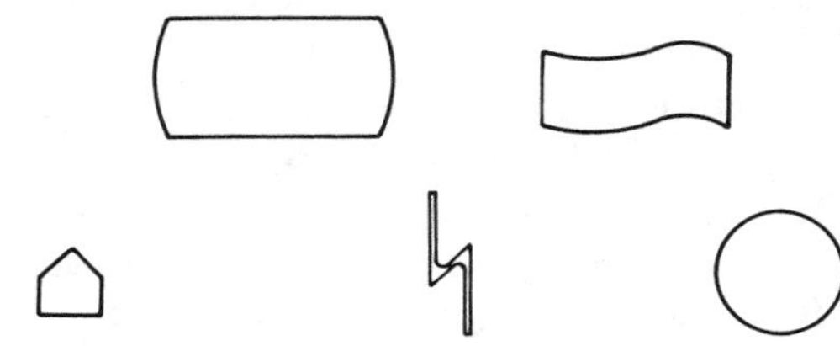

Transformations

Consider the triangles shown in Figure 15.6. How can it be determined if triangle A is congruent to triangle B? One method is to trace triangle A, then see if the traced figure can be made to coincide with triangle B. The traced figure would need to be flipped to establish congruence. We can speak of this as *reflecting* figure A around line *l*. Line *l* is called the line of reflection. A reflection is informally referred to as a *flip*.

Figure 15.7 contains other plane figures, each with an indicated line of reflection. Notice

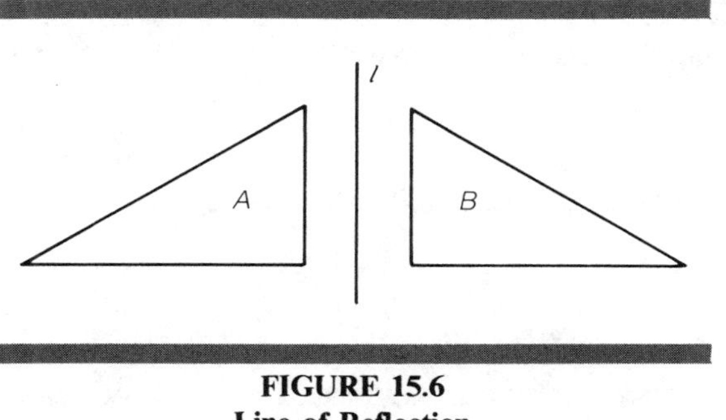

FIGURE 15.6
Line of Reflection

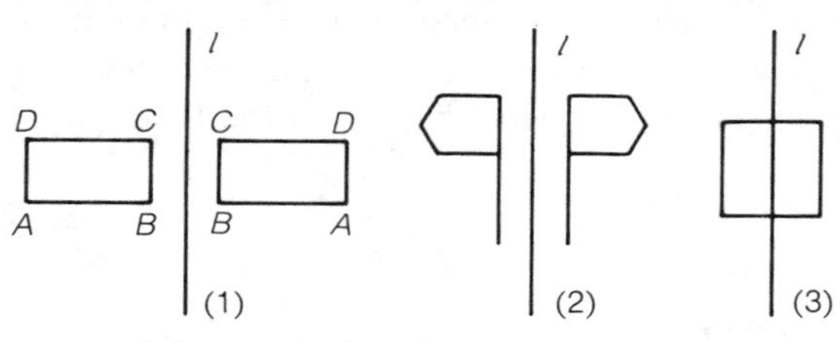

FIGURE 15.7
Figures with Lines of Reflection

that a reflection reverses orientation. A line of reflection can even be contained in the figure itself, as indicated in Figure 15.7 (3).

Figure 15.8 (1) shows two figures that cannot be made to coincide with a single flip. A tracing of triangle A would need to be rotated to make it fit on triangle B. Figure 15.8 (2) suggests the direction of the *rotation* or *turn*. Any rotation is carried out around a point, and in this case it is point X that is the *turn center*—the point around which the turn is made. The turn point may be in either the interior or the exterior of a figure. In Figure 15.8 (3), the turn point is in the interior, and the figure coincides with itself every 1/6th of a complete rotation as it is turned about point K.

A third useful transformation is a *translation* or *slide*. A slide suggests moving or pushing a figure a certain distance in a given direction through a plane. In Figure 15.9, ray *XY* indicates the direction that triangle M must be moved to fit on triangle N, its image. Triangles M and N are congruent since they can be made to coincide. A slide, unlike the turn or flip, retains the original spatial orientation.

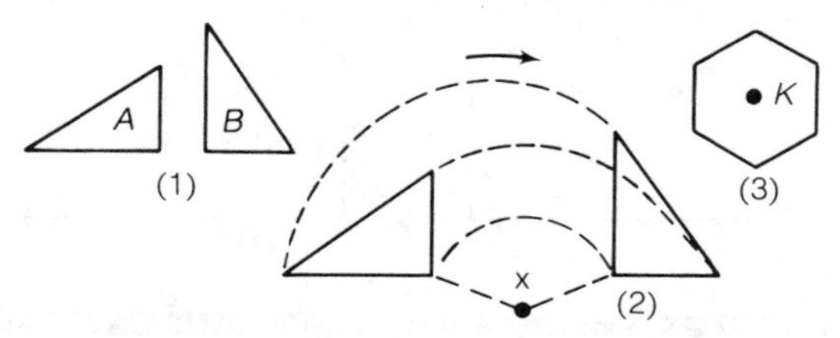

FIGURE 15.8
Rotation Around a Point

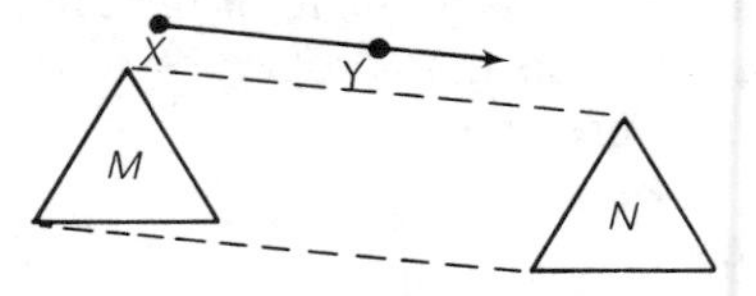

FIGURE 15.9
A Translation

Geometry in Space

Many of the ideas introduced as children study two-dimensional (plane) geometry can be extended to three-dimensional (space) geometry. In addition to points and lines of symmetry, there are planes of symmetry. Three-dimensional figures can also be identified, studied and classified according to their properties and their characteristics.

Polyhedra

One major class of three-dimensional figures is labeled *polyhedra*. A polyhedron is a three-dimensional figure consisting of a collection of polygonal surfaces joined at the edges and vertices of the surfaces. The surfaces are called *faces*. An example of a polyhedron, along with its parts identified, is pictured as follows. Refer to faces, edges and vertices when studying polyhedra.

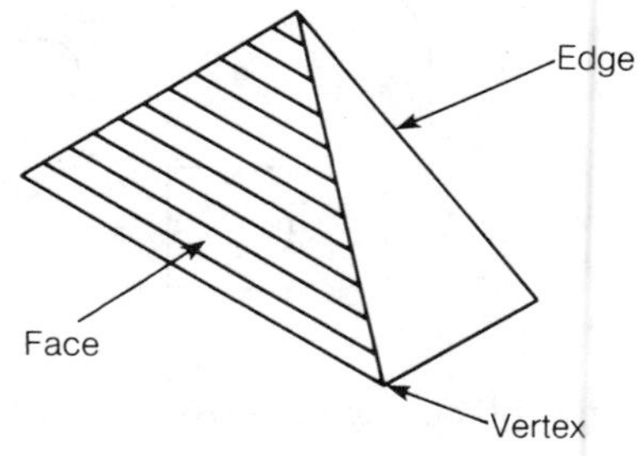

Many polyhedra are categorized as prisms or as pyramids.

> **15.17**
> A *prism* is a polyhedron having *N* faces. At least two faces must be parallel, and at least *N*–2 faces must be parallelograms.

The name of a prism is determined by the shape of the polygons in parallel planes. Prisms can be classified as triangular, rectangular, pentagonal, etc. Examples of prisms follow.

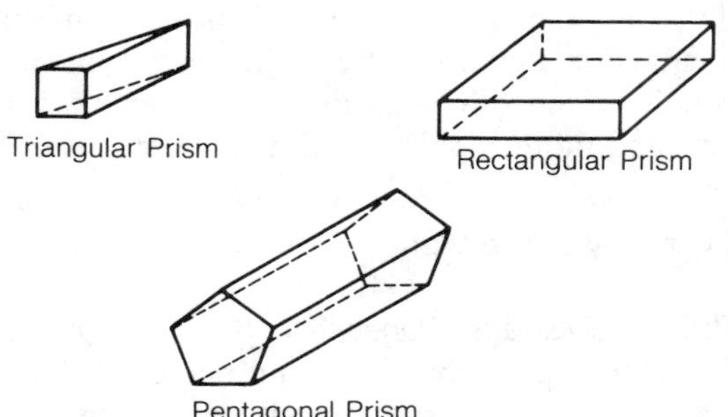

15.18
A *pyramid* is a polyhedron having all but one surface bounded by triangles that intersect at one point.

Thus, a pyramid can be thought of as having a base that is a polygonal region. The other faces are the intersecting triangles. Pyramids are named according to the shape of the polygonal base, that is, they are triangular, rectangular, etc. Examples of pyramids follow.

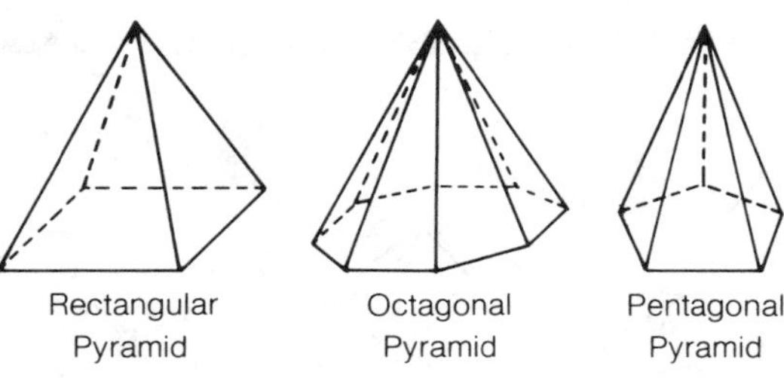

15.19
A *regular polyhedron* is a polyhedron with equal surfaces; each surface is congruent to every other surface.

Some polyhedra are neither prisms nor pyramids. Figure 15.10 shows an example.

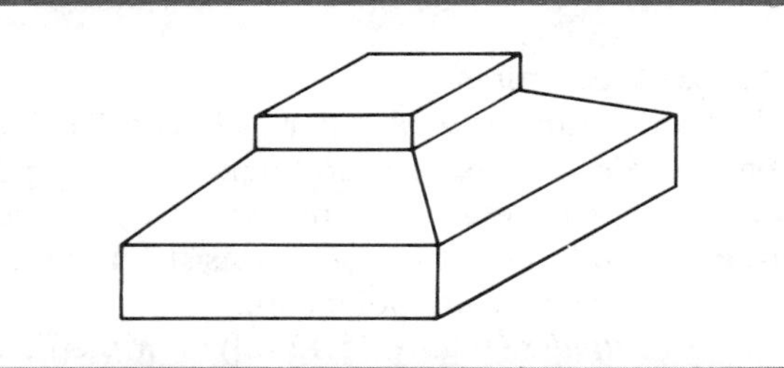

FIGURE 15.10
A Polyhedron

EXCURSION
Euler's Formula Applied to Polyhedra

Euler's formula for networks, $V + R - P = 2$, states a relationship among the number of vertices, paths, and regions. This formula can also be applied to simple polyhedra, where V becomes the number of vertices (V), R becomes the number of Faces (F) and P becomes the number of edges (E). Hence, for simple polyhedra the formula becomes $V + F - E = 2$.

Complete the chart below and compare $F + V$ and E to see if you find the same relationship as Euler.

Polyhedra	*Number of Vertices V*	*Number of Faces F*	*V + F*	*Number of Edges E*	*Comparison of F + V and E*
Triangular Prism	6	5	11		
Rectangular Prism				12	
Tetrahedron					
Rectangular Pyramid					

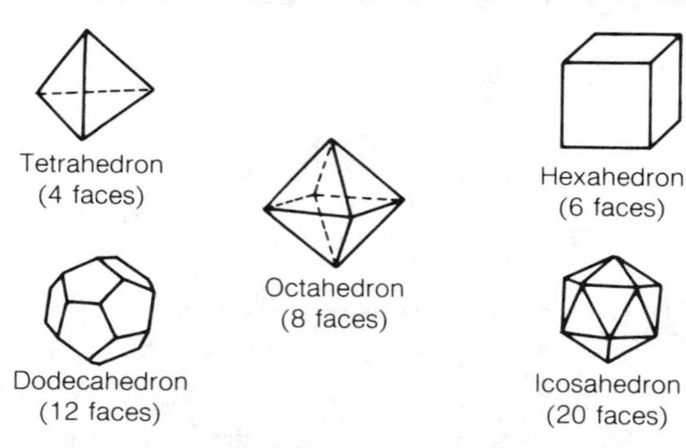

FIGURE 15.11
The Five Regular Polyhedra

While there are many regular polygons, there are only five regular polyhedra. (See Figure 15.11.)

Other Space Figures

Other three-dimensional figures include those having some surfaces that are plane regions and othes that are curved, and those with only curved surfaces. Both classifications consist of space figures, but neither are polyhedra.

A *cylinder* (Figure 15.12 [a]) is in the first category. It has two surfaces that are plane circular regions, and another that is curved. Figure 15.12 (b) illustrates a *cone*, another example in this category. A cone consists of one circular plane region, called the base, and a lateral curved surface. A *sphere* is an example of a space figure belonging to the second category.

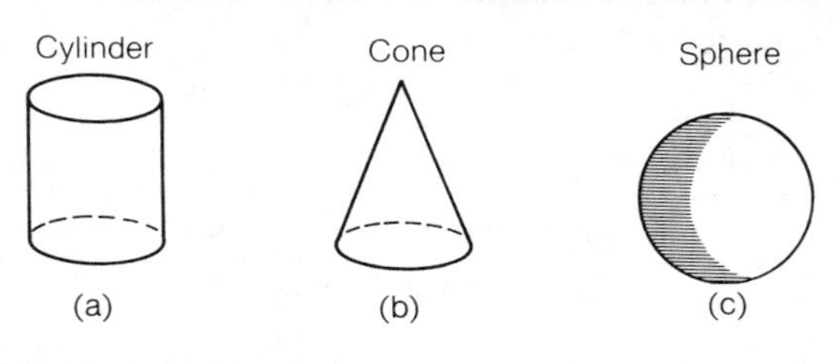

FIGURE 15.12
Space Figures

PLANNING INSTRUCTION

Children need intuitive experiences with geometric patterns and relationships. These experiences begin at the preschool level, as they observe and describe objects within their immediate environment. For older children, the environment is equally rich with situations and exemplars for teaching more sophisticated geometric ideas.

Geometry in the Plane

Points, Lines and Planes

The concepts of point, line and plane are easy to teach because children intuitively sense them. When asked to define a point, most children will identify a dot, a small mark on the desk, or the point of a pencil. These are excellent exemplars, but each child should eventually come to think of a point as a location, a specific position in the plane. We can emphasize this concept as we represent points in various drawings on a chalkboard.

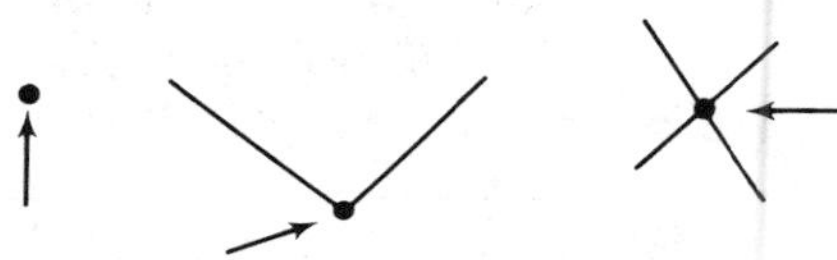

Representations of points can be shown easily with a 5-by-5 geoboard, an exemplar with many uses in nonmetric geometry.

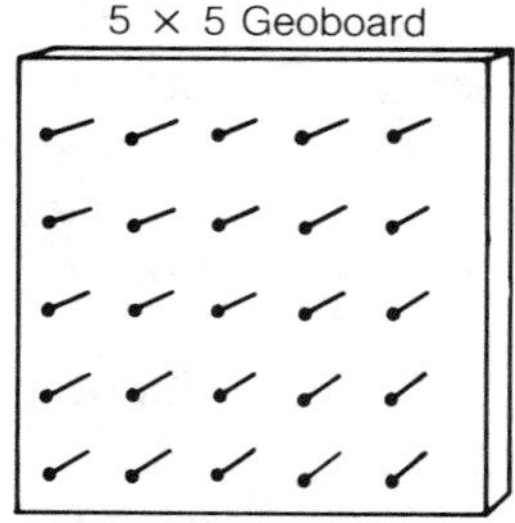

To develop the concept of a line, relate instruction to children's experiences in the world about them. Highways have lines, wires strung between telephone poles give the impression of continuing on and on, and marks on a paper can also represent lines. Representations for both straight and curved lines can be found in everyday surroundings.

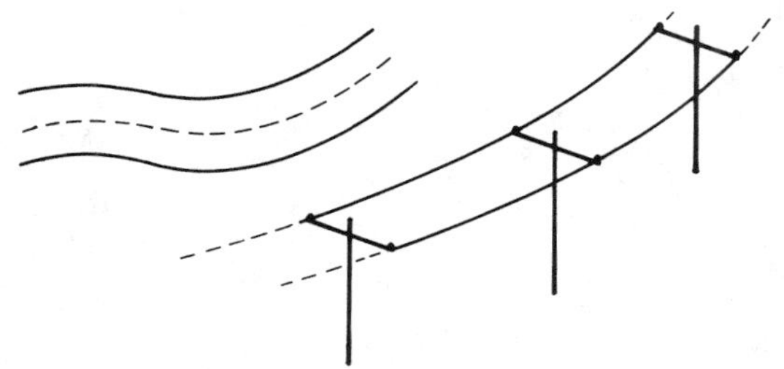

Line segments and rays can also be represented in many ways. Whereas a line segment is a part of a line, including both endpoints, a ray contains only one endpoint and continues indefinitely in one direction. The light beam from a flashlight is one example.

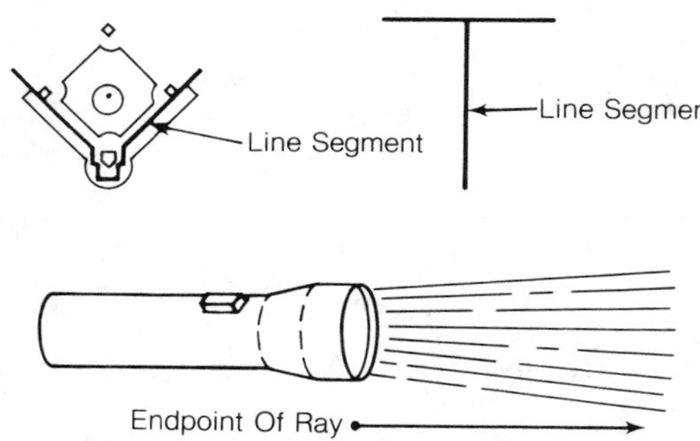

Figure 15.13 is an activity plan involving lines, line segments, points, and rays. Because these ideas are interrelated, they should be developed at about the same time.

Many surfaces can be used to develop the idea of a plane. Tops of tables, the chalkboard, textbook covers, and the gymnasium floor are all plane surfaces. Even the classroom can be used: the walls suggest planes, the intersection of the walls suggests a line, and the corners of the floor and ceiling suggest points.

Curves

Once a child understands what a line is, many related ideas can be developed. Among them is the notion of a *curve*. Within a plane, any figure whose picture can be drawn by moving a pencil from one point to another without lifting it from the surface is called a *plane curve*. A plane curve does not have to be a curved line. A curved line *is* an example of a curve; however, a straight line is also a plane curve. Children need to see lines in a plane named as curves.

Curves can be classified as open or closed; and if closed, either as simple or nonsimple. A curve is referred to as *open* if, when you begin to trace it, you cannot return to the beginning point without tracing some part twice. The letters N and M are examples of open curves, while the letter D and the digit 8 are examples of closed curves. *Closed* curves can be either simple or nonsimple. A closed curve is simple if it does not cross itself; it is nonsimple if it does.

Young children can explore curves and their characteristics on a geoboard.[10] Read through the activity plan in Figure 15.14.

Paper and pencil or chalkboard can also be used to develop these ideas. Dot-to-dot activities, for instance, can help students differentiate between open and closed curves. Which are open and closed in the following?

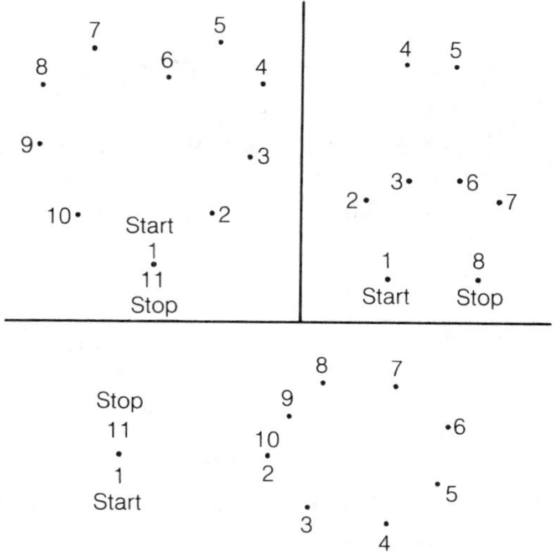

10. See Joan Holcomb, "Using Geoboards in the Primary Grades," *The Arithmetic Teacher* 27, no. 8 (April 1980): 22–25.

ACTIVITY PLAN

Nonmetric Geometry: Points, lines, planes

3-D/Rep→ Symbolic

Abstracting

Content objectives

15.1, 15.2, 15.3

Exemplars/Materials

Waxed paper

Behavioral indicators

Given a sheet of waxed paper, the child will:

1. Identify the crease at a fold as a line segment.
2. Identify the point of intersection of two line segments.
3. Use the figures formed to suggest rays and lines.

Procedures

1. Give each child a sheet of waxed paper and point out that it is an example of a plane.
2. Ask each child to fold the paper once; have children use their fingers to crease the paper at the point where it is folded.
3. Tell each child to open his paper and describe what is seen. (Each paper should picture a line segment.)

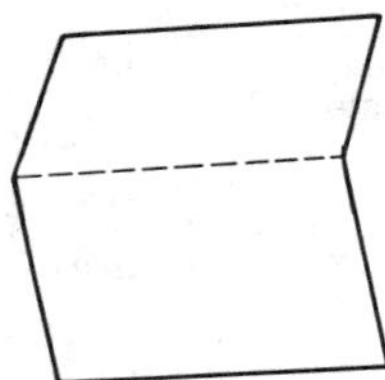

4. Ask each child to fold the paper another way. Demonstrate if necessary. Again make sure the paper is creased.
5. Have children open their papers and describe what they see. Lead children to see that they now have two line segments that intersect at a point. Have the children identify the point of intersection.

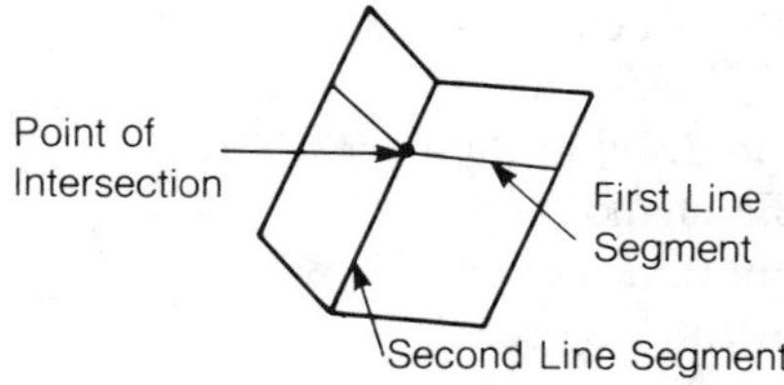

6. Continue folding the paper calling attention to line segments formed and points of intersection.
7. Ask children to choose a point of intersection anywhere on the paper. Have them place their finger there. Say, "Begin at this point and move out along one of the line segments. Continue on and on beyond the edge of the paper. This is an example of a ray. Find other examples of rays."
8. Continue, extending the idea of line segments to the idea of a line.

FIGURE 15.13
Abstracting Activity for Line Segment, Point of Intersection, and Ray

ACTIVITY PLAN

3-D/Rep→ 3-D/Rep

Nonmetric Geometry: Curves

Initiating/Abstracting/ Schematizing

Content objectives

15.5, 15.6, 15.7

Exemplars/Materials

Geoboards; rubber bands

Behavioral indicators

Given a geoboard and a collection of rubber bands, the child will:

1. Make examples of curves.
2. Identify which curves are open and which are closed.
3. Identify which closed curves are simple and nonsimple.

Procedures

Each child will need a collection of different sized rubber bands. If this is his first experience with a geoboard, you may need to demonstrate how to make a figure.

1. Give a geoboard and rubber bands to each child. Allow time for students to become familiar with the process of stretching a band from one nail to another.
2. Ask students to make an upper case (capital) "A." Look for responses such as these:

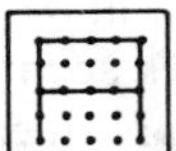 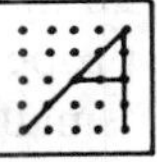  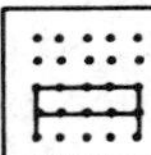 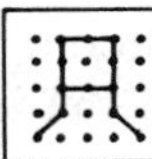

3. Ask students to make other letters, and have them share the many ways they form letters.
4. Ask each child to make the numeral 3. Again look for variety and allow children to share their products.
5. Introduce the concept of a curve as children use their fingers to trace the 3. Say, "When you formed the 3 you formed a curve. Trace the curve with your finger."
6. Have them make letters that consist of line segments (L and M are examples) and have them trace these curves.
7. Once children are comfortable with the concept of a curve, introduce the ideas of open and closed curves. Use letters such as T, V, W, and S as examples of open curves; and O and D as examples of closed curves. Move away from letters and numerals to plane figures in general, and continue to make open and closed curves on the geoboards.
8. After much exploration, ask students to define an open and a closed curve.
9. Again form letters and numerals. As children trace their designs ask, "Are you tracing over any part twice?" "Can you return to the beginning point without going back over any part?" Move to a definition of a simple closed curve and a nonsimple closed curve.
10. Continue this activity by asking each child to use the geoboard to make pictures of objects within the classroom. Decide if their diagrams are examples of open or closed figures. If closed, decide if they are simple or nonsimple.

FIGURE 15.14
An Activity Plan for Initiating, Abstracting and Schematizing the Concept of a Curve

If a curve is closed, it encloses a part of the plane. The plane is partitioned into the curve itself; the inside of the curve (interior), and the outside of the curve (exterior). Materials made from wire as shown allow us to represent the interior and the exterior of the curve, as well as the curve itself. Exemplars can be constructed from clothes hangers or from wire that is easily bent but holds its shape.

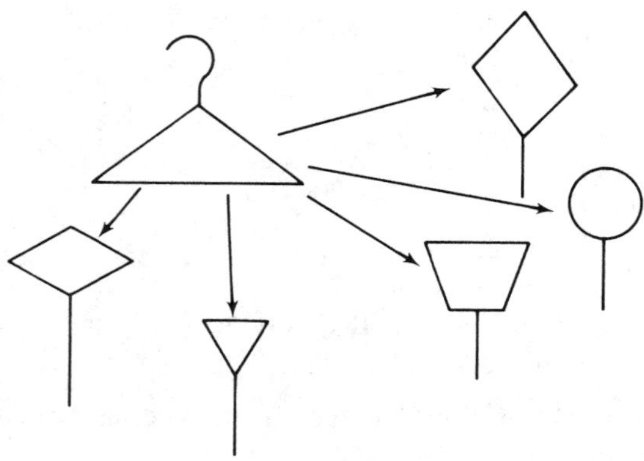

Teachers frequently make figures with masking tape on the classroom floor so children can stand on the curve, in the interior, or in the exterior of the curve.

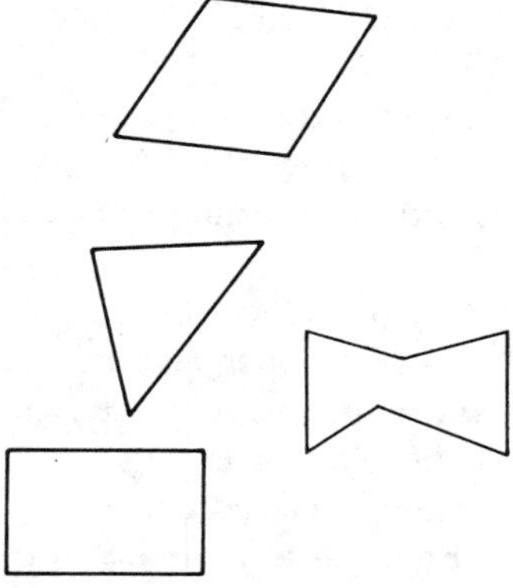

The concepts inside and outside can also be developed by having a group of children join hands and form a simple closed curve. Have at least one child stand inside the curve. Then identify children (or locations) inside and outside. Also give children simple closed curves with identified points for marking inside with an I, on the curve with an X, and outside with an O.

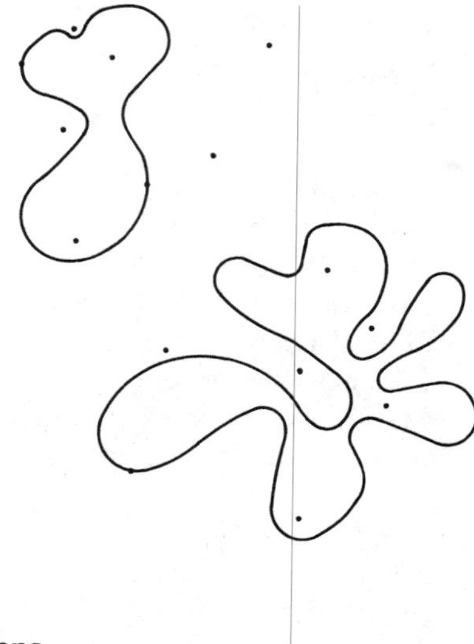

Polygons

Intuitive ideas related to polygons begin at the preschool level, before children begin a more structured mathematics program. As toys, stuffed animals, and playing blocks are inspected, properties such as roundness (flatness) and large (small) are identified and discussed. Attribute blocks are good for developing recognition of shapes and the attributes mentioned in Chapter 4. Although the blocks are three-dimensional, their faces suggest plane figures that can be identified as triangular, rectangular, square, etc.

Children enjoy making polygons on a geoboard. To begin, ask each child to form figures with three sides. Children will note that triangles can be formed in a variety of ways. Focus discussion on what is common to all shapes formed; they all have three sides.

A similar approach can be taken with quadrilaterals (four-sided plane figures). Different quadrilaterals can be shown: a square, a rectangle, a parallelogram, a rhombus, and a trapezoid. In a nonmetric approach, the discussion of these figures will focus on the similarities and differences of shapes such as squares and rectangles rather than on the actual measures of sides. Encourage children to inspect distinctive categories of figures and identify the similarities and differences. Note that a square is an example of a rectangle.[11] Sets of Tangram Pieces can help children become familiar with polygonal figures.[12]

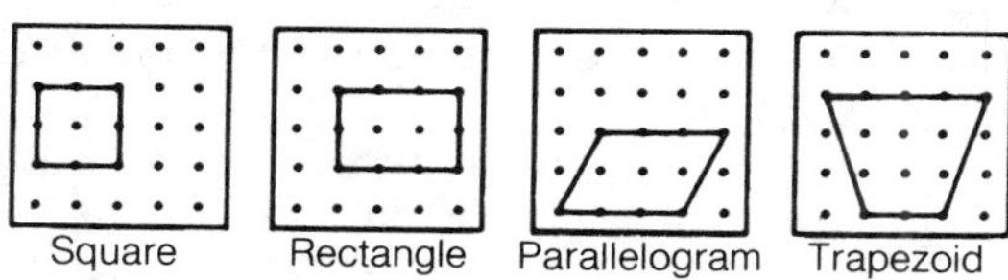

Plane figures other than triangles and quadrilaterals should be investigated; these include pentagons, hexagons, and octagons. Although not a polygon, a circle is a closed curve and has many interesting properties. Study the circle along with other plane figures.

As the informal study of polygons continues, include in- or out-of-class activities such as the following.

1. Make a collage of triangles, squares, rectangles, and so on, possibly with clippings from magazines. Display the finished product.

2. Plan a field trip (or a walk outside your school) to identify ways geometric shapes are used every day. Children will observe the way buildings are constructed, how sidewalks and streets are designed, the shapes of signs, and so on.

The study of polygons can be further extended by finding out which polygons tessellate a plane, that is, by determining which figures have the effect of filling a plane. (Tessellations were discussed briefly in Chapter 11.) Have children consider the shapes shown below for making tessellations, then have them explain which shapes will fill a plane. They can also make their own shapes for investigating tessellations.

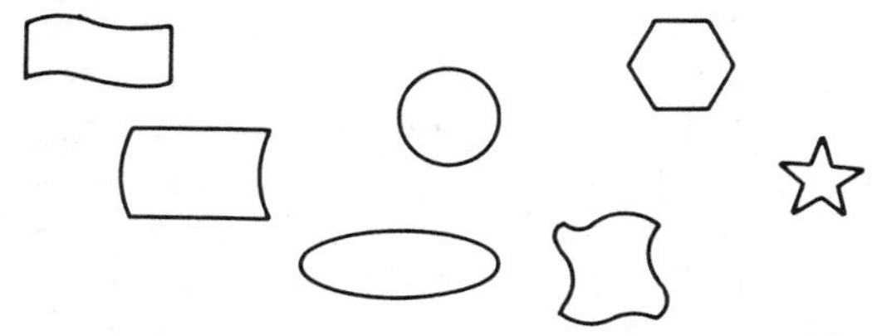

Symmetry and Transformations

Transformational geometry, sometimes called motion geometry, has become increasingly popular in elementary schools. Children study ways figures in a plane can be moved or transformed in order to make them coincide with other figures in the plane. This can be fun but it requires a high level of abstract thinking.

Before ways to transform figures are studied, give some time to concepts of symmetry. Children need to examine many classroom and everyday figures and decide which ones have lines of symmetry. Often, figures drawn on paper can be inspected with a mirror to determine if a line of symmetry is present. Commercial materials such as Mirror Cards are extremely valuable for determining this.[13] The figures below can be

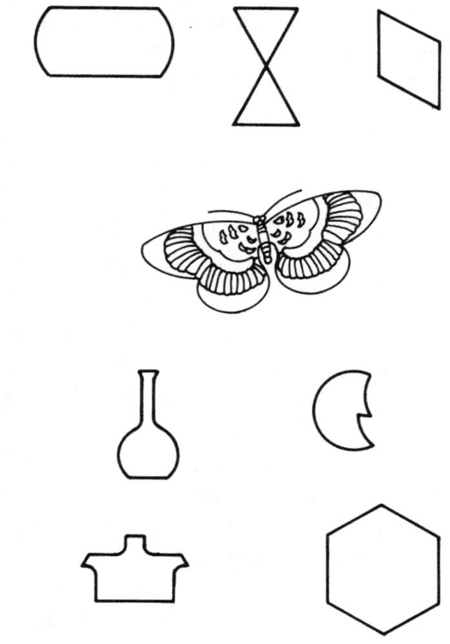

11. See Helen F. Alexick and F. Richard Kidder, "Why is a Rectangle not a Square?" *The Arithmetic Teacher* 27, no. 4 (December 1979): 26–27.

12. Tangram Pieces are available from McGraw-Hill Book Company, 1221 Ave. of the Americas, New York, NY 10020. They were developed by the Educational Development Center, Inc., 1968.

13. Mirror Cards are also available from McGraw-Hill, New York. They were developed by the Educational Development Center, Inc., 1967.

examined; some have more than one line of symmetry.

Many plane figures also have a point of symmetry, that is, a point around which the figure can be turned in order to coincide with itself. Each of the figures below has a point of symmetry.

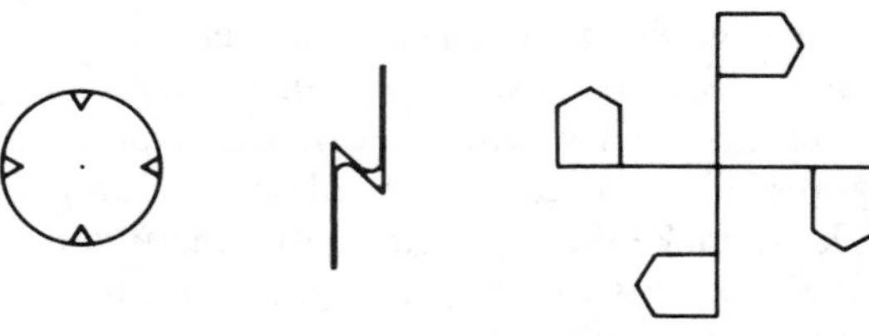

To begin the study of transformational geometry, draw two congruent figures adjacent to each other on a transparency. When the figures are projected, pose such questions as: What is the name of each figure? Are these figures the same size and shape? Can we move one figure to see if it will fit exactly on top of the other? What might we do? A child may suggest that one figure be picked up and placed on the other, but this cannot be done physically. Yet this is really what we would like to be able to do in order to verify congruence.

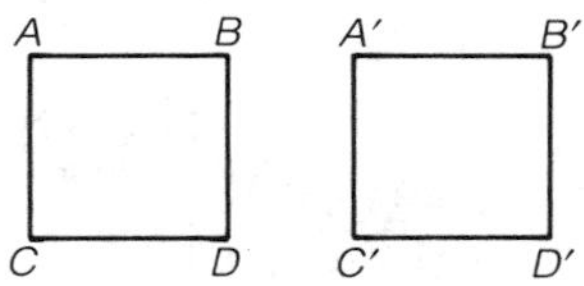

Next, place a transparency over the two figures. One figure can be traced on the transparency, and then moved to see if the tracing fits the second. This strategy should be repeated often. When children understand the idea of tracing, focus on how the tracing paper is moved when trying to get the figures to coincide. Emphasize three motions.

1. *Slide*. When two figures similar to the following are considered, the one (a) can be traced. The tracing should be moved onto the other figure (b) staying within the same plane, that is, without picking up the paper. The tracing can be moved by simply sliding it. This motion, a slide or, more formally, a *translation*, is one of the basic motions used when determining if two plane figures are congruent. When picturing a slide, dotted lines can be used to show the direction of the slide.

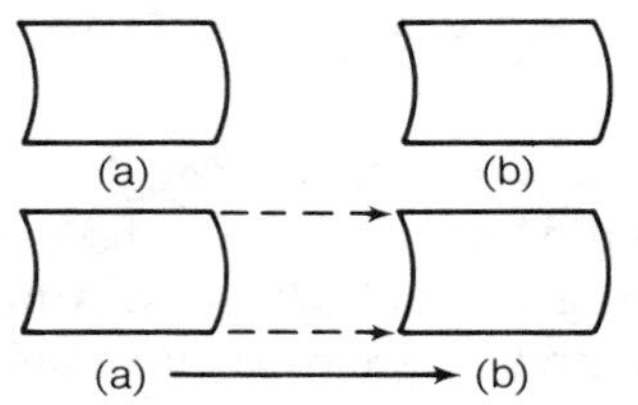

Pairs of drawings similar to the following should be considered, and congruence determined by using a slide motion to make one coincide with the other. If this cannot be done, another motion may be needed.

2. *Turn*. A turn motion employs the idea of a point of symmetry. Again, tracing can be used to see if one drawing can fit on another by simply turning the paper within the plane. This is diagrammed as follows. For the two plane figures (a), the lines in (b) show how a tracing was turned about a point of symmetry to make the first figure fit onto the second.

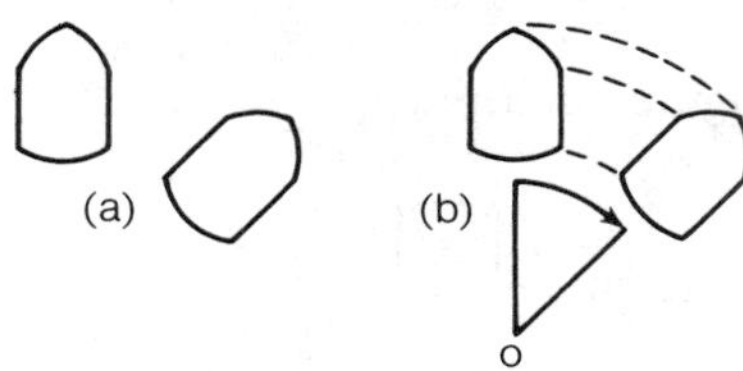

Whenever children use a turn motion to see if two figures coincide, be sure they look for a turn point both inside and outside the figure. Some of the figures that follow can be made to coincide using a turn motion or *rotation*; children can be asked to find the turn points.

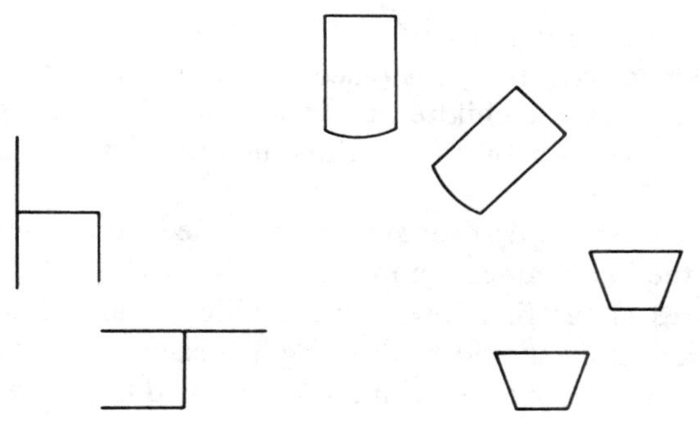

3. *Flip.* The cards below have been flipped from left to right in (a) and from bottom to top in (b). The flip motion is formally called a *reflection*; one object is the mirror image or reflection of the other.

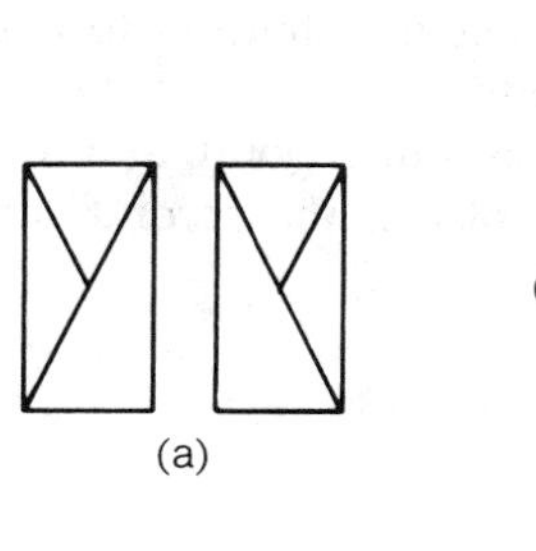

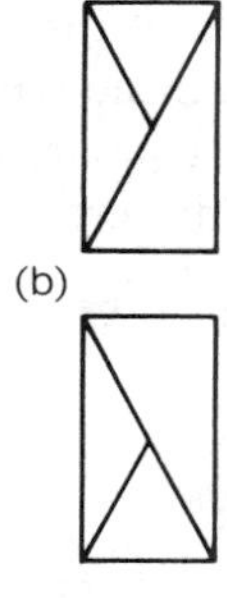

Children should be asked to find the line of reflection, or the *flip line* around which the flip is made so the two figures coincide. For example, a line of reflection is indicated for the first two pairs of the following figures. Only some of the remaining pairs can coincide with a flip.

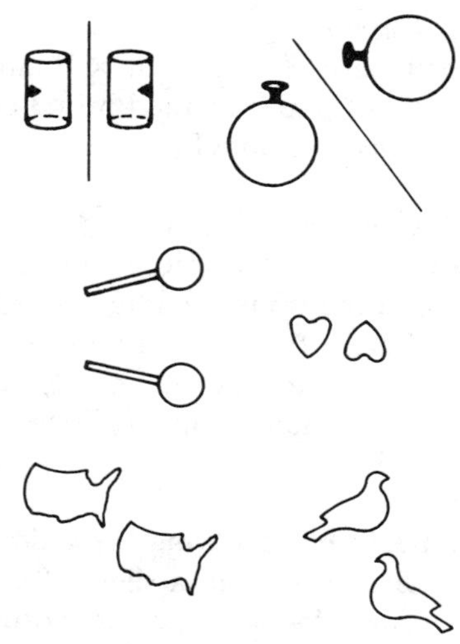

Many activities can incorporate slides, turns, and flips. The activity in Figure 15.15 can help children use motions in generating geometric patterns.

The use of varied materials will help clarify these ideas. As children use the Mirror Cards, for example, they determine which designs can be formed with a mirror and investigate why others cannot be made. This leads to the conclusion that the mirror will reflect, but not slide or turn, the image.

Marion Walter's *Make a Bigger Puddle, Make a Smaller Worm* and *Look at Annette* require that a mirror be used as the child reads each page of the book.[14] Another application of motion geometry can be found in wallpaper designs which frequently incorporate slides, flips, and turns.

---

14. Marion Walter, *Make a Bigger Puddle, Make a Smaller Worm* (London: Andre Deutsch Limited, 1971); Walter, *Look at Annette* (New York: M. Evans, 1971).

ACTIVITY PLAN

Nonmetric Geometry: Transformations | *3-D/Rep→ 2-D/Rep* | *Consolidating*

Content objective

Slides, flips, and turns are basic transformations of figures in a plane.

Exemplars/Materials

Cardboard pieces, 3 cm × 3 cm, with the following pattern: 
Sheets of paper marked into 3 cm × 3 cm squares
Crayons or colored pencils

Behavioral indicators

Given the pattern piece and squared paper, each child will:

1. Use the transformation *slide* to generate patterns.
2. Use the transformation *turn* to generate patterns.
3. Use the transformation *flip* to generate patterns.
4. Use combinations of slides, turns, and flips to generate patterns.

Procedure

1. Begin by reviewing the slide, flip and turn transformations.
2. Give each child a sheet of squared paper, then have the class place a pattern piece in the square in the upper left corner on the paper. Ask the children to notice how the line segments are marked on the pattern piece, have them remove it and draw line segments in the square as they were on the pattern piece.
3. Instruct the children to place the pattern piece on the first square, and to slide it to the next square in the first row. Again notice how the line segments appear.
4. Encourage the children to fill in the squares in the first row, using a slide motion of the pattern piece when needed. At this point each child should be drawing line segments in the squares without relying completely on the pattern. Continue with the second and third rows. When completed, the first three rows should look like this.

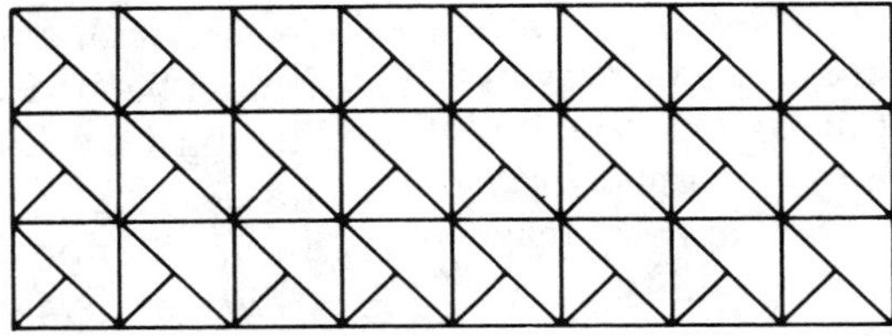

5. Give each child a fresh piece of squared paper, and instruct the children to begin in the upper left corner. Again they should draw in the basic pattern.
6. Have children use the lower right corner of the first square as a turn point, and rotate $\frac{1}{4}$ turn around this point. After they observe how the pattern looks in the second square, they should draw in the design.

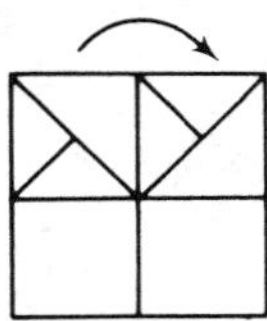

7. Have the children make another $\frac{1}{4}$ turn, and draw in the design.
8. Finally, another $\frac{1}{4}$ turn should be made and the design drawn.

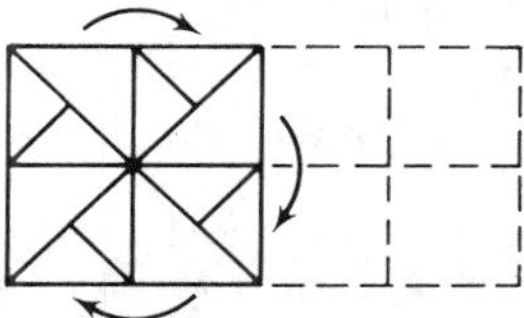

9. Instruct children to move the pattern piece to the *third* square in row one and repeat the turning motion. Continue. The first four rows should look like this when completed.

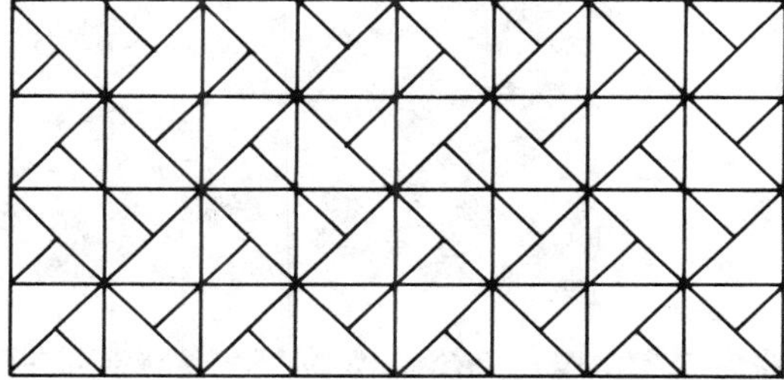

10. Each child should begin with a new piece of squared paper. Instruct children to flip the pattern piece across the row and down, and out of the row: they should draw down each column. When finished, the first four rows should look like this.

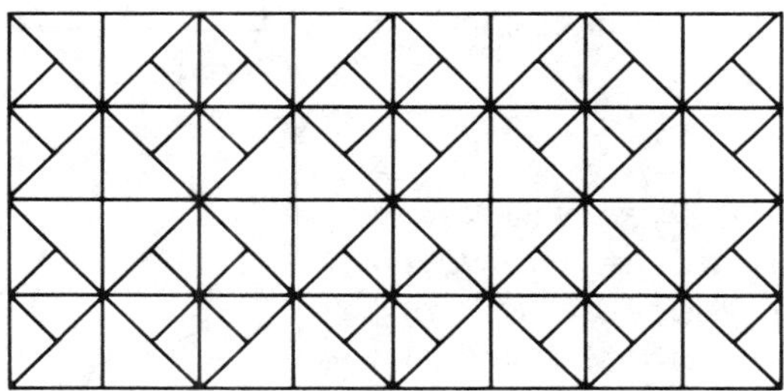

11. Motions can be combined for similar activities. For example, say, "Flip down out of the row, then slide over to the next column . . . and repeat." By varying the rules, different patterns are generated.
12. Finally give each child crayons or colored pencils, and ask the class to color in the designs however they wish.
13. Extend this activity with different pattern pieces.

FIGURE 15.15
A Consolidating Activity for the Slide, Flip and Turn Transformations[15]

15. Adapted from Martin L. Johnson, "Generating Patterns from Transformations," *The Arithmetic Teacher* 24, no. 3 (March 1977): 191–95.

EXCURSION
The Art Work of M. C. Escher

Maurits Escher was born in Holland, and was one of Europe's most original graphic artists. His woodcuts, engravings, and lithographs involve ideas from mathematics, but none as striking or more dramatic than those involving figures that tessellate a plane. The following sample of Escher's work is from *The Graphic Work of M. C. Escher.*[16]

Geometry in Space

Space figures help describe the physical world. Everyday objects such as boxes and basketballs can be used to define properties of more common space figures.

Polyhedra

A model of each of the 5 known regular polyhedra can be constructed by cutting and folding paper, possibly by using the following patterns. Pattern (a) is constructed of squares and will fold to make a cube (hexahedron). Pattern (b) is constructed from isosceles triangles and will fold to make a tetrahedron. Patterns (c) and (d) are constructed from equilateral triangles and will fold to make an octahedron and icosahedron, respectively. Pattern (e) is constructed of pentagons and will fold to make a dodecahedron.

(a) (b) (c) (d) (e)

Many commercial materials, such as kits with cardboard faces that fit together with rubber bands and straws or plastic tubes with strings or flexible joiners, are available. Polyhedra can even be modeled with Tinker Toys.

Prisms can be illustrated with solids; for instance, small boxes can exemplify the idea of a rectangular prism. Solids such as those found in Geoblocks and in Geometric Figures and Solids[17] can help children determine prism characteristics.

16. M. C. Escher, *The Graphic Work of M.C. Escher,* rev. ed. (New York: Hawthorne Books, 1967), p. 12. *Study of Regular Division of Plane with Birds* (detail), © BEELDRECHT, Amsterdam/V.A.G.A., New York, Collection Haags Gemeentemuseum–The Hague, 1981.

17. Geoblocks are a set of geometric solids and activity cards developed by the Elementary Science Study. Geometric Figures and Solids are available from Creative Publications, Palo Alto, California.

Children should handle these and other models for polyhedra.

Examples of *pyramids* can often be found in classrooms. Children should be asked to inspect various prisms and pyramids and describe them by noting the number of faces, whether two faces are in parallel planes, and so on. They should also be taught the correct names; however, extensive classification systems are probably not feasible in the elementary school.

Other Space Figures

A can of tomato juice can be used to illustrate a certain type of *cylinder*. The ends of the cylinder (bases) are congruent simple closed regions from parallel planes, and the lateral face contains at least one curved surface. Many examples of cylinders can be found in the environment. Children should see that the bases of a cylinder differ from those of a prism as they are not polygonal. *Circular cylinders* have bases that are circular regions, as illustrated in the following.

Examples of *spheres* also can be found in the environment: a globe, a basketball, a tennis ball. In addition, children can investigate *cones*. They know what an ice cream cone is (although it is not always in a conical shape). When children find objects they think are cones, have them focus on the base. If the base is a polygonal region, the space figure is a pyramid rather than a cone. If the base is a circular region, the cone is a *circular cone.*

Children need varied experiences with three-dimensional objects in which they contrast these shapes with polyhedra.

- From a bucket of solids, ask children to pick out all prisms, then pyramids. They should also find the cylinders, cones, and spheres. As this is done, have children correctly name type of solid, vertices, shape of faces, and so on.
- Let each child construct models of space figures from cardboard or paper cutouts. Patterns can be found in sources such as *The Arithmetic Teacher.*[18]

PLANNING ASSESSMENT

At any point in time a child's concepts and skills can be determined in different ways. Formal tests and interviews, in addition to a thorough inspection of pupil products, are helpful. For nonmetric geometry, decide which diagnostic procedures provide the most useful information.

Formal tests usually involve only paper-and-pencil tasks and, hence, test for concept knowledge primarily at the two-dimensional representation and symbolic stages. However, some information concerning identification of polyhedra and polygons and knowledge of symbolism can be evaluated through these tests.

Valuable information concerning mathematical relationships can be obtained through informal interviews. A child can respond to questions such as: What is the difference between a triangle and a rectangle? A cylinder and a pyramid? Is a triangle always equilateral? While conducting interviews, utilize three-dimensional, two-dimensional and symbolic exemplars.

Most likely, we will each need to plan our own interview protocols. A protocol should be constructed so we can infer level of knowledge by observing the child's actions. Many examples of protocols can be found in Copeland's text.[19] The protocol in Figure 15.16 could be used to determine a child's knowledge of geometric solids.

Diagnosis must be a continuous process. Observe how children construct figures on the geoboard, how they describe solids, and so on. Inability to form a figure on a geoboard can have

18. See Anthony J. Laffan, "Polyhedron Candles: Mathematics and Craft," *The Arithmetic Teacher* 28, no. 3 (November 1980): 18–19.
19. Richard Copeland, *Diagnostic and Learning Activities in Mathematics for Children* (New York: Macmillan, 1974).

ACTIVITY PLAN

Nonmetric Geometry: Space figures (solids) — *3-D/Rep→ Symbolic* — *Diagnosing*

Content objectives

15.17, 15.18, 15.19
Prisms, pyramids, cylinders, cones and spheres are space figures.
Cylinders and cones consist of curved surfaces and nonpolygonal bases.
Spheres are simple closed surfaces.

Materials/Exemplars

Sets of geometric solids (wood or plastic)
Material from the environment such as balls, cans of various sizes, boxes of assorted sizes, and ice cream cones.

Behavioral indicators

1. Given a set of geometric solids, the child will:
   a. Identify a prism, a pyramid, a cylinder, a cone and a sphere.
   b. Describe each solid in terms of faces, edges, vertices and/or lateral faces and bases.
2. Given everyday objects, the child will identify those objects that have the shape of the solid figure named.

Procedure

1. Place a collection of geometric solids on a table. Be sure to include examples of all types.
2. Ask the child to find the prism. When this is accomplished ask the child about attributes of this solid. See if the child can identify faces, edges, and vertices.
3. Ask the child to identify the pyramid. Repeat in-depth probing.
4. Repeat this procedure with each of the other solids.
5. Now place everyday objects on the table. Ask the child to find all the objects that are spheres. Then ask for all cylinders, and so on. As the child selects objects, discuss why each was identified.

FIGURE 15.16
A Diagnostic Activity for Identifying Geometric Solids

many causes: inadequate understanding of the concept, a spatial deficit, lack of fine motor coordination, or a specific learning disability. Much of the content is visual, so constantly look for problems related to visual discrimination. We also need to consider developmental stages.

EXCURSION
Tips on Managing the Classroom

Introductory activities in geometry usually involve exploring with interesting objects. Children can touch, move and handle the objects. As we plan instruction, we should keep these ideas in mind.

1. Make sure that the materials children manipulate are able to withstand prolonged use. For example, pieces cut from paper are not durable.
2. Make sure that sufficient quantities of materials are available before beginning an activity. If commercial materials are not available, use cans, balls, toys, and other everyday items.
3. Let each child become familiar with the material; allow a play time. Expect to lose small items such as rubber bands, straws, and pegs.
4. Look for different ways to present activities. One popular strategy is to use an independent learning center in which a child can work without strict teacher supervision. Basic guidelines for developing learning centers include:
   a. The content should be developed through exploration, preferably through independent work.
   b. Determine a format for the learning center. Most centers consist of a set of subunits called *learning stations.* Each learning station contains cards or sheets giving instructions and whatever manipulatives are needed for the activity. Decide if the stations are arranged in a specific order, and if so, how this will be communicated to students.
   c. Determine a system for evaluation. Will each child's work be checked by the teacher or by students? Consider the different items used for evaluation: pupil products, worksheets, and student reports.

   Learning centers can range from simple shoebox packages to large posters on bulletin boards. Many involve gamelike activities. Good ideas for mathematics learning centers are available in periodicals such as *The Arithmetic Teacher.*[20]

5. In teacher-directed activities, ask questions requiring more than a yes-no answer. Such questions are often called *higher order questions* because a child must reflect on basic relationships and form logical deductions in answering.
6. Show children that you love mathematics and believe geometry is an important area of study. If geometry is only taught when number work is completed or just before the end of the school year, children may regard it as somewhat insignificant.

20. See Mary M. Lindquist and Marcia E. Dana, "Independent Mathematics Center for Everyone–Let's Do It," *The Arithmetic Teacher* 26, no. 7 (March 1979): 8–12.

16

Early Experiences with Measurement

Measuring is a basic, everyday activity involving time, land, food, and innumerable other things. Many experiences with measurement take place intuitively; at other times, formal instruction in methods and units is required.

Measurement is often taught in relation to geometry. However, it is sometimes used to develop numerical ideas; the real-life use and importance of numbers needed to express measure of quantity is stressed. This is called a *measurement approach* to mathematics instruction.

Children have intuitive ideas about measuring and what it means; determine what they already know when planning instruction. Kindergartners notice that one string is longer than another or that Mary is taller than John; such insights suggest a level of knowledge concerning length and height. A different level of knowledge is indicated when older children use numbers to make comparisons; such as, 5 feet is longer than 3 feet.

This chapter focuses on those attributes most commonly studied in elementary schools. Often, instruction begins with length measurement in kindergarten, and proceeds to other attributes such as weight, area, and volume. The measurement of other attributes such as speed, velocity, and acceleration are more complex and hence come much later. A preferred sequence based on knowledge of mathematical skills needed to work meaningfully with each measure is presented in most elementary textbook series.

Also, consider a child's mental development, because many attributes—such as weight and time—cannot be perceived by seeing and hearing.

Once an attribute is chosen, select units for measuring. For example, length can be measured in centimeters, inches, yards, meters, and so on. Units such as square meters, cubic yards, and kilometers per hour are constructed from simpler units; very sophisticated levels of thinking are involved. The mathematics knowledge and mental maturity of each child must enter into these decisions.

Instruction should focus on both measurement processes and the selection of appropriate units. Interpretation of relative measures and the application of measurement to real-life situations should also be included.

PLANNING THE CONTENT

The content has two interrelated parts: knowledge of units of measure, and knowledge of the measurement process itself. Development of the latter requires performing actual measurements through a hands-on approach, and careful analysis of how the child is measuring.

At an intuitive level, measurement begins when a person identifies an attribute of an object, and then attempts to determine how much of this attribute the object possesses. Measurement involves knowing both a measurable attribute, and finding an appropriate unit to use.

16.1
A *measurable attribute* of an object is a characteristic that can be quantified by comparing it to some standard unit.[1]

Some attribute of any object encountered can be identified to measure. Obvious attributes include length, size, weight, and volume. Other attributes, such as color intensity and hardness, can be measured, but are not usually part of an elementary school curriculum.

16.2
Measuring is the process of comparing an attribute of a physical object to some unit selected to quantify that attribute.

16.3
A *unit* is a fixed quantity, value, or size and need not have physical substance.

16.4
A *standard* is a physical representation of a unit.

16.5
A *measurement* is the result of measuring. As such, a measurement involves both a number and the unit, whereas a *measure* is a number alone.

1. George Bright, "Estimation as Part of Learning to Measure," in *Measurement in School Mathematics,* 1976 Yearbook, ed. Doyle Nelson (Reston, Va.: National Council of Teachers of Mathematics, 1976), p. 88.

EXCURSION
Historical Units

Throughout history people have attempted to measure.[2] Some units were related to parts of the body or an occupation. For instance:

- Cubit: The length of a forearm from elbow to the tip of the middle finger. Used by the Babylonians and Egyptians, the cubit was the first recorded unit of measurement.
- Yard: The circumference of a person's waist, or from the nose to the tip of the middle finger.
- Inch: The length of three round and dry barleycorns taken from the center of the ear and laid end-to-end.
- Furlong: Originally defined as the length of a furrow plowed by a farmer. It was defined later to be 40 yards long.
- Fathom: The length across a man's two outstretched arms.

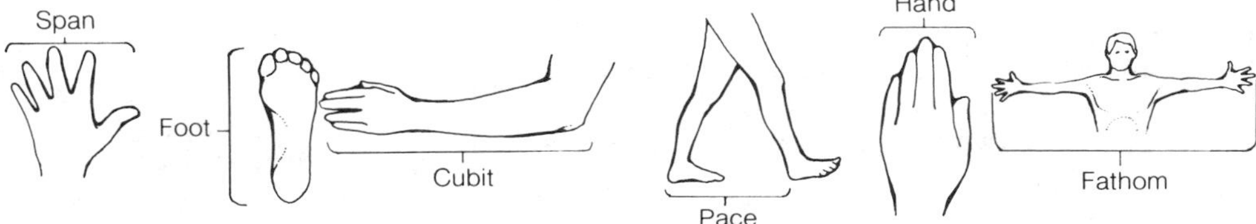

Some relationships established among units were as follows: 9 inches = 1 span; $\frac{2}{3}$ inch = 1 digit; 4 inches = 1 hand; 6 feet = 1 fathom. Can you see any problems with using these units in your everyday measurement? What value would a knowledge of historical units be to your students?

A statement such as "the tree is *37 meters* high" is a measurement; it contains both a unit and a number. Measuring involves making a comparison with a representation of a unit, such as a meter stick or a ruler. Measurement of *continuous* quantities is subject to error: physical representations of units, such as meter sticks, may differ in actual length, or the people who measure may use an instrument differently.

In contrast, *discrete* quantities can be measured by counting. The number of objects, units, or value can be exactly determined. Examples include 7 chairs, 5 cartons, and 10 pennies.

16.6
Accuracy refers to the discrepancy between the true value and the result obtained by measurement. Accuracy is affected by the measuring instrument and the way the instrument is used.

16.7
Precision refers to the agreement among repeated measures of the same physical quantity.

If repeated measures vary greatly, there may be a problem with the instrument or with the procedure being used.

The precision of a measurement increases with smaller and smaller units: measuring length

2. See James Cunningham, *Teaching Metrics Simplified* (Englewood Cliffs, N.J.: Prentice-Hall, 1976).

TABLE 16.1
Partial Listing of Units from the English System of Measurement

Length	*Area*	*Volume (Solid)*	*Capacity (Liquid)*	*Capacity (Dry)*	*Weight*
Inch	Square inch	Cubic inch	Gill	Pint	Ounce
Feet	Square feet	Cubic foot	Quart	Quart	Pound
Yard	Square yard	Cubic yard	Gallon	Peck	Hundred-weight
Rod	Square rod	Load		Bushel	Ton
Mile	Acre	Perch			
	Square mile	Cord			
	Section				
	Township				

in millimeters is more precise than centimeters which in turn is more precise than inches. Generally, the more precise measurements are more accurate.

Many units for measuring have international acceptance and are called *standard* units. Objects such as paper clips or beans are called *nonstandard* units.

Well-defined systems of measurement are used throughout the world. In the past, the United States used the English System of Measurement. Since 1975, however, the U.S. has moved toward the Metric System of Measurement. Both the English System (also called the Customary System of Measurement) and the Metric System contain units relevant for measuring attributes and quantities in elementary schools.

English System of Measurement

Table 16.1 contains many commonly used units in the English System of Measurement, only part of which are considered in the typical elementary program.

TABLE 16.2
Length Measure

12 inches = 1 foot (ft.)
3 feet = 1 yard (yd.)
$5\frac{1}{2}$ yards = 1 rod (rd.)
320 rods, 5280 feet = 1 mile (mi.)

Many relationships exist among the units in Table 16.1. Some of these are indicated in Tables 16.2 through 16.7.

TABLE 16.3
Area Measure

144 square inches = 1 sq. ft.
9 square feet = 1 sq. yd.
$30\frac{1}{4}$ square yards = 1 sq. rd.
160 square rods = 1 acre (A.)
640 acres = 1 square mile
1 square mile = 1 section
36 square miles = 1 township

TABLE 16.4
Volume (Solid) Measure

1728 cubic inches = 1 cubic foot
27 cubic feet = 1 cubic yard
1 cubic yard = 1 load (earth)
$24\frac{3}{4}$ cubic feet = 1 perch (stone)
128 cubic feet = 1 cord (wood)

TABLE 16.5
Capacity (Liquid) Measure

4 gills = 1 pint (pt.)
2 pints = 1 quart (qt.)
4 quarts = 1 gallon (gal.)

EXCURSION
One Pound = One Pound ?

Have you ever been asked, "Which is heavier, a pound of feathers or a pound of gold?" Did you say they are the same?

The correct answer is that a pound of feathers weighs more than a pound of gold. In the United States, there are two units named pound, Troy and Avoirdupois. Feathers are measured in an Avoirdupois pound, which equals 453.59 grams; while gold is measured in a Troy pound, which equals 373.24 grams.

Metric System of Measurement

In 1975, President Ford signed Public Law 94–168 declaring the International Metric System as the official measurement system of the United States. This law was the culmination of many years of effort by educators and others.[3]

The metric system has been recognized as a legal system of measurement in the U. S. since 1866. In recent years, there has been heated debate over making the metric system the *primary* system. Although the issue of conversion has basically been settled, arguments advanced for each system make interesting reading.[4]

TABLE 16.6
Capacity (Dry) Measure

2 pints = 1 quart (qt.)
8 quarts = 1 peck (pk.)
4 pecks = 1 bushel (bu.)

TABLE 16.7
Weight Measure

16 ounces (oz.) = 1 pound (lb.)
100 pounds = 1 hundred-weight (cwt.)
2000 pounds = 1 ton (T.) (short ton)

The metric system of measurement consists of a set of base units and prefixes that define multiples and/or fractional parts of those base units. Base units applicable to elementary school are listed in Table 16.8; commonly used prefixes appear in Table 16.9.

The prefixes in Table 16.9 are combined with the units given in Table 16.8 to name both larger and smaller units for measuring a particular attribute. Tables 16.10 through 16.13 present many interrelationships that result.

In the metric system, the unit of measure for temperature is the degree Celsius. The Celsius scale is quite different than the Fahrenheit scale. Some of the specifics of the Celsius scale are given in Table 16.14.

Measurement of Time

Time may be studied from two perspectives: time as duration, and telling the time shown on a clock or a watch. In some languages, such as German, separate words are used for the two concepts of time.

3. See Marilyn Suydam, "Historical Steps Toward Metrication," in *A Metric Handbook for Teachers* (Reston, Va.: National Council of Teachers of Mathematics, 1974), pp. 26–27.
4. See Cunningham, *Teaching Metrics Simplified,* pp. 4–6.

TABLE 16.8
A Partial Listing of Base Units from the Metric System of Measurement

Length	*Mass*	*Volume*	*Time*	*Temperature*
Meter	Gram*	Liter	Second	Degree (Celsius)

*Officially, the kilogram is listed as the base unit of mass. For consistency and for introduction of the system in the elementary school, the use of gram as a base unit is preferred.

TABLE 16.9
Commonly Used Prefixes in the Metric System

Prefix	*Symbol*	*Prefix Times the Base Unit*
mego	M	10^6 (one-million times)
kilo	k	10^3 (one-thousand times)
hecto	h	10^2 (one-hundred times)
deka (deca)	da	10^1 (ten times)
deci	d	10^{-1} (one-tenth of)
centi	c	10^{-2} (one-hundredth of)
milli	m	10^{-3} (one-thousandth of)
micro	$\mathcal{M}$	10^{-6} (one-millionth of)

TABLE 16.10
Linear Measure

10	millimeters (mm)	=	1 centimeter (cm)
1000	millimeters	=	1 meter (m)
10	centimeters	=	1 decimeter (dm)
100	centimeters	=	1 meter (m)
10	decimeters	=	1 meter
10	meters	=	1 decameter (dam)
10	decameters	=	1 hectometer (hm)
10	hectometers	=	1 kilometer (km)
1000	meters	=	1 kilometer

TABLE 16.11
Area Measure

10	square millimeters (mm^2)	=	1 square centimeter (cm^2)
100	square centimeters	=	1 square decimeter (dm^2)
100	square decimeter	=	1 square meter (m^2)
1 000 000	square meters	=	1 square kilometer (km^2)

Units of time commonly studied are given in Table 16.15. Additional ideas relating to time as duration are found in content objectives 16.8 through 16.11.

16.8
A year is the period of time required for the earth to make one complete revolution around the sun.

16.9
A month is approximately the period of time it takes for the moon to make one revolution around the earth.

16.10
A day is the period of time that it takes for the earth to make one complete revolution on its axis.

16.11
AM and PM refer to before noon and after noon, respectively.

TABLE 16.12
Mass Measure

10	milligrams (mg)	=	1 centigram (cg)
1000	milligrams	=	1 gram (g)
10	centigrams	=	1 decigram (dg)
100	centigrams	=	1 gram
10	decigrams	=	1 gram
10	grams	=	1 decagram (dag)
10	decagrams	=	1 hectogram (hg)
10	hectograms	=	1 kilogram (kg)
1000	grams	=	1 kilogram

TABLE 16.13
Volume and Capacity Measure

Volume	1	cubic centimeter (cm^3)	=	1 milliliter (ml)
	1	cubic decimeter (dm^3)	=	1 liter (1)
	1000	liters	=	1 cubic meter (m^3)
Capacity	10	milliliters (ml)	=	1 centiliter (cl)
	10	centiliters	=	1 deciliter (dl)
	10	deciliters	=	1 liter (L)
	10	liters	=	1 decaliter (dal)
	10	decaliters	=	1 hectoliter (hl)
	10	hectoliters	=	1 kiloliter (kl)
	1000	liters	=	1 kiloliter

The following additional ideas are related to telling time on a clock. These objectives refer to telling time on a traditional and not on a digital clock.

16.12
The face of the clock is the area that has the twelve numerals and the hands.

16.13
The short hand of the clock shows the hours.

16.14
The long hand of the clock shows minutes.

16.15
When the minute hand points to the numeral 12, the time is at the hour; when the hour hand points directly to a numeral, the hour of the day is indicated.

TABLE 16.14
Celsius Temperature

Freezing Point of Water:	0° Celsius
Boiling Point of Water:	100° Celsius
Body Temperature:	37° Celsius

TABLE 16.15
Time Measurement

60	seconds (sec.)	=	1 minute (min.)
60	minutes	=	1 hour (hr.)
24	hours	=	1 day
365	days	=	1 year (yr.)
10	years	=	1 decade
100	years	=	1 century

EXCURSION
Guidelines for Writing Metric Measurements[5]

1. When the names of units are written in full, they usually start with a lowercase letter. (The modifier in "degree Celsius" is capitalized because it is derived from the Swedish astronomer, Anders Celsius.) Unit symbols are also written in this manner: prefixes typically begin with a lowercase letter, as do their symbols, except for tera (T), giga (G), and mega (M).
2. The names of units are made plural when appropriate; however, do not add an *s* to the symbol to indicate a plural.
3. A period is *not* used after a symbol, except at the end of a sentence.
4. A zero must be written before the decimal point for numbers less than 1.
5. Digits are shown in groups of threes (counting from the decimal point). A comma is *not* used; rather a space is left between the groups (except for four-digit numerals).

16.16
When the minute hand points to the numeral 6, the time is at the half hour.

16.17
As the long hand moves from numeral to numeral, the time lapse is five minutes.

16.18
When there is a second hand, the time lapse for one complete revolution around the clock is one minute.

16.19
The time lapse for the short hand to move from one numeral to the next is one hour.

5. Ibid., pp. 177–79.

Money

Many educators view an understanding of the monetary system as essential. Often, such knowledge comes under the heading of functional mathematics.

The monetary system currently in use contains both coins and bills. The coins represent fractional parts of the base unit, one dollar, while bills represent multiples of the one-dollar base unit. Most instruction concerning money focuses on the fractional parts of one dollar; Table 16.16 outlines many of the interrelationships. Denominations of bills include a two-dollar bill, a five-dollar bill, a ten-dollar bill, a twenty-dollar bill, a fifty-dollar bill, a hundred-dollar bill, a five-hundred dollar bill and a one-thousand dollar bill.

TABLE 16.16
Money Relationships

Penny	*Nickel*	*Dime*	*Quarter*	*Half-dollar*	*Dollar*
1 cent	5 cents (pennies)	10 cents 2 nickels	25 cents 5 nickels	50 cents 10 nickels 5 dimes 2 quarters	100 cents 20 nickels 10 dimes 4 quarters 2 half-dollars

PLANNING INSTRUCTION

Because judgment must be used when deciding the most appropriate unit for a particular measurement situation, we provide experiences that help children become skilled at *selecting* such units. This skill requires both knowledge of available units and of how they relate to each other.

Discussion of a topic as comprehensive as measurement must be limited in a methods textbook; only sample activities that are usually effective for measurement topics included in elementary programs are presented: length, area, volume and capacity, mass and weight, temperature, time and money. One common thread woven throughout all activities is hands-on experience. *Children learn to measure by measuring.*

Measurement begins at an intuitive level as a child notices specific attributes of an object: e.g., how long, how large, or color. Help the child focus on a particular attribute and describe it as clearly as possible.

When the child can identify an attribute, begin comparing more than one object regarding this particular attribute.

Comparisons may take place in many ways. If length is the attribute being compared, objects can be placed adjacent to each other to determine which is longer or shorter.

Eventually determine *how much* of a particular attribute an object has. This involves deciding on a way to measure the object. How this decision is made reflects the child's understanding of measurement as a process. At first, many units may be tried to see if they are appropriate for measuring the attribute chosen. For instance, length could be measured with toothpicks, paper clips, straws, strings and other objects; time could be measured in years, months, days, hours, minutes or seconds. Competence in choosing the most appropriate unit should develop with experience.

Figure 16.1 contains a model describing how a person may progress in knowledge of measurement as a process. Notice that many of the activities for identifying and comparing attributes are called *premeasurement* activities; they do not produce a number of units. Much of the measurement program in the primary grades will involve premeasurement activities.

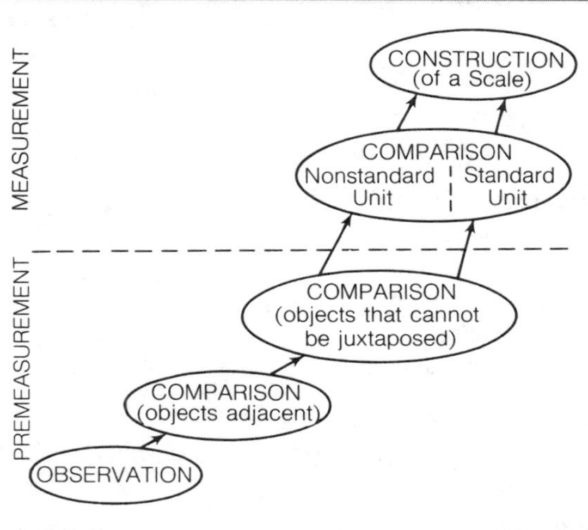

FIGURE 16.1
A Model for Measurement Instruction

Children should *estimate* the number of units before a measure is actually taken, thereby giving us insight into what they know about a particular unit. An estimate also enables us to see if the child is learning to choose an appropriate unit and to determine the number of units.

While whole numbers are used in many of our measurements, the measurements themselves are only approximations of the true measure. The technical aspects of precision and accuracy can be introduced to intermediate grade students. Children can determine the effect that size of a unit has on the accuracy and/or precision of a measurement.

Linear Measurement

Begin with activities that emphasize the role of a unit in the measuring process. Nonstandard units such as paper clips, beans, toothpicks or even a child's thumb can be used to measure a variety of linear objects. Encourage each child to express the measurement as ________ paperclips, ________ thumbs, and so on. (See the activity described in Figure 16.2.)

Before introducing standard units, we must choose whether to teach meters or yards,

ACTIVITY PLAN

Measurement: Length — ***3D→Symbolic*** — ***Initiating/Abstracting***

Content objectives

16.2 through 16.5

Materials/Exemplars

Pieces of string
Paper clips
Unifix cubes
Beans
Line segments on paper

Behavioral indicators

Given a linear object, the child will determine its length in paper clips, unifix cubes, beans, and thumbs.

Procedure

1. Give each child a piece of string and ask her to measure it with paper clips. You may need to show how to do it.

Say, "Eleven paper clips."
2. Repeat and ask for the measure in Unifix cubes.
Ask, "What is the measuring unit?" "How many units long is this string?"
3. Continue measuring linear objects with nonstandard units. Focus attention on the fact that as we measure, we determine how many units are needed.
4. Give each child a piece of paper on which you have drawn a short line segment. Make sure all segments are the same length. Ask each child to measure the line segment using the beans. Again ask for a report of the number of beans.
5. Repeat Step 4, with each child using her thumb as the measuring unit. As the measurements are reported, ask why different measures are being given when each person measured line segments of the same length. Discuss in detail.

FIGURE 16.2
Measurement of Length

centimeters or inches, and so on. The metric system of measurement is usually the primary system for instruction, but the school board or mathematics supervisor may ask you to include English units such as foot, inch, and yard. The Metric Implementation Committee of the National Council of Teachers of Mathematics has prepared guidelines that include suggestions on grade placement of metric topics.[6]

6. See NCTM Metric Implementation Committee, "Metric: Not *If,* But *How,*" *The Arithmetic Teacher* 21, no. 5 (May 1974): 366–69.

Regardless of the unit, children must be allowed to move around and measure objects found in their environment. Ask questions similar to the following:

- Can you find objects that are longer than a decimeter?
- What objects are shorter than a meter?
- What is the distance from the teacher's desk to the door? (Use a trundle wheel to find out.)

Learning centers can involve children with stations on attributes such as length, area, weight, and time. Activity cards focusing on measurement with different units are illustrated in Figure 16.3. Develop sets of activity cards for learning centers and prepare a management plan that allows easy movement through the sequence of activities.

Many materials help children understand linear measurement. Cuisenaire rods are constructed on a metric scale, and can be used for measuring centimeters and decimeters. Base 10 blocks are also available in metric dimensions. Of course, metric measuring instruments such as tape measures, meter sticks, and trundle wheels are commercially available.

Children in the intermediate grades should study the ways linear units are related; for example, the number of centimeters contained in a decimeter and a meter. The meanings of prefixes should be discussed and learned (*milli* means one-thousandth, *centi* means one-hundredth, and so on). Children should be able to convert one linear unit to another.

The schematizing activity in Figure 16.4 provides a systematic procedure for converting from one unit to another within the metric system. However, do not emphasize conversion activities so much that children begin to view this as the most important aspect of measurement.

When children are ready for consolidating activities, games such as those described in Figures 16.5 (page 386) and 16.6 (page 387) can be used. These games can be adapted to measurement of other attributes.

Area Measurement

Measurement in two dimensions seeks to determine how much surface is enclosed within a closed plane curve. As with linear measurement, the initial concern of area measurement is to find an *appropriate* unit. Nonstandard units such as circular discs, cards, books, and magazines can be used. Ask, "Will this unit cover the surface without any space remaining uncovered?"

A geoboard (see page 385) can be used to make the transition to a standard unit. Begin by enclosing a rectangular region with a rubber band. Then use rubber bands of a different color to enclose unit square regions to determine the number of square units contained in the rectangle.

Card 1—Centimeter Tape Measure

	You	*Friend 1*	*Friend 2*
Wrist			
Arm			
Height			
Around Head			
Smile			

Card 2—Decimeter Rods

Find 5 things longer than a decimeter.

____
____
____
____
____

Find 5 things shorter than a decimeter.

____
____
____
____
____

Card 3—Centimeter Tape Measure

1. Take 10 steps. Measure in centimeters. _____ cm
2. Stand with both feet together. Jump. Measure your jump in centimeters. _____ cm
3. Measure length of one step in centimeters. _____ cm

Card 4—Trundle Wheel

Find the distance around the room. _____ m
Find the distance from the door to your desk. _____ m
Go outside and find the distance around the playground. _____ m

FIGURE 16.3
Activity Cards for a Learning Center

ACTIVITY PLAN

Measurement: Length | *Symbolic→Symbolic* | *Schematizing*

Content objectives

The units of the metric system are interrelated, i.e., 10 mm = 1 cm, 10 cm = 1 dm, 10 dm = 1 m, etc.

Materials/Exemplars

Paper and pencil

Behavioral indicator

Given a length measurement, the child can convert it to other linear units within the metric system.

Procedure

1. Present a measurement such as 1 meter and ask, "How many decimeters is this equivalent to?" Continue to ask questions, such as: "Which unit is larger, meter or decimeter?" "If a meter is larger than a decimeter, would you have more meters or more decimeters?"
2. Place emphasis on the relationship 1 meter = 10 decimeters. Ask "How many decimeters would be equivalent to 6 meters?" (60) "7 meters?" (70) and so on. Conclude that we multiply the number of meters by 10 to find the number of decimeters.
3. Relate meter to centimeter by following a similar procedure, emphasizing that we multiply the number of meters by 100. (1 meter = 100 cm)
4. Introduce the following diagram and explain that it can be used for converting.

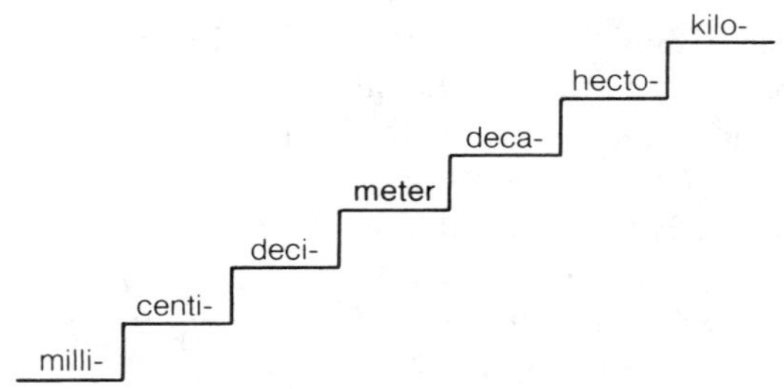

Each step represents a linear unit, millimeter to kilometer. To change meters to decimeters, take one step down the stairs. To change millimeters to meters, take three steps up the stairs, and so on.

5. Place the number 10 at each step as shown.

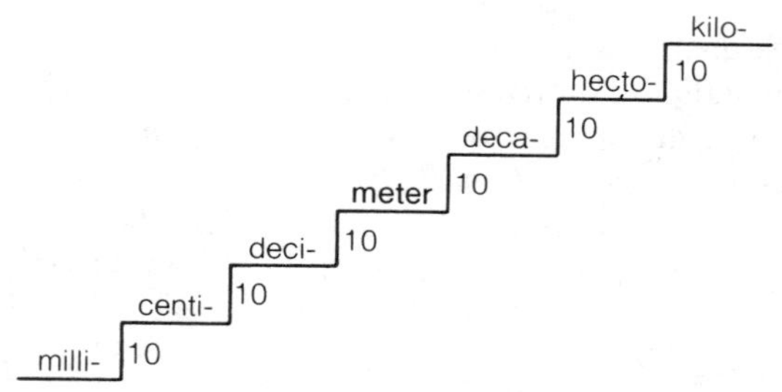

Show children that when they wish to change from meters to decimeters, they can take one step down and multiply by 10. From meters to millimeters they take three steps down, and multiply by 10 × 10 × 10 or (10^3).

6. Finally, review the reasoning for multiplying or dividing when changing from one unit to another. Emphasize that if you change from a smaller to a larger unit, you use *fewer* of the larger units. Hence, you will divide. If you convert in the other direction, larger to smaller, you will have more units and therefore multiply. The multiple of 10 used for multiplying or dividing depends on the number of steps taken on the stairs. Your final diagram should look like this:

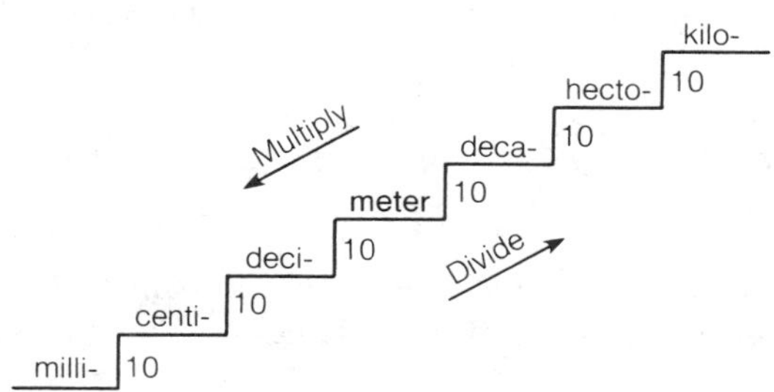

Note: Similar diagrams can be used for converting when using grams or liters.

FIGURE 16.4
A Diagram for Converting Within the Metric System

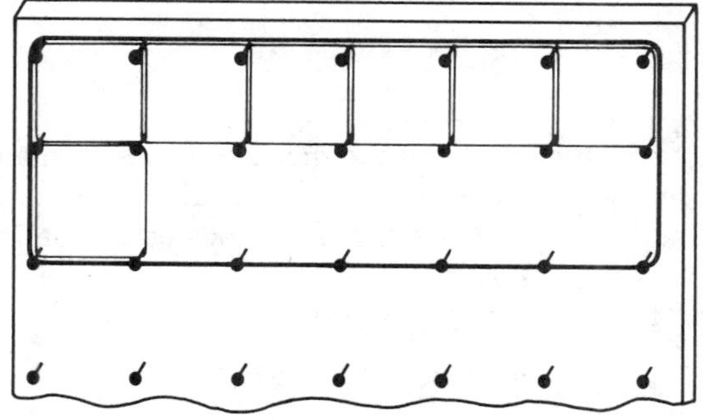

At this point, move to centimeter graph paper. Help children identify the unit, a square centimeter. Have them color in a rectangle 2 centimeters long and 3 centimeters wide, then count the square centimeters. Initially, do not emphasize a formula for computation; instead, stress units and their size. Paper or cloth cut-outs for standard units often help a child comprehend the difficult concept of area measurement.

Eventually observe that the number of units in a rectangular region is equal to the number of units in the length of the array times the number of units in the width of the array. A child applies what she already knows about an array in learning a formula for the area of a rectangular region. Have children find the areas of table tops, chalkboards, the classroom floor, and part of the playground. As children move into the intermediate grades, other units can be studied such as square meters, square kilometers and square millimeters.

Volume and Capacity Measurement

Even primary grade children want to know how much a container holds. They can informally determine the capacity of a container by filling it with sand or water using a smaller container, counting as material is poured. A juice can, for example, can be a nonstandard unit for determining the capacity of a bowl. Measuring containers for one cup, one quart, and one liter can also be used. Many teachers use simple recipes for a cooking lesson when teaching direct measurement of capacity.

Indirect measurement of volume is usually introduced in the intermediate grades, after children have shown some knowledge of both linear and area measure. The transition from the direct measurement of earlier activities to indirect measurement can be accomplished by asking children to determine how many cubic centimeters (cm^3) are in a small box. Make sure each child understands the dimensions of length, width, and

ACTIVITY PLAN

Measurement: Length — *3D → 3D* — *Consolidating*

Content objectives

The meter, decimeter and centimeter are units of length in the metric system.

10 decimeters = 1 meter 10 centimeters = 1 decimeter 100 centimeters = 1 meter

Materials/Exemplars

Meter stick	Centimeter cubes
Decimeter rods	Dice

Behavioral indicators

1. Given 10 or more centimeter cubes, the child will exchange 10 cubes for 1 decimeter rod.
2. The child will place 10 decimeter rods beside 1 meter stick.

Procedure

1. Group the children in pairs. On desk tops, tape two meter sticks as shown for each pair of children.

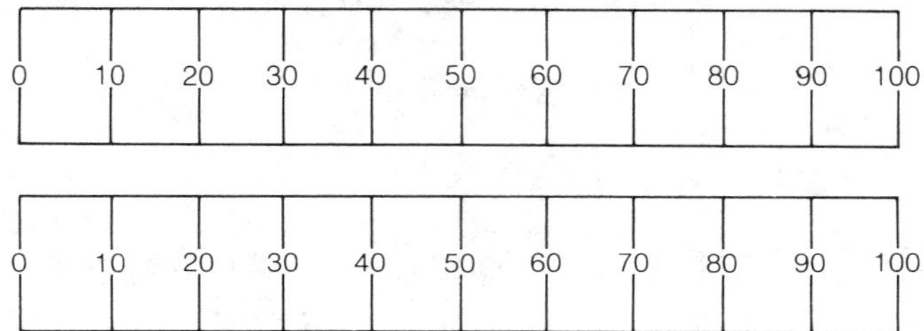

2. Explain how the game is played. Each child takes turns tossing the dice and takes as many centimeter cubes as the sum thrown. The centimeter cubes are matched against the meter stick. Suppose the first roll is 6 and 6; 12 centimeter cubes are taken and placed beside the meter sticks.

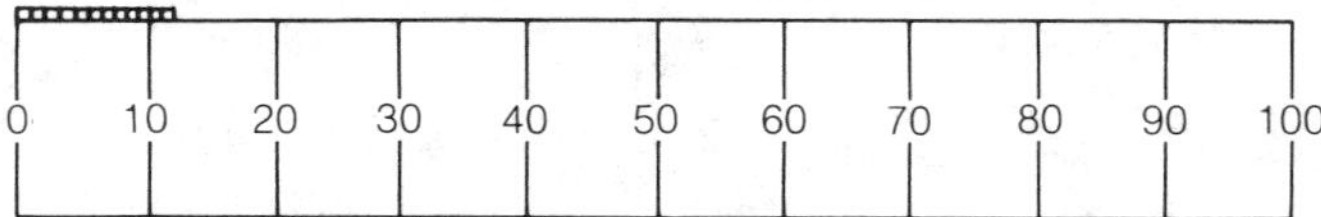

3. A child must exchange 10 centimeters for 1 decimeter whenever 10 centimeters in length are accumulated. At some point the desk top may look like this.

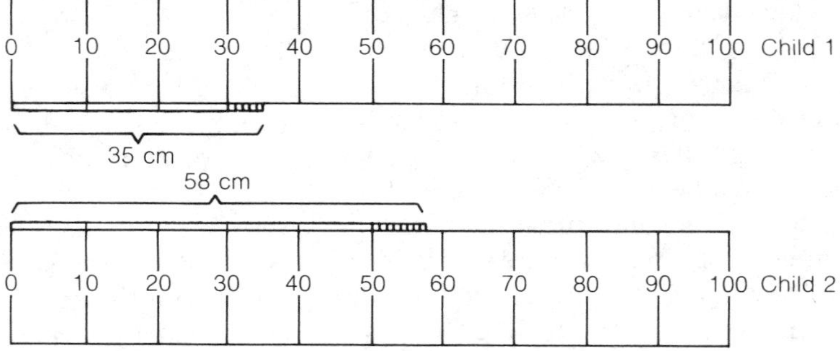

4. The object of the game is to see who reaches 100 centimeters first.

FIGURE 16.5
A Metric Consolidating Activity

ACTIVITY PLAN

Measurement: Length | *Symbolic →Symbolic* | *Consolidating*

Content objectives

1 kilometer (km) = 10 hectometers
1 hectometer (hm) = 10 decameters
1 decameter (dam) = 10 meters
1 meter (m) = 10 decimeters
1 decimeter (dm) = 10 centimeters
1 centimeter (cm) = 10 millimeters

Materials / Exemplars

Paper with 5 × 5 rectangular grid
Paper squares with linear measurements
Items to cover squares

Behavioral indicator

Given a linear measure written as a unit times 10 or times a multiple of 10, the child will identify an equivalent unit.

Procedure

1. Give each child a sheet of paper containing a 5 × 5 rectangular grid.

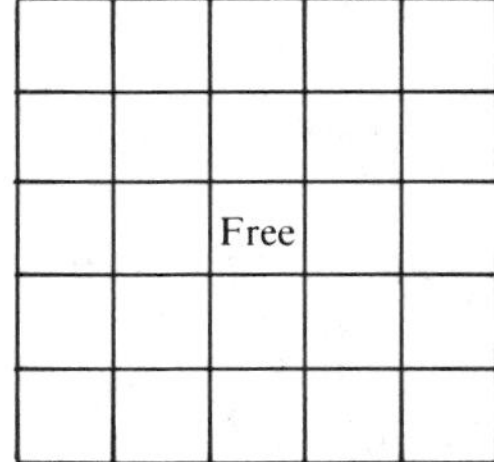

2. Tell each child to write the symbols *km, hm, dam, m, dm, cm,* and *mm* in the remaining 24 blocks. The symbols should be written in random order; some will need to be repeated.
3. Explain that the game will be played following the rules of Bingo. The symbols represent *one* millimeter, *one* centimeter, and so on.
4. Pick up a paper square and call out the linear measurement. Say, "Suppose it is 100 meters. What measurement on the card is equal to 100 meters?" (one hectometer). Say, "Cover a block having *hm* in it."
5. Continue calling number values until someone has metric Bingo.

FIGURE 16.6
A Metric Consolidating Activity

height (depth). Children can fill the box with centimeter cubes or construct a cubic decimeter and attempt to fill it with cubic centimeters. The large thousands block in a set of metric base 10 is a good model (see page 388, top left).

Although volume is measured formally in cubic units, have children determine the *capacity* of containers (how much each will hold) by filling them with liquid. Volume and capacity are related—a capacity measure of 1 milliliter of

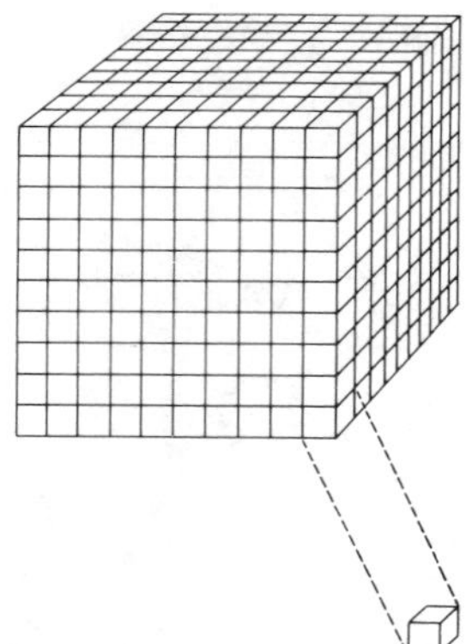

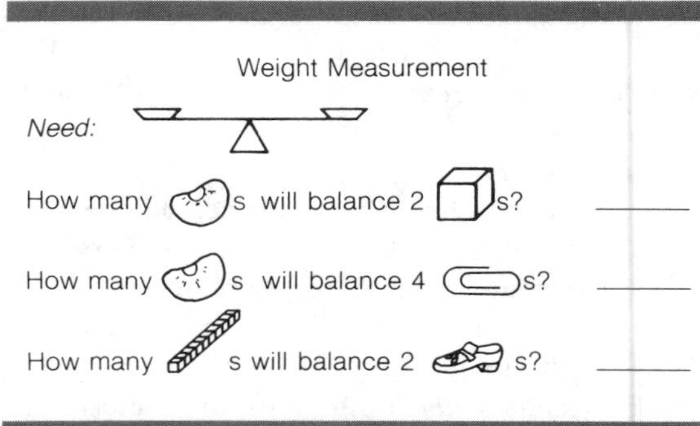

FIGURE 16.7
Activity Card for Weight Measurement

water has a volume of 1 cubic centimeter (sometimes referred to as a cc). In the metric system, the base unit for capacity measure is the liter, a unit being slightly larger than the quart. Children should become familiar with the liter, the deciliter and the milliliter as they determine the volumes of shoeboxes, lunchboxes, and classrooms. For capacity measure, use soft-drink bottles, coffee cups, milk jars and food cans.

Weight and Mass Measurement

The *mass* of an object–the amount of matter–will remain the same regardless of where it is moved. Weight, on the other hand, is the amount of gravitational pull exerted on an object. Thus, the weight of an object may change, depending upon how far the object is from the earth's surface. Also, a person's weight on the moon is about one-sixth of her weight on earth. (The difference between a person's mass and weight on earth is negligible, so in elementary school little time is spent making the distinction.)

Not surprisingly, many children have difficulty with the concept of weight. The weight of an object can be sensed by holding it in your hand, but this is confusing when a small object weighs more than an object 10 times its size.

The two skills critical to other work with measurement are central to weight as well: determining appropriate units and estimating the measure of the attribute. Both are best developed through activities in which children actually weigh many objects, using a variety of units. Beginning in kindergarten and throughout subsequent elementary grades, pose questions such as: How many paper clips will balance a Unifix cube? How many cubes will balance 3 navy beans? A wide variety of objects can be weighed and items used for a unit. An activity card similar to Figure 16.7 can be used with children in the primary grades.

As children enter the intermediate grades, move toward use of standard units: grams, kilograms, decigrams, decagrams and hectograms. Thoroughly develop awareness of frequently used units, such as the gram and the kilogram. Children can use a balance to find objects weighing 1 gram, more than 1 gram and, if possible, less than 1 gram. A paper clip, a dollar bill, and an aspirin tablet are good examples of objects weighing about 1 gram. An activity focusing on the kilogram is outlined in the handbook accompanying this text.

Temperature Measurement

Although children experience sensations of heat and cold daily, a thermometer is required to actually determine the temperature. The metric unit of measure for temperature, the degree Celsius, is written as °C.

The ability to estimate a temperature requires skills not usually found among elementary school children. There are, however, a few Celsius temperatures we can focus upon. These are named on the thermometer.

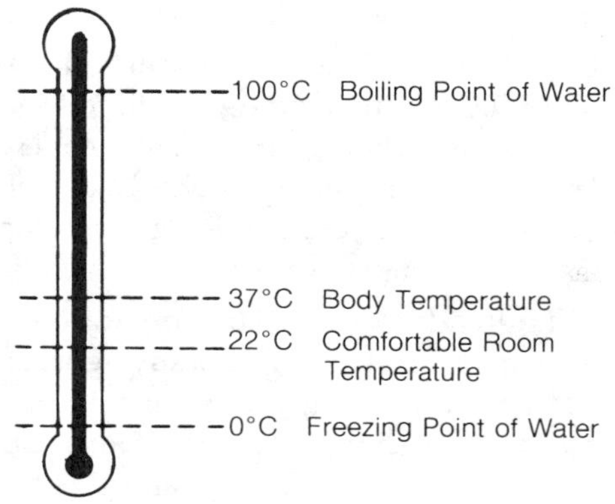

The following activities can help develop awareness of temperature:

1. Place a Celsius thermometer in a pan of ice. Leave it there for five minutes. What is the temperature?
2. Place a Celsius thermometer in a sunny window. Leave it there for five minutes. What is the temperature?
3. Listen for the daily temperature given in Celsius from a local weatherperson. Make a chart for a two-week period showing daily temperatures for a set time of day.

Time Measurement

Most children have minimal awareness of time as duration before they are 8 or 9 years old. Most effort expended in this area involves teaching them how to read a clock, at first to the hour, then to the half-hour, to the quarter-hour, and finally to the minute.

Reading a clock requires that a child recognize the two (possibly three) hands, and know what each hand represents. Begin instruction by asking them to determine time to the hour. Emphasize the fact that the long hand points directly to twelve and the short hand points directly to the name of the hour. Use a real clock; paper clocks often display the time inaccurately.

After children can tell time to the hour, introduce the half-hour. A real clock will show clearly what happens to the short hand whenever the long hand points to 6.

30 minutes past 2,
or half-past two
or two thirty
or 2:30

As children gain skill, they should be encouraged to read the measure written different ways. The symbolism 2:30, for example, should be introduced for 30 minutes past 2. Many children will have already encountered this in digital clocks or watches.

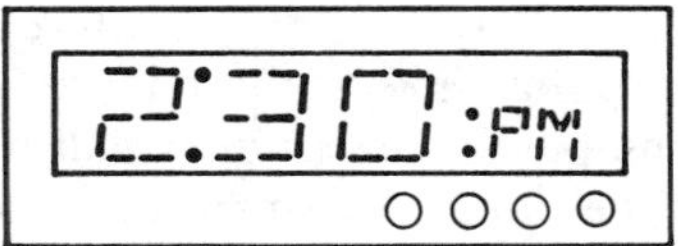

Reading time to the quarter-hour and the minute requires that children see the numbered part of the clock face as a circular number line from 1 to 60. Children who count well by fives can be encouraged to work on a clock marked off in units of 5 minutes. By using just the long hand, reading minutes can be practiced: 10 minutes, 15 minutes, etc.

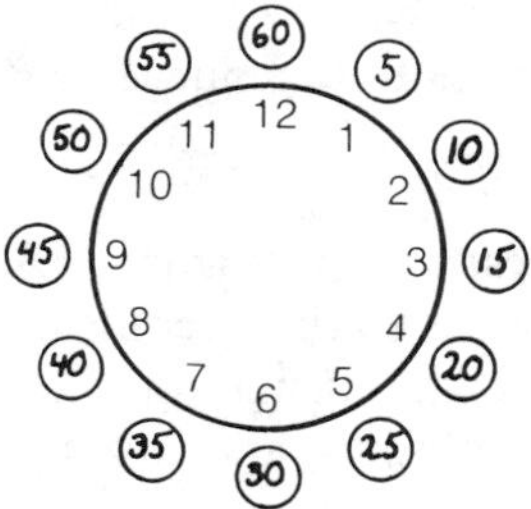

Next, move a real clock through one hour's revolution while focusing on the position of the minute and hour hands. Start with the clock positioned as in (a) in the following. Then, as the minute hand moves, stop at each five minutes and read the time as shown in (b)–(e).

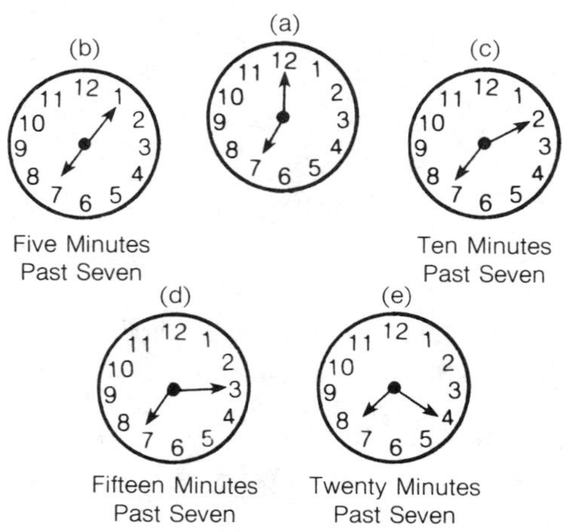

Some teachers prefer to start with the opposite: reading time to the minute, then progress to 5 minutes, quarter-hours, half-hours, and finally to the hour. Reisman has prepared a detailed description of this approach.[7]

Teachers often do the following:

- Have a child say and then show the time she goes to bed or rises in the morning.
- Bring in a digital clock. Have each child show the corresponding time on a standard clock.
- As children move to the intermediate grades, emphasize time as duration by making schedules to be followed for days and weeks.
- Emphasize how to tell time while teaching other content areas. Many situations in social studies, science, and language arts involve telling time.

The activity described in Figure 16.8 can identify the best unit to use when reporting duration of time. It also provides insight into a child's knowledge of hours, minutes, and seconds.

7. Fredricka K. Reisman, *A Guide to the Diagnostic Teaching of Arithmetic* (Columbus, Ohio: Charles E. Merrill, 1972), pp. 27–29.

Money

The names, values, and relationships between coins and bills are the three major topics taught regarding money. Children can usually name several coins when they come to school. Therefore, focus on the values of each and begin developing relationships among coins.

Children learn to make many-for-one and one-for-many exchanges when studying numeration. Money relationships also require that a child count and exchange when the count reaches a designated amount. Unlike exchanges with base blocks, size is irrelevant when exchanging coins. Each child must learn that a coin has a value assigned to it, and the value does *not* necessarily relate to the size.

Begin by designating the value of a penny as 1 cent. The other coin values to be learned in the primary grades are nickel (5 cents), dime (10 cents), quarter (25 cents), half-dollar (50 cents), and one-dollar bill (100 cents).

Have each child count sets of pennies and trade for other coins. Five pennies can be traded for 1 nickel, 10 pennies for 1 dime, and 25 pennies for 1 quarter. The exchanges can also be illustrated on a chalkboard or a flannel board.

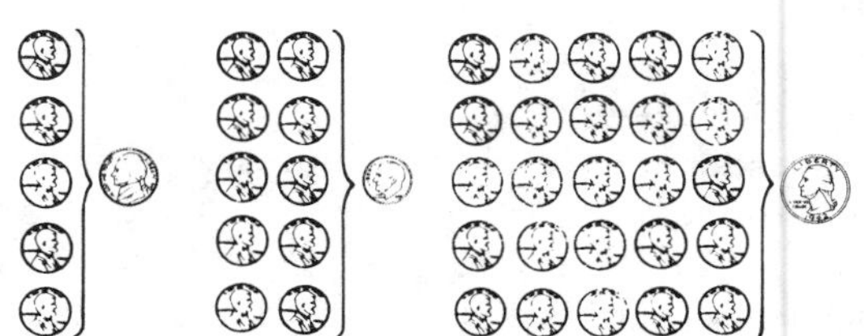

Grouping pennies in clusters of five also shows other relationships: 2 groups of 5 pennies (or 2 nickels) can be traded for 1 dime; 5 nickels (or 5 groups of 5 pennies) can be traded for 1 quarter. Emphasize these relationships: trade coins of less value for coins of greater value, and vice-versa. Some teachers make these activities especially meaningful by using real money.

As with all measurement, children should be encouraged to use the most appropriate units; they should use the most appropriate coins for a given amount of money. Also, have children

ACTIVITY PLAN

Measurement: Time | *Symbolic→Symbolic* | *Diagnosing*

Content objectives

16.13 through 16.19, and definitions of a day or less as found in Table 16.15

Materials/Exemplars

Paper and pencil

Behavioral indicators

1. Given a stated period of time, the child will identify the most appropriate unit (days, hours, minutes, or seconds) to measure the time.
2. Given a stated period of time, the child will justify why she chose a particular unit to measure the time.

Procedure

Have children indicate in writing the best unit to use for measuring each period of time. Probe to see if each child can justify her answers.

Questions:

1. How much time is needed to walk to school?
2. How much time is needed to pour milk in a glass?
3. How much time is needed to drive from Washington, D.C. to New York City?
4. How much time does it take to throw a ball across the room?
5. How much time is needed to eat breakfast?
6. How much time do you stay in school each day?
7. What is the length of a TV commercial?
8. What is the length of time you sleep at night?
9. What is the length of a baseball game?

FIGURE 16.8
Sample Diagnosing Activity for Time Measurement

show a given amount in different ways. For instance, how many ways can 47 cents be shown? Possible combinations include:

- Forty-seven pennies
- Nine nickels and two pennies
- Four dimes, one nickel, and two pennies
- One quarter, two dimes, and two pennies

A classroom store where children make purchases and count out whatever change is due is a popular activity. One child can act as the storekeeper while others make purchases. The storekeeper must make the correct change. Children take turns at being storekeeper so that each child can practice skills.

When children understand decimals, they need to learn that the dollar is a standard unit for money; a cent is one-hundredth of a dollar. Children should not confuse one dollar with the hundred's place in an expression like $1.65.

PLANNING ASSESSMENT

Determine each child's understanding of a unit and the names of units already known. Can each child actually carry out a measurement? What, if any, relationships among units are known and understood?

Many types of diagnosis are required to gather such information.

1. A paper-and-pencil test, along with an interview, if possible, can determine if a child knows the names of units such as meter, gram, and liter.

2. A performance test can determine if a child is able to carry out a measurement with instruments such as a meter stick, a centimeter ruler, or a graduated container.

3. An interview can determine if a child understands relationships among units. Does she understand, for example, how a gram relates to a kilogram?

Inferences about real understanding must be based on more than paper-and-pencil performance. Much of the Piagetian research on measurement clearly demonstrates that children will measure as the teacher instructs, yet may have little understanding of the unit used. For example, they may believe that if a ruler is oriented differently, the unit changes length. The level of a child's understanding can only be diagnosed by using a variety of procedures.

EXCURSION
Tips on Managing the Classroom

A good measurement program requires that children become physically involved in activities. As you plan, consider the following:

1. If you are using an unfamiliar 3–D exemplar, spend time explaining the proper method of use. In the intermediate grades, instructions can be given on a cassette tape or on a carefully written reference card.
2. Children can make measuring devices such as centimeter tapes and meter sticks. Each child has her own device and also learns from the actual construction.
3. Have a collection of measuring instruments. Anticipate the number needed for small group and independent work.
4. Start collecting objects from home and stores early in the year. Also have children bring in cans, boxes, counters, and things to weigh.
5. Attempt a good mixture of hands-on and paper-and-pencil activities. Use learning centers whenever possible to provide opportunities for children to work independently. Have them discuss their findings with peers.

17

Measurement and Metric Geometry

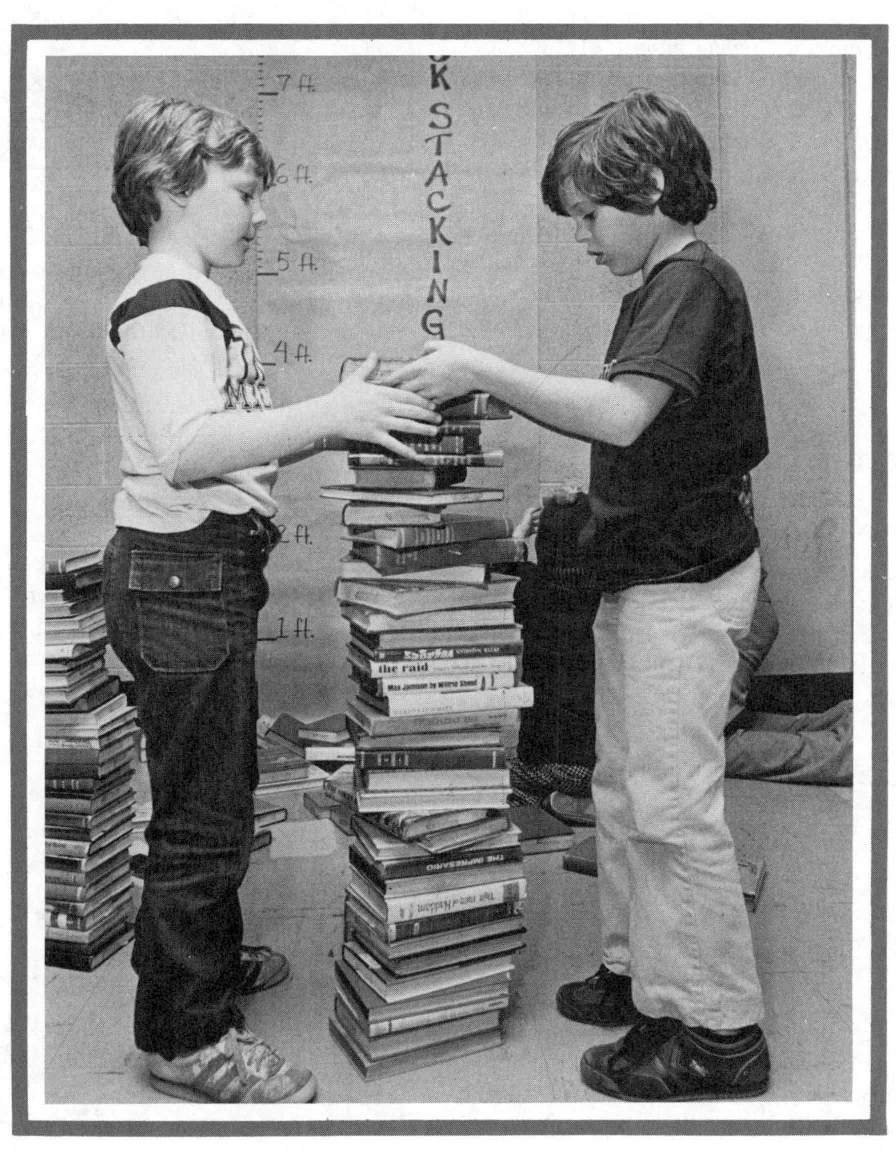

Children who look for patterns and relationships during their many experiences with informal geometry and also learn to use units of measure are soon ready to begin a more formal study of geometry.

Geometric topics in elementary school mathematics include measurement of line segments, angles, perimeter, area, and volume. Even topics not discussed in the following section are sometimes taught: constructions, scale drawings, and more formal treatments of coordinate and transformational geometry.

Each topic involves many concepts and skills, so we must make judgments as we select concepts and skills to emphasize. We are likely to teach some topics only as enrichment material.

PLANNING THE CONTENT

Lines and Line Segments

To measure a line segment, choose a unit of measure (possibly a centimeter) and determine the number of centimeters contained in $\overline{AB}$. This can be shown as follows:

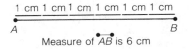

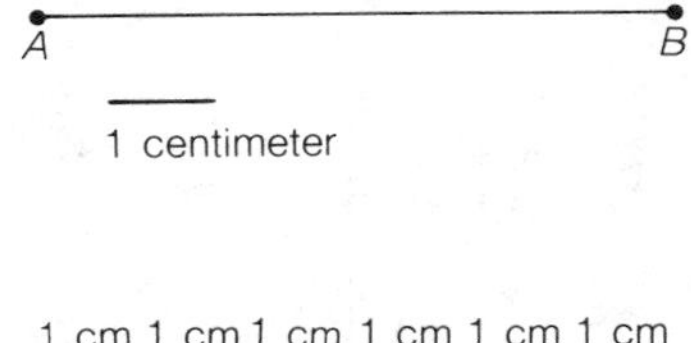

When the measure is determined to be 6 centimeters, it can be written $m(\overline{AB}) = 6$ cm. You can compare the lengths of two or more line segments as long as the same unit of measure is used.

> **17.1**
> Two line segments are defined as being *congruent* if their measures are equal.
>
> **17.2**
> The symbol $\cong$ is used to indicate congruence, eg., $\overleftrightarrow{AB} \cong \overleftrightarrow{CD}$.

Angles

An angle is defined as the union to two noncollinear rays having a common endpoint. The common endpoint is called the *vertex* and the rays are called the *sides* of the angle.

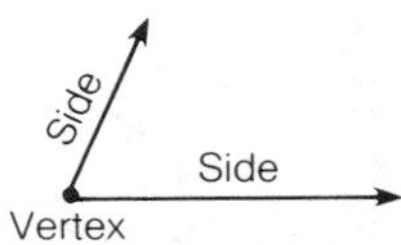

The basic unit for angle measure is the *degree*. The measure of an angle is read from a protractor. We see that $\angle BAC$ has a measure of 60°. We could also write $m(\angle BAC) = 60°$.

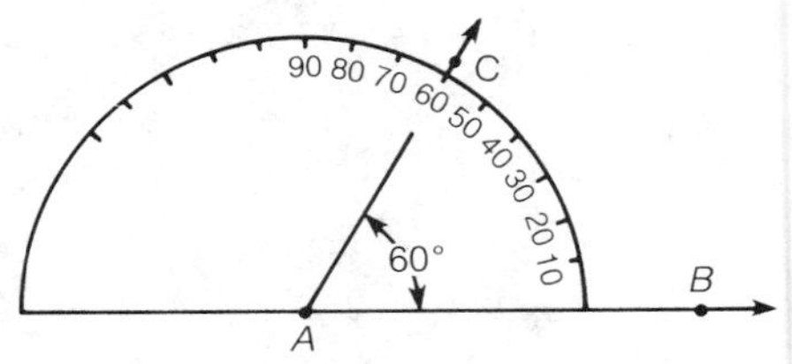

> **17.3**
> Two angles are defined as being *congruent* if their measures are equal.

17.4
An angle having a measure of 90° is a *right angle.*

17.5
An angle having a measure less than 90° is an *acute angle.*

17.6
An angle having a measure greater than 90° but less than 180° is an *obtuse angle.*

17.7
An angle having a measure of 180° is a *straight angle* (also named a straight line).

17.8
Two angles having the same vertex and a common side are *adjacent angles.*

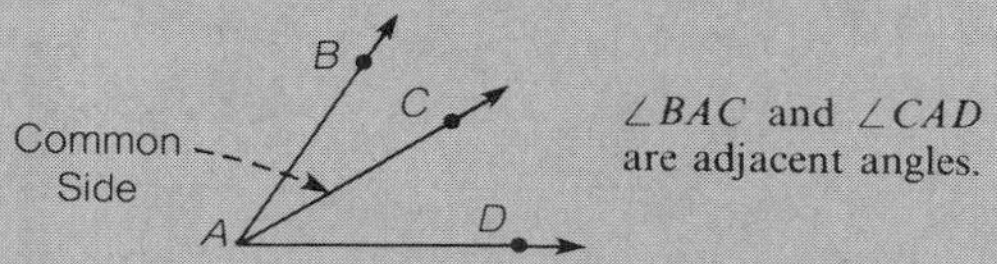

17.9
Angle measurement is additive, that is, $m(\angle BAC) + m(\angle CAD) = m(\angle BAD)$

17.10
Two angles are defined as *complementary* if the sum of their measures equals 90°.

17.11
Two angles are defined as *supplementary* if the sum of their measures equals 180°.

Lines in a plane may or may not intersect. Whenever they do intersect, angles are formed.

17.12
If the angle formed when two lines intersect is equal to 90°, the lines are *perpendicular.*

17.13
Two lines that do not intersect within a plane are *parallel.*

17.14
Two lines that do not intersect within different planes are *skew lines.*

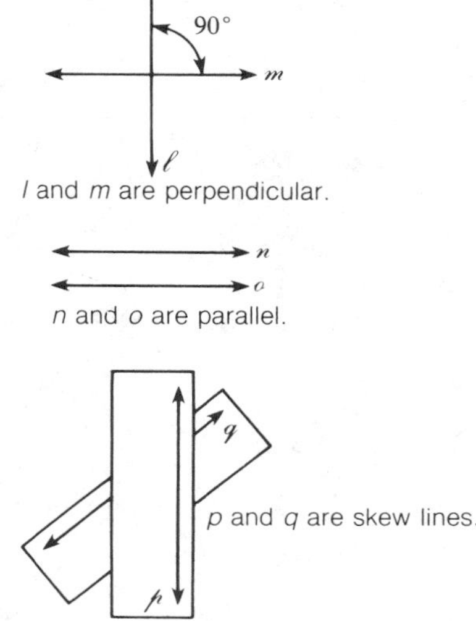

Polygons

In Chapter 15, definitions of polygons simply considered the number of sides. When a child understands how line segments and angles are measured, we can define polygons by referring to the measures of sides and angles.

Types of Triangles

Triangles may be classified into types according to their angle measures and the measures of their sides.

Objectives 17.16 through 17.24 identify the major categories of triangles usually covered

17.15
The sum of the measures of the three interior angles of a triangle is 180°.

17.16
A triangle having a 90° angle (a right angle) is called a *right triangle.*

17.17
A triangle having an obtuse angle is an *obtuse triangle.*

17.18
A triangle in which all three angles are acute is an *acute triangle.*

17.19
A triangle having all three angles equal is an *equiangular triangle.*

17.20
A triangle having all three sides equal is an *equilateral triangle.*

17.21
A triangle having at least two sides congruent is an *isosceles triangle.*

17.22
A triangle in which all sides differ in length is a *scalene triangle.*

17.23
Two triangles are congruent if corresponding sides and angles are congruent. For example, $\triangle ABC$ is congruent to $\triangle DEF$ if $\angle A \cong \angle D$, $\angle B \cong \angle E$, $\angle C \cong \angle F$ and $\overline{AB} \cong \overline{DE}$, $\overline{AC} \cong \overline{DF}$, $\overline{BC} \cong \overline{EF}$.

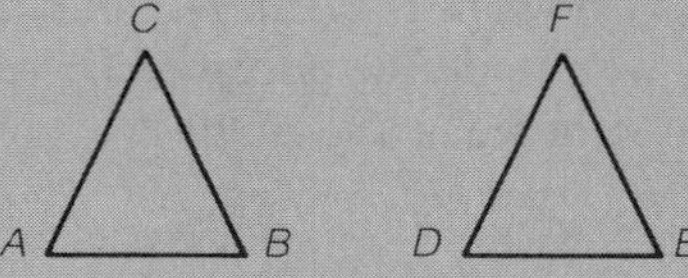

17.24
Two triangles are *similar* if corresponding angles are congruent; sides may or may not be congruent.

in elementary schools. Knowledge of measures of line segments and angles is needed for this classification to be meaningful. Examples of the categories follow. (See top, next column.)

A triangle may fit into more than one category. A similar classification system can be used with quadrilaterals and other polygons.

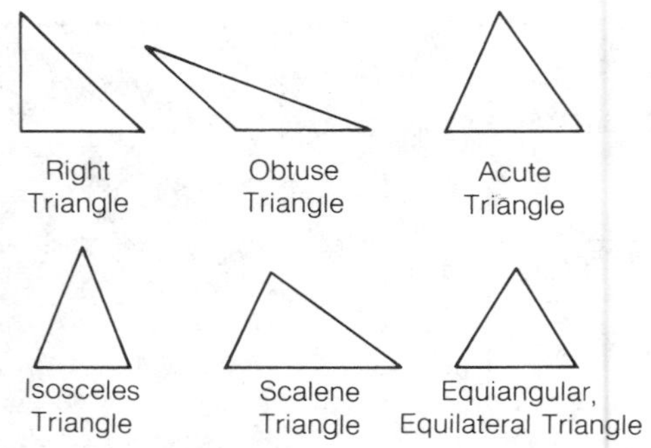

Types of Quadrilaterals

A quadrilateral is a polygon with four sides. Many different quadrilaterals are studied in elementary schools.

Some examples of the quadrilaterals named below are given here; some fit into more than one category.

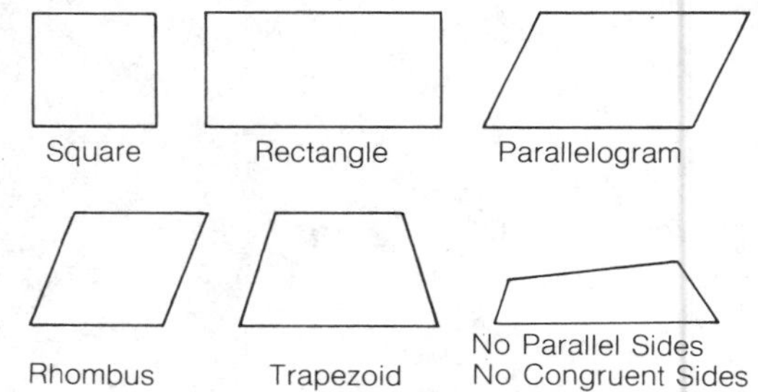

17.25
A quadrilateral having four right angles *and* four congruent sides is a *square.*

17.26
A quadrilateral having four right angles and opposite sides congruent is a *rectangle.*

17.27
A quadrilateral having opposite sides congruent and parallel is a *parallelogram.*

17.28
A quadrilateral with all sides congruent is a *rhombus.*

17.29
A quadrilateral having two sides parallel is a *trapezoid.*

17.30
A quadrilateral may have no parallel or congruent sides.

Other Polygons

Other polygons can be classified in a similar manner, but this is largely beyond the scope of the elementary school program. The idea of a regular polygon should be emphasized, however.

17.31
A polygon having all sides and angles congruent is a *regular polygon* (regular pentagon, regular hexagon, etc.).

Perimeter

The perimeter of a closed plane figure is the distance around it. As a student studies this concept using a ruler with models for the figure, several basic patterns are revealed.

17.32
The perimeter of a square can be found by using the formula $P = 4s$, where s is the length of a side.

17.33
The perimeter of a rectangle can be found using the formula $P = 2l + 2w$, where l is the length and w is the width.

17.34
Generally, the perimeter of a polygonal figure is found by using the formula $P = s_1 + s_2 + \ldots + s_n$, where s is the length of a side and n is the number of sides.

Area

Area refers to the amount of surface inside a closed curve. Many important relationships can be learned as children determine the areas of commonly studied plane figures.

17.35
The area of a square or a rectangle is found by using the formula $A = lw$, where l is the length and w is the width.

17.36
The area of a rhombus or a parallelogram is found by using the formula $A = bh$, where b is the base and h is the height or altitude.

17.37
The area of a triangle is found by using the formula, $A = \frac{1}{2}bh$, where b is the length of the base and h is the height.

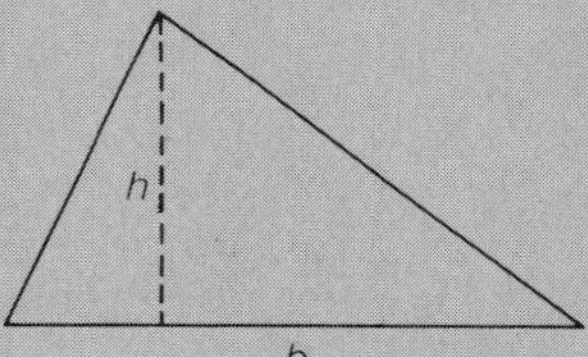

17.38
The area of a trapezoid is found by using the formula $A = \frac{1}{2}h(b_1 + b_2)$, where h is the height and b_1 and b_2 are the bases, respectively.

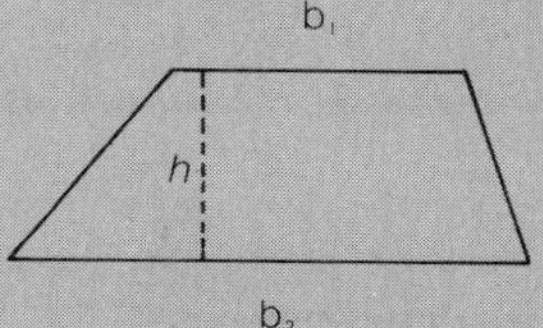

17.39
The area of a circle is found by using the formula $A = \pi r^2$, where π is a constant (about 3.14) and r is the radius of the circle.

Some elementary programs include surface areas of three-dimensional or space figures. Surface area is defined as the total area of all surfaces in a space figure. Formulas for computation of surface area are not included here, but can be found in most junior high school mathematics textbooks.

Volume

Intuitively, the volume of a space figure is the amount of space contained within the figure. We often think of taking a space figure, such as a rectangular prism, and filling it with units, such as cubic centimeters.

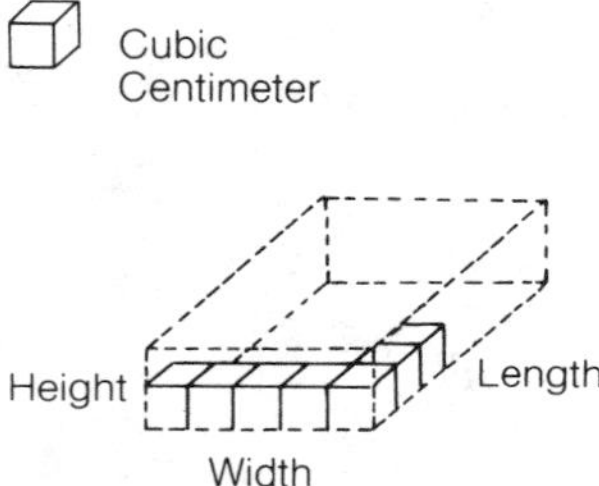

17.40
The volume of a rectangular prism is found by using the formula, $V = lwh$, where l is length, w is the width and h is the height. This definition can be shortened to $V = Bh$, where B is the area of a base and h is the height.

PLANNING INSTRUCTION

Provide experiences in geometry that are distinct from the study of operations on numbers. Each child should notice patterns and identify logical relationships. Many children develop and use mathematical formulas for the first time as they describe relationships that define perimeter, area, and volume. The study of geometry also provides opportunities to apply and extend measurement concepts and previously learned skills.

Basic Measurement of Line Segments and Angles

After children have many informal experiences with length measurement, introduce the ruler. Have children make their own rulers and mark them in centimeters or in inches, or even with a nonstandard unit. A standard unit should be used when we begin to discuss two segments having the same measure, thereby pointing toward congruence. In the following drawing, line segments AB and CD both have equal measures of 8 cm, and are defined as being congruent. Children should measure many line segments to determine whether one is longer or shorter than the other.

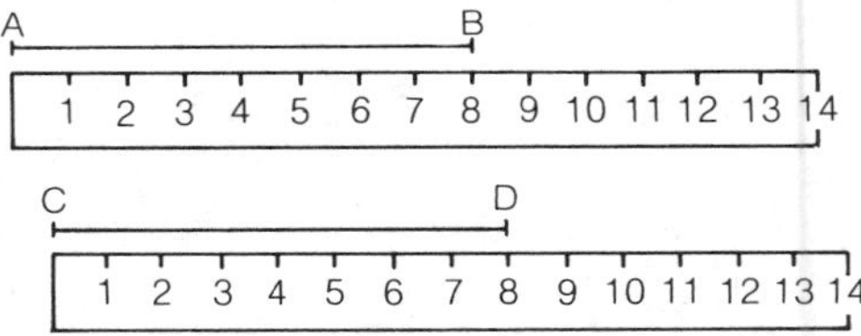

The concept of an angle and its measure can be developed intuitively. Begin by focusing on the concept of rotation. Then, fasten two strips of cardboard together with a paper fastener as shown. Start with both strips in position (a), and rotate strip n gradually to positions (b), (c), and (d), focusing discussion on the amount of spread—the distance moved from the starting position.

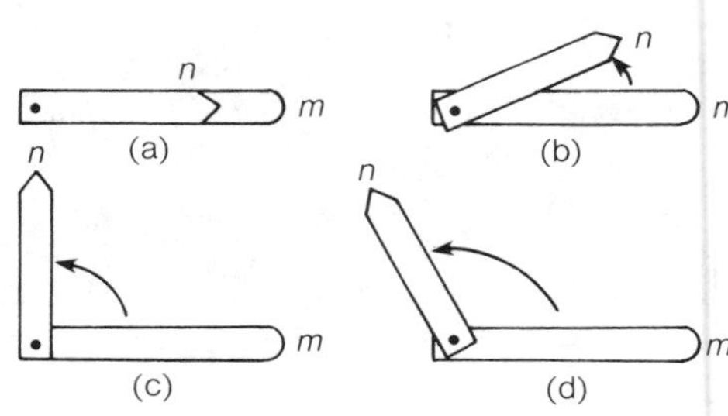

You may also introduce measurement of an angle by using a nonstandard unit. As illustrated, place copies of a pie-shaped wedge between the strips to see how many wedges it contains. Later, compare the wedge to one degree.

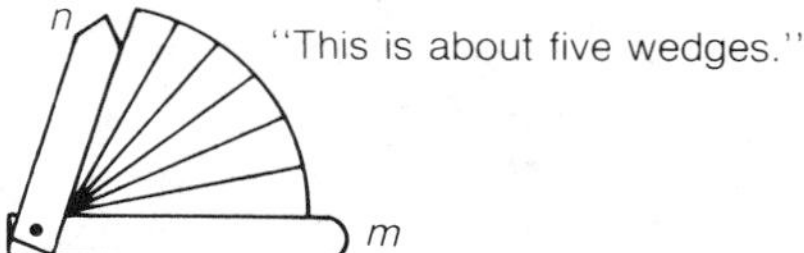

To introduce a *protractor,* an instrument used to measure an angle, prepare a scale in the shape of a semicircle on the chalkboard. Place cardboard strips as illustrated, and show how angles are measured in degrees. Compare angles as to size, focusing on angles of 30°, 45° and 90°.

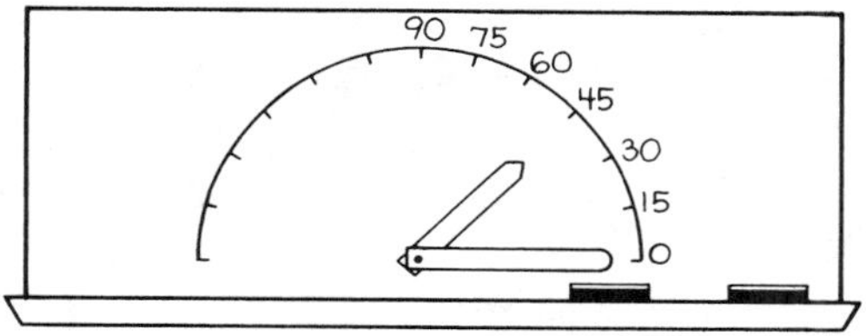

After initial work with the concepts of degree and angle measure, you can teach children how to use a standard protractor. Stress that the vertex of the angle should be lined up with the designated point on the protractor, and that one side of the angle should be lined up with the zero degree mark.

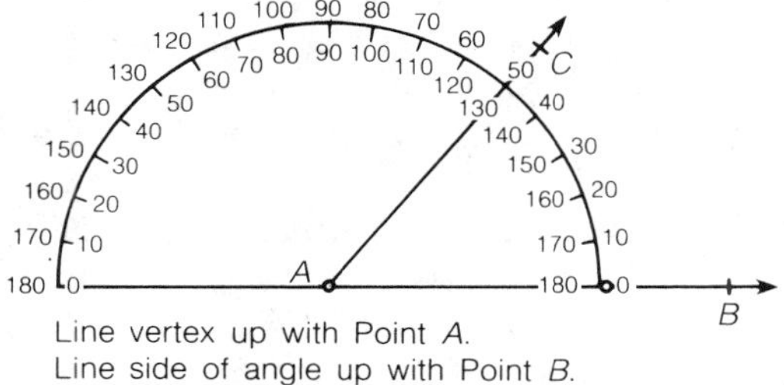

Line vertex up with Point *A*.
Line side of angle up with Point *B*.

The measure of ∠CAB is about 50°. Notice that this protractor has both an inner and an outer scale, making it easy to measure angles larger than 90°. It can even be used to help illustrate concepts such as, the measure of a *straight angle* is 180°, and, 180° defines a straight line. If the three vertices of a paper triangular region are cut off and arranged as follows, the total degree measure is obviously 180°, again showing that a 180° angle is essentially a straight line.

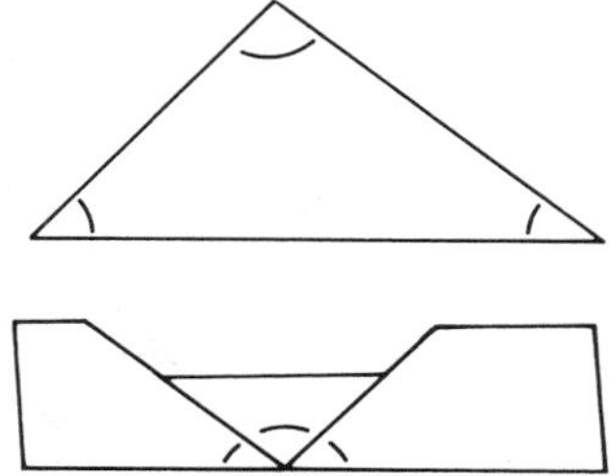

Once children are able to use a protractor, have them measure angles and classify them according to size. (A right angle is 90°, an acute angle is less than 90°, and an obtuse angle is greater than 90° but less than 180°.) The concept of perpendicular lines can be developed as children find intersections that form 90° angles. Many examples of perpendicular can be cited in your classroom: lines that intersect along the walls, at the corners, and at the ceiling. Have children find other examples in their environment.

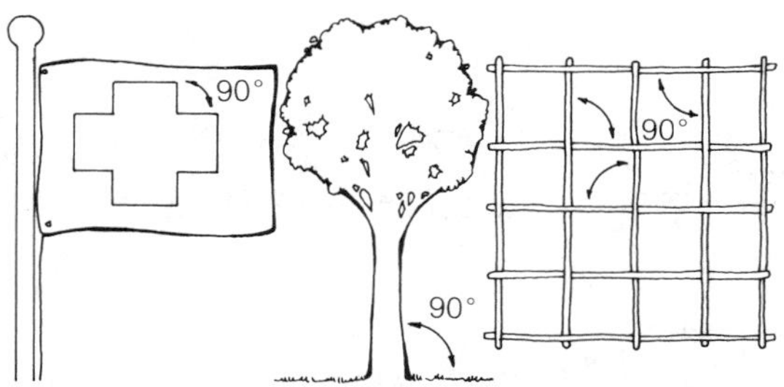

Types of Polygons

Objectives 17.15 through 17.30 give criteria for specific plane figures; concepts of linear measure or angle measure are involved in each definition. These definitions can be taught and plane figures classified as soon as children understand and can measure length and angles. Highly contrasted figures should be presented at first; but as children gain competence, attempt to help each child understand the general classes and their relation-

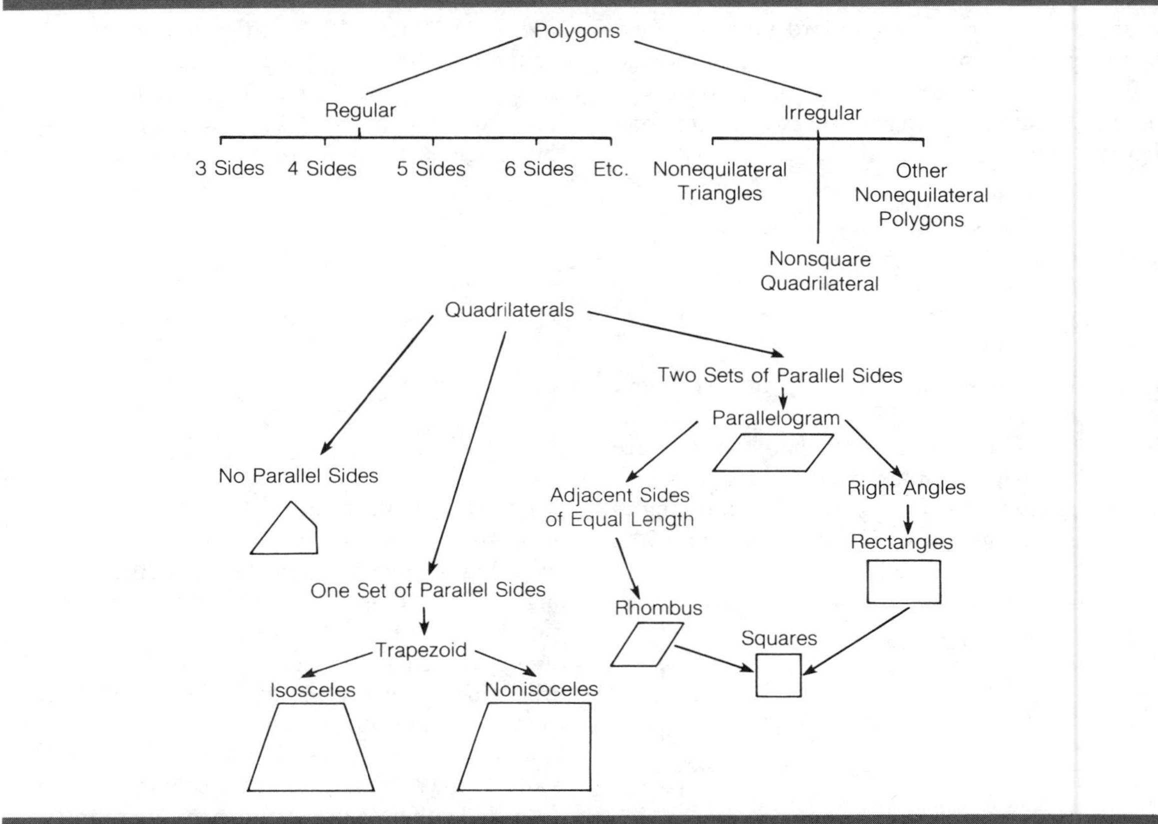

FIGURE 17.1
Classification of Polygonal Figures

ship to each other. Figure 17.1 gives a useful schematic of relationships among polygons.[1]

Perimeter

One immediate application of linear measurement is the study of distances around plane figures. Make sure children understand what perimeter is before attempting to develop a formula.

Begin with situations in the environment. For instance, have children determine the distance around a desk top, the classroom, or a playground. Discuss the fact that each total is found by adding the measures of all sides. Because plane figures can have many sides, it is difficult to establish a general formula. Yet children can develop formulas for some polygons, such as squares and rectangles.

Give children plane figures and ask them to find the perimeter. At first, provide measurements, then ask each child to measure. Make sure that the initial work with linear measure involves figures drawn to a scale. Teachers sometimes carelessly present examples that do not conform to any scale. Measures and visual inspection of how long should be in agreement.

Area of a Polygon

Formulas for areas of squares and rectangles can be developed with geoboards. Figure 17.2 (pages 401–2) describes one plan for developing the formula for area of a rectangle.

1. Robert Underhill, *Teaching Elementary School Mathematics,* 3rd ed. (Columbus, Ohio: Charles E. Merrill, 1981), pp. 356–57.

ACTIVITY PLAN

Geometry: Area of a Rectangle — *3-D → Symbolic* — *Abstracting/Schematizing*

Content objective

17.35

Materials/Exemplars

Geoboards
Rubber bands
Paper and pencil
Overhead projector
Transparent 5 × 5 geoboard

Behavioral indicator

Given the length and width of a rectangular region, the child will compute its area by using the formula, Area = length × width.

Procedure

1. Give each child a 5 × 5 geoboard and a supply of rubber bands. Use a transparent 5 × 5 geoboard on the overhead projector to demonstrate how to use rubber bands to form the required figures.
2. Show a square unit.

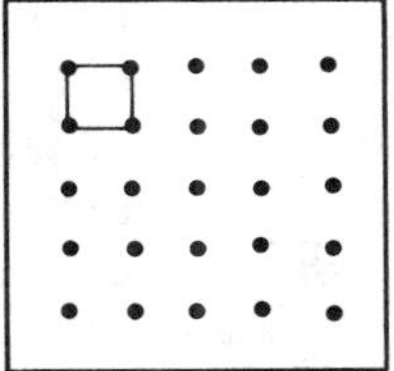

Ask children to make a similar unit on their geoboards. Discuss the length of the figure (1 unit), the width of the figure (1 unit) and the total area (1 square unit).

3. Draw attention to the areas of various shapes and make a chart similar to the following.

	Length	*Width*	*Area*
a.	2 units	1 unit	2 square units
b.	3 units	1 unit	3 square units
c.	3 units	2 units	6 square units
d.	4 units	2 units	8 square units

(a) (b)

(c) (d)

4. Continue the previous activity as long as necessary. Then focus attention on the entries in the table. Ask, "How are the numbers related? Can you see a way to find the area using the length and the width of the figure?" Arrive at the generalization that the number of square units in the area can be found by multiplying the number of units in the length by the number of units in the width.
5. Present problems that allow children to apply what they have discovered. For example:
   a. How many floor tiles measuring 1 foot square are needed to cover the floor of a room 15 feet long and 12 feet wide?
   b. A playing field is 100 meters long and 50 meters wide. What is the area of the field?

FIGURE 17.2
Abstracting/Schematizing Activity for Area of a Rectangle

Centimeter graph paper can also be used. Have children color in rectangles of various dimensions. Then focus on the length and width, and relate them to total number of units as determined by counting. Children will observe that the total number of units of area is the same as the product of measures for length and width. The more general terms *base* and *height* are used in the formula: Area = base × height.

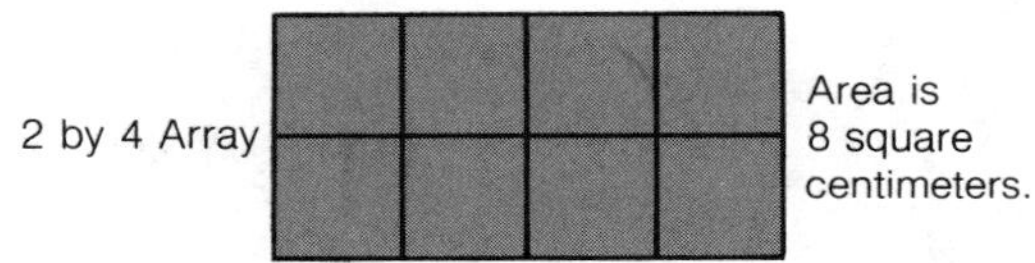

Once children understand this formula, they can discover area formulas for other quadrilaterals, triangles, and circles. The sequence described in Figure 17.3 illustrates how to move on to area of a parallelogram using two-dimensional exemplars. You may also want to actually use paper cut outs, a 3-D representation.

Once children know how to find the area of a parallelogram, they are ready to study the area of a triangle. Begin by focusing on figures formed whenever a diagonal is drawn in a rectangle or in other parallelograms.

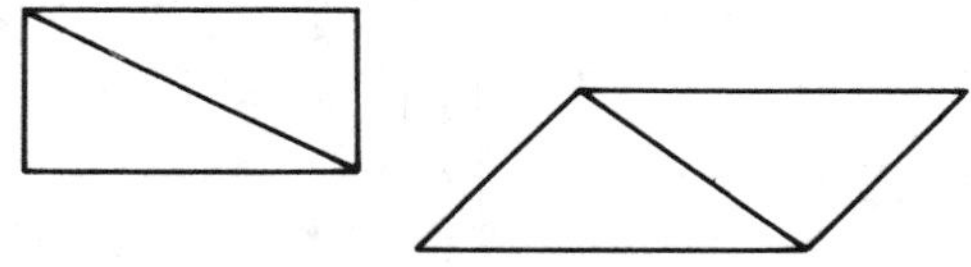

What happens when diagonals are drawn? Do children see that two triangles are formed? If so, what do they believe the area of each triangle to be? Try to reason it out. Since the area of the entire parallelogram is found by the formula Area = base × height, the area of each triangle should be one-half the area of the parallelogram. Thus, the area of each triangle is found by the formula: $A = \frac{1}{2}bh$. Try this formula with a few examples and verify the results by using a copy of the triangle and the original triangle to make a parallelogram.

Areas of other polygonal figures can be found by partitioning a figure into triangles, rectangles or parallelograms, then finding the area of each and computing the sum. To find the area of this figure, for example, mark it off as indicated, find the area of *A, B,* and *C*, and add to get the total area.

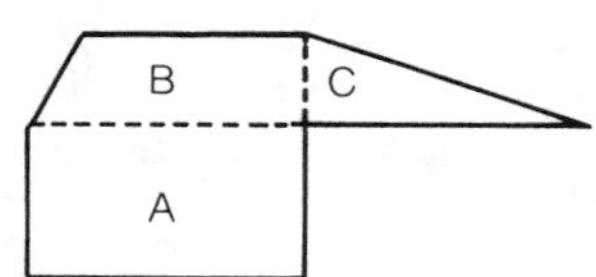

Area of a Circle

Many students can state the formula for area of a circle ($A = \pi r^2$) but do not really understand it. Develop the formula carefully, beginning with a review of basic circle terminology, including circumference, radius, and diameter. To develop the notion of pi (π), have children look for relationships between the circumference and the diameter of various sized circles. In each case, the circumference is a little more than 3 times the diameter. This constant relationship, named pi

Area of a Parallelogram

Draw a 2 × 3 rectangle and review how the area is determined. Recall the formula: Area = length × width. So, 3 × 2 = 6. The area is 6 square units.

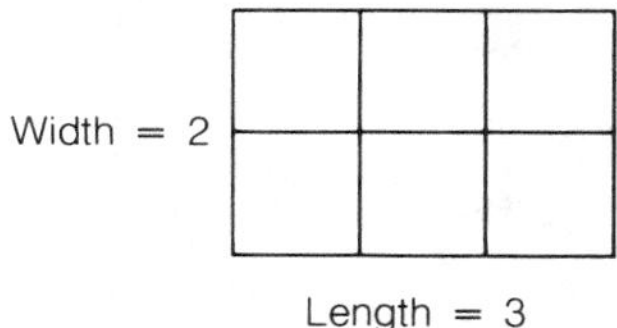

Next, consider the figure drawn on centimeter grid paper.

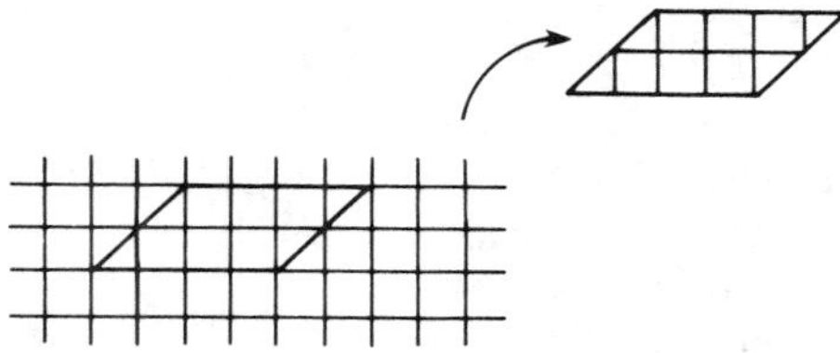

Ask, "How could we find the area of this new figure, a parallelogram?" Shade part of the figure as shown, and ask children to think of moving the shaded piece to the other end of the figure.

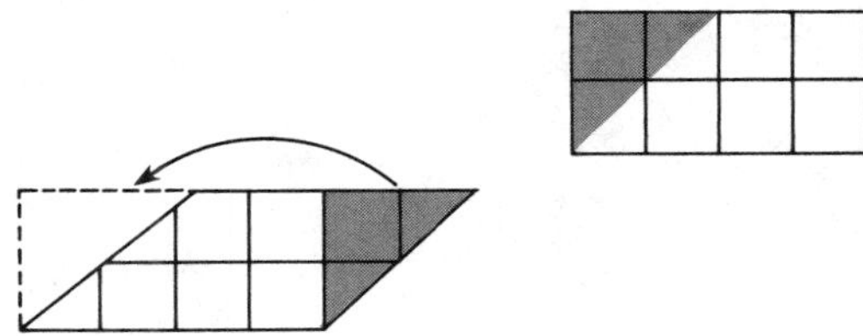

Observe, "Now we have a rectangle, and we already know how to find the area of a rectangle."

Focus on the formula A = lw, and introduce terminology commonly used when studying parallelograms, *base* and *height*. Relate length to base, and width to height.

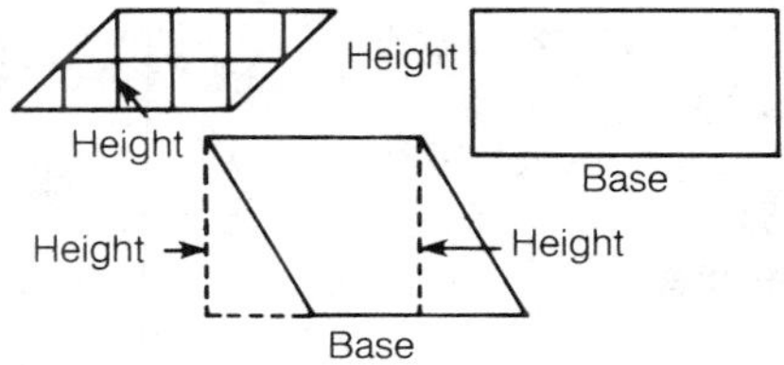

Thus, the formula for area of a parallelogram is: Area = base × height.

FIGURE 17.3
Sequence for Transition from Area of a Rectangle to Area of a Parallelogram

(pronounced "pie"), is represented by the Greek symbol π, and is given a numerical value of approximately $\frac{22}{7}$ (about 3.14). Thus, $C = \pi d$.

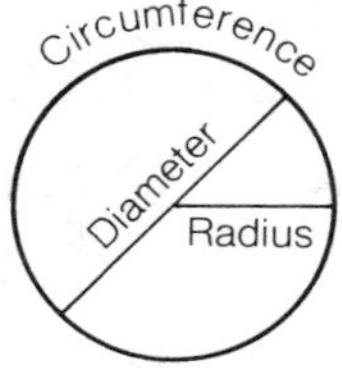

This formula must be understood before going on to the formula for the area of a circle.

Figure 17.4 describes an approach for teaching the area of a circle that involves successive approximations to a rectangle.

Volume

The study of volume always relates to how much something will hold. In the primary grades, capacity units such as cupfuls, pints, and liters are used to answer this question. The formal study of volume begins by extending two-dimensional space, in which area was considered, to three-dimensional space. Exemplifications abound, such as boxes, blocks, buildings and the classroom. Introduce the words *length, width,* and *height* in examining these exemplars.

Then give each child a set of centimeter cubes. Have children note the length, width, and height of each cube.

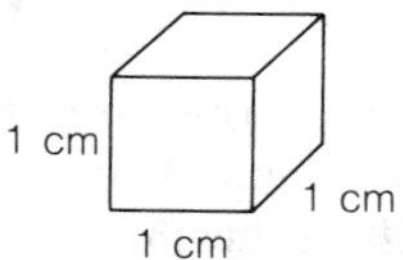

With the 1 cm cube as the unit, have children fill containers, such as small, shallow boxes, with cubes and record the count. The children can conclude, for instance, that a container has a volume of about 45 cubic centimeters.

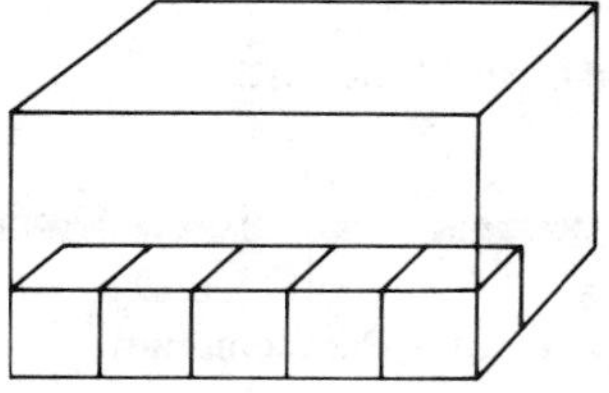

For a formal representation of this idea, use a sheet of paper marked in square centimeters as shown in (a). Cut the paper so that it looks like (b). Then, fold so that the paper looks like a box without a lid as in (c).

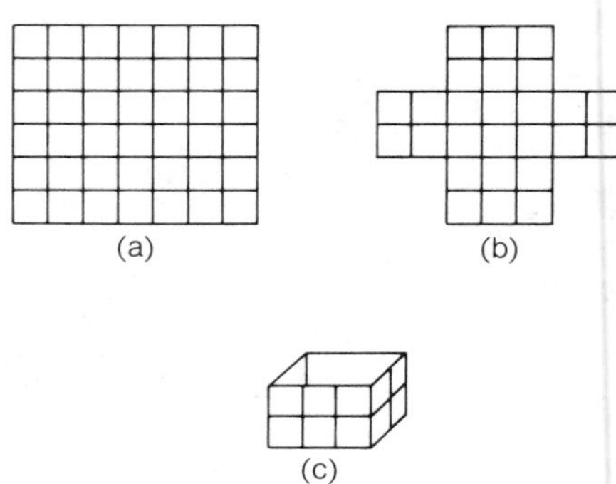

Record the length, width and height of this box.

Length	*Width*	*Height*
3 cm.	2 cm.	2 cm.

Have a child fill the box with centimeter cubes; exactly 12 will fit. Then seek to relate the dimensions of the box to its volume. At first focus on the area of the floor (2 cm $\times$ 3 cm = 6 sq. cm), then on the height (or depth) as a third dimension. Layers of cubes make it clear that the measure of the height can be multiplied times the area of the floor or base. Therefore, Volume = (area of base) $\times$ height, or $V = lwh$.

Have children test the formula to see if it can be used to determine the volume of a rectangular solid such as a block of wood. Eventually, incorporate purely symbolic situations that apply the concept of volume.

PLANNING ASSESSMENT

We need to be aware of developmental, physical, and mathematical factors that may inhibit learning of the content. The ideas that follow can help us plan whatever diagnosis is needed.

1. When using materials that involve motor coordination (such as forming shapes on a geoboard or using a protractor or ruler) note which children experience difficulty. Different exemplars or a protractor with larger numerals may eliminate problems.

Area of a Circle

Suppose we cut a circular region as shown in (a) and pull it apart as in (b) and (c).

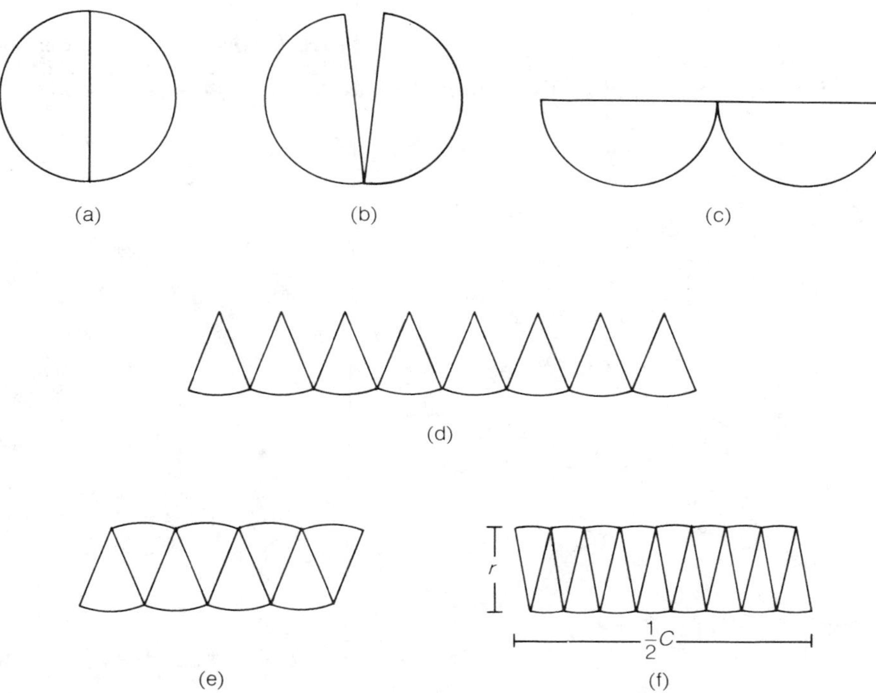

Take each half and cut it into equal sized pieces as shown in (d), then fit the pieces together as in (e). Say, "Look for the radius of the original circle. Also, see if you can find the original circumference."

Observe that, as the circular region is cut into smaller and smaller pieces of equal size, we have a figure that begins to look like a rectangle. The base is one-half of the original circumference, and the height is the radius of the original circle. See (f).

By using what we already know about the area of a rectangle, we can organize this information as follows:

Area of a circle	$= (\frac{1}{2}C)r$	Remember: lw
	$= (\frac{1}{2}\pi d)r$	Remember: $C = \pi d$
	$= (\frac{1}{2}\pi 2r)r$	Remember: $d = 2r$
	$= \frac{1}{2}(2\pi r)r$	
	$= (\pi r)r$	
	$= \pi r^2$	
Area of a circle	$= \pi r^2$	

Figure 17.4
Sequence for Developing Area of a Circle

2. Often failure to perform well in geometry is a reflection of weaknesses in arithmetic or an inadequate sense of how number ideas can be related to mathematical expression. Formal and informal assessment can help us gain insight into the causes of poor performance.

3. Much of formal Euclidean geometry involves visualization of plane and spatial figures. Some students find it extremely difficult to conceptualize three-dimensional figures, especially if they are pictured in two dimensions. Formal assessment instruments can help identify this type of learning difficulty.[2]

---

2. See Hilda Lewis, "Spatial Representation in Drawing As a Correlate of Development and a Basis for Picture Preferences," *Journal of Genetic Psychology* 102 (1963): 95–107.

Teaching Other Topics

V

18

Early Experiences with Integers

Some elementary school textbooks introduce integers as early as the fourth grade. The study of integers typically involves discovery-type experiences designed to help children understand in an intuitive way. A more formal approach is attempted in the eighth or ninth grade. A child's exposure to situations requiring the use of integers is limited, but a few situations do occur: measuring temperature, reading altitude with reference to sea level, reading about yardage gained or lost by football teams, and keeping score for games in which gains and losses are recorded.[1]

PLANNING THE CONTENT

Some elementary teachers see the study of integers only as a unit to be used for enrichment purposes, or for gifted children. Others feel that integers have a rightful place in the regular mathematics program. There are several reasons for teaching integers in elementary schools:

1. The fundamental ideas associated with integers can be understood by most children.
2. The study of integers provides children with opportunities for discovery and problem solving.
3. Integers offer an opportunity for children to practice computation with whole numbers and rational numbers in an interesting manner.
4. Introduction of integers can provide a good foundation for a formal study at the secondary level. Teaching a topic over a period of years is more effective than teaching it all at once during a short time.

1. Leonard M. Kennedy, *Guiding Children to Mathematical Discovery*, 3rd ed. (Belmont, Calif.: Wadsworth, 1980), p. 140.

Although textbooks and curriculum guides are excellent instructional aids, we make the final decision about what to teach each child. Selections are often based on individual needs; not all the following content objectives need to be used. Think of instructional activities to help each child understand the ideas. Also think of behaviors that might indicate a child understands.

Number and Notation

The following focuses on the meaning, notation and models associated with integers.

18.1
The set of integers contains the set of natural numbers, their opposites and zero.

18.2
Integers may be associated with equidistant points on a line.

18.3
Numbers to the right of zero are called positive numbers or positive integers. Numbers to the left of zero are called negative numbers or negative integers.

18.4
A + (read *positive*) placed before a numeral denotes a positive number, and − (read *negative*) placed before a numeral denotes a negative number.

18.5
An integer is greater than a given integer if its position on the number line is to the right of the given integer.

The set of integers is useful for many applications; for example, when we subtract cer-

EXCURSION
Wind Chill Temperature

Temperature is a measure of hot and cold. How cold we *feel* is determined by the wind as well as the temperature. This chart shows the chilling effect of wind.

Wind Chill Temperature in °F

	Actual Thermometer Reading in °F								
Wind Speed	*50*	*40*	*30*	*20*	*10*	*0*	*−10*	*−20*	*−30*
Calm	50	40	30	20	10	0	−10	−20	−30
5 MPH	48	37	27	16	6	−5	−15	−26	−36
10 MPH	40	28	16	4	−9	−21	−33	−46	−58
15 MPH	36	22	9	−5	−18	−36	−45	−58	−72
20 MPH	32	18	4	−10	−25	−39	−53	−67	−82
25 MPH	30	16	0	−15	−29	−44	−59	−74	−88
30 MPH	28	13	−2	−18	−33	−48	−63	−79	−94
35 MPH	27	11	−4	−20	−35	−49	−67	−82	−98
40 MPH	26	10	−6	−21	−37	−53	−69	−85	−100

(Wind speeds greater than 40 mph have little additional effect.)

- What is the wind chill temperature when the actual temperature is 30°F and the wind speed is 20 mph?
  (Locate 20 mph on the scale on the left of the chart. Look across until you are under 30°F. The wind chill temperature is 4°F.)
- What is the wind chill temperature if the actual temperature is 10°F and the wind speed is 20 mph?

tain whole numbers such as $4-6$, we find that the answer is not a whole number; but, it *is* an integer. The set of integers can also be used for expressing direction. Directional numbers are needed to express the positions of above and below, to the right of and left of, and gains and losses.

The set of integers may be thought of as the union of three disjoint sets:

The positive integers: $\{^{+}1, ^{+}2, ^{+}3, ^{+}4 \ldots\}$
The negative integers: $\{\ldots, ^{-}4, ^{-}3, ^{-}2, ^{-}1\}$
Zero: $\{0\}$

It is symbolized as:

$$\{\ldots, ^{-}4, ^{-}3, ^{-}2, ^{-}1, 0, ^{+}1, ^{+}2, ^{+}3, ^{+}4, \ldots\}$$

When integers are associated with equidistant points on a number line, it becomes obvious that each integer (except zero) has an opposite.

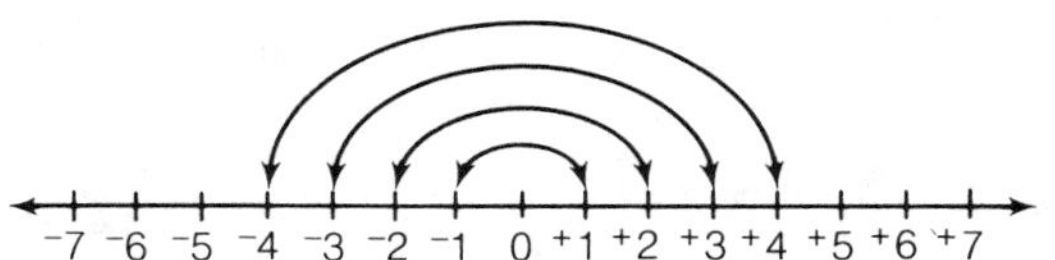

Zero is often misunderstood to mean "nothing." However, it corresponds to a definite point on the number line like each positive and negative number. Zero is neither positive nor negative, but it is an integer. Numbers other than zero without a positive or negative symbol are considered positive.

For every point on the line (other than zero) for which a number has been assigned, there is a number assigned to a second point on the line the same distance from zero but in the opposite direction. This other number can be thought of as the opposite of the given number

EXCURSION
Absolute Value

When a number and its opposite are compared, the greater in value is called the *absolute value* of either number. The absolute value of a number is denoted by placing the number between a pair of vertical bars | |.

The absolute value of any number and its opposite are equal. For example, $|^-5| = |^+5| = {}^+5$, which is read, "The absolute value of negative 5 equals the absolute value of positive 5, equals positive 5." Similarly, on a number line, the distances from a given number to zero and from its opposite to zero are the same. The absolute value of zero is zero.

and is called its *additive inverse*. A number plus its additive inverse is always zero. For example, the opposite of $^+7$ is $^-7$, and the sum of $^+7$ and $^-7$ is 0. The opposite of $^-3$ is $^+3$, and the sum of $^-3$ and $^+3$ is 0.

Positive integers are often assumed to be the same as natural numbers, that is, we write positive 7 as either 7 or $^+7$. The signs for positive and negative are raised when writing integers to distinguish them from the operational signs for addition and subtraction.

The number line provides an excellent way to visualize relationships among the integers. On the whole number line, the number to the right of another number is always greater. This same order relationship holds for the set of integers. Thus:

$^+6 > {}^+5 \quad {}^+2 > 0 \quad 0 > {}^-1 \quad {}^-3 > {}^-4 \quad {}^+2 > {}^-6$

Operations

Children need to carefully consider what happens when we perform operations with integers.

Addition and Subtraction

Important content objectives related to addition of integers include:

18.6
Addition of integers is a binary operation that associates each pair of integers with a unique integer.

18.7
To add two positive integers, add the absolute values. The sum is positive.

18.8
To add two negative integers, add the absolute values. The sum is negative.

18.9
To add a positive and a negative integer, subtract to find the difference between their absolute values. The sign for the sum of the integers is the same as the sign of the integer with the greater absolute value.

When two nonzero integers are added, five possible conditions can occur:

1. Addition of two positive numbers. The sum is positive.

$$(^+2) + (^+3) = {}^+5$$

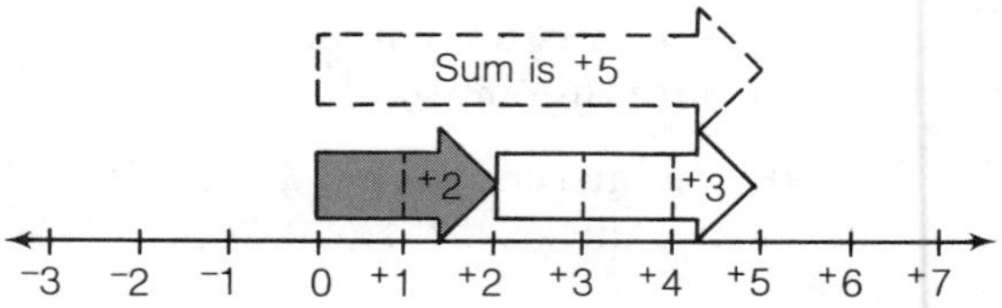

2. Addition of two negative numbers. The sum is negative.

$$(^-3) + (^-4) = (^-7)$$

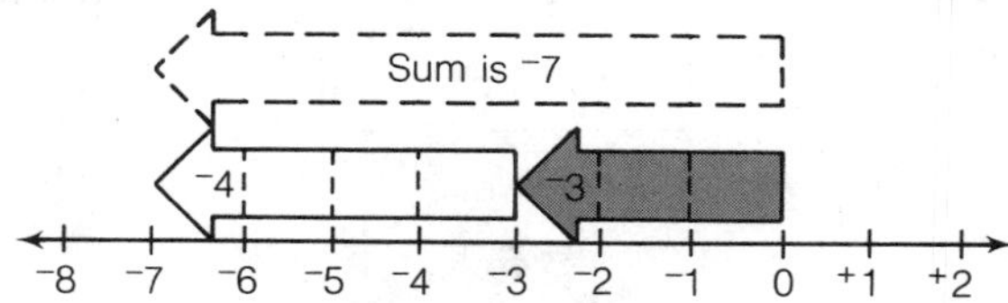

3. Addition of a positive and negative number. The sum is positive.

$$(^{-}4) + (^{+}6) = {}^{+}2$$

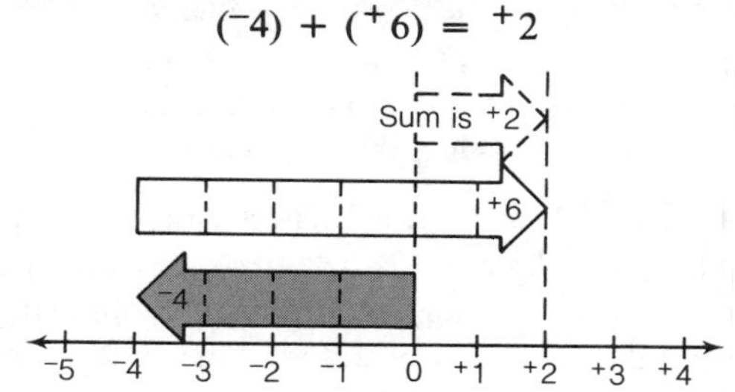

4. Addition of a positive and negative number. The sum is negative.

$$(^{-}7) + (^{+}4) = {}^{-}3$$

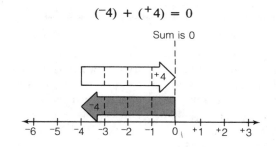

5. Addition of a number and its opposite. The sum is zero.

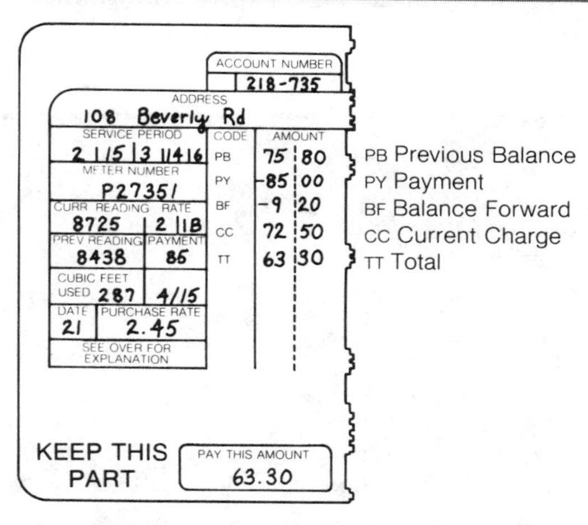
ACCOUNT NUMBER 218-735
ADDRESS 108 Beverly Rd
SERVICE PERIOD 2 1/5 3 1/4 16
METER NUMBER P27351
CURR READING 8725 RATE 2 11B
PREV READING 8438 PAYMENT 86
CUBIC FEET USED 287 4/15
DATE 21 PURCHASE RATE 2.45
SEE OVER FOR EXPLANATION

CODE	AMOUNT
PB	75 80
PY	-85 00
BF	-9 20
CC	72 50
TT	63 30

PB Previous Balance
PY Payment
BF Balance Forward
CC Current Charge
TT Total

KEEP THIS PART
PAY THIS AMOUNT 63.30

FIGURE 18.1
Credits and Debits on a Gas Bill

$$(^{-}4) + (^{+}4) = 0$$

Children have more difficulty with subtraction of integers than with addition, largely because it is unlike the usual take away concept of subtraction. The interpretation we want to

EXCURSION
Credits and Debits

Positive and negative numbers are often used to show the current status of an account. For example, on the gas bill in Figure 18.1, positive numbers show how much is owed. A negative number indicates that a payment has been made or that the account is overpaid. The total on the bill is the sum of the balance forward—the sum of the previous balance and the payment—and the current charges. If a check for $70.00 is sent in payment, what will the balance forward be on the next gas bill?

convey for subtraction is expressed in these content objectives:

18.10
Subtraction of integers is a binary operation that associates each pair of integers with a unique integer.

18.11
Subtraction is the inverse of addition.

18.12
Subtraction may be thought of as an operation used to find one addend when the sum and the other addend are known.

18.13
To subtract one integer from another, add the opposite of the integer to be subtracted (subtrahend or known addend) to the other integer (minuend or sum).

When 4 is to be subtracted from 1, a person cannot take away more than there is. Instead, think of subtraction as the process of finding a missing addend. Thus, we might rewrite the equation $(^+1) - (^+4) = \square$ as $(^+4) + \square = {}^+1$. While no number for $\square$ exists in the set of whole numbers, the integer $^-3$ *will* complete this number sentence; that is, $(^+4) + (^-3) = {}^+1$.

Subtraction is so closely related to the operation of addition that, with an additive inverse (or opposite) for every integer, a difference can be written in equivalent form as a sum. For example, $(^+6) - (^+2) = \square$ becomes $(^+2) + \square = {}^+6$. On the number line in Figure 18.2, the missing addend is represented by the distance from $^+2$ to $^+6$. Thus, the missing

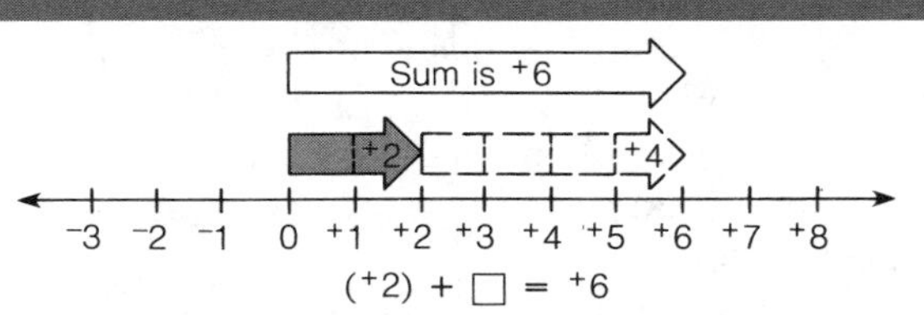

FIGURE 18.2
Number Line Showing Missing Addend

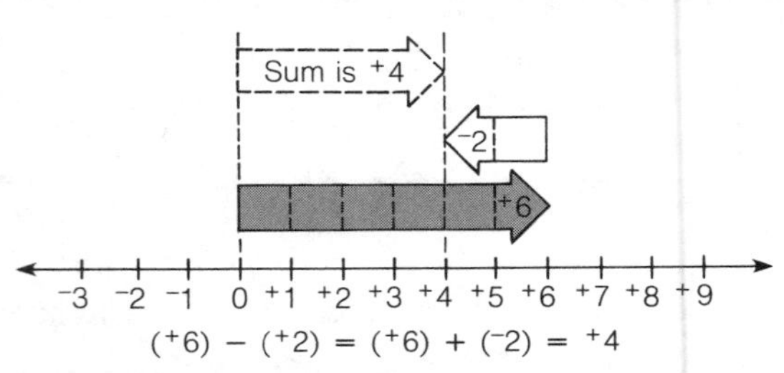

FIGURE 18.3
Number Line Showing Subtraction by Adding the Opposite

addend is $^+4$. The same answer can be obtained by applying objective 18.13; that is, subtract one integer from another by adding the opposite of the number to be subtracted. This is illustrated on the number line in Figure 18.3.

Subtraction can be accomplished similarly with other possible combinations of integers.

1. Subtracting a negative integer from a positive integer.

$$(^+2) - (^-5) = \square \longrightarrow (^-5) + \square = {}^+2$$

2. Subtracting a negative integer from a negative integer.

$$(^-3) - (^-5) = \square \longrightarrow (^-5) + \square = {}^-3$$

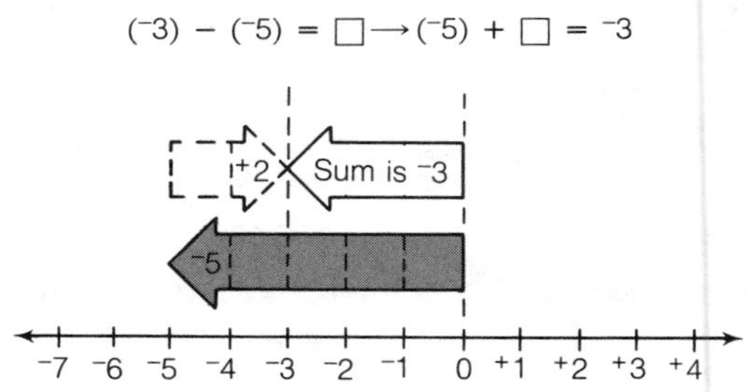

3. Subtracting a positive integer from a negative integer.

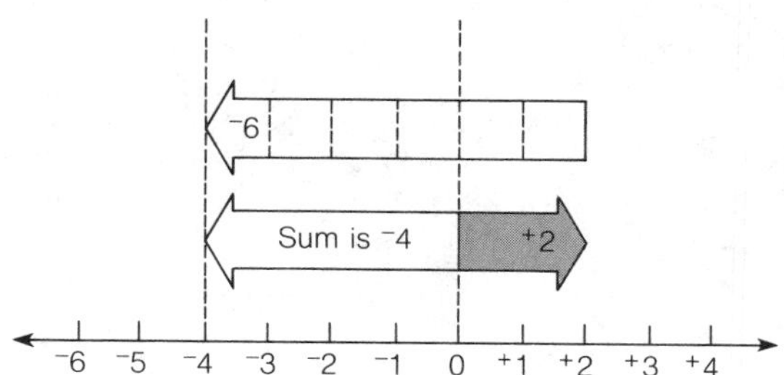

Generally, if a and b are integers, then $a - b = a + (^-b)$ for any subtraction situation. Other models also represent addition and subtraction with integers. However, the models pictured here are similar to those used for whole numbers, and build on what each child has already learned.

Multiplication and Division

Important content objectives associated with multiplication are as follows:

18.14
Multiplication of integers is a binary operation that associates each pair of integers with a third integer.

18.15
Multiplication is used to find the product when its factors are known.

18.16
To multiply two positive or two negative integers, multiply their absolute values. The product is positive.

18.17
To multiply a positive and a negative integer, multiply their absolute values. The product is negative.

Multiplication of integers requires abstract thought and therefore is not treated fully in some elementary programs.

If children already understand multiplication of whole numbers, they should have little difficulty with multiplication of positive integers. The multiplication of a positive integer times a negative integer, which causes more difficulty, can be pictured on a number line. Figure 18.4 shows clearly that the product of 3 (⁻4) is ⁻12.

Children will have little use for multiplying a negative integer times a positive integer, for example, (⁻4) (3). Practical situations that show the meaning are scarce.[2] A child can find the product by applying the commutative property if needed.

The multiplication of a negative integer times a negative integer cannot be shown on a number line. To present this concept, extend the pattern approach illustrated in the Excursion *Counting Backwards* by starting with the products of positive whole numbers that are known, and extending the pattern to other rows and columns (see Figure 18.5). The fact that two negative factors have a positive product soon becomes apparent.

Most elementary school children will not need division involving integers, but if it is taught, consider these content objectives:

18.18
Division of integers is a binary operation that associates a pair of integers with a third integer.

2. Esther J. Swenson, *Teaching Mathematics to Children,* 2nd ed. (New York: Macmillan, 1973), p. 422.

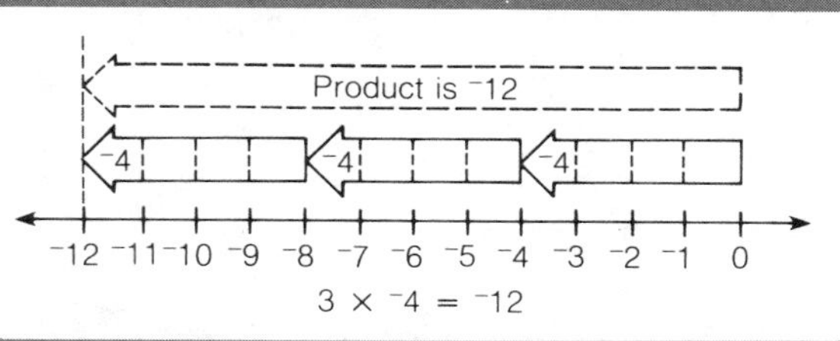

FIGURE 18.4
Number Line Showing a Positive Integer Times a Negative Integer

×	$^{+}4$	$^{+}3$	$^{+}2$	$^{+}1$	0	$^{-}1$	$^{-}2$	$^{-}3$	$^{-}4$
$^{+}4$	$^{+}16$	$^{+}12$	$^{+}8$	$^{+}4$	0	$^{-}4$	$^{-}8$	$^{-}12$	$^{-}16$
$^{+}3$	$^{+}12$	$^{+}9$	$^{+}6$	$^{+}3$	0	$^{-}3$	$^{-}6$	$^{-}9$	$^{-}12$
$^{+}2$	$^{+}8$	$^{+}6$	$^{+}4$	$^{+}2$	0	$^{-}2$	$^{-}4$	$^{-}6$	$^{-}8$
$^{+}1$	$^{+}4$	$^{+}3$	$^{+}2$	$^{+}1$	0	$^{-}1$	$^{-}2$	$^{-}3$	$^{-}4$
0	0	0	0	0	0	0	0	0	0
$^{-}1$	$^{-}4$	$^{-}3$	$^{-}2$	$^{-}1$	0	$^{+}1$	$^{+}2$	$^{+}3$	$^{+}4$
$^{-}2$	$^{-}8$	$^{-}6$	$^{-}4$	$^{-}2$	0	$^{+}2$	$^{+}4$	$^{+}6$	$^{+}8$
$^{-}3$	$^{-}12$	$^{-}9$	$^{-}6$	$^{-}3$	0	$^{+}3$	$^{+}6$	$^{+}9$	$^{+}12$
$^{-}4$	$^{-}16$	$^{-}12$	$^{-}8$	$^{-}4$	0	$^{+}4$	$^{+}8$	$^{+}12$	$^{+}16$

FIGURE 18.5
Patterns Extended to Complete a Table

18.19
Division is used to find a missing factor (quotient) when the other factor and the product are known.

18.20
To divide two positive or two negative integers, divide their absolute values. The quotient is positive.

18.21
To divide a negative integer by a positive integer, or to divide a positive integer by a negative integer, divide their absolute values. The quotient is negative.

As with subtraction and addition, division is defined in terms of multiplication. For example, $^{+}6 \div {}^{-}3 = \square$ can be viewed as $^{-}3 \times \square = {}^{+}6$. Thus, the quotient can be thought of as a missing factor just as the difference is considered to be a missing addend.

Division of integers and whole numbers parallel, except when determining if the quotient is positive or a negative. This variation is easily resolved by referring to previous knowledge of finding the products of two integers. For example, the division statement $^{-}12 \div {}^{+}4 = \square$ is

EXCURSION
Counting Backwards

Although negative integers are not counting numbers, we can use them to count backwards:

$$4, 3, 2, 1, 0, {}^{-}1, {}^{-}2, {}^{-}3, {}^{-}4, \text{etc.}$$

We can also count backwards by threes:

$$12, 9, 6, 3, 0, {}^{-}3, {}^{-}6, {}^{-}9, {}^{-}12$$

Put these two ideas together and observe the pattern associated with multiplication by an integer:

$$\begin{aligned} 12 &= 4 \times 3 \\ 9 &= 3 \times 3 \\ 6 &= 2 \times 3 \\ 3 &= 1 \times 3 \\ 0 &= 0 \times 3 \\ {}^{-}3 &= {}^{-}1 \times 3 \\ {}^{-}6 &= {}^{-}2 \times 3 \\ {}^{-}9 &= {}^{-}3 \times 3 \\ ? &= {}^{-}4 \times 3 \end{aligned}$$

This approach usually makes sense to children.

equivalent to the multiplication statement $^+4 \times \square = {}^-12$. Thus, the quotient is $^-3$, since $^+4 \times {}^-3 = {}^-12$. Children soon learn that the product (quotient) is negative when only one of the integers is negative.

Properties

Since the set of whole numbers is a subset of the set of integers, integers have the same properties of closure, commutativity, and associativity. Zero also serves as the identity element for addition. Furthermore, the additive inverse property, a new property, can be observed.

> **18.22**
> *Additive Inverse*—If a is an integer, then there exists a unique integer, ^-a, such that $a + {}^-a = 0$.

Each of the integers a and ^-a is called the opposite or additive inverse of the other. The additive inverse (opposite) of $^+3$ is $^-3$.

Because addition and subtraction are inverse operations, they can undo each other.

> **18.23**
> If a, b and c are integers, then $a - b = c$ if and only if $b + c = a$.

Interpretation of subtraction in terms of addition of the opposite results in the closure property for subtraction of integers.

> **18.24**
> *Closure for Subtraction*—If a and b are integers, then $a - b$ is an integer.

For example, if we subtract $6 - 8$, the remainder, $^-2$, is also an integer.

Earlier we listed the closure, commutative and associative properties of multiplication with whole numbers, noted that 1 is the multiplicative identity element, and pointed out that multiplication distributes over addition and subtraction for whole numbers. Multiplication in the set of integers is defined in such a way that all of these properties also hold for integers.

Just as subtraction can be defined in terms of addition, so may division be defined in terms of multiplication.

> **18.25**
> If a, b and c are integers, and $b \neq 0$, then $a \div b = c$ if and only if $b \times c = a$.

Ordering the Content

When introducing integers, begin with understanding and recording number concepts. For some children, this may be enough; for others, you may want to introduce addition and subtraction of integers. The latter two concepts should be taught together due to their close, inverse relationship. The operations of multiplication and division can be taught together at a later time.

PLANNING INSTRUCTION

Remember the cycle of instructional activities, what we know about each child, and what has already been learned when planning instruction. This way, we can build upon each child's successes.

Initiating, Abstracting and Schematizing Activities

Initial instruction should be similar in character to procedures used when introducing non-negative numbers. Use physical representations to depict numbers and their operations; then, as notation is introduced, make records of these experiences. Children can notice patterns and hypothesize rules, testing the rules in varied situations to see if they always hold.[3]

3. Robert B. Ashlock and Tommie A. West, "Physical Representations for Signed-Number Operations," *The Arithmetic Teacher* 14, no. 7 (November 1967): 549.

TABLE 18.1
Uses of Positive and Negative Numbers[4]

Situations	*Positive Interpretations*	*Negative Interpretations*
Altitude	Above sea level	Below sea level
Riding elevator	Riding upward	Riding downward
Temperature	Above zero	Below zero
Games	Points made	Points lost (penalized)
Time	Forward	Backward
Weights	Gaining weight	Losing weight
Walking	Forward	Backward
Money dealings	Received	Spent
	Gains	Losses
	In the black	In the red

Number and Notation

Special notations may make integers more meaningful, at least initially. In order to avoid confusing the operational signs minus and plus with those for negative and positive, designate the latter two by writing the sign that precedes the numeral in a raised position ($^-2$ and $^+2$). Later children can learn to write the symbols in conventional algebraic format.

Plan initiating activities that involve each child in situations in which positive and negative numbers are used as in Table 18.1.

A large number line can be taped on the floor. Give each child a card with an integer written on it, then ask the children to arrange themselves along the line as indicated by the positive and negative numerals.

A homemade thermometer can also be used to illustrate integers ordered in a line. Have children select numeral cards for specified temperatures and place them in appropriate positions along the thermometer.

More abstract activities can follow as children grasp concepts of order. With a number line for reference, paper-and-pencil activities such as the following can be useful.

Write the symbol > or < to make each statement true.

$^+3$ _____ $^+2$ $\quad$ $^-2$ _____ $^-3$

$^-3$ _____ $^+1$ $\quad$ 0 _____ $^-2$

Operations

By fourth or fifth grade, children usually understand what it means to owe 10¢, or to be in the hole in a game of cards. These concepts, while

4. Table adapted from Swenson, *Teaching Mathematics to Children,* 2nd ed. by Esther J. Swenson (New York, Macmillan, 1973), p. 417. Reprinted with permission of Macmillan Publishing Co., Inc. Copyright © 1973 by Esther J. Swenson.

EXCURSION
Tolerance

Machinists make tools and metal parts. Since all measurements are approximations, the size of each metal part can vary by a certain amount. This variance is called the *tolerance*.

Measurement: 6.72 ± 0.05 cm
Tolerance: ± 0.05 cm

The tolerance is read, "Plus or minus 0.05 centimeters." The measurement 6.72 ± 0.05 centimeters means that the part can be as long as $6.72 + .05$ centimeters and as short as $6.72 - .05$ centimeters. Thus, the longest acceptable length is 6.77 cm and the shortest acceptable length is 6.67 cm.

A blueprint gives a measurement of $4.172 \pm .005$ millimeters. What are the longest and shortest acceptable measurements?

rudimentary and undeveloped, form a basis for introducing operations with integers.

Skill in computing with integers is usually reserved for the study of algebra. What we want to do at the elementary level, however, is help children become familiar with integers for use in everyday situations, and provide a foundation for later study.

There are two strategies for teaching computation with integers: one uses the number line and measurement representations, and the other is based on counting.

Addition and Subtraction. Probably the most effective way to introduce addition of integers is with a number line drawn on the chalkboard or on paper, or shown on a flannel or a magnetic board. Cricket or frog hops may be used to illustrate the operations, making them seem like moves in a game. Children soon learn to interpret a positive integer as a move to the right and a negative integer as one to the left. Addition of integers can be represented as shown in Figure 18.6. When children compute, they can use a number line to verify their answers.

Encourage children to make up stories to illustrate the sentences shown on their number lines. As they work with examples, they soon discover the rules for adding integers.

The number line can also be helpful when introducing subtraction of integers. As with addition, begin with familiar situations. For example, when working with whole numbers, use problems such as $(^{+}6) - (^{+}2) = {}^{+}4$, or $6 - 2 = 4$.

Show children a number line and ask, "How far is it from $^{+}2$ to $^{+}6$?" Have them draw a ray from $^{+}2$ to $^{+}6$, illustrating that the answer is $^{+}4$. Then, as the child studies examples such as Figure 18.7, he can observe the patterns involved in subtracting integers.

We may also want to have children search for patterns in a chart similar to Figure 18.8. The chart can be easily adapted to show that any difference can be written as a sum; that is, subtraction can be shown by finding a missing addend.

Another way to introduce operations with integers is with counting. One interesting counting method involves a postman who delivers checks (+) and bills (−), and takes them back

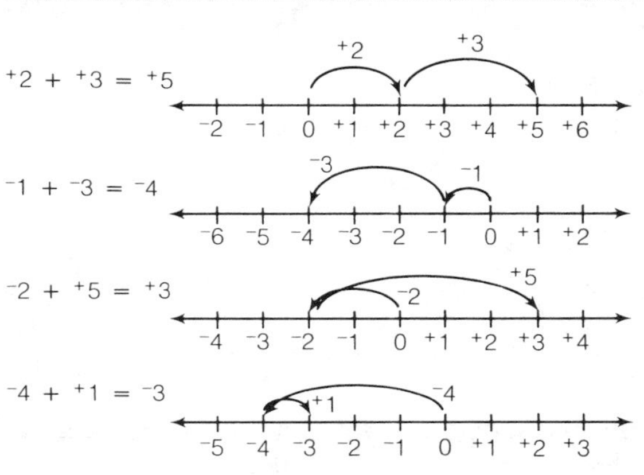

FIGURE 18.6
Adding Integers with a Number Line

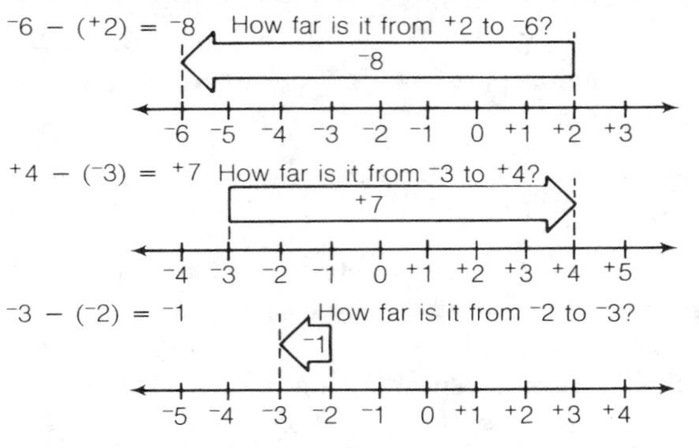

$^-6 - (^+2) = {}^-8$ How far is it from $^+2$ to $^-6$?

$^+4 - (^-3) = {}^+7$ How far is it from $^-3$ to $^+4$?

$^-3 - (^-2) = {}^-1$ How far is it from $^-2$ to $^-3$?

FIGURE 18.7
Subtracting Integers with a Number Line

because he left them at the wrong address and discovered his mistake. For example, delivering a check for $5.00 is recorded as $^+5$. Delivering a bill is recorded as $^-5$. Taking back a $5.00 check is a $^-5$ because the person is now $5.00 poorer, and taking back a $5.00 bill is a $^+5$ because the person is $5.00 richer since the bill does not have to be paid. Stories such as the following can be made up involving deliveries and taking back bills and checks.[5]

5. Louis S. Cohen, "A Rationale in Working With Signed Numbers," *The Arithmetic Teacher* 12, no. 7 (November 1965): 563–67.

Subtract:

5	5	5	5	5	5	5	5
3	2	1	0	$^-1$	$^-2$	$^-3$	$^-4$
2	3	4	5	?	?	?	?

Subtract:

$^-5$	$^-5$	$^-5$	$^-5$	$^-5$	$^-5$	$^-5$	$^-5$
$^-5$	$^-4$	$^-3$	$^-2$	$^-1$	0	1	2
0	$^-1$	$^-2$	$^-3$	?	?	?	?

FIGURE 18.8
Observing Patterns for Subtraction of Integers

PROBLEMS

A postman delivers a $5.00 check and an $8.00 bill. When the check is cashed and the bill paid, how much richer or poorer is Mrs. Jones?

$$^+5 + (^-8) = {}^-3$$

Mr. Smith believes he will have $10.00 left after he pays his bills. The postman discovers that he has delivered a $4.00 bill by mistake. After the postman takes back the bill, how much money will Mr. Smith have left?

$$^+10 - (^-4) = {}^+14$$

An important discovery for children is that of the additive inverse relationship. A child's understanding of the additive inverse depends upon attainment of the concept of reversibility as described by Piaget. The child must realize that a one-to-one relationship exists between a number and its inverse, and that the effect of adding a quantity can be nullified by moving in reverse; that is, by adding a negative quantity.[6] Children can easily demonstrate that any number plus its inverse equals 0 on a number line.

Multiplication and Division. If we use a number line to illustrate multiplication, the mode of representation has to be extended. Even so, it is not always effective.

For a positive integer times a positive integer, the number line representation is similar to that for whole numbers. Thus, it is easy to establish the rule that the product of two positive integers is positive.

In the case of multiplying a positive integer times a negative integer, we can refer to the repeated addition interpretation learned earlier: 4×2 can be interpreted as $2 + 2 + 2 + 2$. Consequently, $4 \times {}^-2$ would mean $^-2 + {}^-2 + {}^-2 + {}^-2$ or $^-8$.

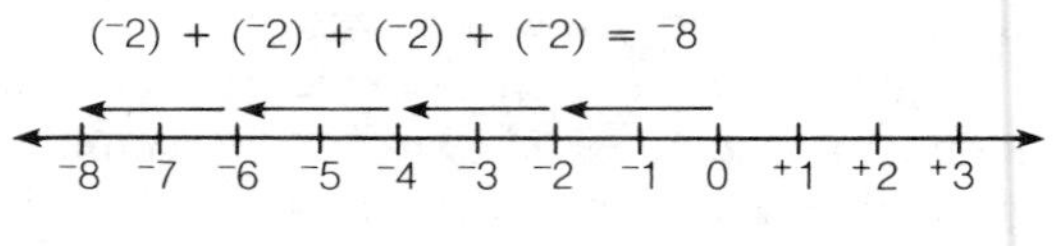

$(^-2) + (^-2) + (^-2) + (^-2) = {}^-8$

6. C. W. Schminke et. al., *Teaching The Child Mathematics,* 2nd ed. (New York: Holt, Rinehart and Winston, 1978), pp. 182–83.

EXCURSION
Using the Additive Inverse

The importance of the additive inverse concept can be shown with the addition and subtraction of integers. For example, to demonstrate that $^{+}6 + (^{-}2) = {}^{+}4$, use 6 blue (positive) chips to represent $^{+}6$, and 2 red (negative) chips to represent $^{-}2$.

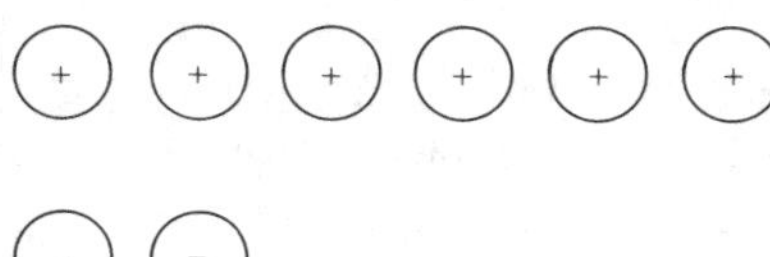

Two blue chips can be matched with 2 red chips. The excess is 4 blue chips. Thus, $^{+}6 + (^{-}2) = {}^{+}4$

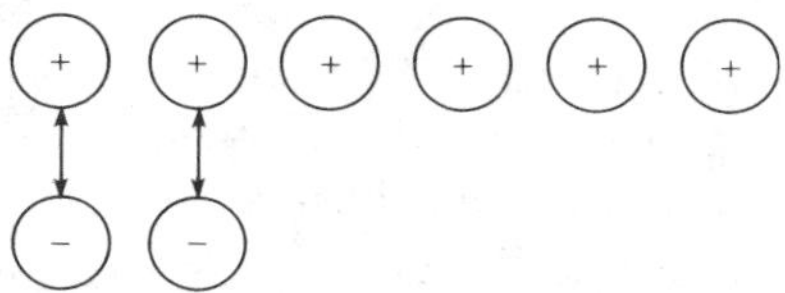

We can justify this mathematically by applying the additive inverse.

$$\begin{aligned} ^{+}6 + (^{-}2) &= (^{+}4 + {}^{+}2) + (^{-}2) \\ &= (^{+}4) + (^{+}2 + {}^{-}2) \\ &= (^{+}4) + 0 \\ &= {}^{+}4 \end{aligned}$$

For the subtraction statement $(^{+}6) - (^{-}2)$, first think, "There are no negative chips to be removed."

Two negative chips must be joined with the others; however, in order not to change the amount, 2 positive chips must also be added to the collection (additive inverse is applied by using $^{-}2 + {}^{+}2 = 0$).

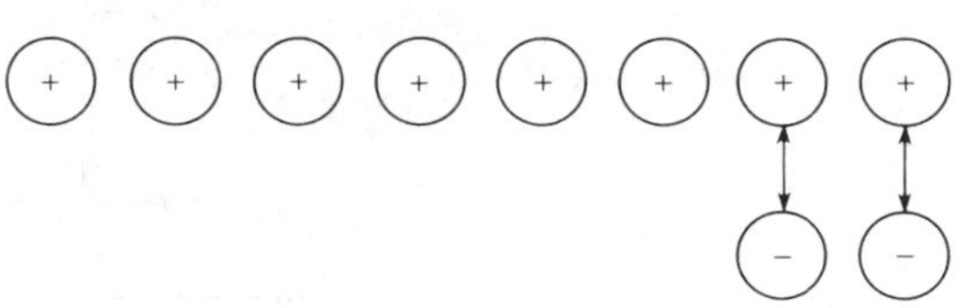

Now remove 2 negative chips.

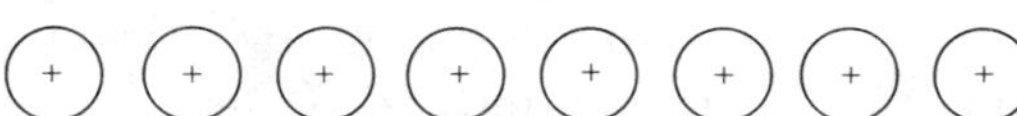

Thus, 8 blue (positive) chips remain, so $^{+}6 - (^{-}2) = {}^{+}8$.

Illustrate $(^{-}2) - (^{-}4)$ using this procedure.

Although the number line cannot be used to explain the other cases of multiplication involving integers, a negative integer times a positive integer provides an excellent opportunity to apply the commutative property of multiplication. Children easily accept the notion that $4 \times {}^-2$ and ${}^-2 \times 4$ have the same product (${}^-8$).

Multiplying a negative times a negative is also difficult to develop. The easiest way for children to see that a negative times a negative is a positive is to use a pattern approach similar to the one described in the excursion *Counting Backwards,* and extend the patterns as has been done in Figure 18.5.

Division with integers can be illustrated on a number line if both dividend and divisor are positive, or both are negative. Since the quotient will always be a positive number, the division can be interpreted as showing the number of times the divisor is contained in the dividend (measurement division). In the following example, ${}^-3$ is contained in ${}^-6$ two times.

Division is often approached as the inverse of multiplication, as finding the missing factor. Thus, if ${}^+24$ is to be divided by ${}^-6$, a child can be encouraged to rewrite the problem as ${}^-6 \times \square = {}^+24$. Then, using the rules for multiplying he already knows, he can conclude ${}^-6 \times {}^-4 = {}^+24$, so ${}^+24 \div {}^-6 = {}^-4$. If division is presented in this fashion for all cases, children soon observe a similarity between rules for multiplying and dividing integers.

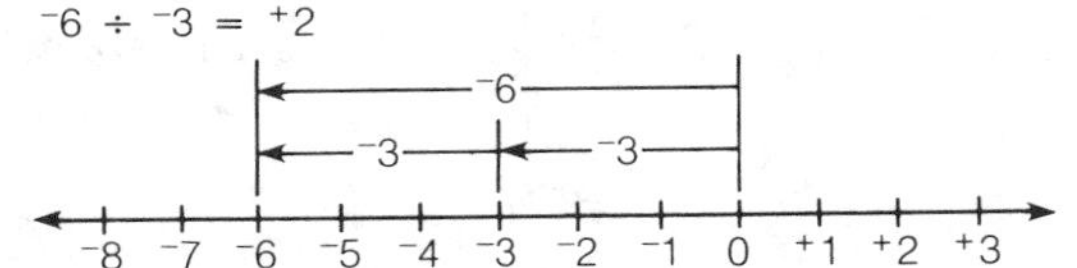

Consolidating and Transferring Activities

Practice activities take many forms, including oral and paper-and-pencil games that provide variations from classroom routine.

Climb the Ladder[7]

This is a board game consisting of ladders that have steps numbered from ${}^-12$ at the bottom to ${}^+12$ at the top. All players start at zero and take turns with the spinner. The numbers tell players

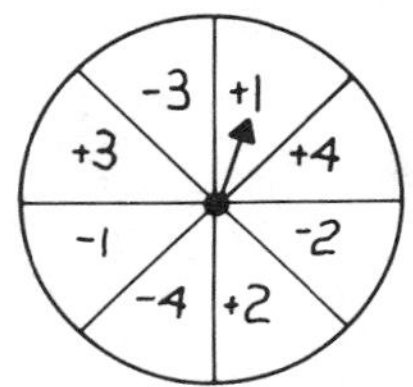

7. Adapted from Albert H. Mauthe, "Climb the Ladder," *The Arithmetic Teacher* 16, no. 5 (May 1969): 354–56.

EXCURSION
Nomograph

Nomographs can be used to find sums and differences.

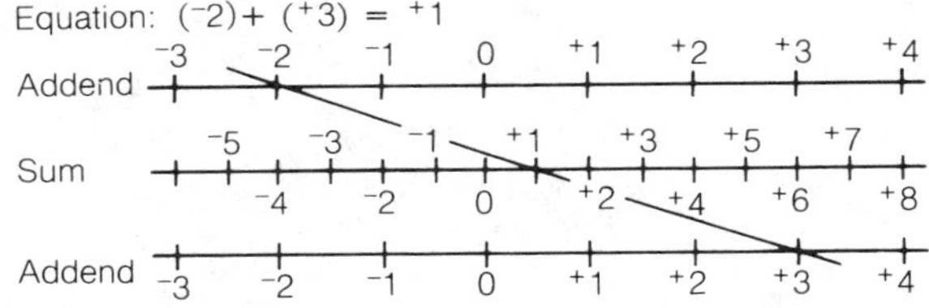

To make a nomograph, draw 3 parallel number lines. Line up the zeros and make the distance between points on the *sum* line half the distance between points on the *addend* line.

- Find the sum of ${}^-3$ and ${}^-1$ using the nomograph.
- Can you use the nomograph to find the difference between ${}^+1$ and ${}^-2$? Hint: Subtraction may be thought of as finding the missing addend.

how many steps to go and if they should climb up or go down. The winner is the first player to get to $^{+}12$ or the last player to stay on the ladder after the others have fallen off.

Integer Shuffleboard

This game is played like shuffleboard, except the board includes negative numbers. A target diagram is drawn on the floor, or on a large sheet of paper and taped to the floor (see Figure 18.9.) Four red and 4 black discs are also needed. Each player tries to either push his disc into one of the scoring areas, or knock the opponent's out of a scoring area and into a penalty area (negative integer). Children should be required to compute their scores. The winner is the player with the largest score for a given number of turns.

Roll and Subtract

A board with a path to a goal, several markers, and a red and a blue cube are needed. The red cube is marked with 0, $^{+}1$, $^{+}2$, $^{+}3$, $^{+}4$, $^{+}5$ and means *add*. Mark the blue cube with $^{+}1$, $^{+}2$, $^{+}3$, $^{-}1$, $^{-}2$, $^{-}3$; it means *subtract*. Numerals on the red and blue cubes can be varied to meet the specific needs of the class. Each player rolls both the red cube then the blue cube per turn. For example, if the red (add) cube is rolled and $^{+}2$ turns up, the player moves 2 places forward. If a $^{-}3$ is rolled on the blue (subtract) cube, the player moves forward another 3 places. That is, $^{+}2 - (^{-}3) = {}^{+}5$. Players soon learn that it is preferable to roll a negative number with the subtract cube if they are to reach the goal first.

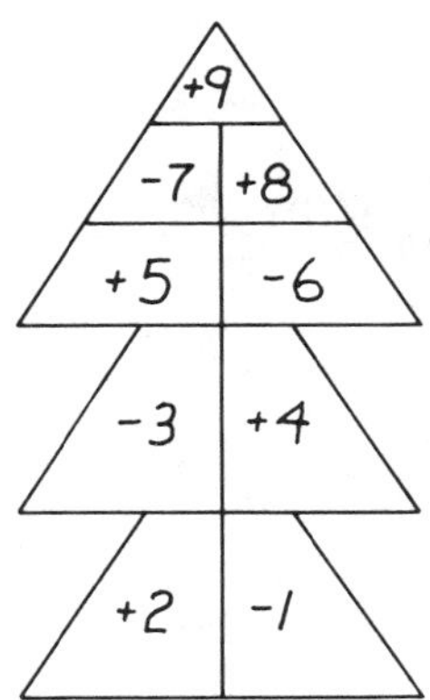

FIGURE 18.9
Gameboard for Integer Shuffleboard

Spin and Multiply[8]

This is a game for two or three children, or for the class divided into two teams. Two spinners are needed: one divided into ten sections and labeled 0, $^{+}1$, $^{+}2$, $^{+}3$, $^{+}4$, $^{-}5$, $^{-}6$, $^{-}7$, $^{-}8$, $^{-}9$; the second divided into seven sections, two of which are labeled as powers (for example, *power of 2*, meaning to raise the number to the second power) and the other five labeled with positive and negative integers. A player spins the first spinner to get the first factor, and the second spinner to get either the second factor or the power to raise the first factor. After the first product is found, each additional product is added to the last product. For example, if the first product is $^{+}12$ and the second product $^{-}9$, the cumulative score is $^{+}3$. The winner is the score farthest from zero (whether positive or negative) for a predetermined number of plays.

At the elementary school level, there are fewer transferring activities for integers than for whole numbers, fractions, and decimals. Transferring activities vary greatly and depend upon other topics being studied and on the age and interest of each child. If possible, relate applications to other parts of the curriculum.

The following number sentences can be used as integer transferring activities.

PROBLEM

Find the value of n in each equation:

$$n + (^{+}6) = {}^{+}9$$

$$^{+}2n = {}^{-}6$$

$$n - (^{+}3) = {}^{+}4$$

$$\frac{n}{^{+}2} = {}^{-}3$$

$$^{+}3n - (^{+}2) = {}^{+}4$$

Verbal problems can also be used. When selecting story problem situations, be sure to

8. Adapted from Esther Milne, "Disguised Practice for Multiplication and Addition of Directed Numbers," *The Arithmetic Teacher* 16, no. 5 (May 1969): 397–98.

include those that require a child to translate from a number sentence, a verbal problem, or a two- or three-dimensional representation to other forms of representation.

PLANNING ASSESSMENT

Assessment begins by focusing on important objectives. For example, consider assessment procedures we might use in relation to content objectives 18.7–9.

What behaviors might we look for as evidence that a child understands this cluster of objectives? More than one of the following behaviors will indicate that the child actually does understand.

1. Given several addition statements involving integers and their representations on a number line, the child matches each statement with the appropriate representation.
2. Given an addition example involving integers, the child finds the unknown sum.
3. Given an addition statement involving integers, the child represents the statement on a number line.
4. Given a story problem involving addition of integers, the child solves the problem.
5. Given an addition statement involving integers, the child constructs a story problem to illustrate the statement.
6. The child can state the rules for adding integers.

By eliciting such behaviors, much can be learned about each child's knowledge of addition with integers. Behavioral indicators and assessments tasks can be prepared in a similar fashion for the other operations.

19

Organizing, Analyzing, and Reporting Data

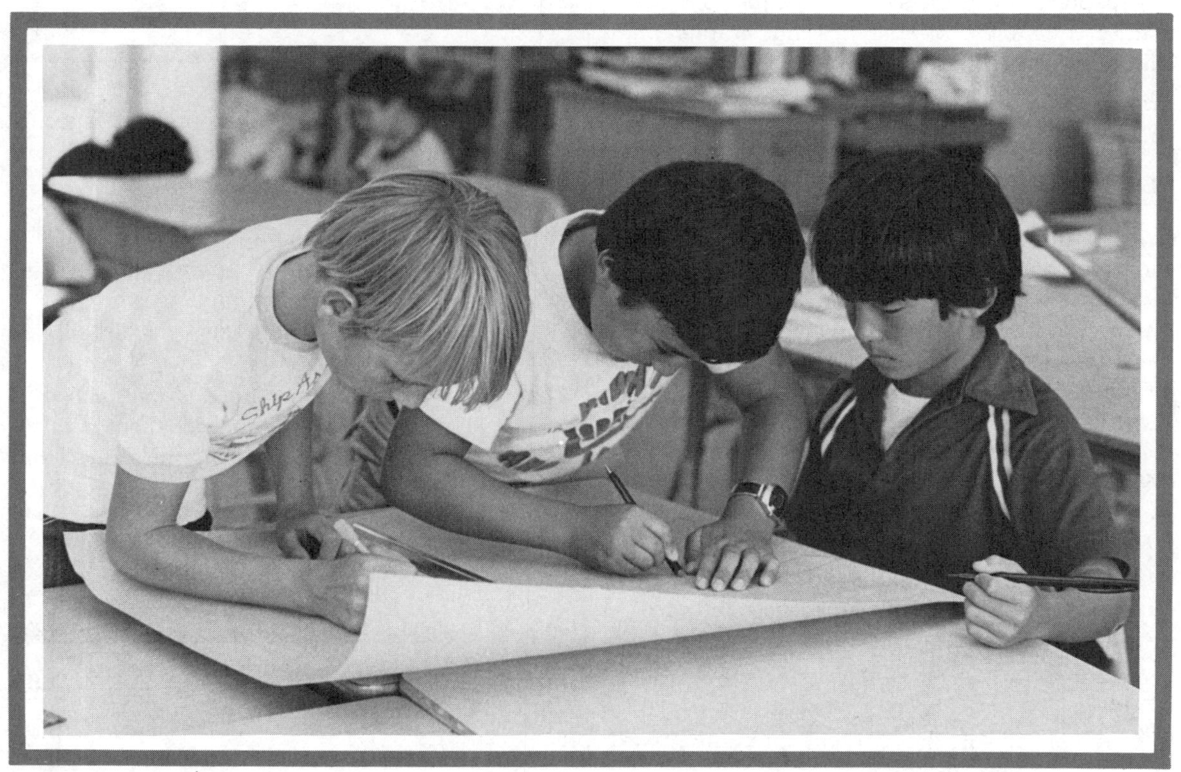

The study of statistics and graphs has become increasingly important, due to the present emphasis on data collection. Advertisements in newspapers, magazines, radio \ and television make frequent use of statistics and graphs, and children will be influenced by the data they hear and see.

PLANNING THE CONTENT

Graphs and statistics should be studied for a variety of reasons:

- Graphs and data are prominent in current publications and on radio and television.
- Graphs are a simple and powerful way of presenting data in a condensed, understandable, and interesting manner.
- Problem-solving skills are enhanced because children are involved in collecting, organizing, presenting, and critically evaluating data.
- Other mathematics skills such as counting, computing, and measuring can be reinforced.
- Children become highly motivated when they collect and organize data, then analyze and report it.

Not all of the content presented here should be taught to all children; we will need to carefully select appropriate content for each child.

Gathering and Organizing Data

The word *data* is derived from a Latin word meaning *something given*, and is generally used to mean given information or facts.

Numerical data should be collected in an organized fashion. This can be done by making a frequency table.

19.1
The number of items in a set of data is called the *frequency of the data*. An organized summary of the data is called a *frequency distribution*.

We often gather data to solve a specific problem, such as examining the scores on a test. Suppose these are the scores.

95, 85, 90, 85, 75, 70, 80, 85, 90, 80
80, 75, 80, 70, 80, 95, 80, 75, 100, 90
90, 65, 95, 85, 80, 75, 60, 75, 85, 80
85, 80, 90, 80, 100, 85, 80, 80, 85, 90

This list of 40 scores does not tell us much. But we can learn more if we summarize them in a frequency distribution by reordering the scores from the highest to the lowest, then showing how often each score occurred, as in Figure 19.1.

Scores	*Tally Marks*	*Frequency*
100	//	2
95	///	3
90	~~////~~ /	6
85	~~////~~ ///	8
80	~~////~~ ~~////~~ //	12
75	~~////~~	5
70	//	2
65	/	1
60	/	1

FIGURE 19.1
A Frequency Distribution for Scores on a Mathematics Test

EXCURSION
Using Statistics to Support a Position

The graph and chart in Figure 19.2 were used to show that women still lag behind men in earnings.[1] The data appear to make a strong case for sex discrimination, but a closer examination of how the statistics were derived is essential. Is the reason that the average female accountant earns only 69% of the income earned by the average male in the same job classification due to the fact that males have been working for a longer period of time? We cannot tell from the information presented. These and other statistics are frequently used to support a position.

Make certain that children learn to ask similar questions when statistics are used in this manner.

Analyzing Data

After data has been collected and organized, it can be numerically described in different ways. One of the most common descriptions of data is an *average* or a *measure of central tendency*. Statisticians use the latter term to designate numbers that tell what is typical. Mean, median and mode are the three different measures of central tendency.

1. Jeffery L. Sheler, "A Fresh Round in the Fight Over Equal Pay," *U.S. News and World Report,* June 22, 1981, p. 81.

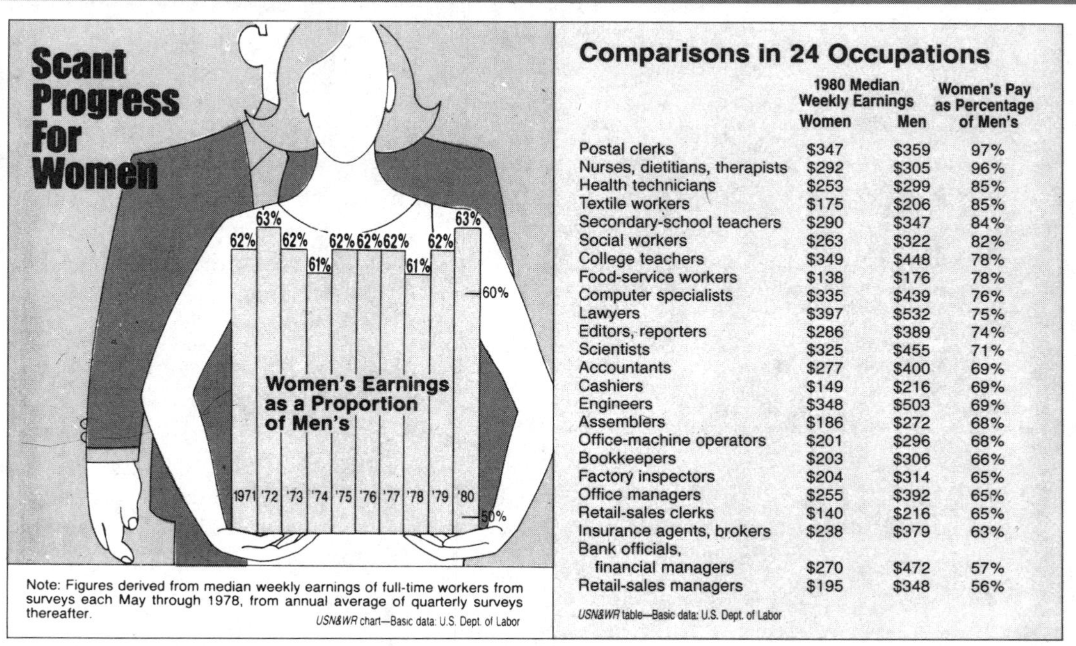

Comparisons in 24 Occupations

	1980 Median Weekly Earnings Women	Men	Women's Pay as Percentage of Men's
Postal clerks	$347	$359	97%
Nurses, dietitians, therapists	$292	$305	96%
Health technicians	$253	$299	85%
Textile workers	$175	$206	85%
Secondary-school teachers	$290	$347	84%
Social workers	$263	$322	82%
College teachers	$349	$448	78%
Food-service workers	$138	$176	78%
Computer specialists	$335	$439	76%
Lawyers	$397	$532	75%
Editors, reporters	$286	$389	74%
Scientists	$325	$455	71%
Accountants	$277	$400	69%
Cashiers	$149	$216	69%
Engineers	$348	$503	69%
Assemblers	$186	$272	68%
Office-machine operators	$201	$296	68%
Bookkeepers	$203	$306	66%
Factory inspectors	$204	$314	65%
Office managers	$255	$392	65%
Retail-sales clerks	$140	$216	65%
Insurance agents, brokers	$238	$379	63%
Bank officials, financial managers	$270	$472	57%
Retail-sales managers	$195	$348	56%

USN&WR table—Basic data: U.S. Dept. of Labor

FIGURE 19.2
Graph and Chart Used to Support a Position

EXCURSION
Computing the Earned Run Average (ERA)

The most important statistic for a pitcher is the earned run average (ERA). As the term denotes, earned run average is based only on earned runs given up by a pitcher. Unearned runs are not counted. The earned run average compares the earned runs given up by a pitcher to the number of innings pitched. The formula is:

$$\text{Earned Run Average} = \frac{\text{Earned Runs} \times 9}{\text{Innings Pitched}}$$

In 1975 Jim Palmer of the Baltimore Orioles had the lowest earned run average in the American League. Palmer gave up 75 earned runs in 323 innings. What was his earned run average?

$$\begin{aligned}\text{Earned Run Average} &= (75 \times 9) \div 323 \\ &= 675 \div 323 \\ &= 2.09\end{aligned}$$

Palmer's earned run average for 1975 was 2.09. An earned run average of 2.09 means that he allowed an average of 2.09 runs for every 9 innings pitched.

19.2
To find the arithmetic *mean* of a set of numbers, divide their sum by the number of numbers in the set.

19.3
To find the *median* of a set of numbers, arrange them in order, smallest to largest, and find the middle number.

19.4
The *mode* for a set of numbers is the number that occurs most frequently.

The most commonly used measure of central tendency is the arithmetic *mean*. To find the average weight (mean) of 5 football players, divide the sum of their weights by the number of weights.

$$\begin{aligned}\text{Mean} &= \frac{195 + 206 + 243 + 220 + 256}{5} \\ &= \frac{1120}{5} = 224\end{aligned}$$

Unfortunately, the means can be considerably affected by one or two extreme (very high or very low) values in the data collected.

When the mean is not the best measure of central tendency, the *median*—which is not affected by extremes—is often used. The median for a set of data is found by arranging the data from smallest to largest, then selecting the middle number. For example, the median of these 7 scores is 75.

65, 65, 70, 75, 80, 80, 95
↑
Median

When there is an even number of scores, think of the median as midway between the two scores nearest the middle. For example, the median for this set of data is 61.

54, 57, 58, 64, 72, 80

$$\frac{58 + 64}{2} = \frac{122}{2} = 61$$

The other measure of central tendency is the *mode*. This is the value that occurs with the greatest frequency.

Reporting Data

Frequently used data is usually presented in charts and tables. It is often easier to get information from tables and charts than from written materials. Statisticians also make use of graphs to report data in a clear and concise manner.

> **19.5**
> *Bar graphs* compare the size or frequency of two or more quantities.
>
> **19.6**
> *Line graphs* show change over time and relationships between two or more variables that involve continuity.
>
> **19.7**
> *Circle graphs* display the relation of parts to the whole and to each other.

Bar graphs consist of a series of bars of uniform width with some form of measurement noted on a vertical or horizontal axis. The vertical bar graph in Figure 19.3 compares the heights of several large mountains, while the horizontal graph in Figure 19.4 shows the running speeds of various animals. The width of the bars or the direction in which bars are drawn (horizontal or vertical) is not significant to the meaning of a graph. However, in a bar graph, the height or length of each bar is drawn in proportion to the number or size of the measures, scores, or percentages. Bars should be drawn of uniform width, and the scale should be complete.

Multiple bar graphs similar to Figure 19.5 show several relationships at once. Divided bar graphs such as Figure 19.6 are used similarly.

A more dramatic device for displaying data is the picture graph or pictograph that consists of rows of small schematic pictures denoting the items being reported. Each picture represents a specific amount. In Figure 19.7 (page 432), how many telephones are there in Dayton, Ohio?

Line graphs (see pages 432–33) show changes over a period of time. Examples include temperature and rainfall variations (Figure 19.8) and trends in population (Figure 19.9). Relationships like Miles Per Gallon (M.P.G.) to speed (Figure 19.10) can also be pictured. One line graph can compare two sets of similar data. Figure 19.11 shows a line graph that compares actual rainfall with predicted rainfall for a given month.

Circle graphs are also introduced in elementary schools. They compare the parts of something to the whole, or one part to another part. Budgets are frequently displayed, as in Figure 19.12 (page 433).

EXCURSION
Misuse of the Mean

If the mean is used as a representative measure, a distorted picture of the set of data may be given. For example:

A small company employs only 6 persons in addition to the 4 officers of the company. These are the yearly earnings of each:

President	$60,000	Production person	$15,000
First Vice-President	$45,000	Production person	$15,000
Second Vice-President	$40,000	Production person	$10,000
Treasurer	$25,000	Production person	$10,000
Secretary	$9,000	Custodian	$8,000

What is the mean income of people working for this company? Is it misleading to tell a person seeking a job what the average income is if you use the mean as the average?

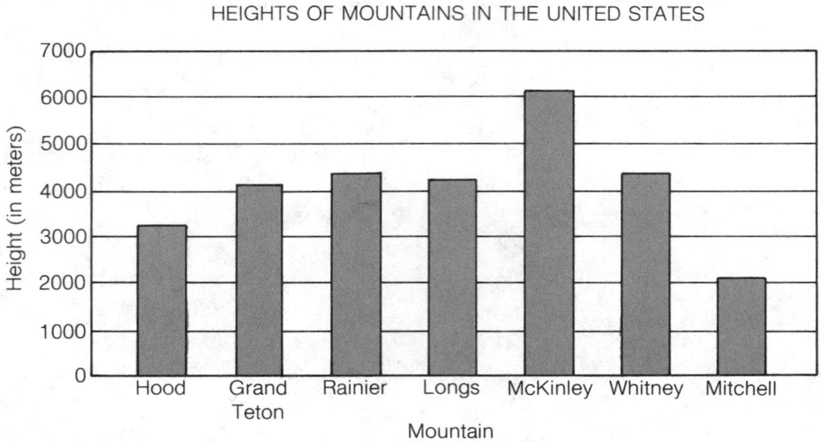

FIGURE 19.3
Vertical Bar Graph

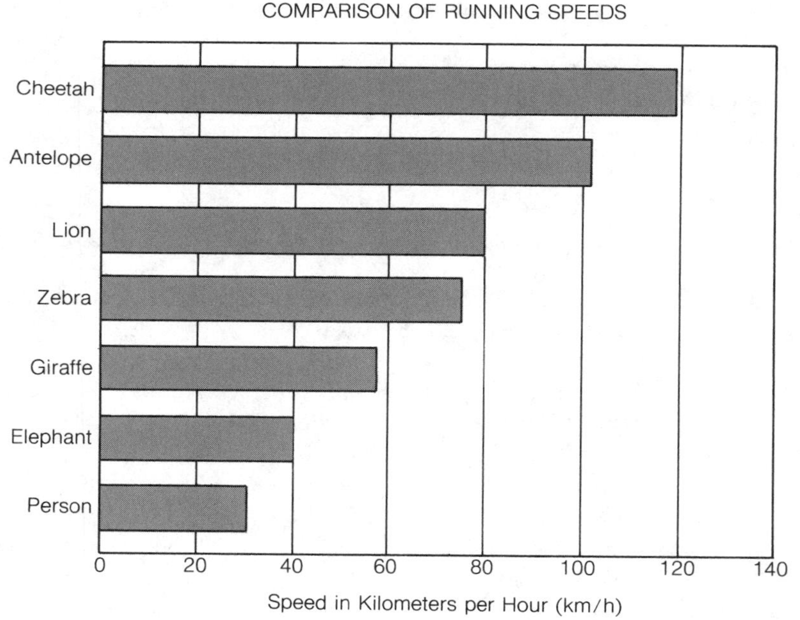

FIGURE 19.4
Horizontal Bar Graph

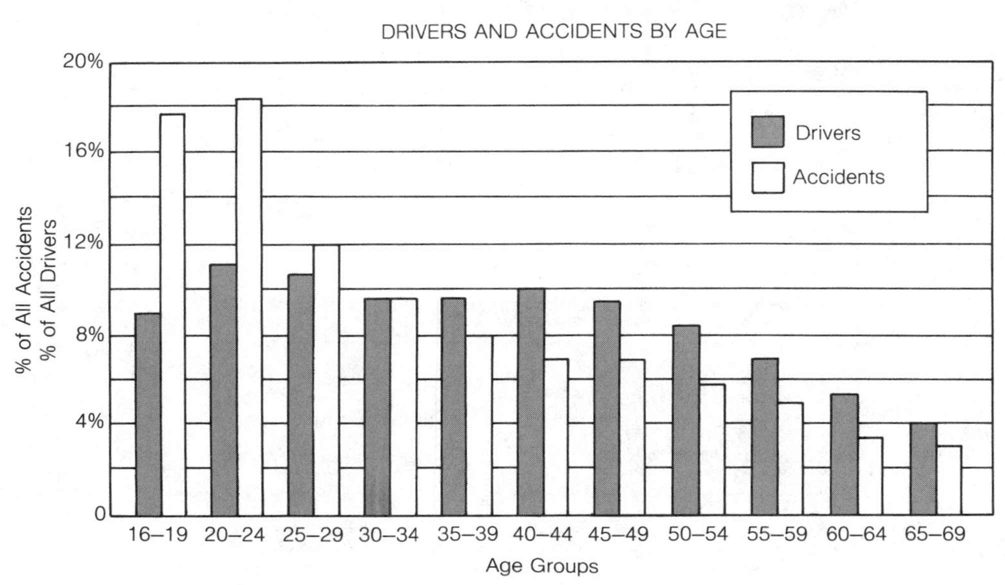

FIGURE 19.5
Vertical Multiple Bar Graph

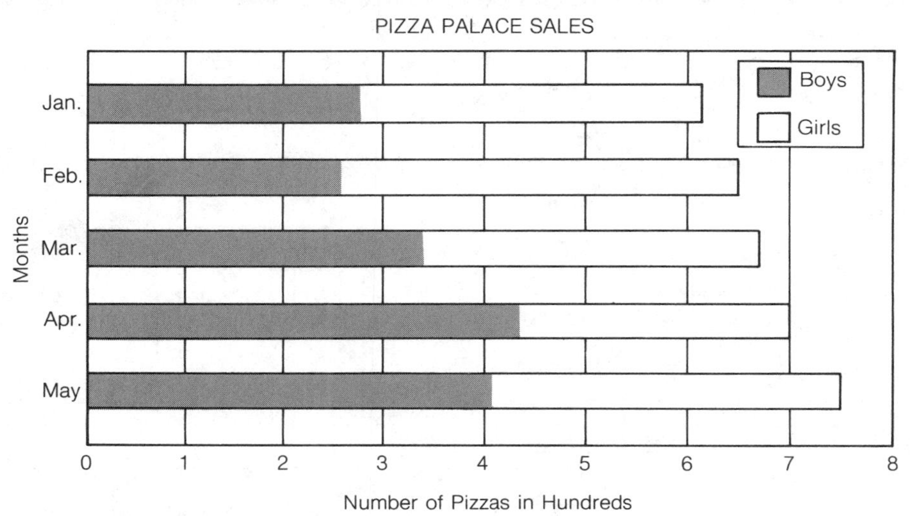

FIGURE 19.6
Divided Bar Graph

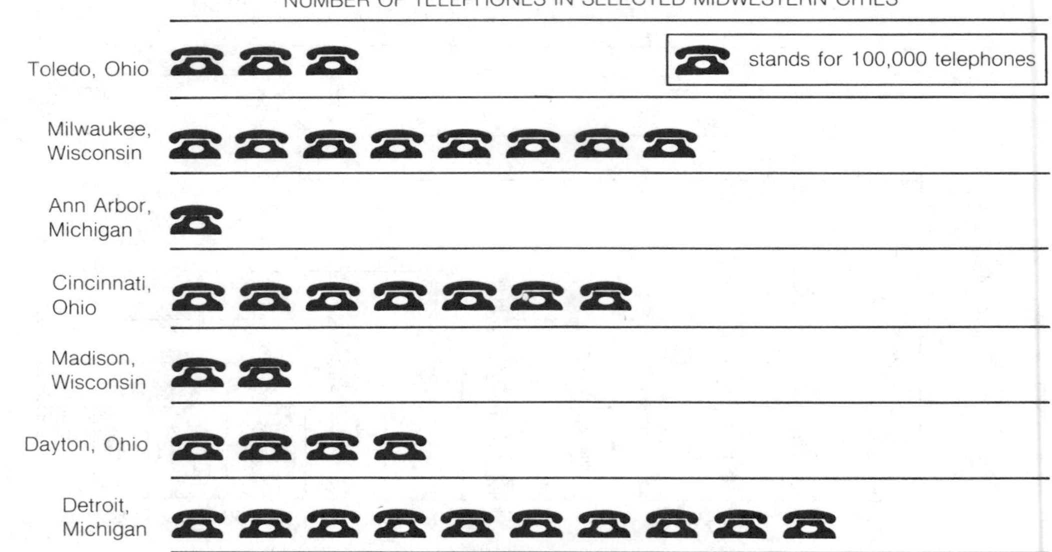

FIGURE 19.7
A Picture Graph

Ordering the Content

Children should observe and collect information even in the earliest years of school. A teacher can work with kindergartners in collecting and organizing data for a simple bar graph. Throughout elementary school, instruction in organizing, analyzing, and reporting should focus primarily on data the children themselves collect, either through direct observation or measurement, or by consulting reference works such as an encyclopedia. As children help to build graphs, they learn to interpret similar data in magazines, newspapers, and books.

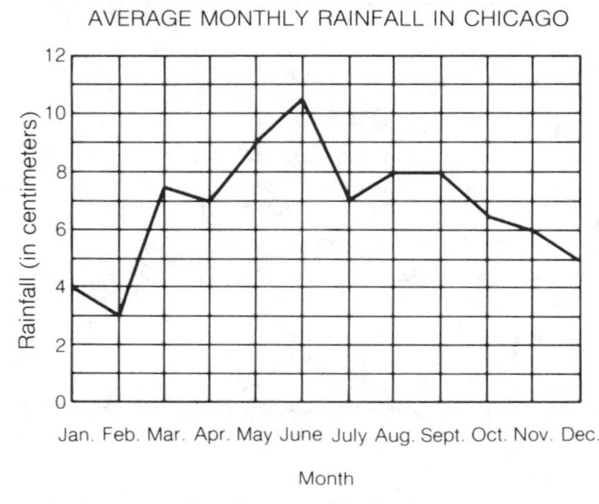

FIGURE 19.8
Line Graph for Variations in Rainfall

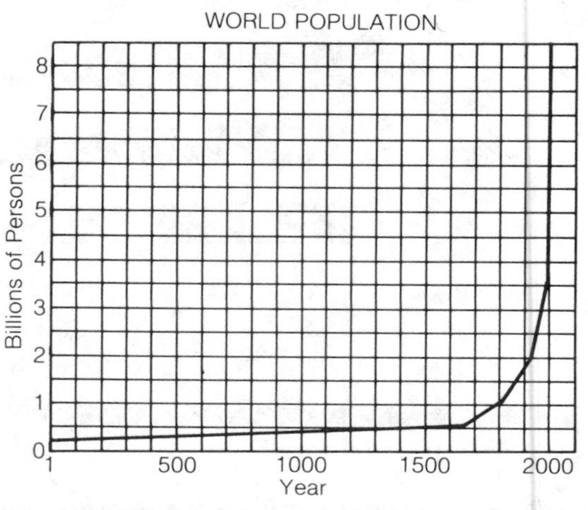

FIGURE 19.9
Line Graph for Population Trends

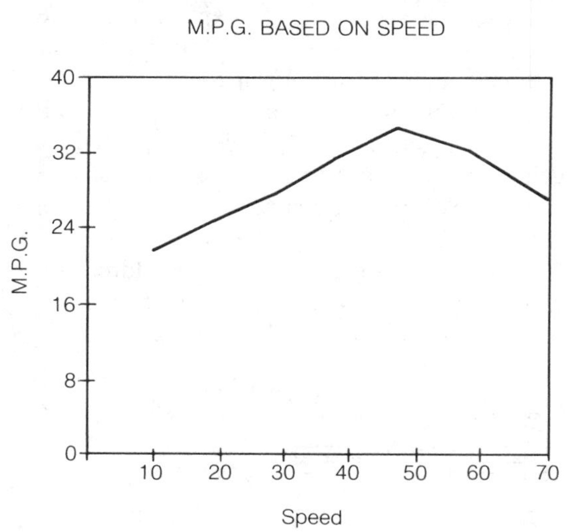

FIGURE 19.10
Line Graph for Relationship of Miles Per Gallon (M.P.G.) to Speed

The construction of simple bar graphs is usually followed by picture, then line graphs. Measures of central tendency can usually be introduced in the late primary or intermediate grades. After children have had experiences with ratio and proportion, angle measurement and percent, more extensive work with picture graphs, line graphs, and statistics can be done. Multiple bar and line graphs can be introduced at this time. With more advanced children, we may also want to present frequency-distribution tables.

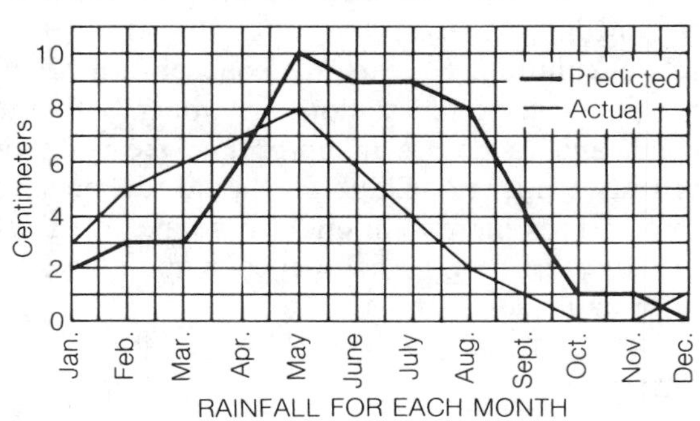

FIGURE 19.11
Line Graph Comparing Two Sets of Data

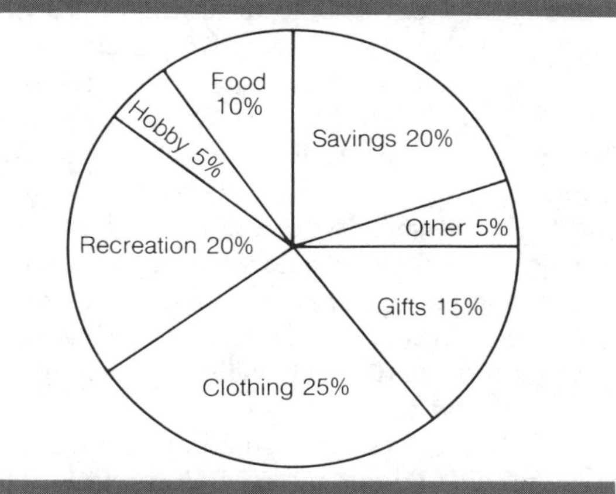

FIGURE 19.12
Circle Graph

Children should have prerequisite concepts and skills whenever a topic is introduced. At the earliest levels, for example, counting and sorting skills are important. Later, make sure children understand percent before they work with circle graphs.

We also need to decide if we want to focus on these objectives during mathematics class alone, or interweave them with social studies, science, and art curricula as well.[2]

PLANNING INSTRUCTION

Before we plan instruction, we need to know what concepts and skills each child has learned already. We must also decide whether we are going to make this a unit within our mathematics program, or integrate the topic with other curricular areas.

Initiating, Abstracting and Schematizing Activities

Although activities for gathering, organizing, analyzing and reporting data are described in separate sections, a classroom activity will usually incorporate each of these.

Gathering and Organizing Data

Children understand graphic displays if they collect data themselves, then organize the informa-

2. Delia Sullivan et. al., "This is Us! Great Graphs for Kids," *The Arithmetic Teacher* 28, no. 1 (September 1980): 14–18.

tion. This process should begin early, as soon as children learn about themselves, their families, and school. Information about birth dates, family size, hair color, height, shoe size, and so on, can be gathered and discussed, counted, and compared.

Early sorting, counting, and recording activities can include collectibles such as buttons, shells, and cut-out shapes. Class records of who goes home for lunch, the number of birthdays each month, and which members own dogs or cats also lead to graphing experiences.

As children understand more about numbers, provide additional experiences with polling and census taking. Polls should focus on topics of interest, such as favorite colors, television shows, and the use of recreational time. Census taking requires more detailed planning because children must decide on topics and questions; they also need to agree on how to collect information. Emphasize the importance of representing cumbersome sets of numbers in a more manageable way using, for example, frequency distribution tables similar to those illustrated earlier in this chapter.

Analyzing Data

Certain everday statistical concepts are easily understood and used by children to make judgments about numerical data. The mean or arithmetic average can be introduced through simple activities with books or blocks. For example, the height of 3 identical stacks of books might be examined.

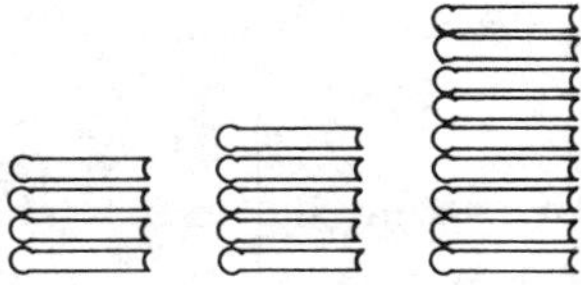

Ask, "If these 3 stacks were all the same height, about how many books would there be in each stack?" Have children move the books about until there are 3 stacks of equal height.

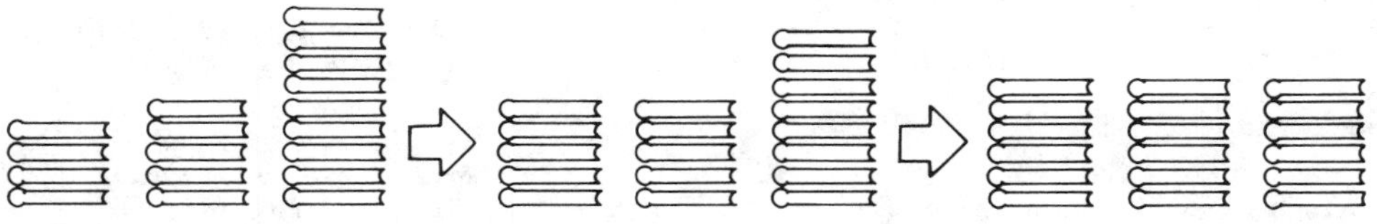

Children will use different methods to get 3 stacks the same height. Explain the different solutions, then illustrate the solution that pictures the algorithm for finding a mean. Stack all the books in one pile, then separate or parcel out the books into 3 stacks of equal size. This leads to the generalization that, to find the mean of a set of numbers, divide their sum by the number of elements. Means can then be computed for daily high temperatures, heights of children in the class, the time it takes each child to run a hundred yards, and so on.

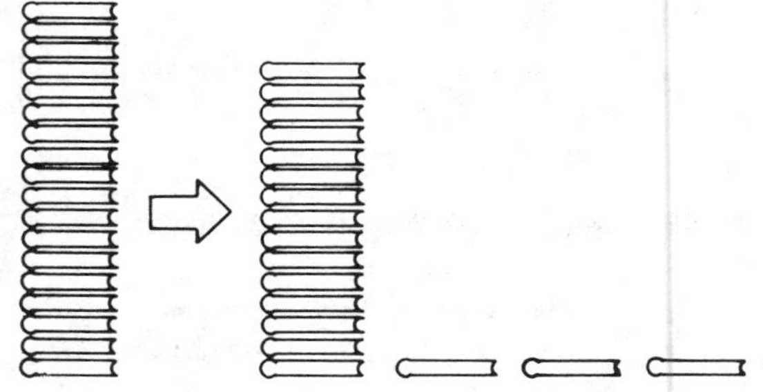

Emphasize that, although the mean is the most commonly used average, it is not always the most appropriate. For example, the chart in Figure 19.13 is a record of the number and sizes of shoes sold by the Easy Walk Shoe Store. In determining the most popular size sold, the store owner would not find the mean size sold to be helpful in placing an order for more shoes. The owner needs to know the mode, the most popular size or sizes. After children learn the appropriate use of a mode, they can easily inspect data to find the most frequently occurring number.

The other measure of central tendency, the median, is used less often by elementary school children. To teach the concept of median, provide several illustrations in which both the mean and mode are inappropriate measures of central tendency. Children should understand that there is more than one way to average a set of data. The best method depends upon how the information is to be used.

Size	*Sold*	*Size*	*Sold*
$6\frac{1}{2}$	2	9	28
7	3	$9\frac{1}{2}$	42
$7\frac{1}{2}$	6	10	10
8	11	$10\frac{1}{2}$	6
$8\frac{1}{2}$	25	11	3

FIGURE 19.13
Shoe Sizes Sold at Easy Walk Shoe Store

Consider this example. A group of 5 children entered a spelling contest. After the results of the contest were tabulated, the announcer said, "On the average, each of the contestants spelled 35 words correctly." The mean is 35, but 35 is not the best way to express their average spelling ability. Only one child spelled more than 35 words correctly.

Alice	65
Mark	18
Janet	12
Susan	6
John	4

Reporting Data

Charts, tables, and graphs organize and display information in a usable form. As children learn to construct graphs, they review and reinforce previously learned skills involving measurement, ratio and proportion, computation, and the arrangement of lines in space. Children seem to enjoy constructing graphs, regardless of age.

When children first collect and organize data for comparison, they sometimes place different-sized objects in rows and compare lengths, thereby fooling themselves as to which row has more. This can be avoided by making sure they understand that equal-sized units must be used. Graphs are first made from collectibles, then with a symbol for each object; later a dimension such as length is introduced.[3]

3. See Elizabeth M. Johnson, "Bar Graphs for First Graders," *The Arithmetic Teacher* 29, no. 4 (December 1981): 30–31.

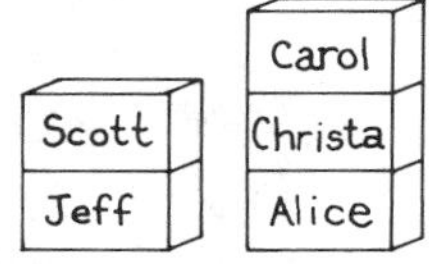

FIGURE 19.14
Boys and Girls Shown with Blocks

One way to introduce simple graphing is to have boys and girls place blocks labeled with their names in two separate piles, as in Figure 19.14.

As children grasp the idea of one-to-one correspondence (one block per child) increase the number of categories. In Figure 19.15, for example, each child has placed a name block on top of the month of her birthday. Important mathematical ideas are involved in an analysis of the results:

- In which month are the *greatest* number of birthdays?
- In which month are the *fewest* number of birthdays?
- How many *more* children have birthdays in July than in May?
- In what month is the number of birthdays *3 fewer* than the number in August?

To introduce more abstract graphs, interlocking blocks and graph paper can be used. In Figure 19.16, Unifix cubes show the family size for each child in the group. The same information

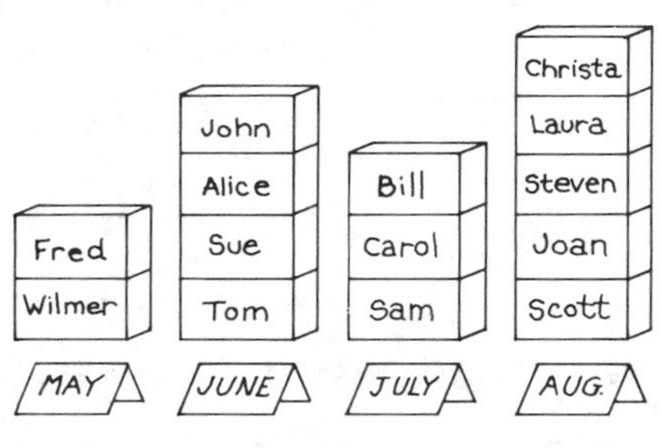

FIGURE 19.15
Birthday Months Shown with Blocks

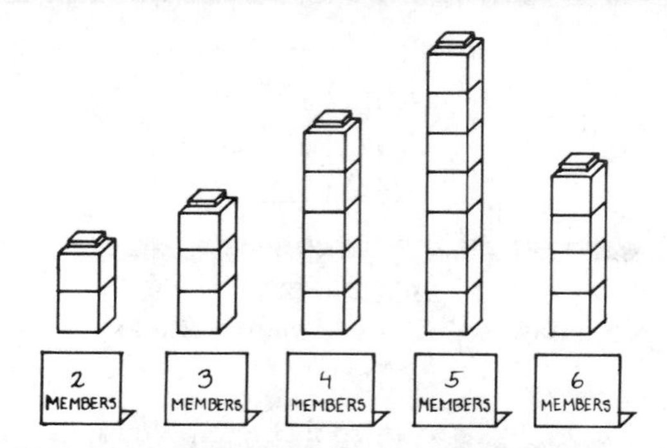

FIGURE 19.16
Interlocking Cubes Used to Show Family Size

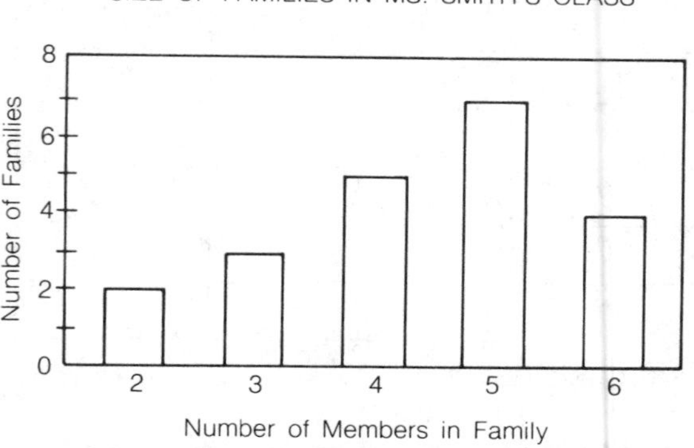

FIGURE 19.18
Graph Used to Show Family Size

can be presented with graph paper (Figure 19.17). Picture graphs can be introduced similarly, each picture representing one unit. Later, each child can express this information in the format seen in Figure 19.18.

After children have used simple bar and picture graphs, one symbol can represent *multiple* units. Children might examine graphs, such as Figures 19.19 and 19.20, to see which is the clearer presentation of the data.

Children are likely to encounter difficulty if they construct picture graphs in which one symbol represents multiple units. Partial figures that show part of a group are often difficult to draw. Picture graphs are easier to construct when regular-shaped regions, such as circles or rectangles, are used.

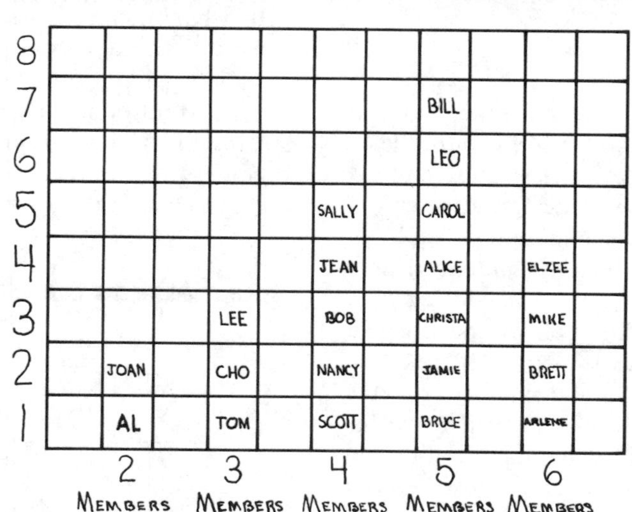

FIGURE 19.17
Graph Paper Used to Show Family Size

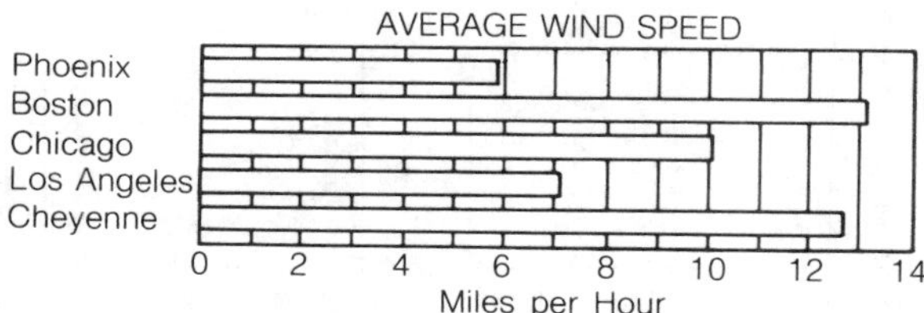

FIGURE 19.19
Bar Graph with Multiple Units

AVERAGE WIND SPEED

Phoenix, Ariz.

Boston, Mass.

Chicago, Ill.

Los Angeles, Calif.

Cheyenne, Wyo.

stands for 2 miles per hour

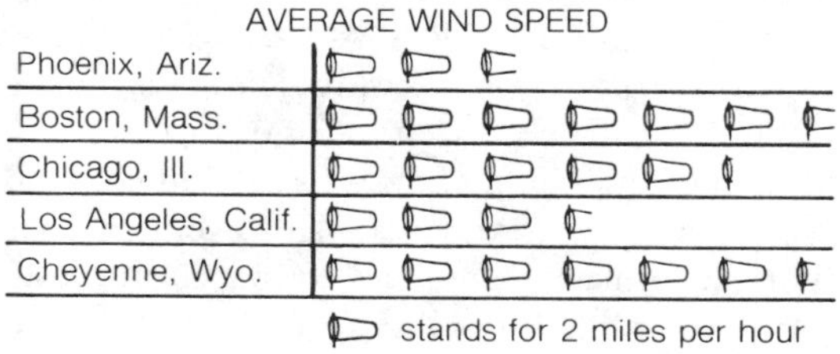

FIGURE 19.20
Picture Graph with Multiple Units

Line graphs can be introduced to children by using selected bar graphs, as shown in Figure 19.22. Simply connect the midpoints of the top of each bar.

Line graphs present data that is continuous rather than discrete. For example, a line graph would provide a better picture of temperature changes over 24 hours than either a picture or bar graph, because temperature is a continuous function that occurs between observed intervals.

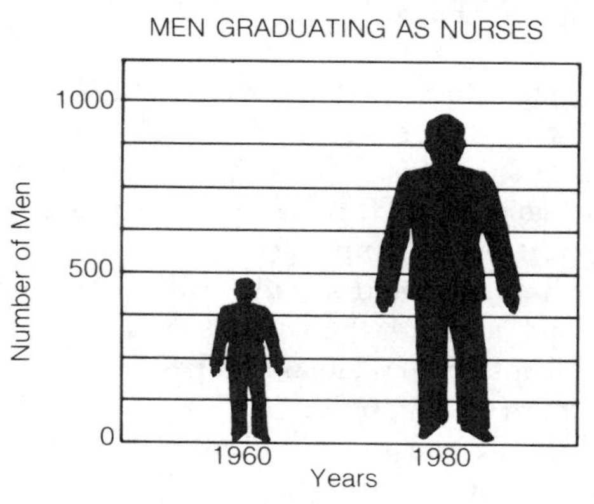

FIGURE 19.21
Misleading Picture Graph

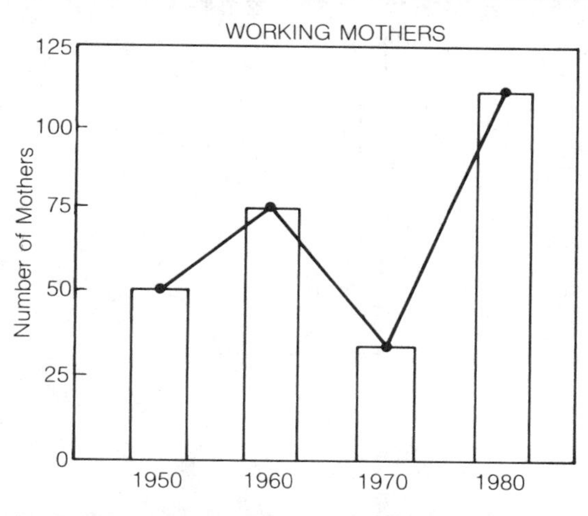

FIGURE 19.22
Transition from Bar to Line Graph

EXCURSION
Misleading Graphs

The size of figures in graphs is sometimes varied in more than one dimension. For example, in Figure 19.21 the man drawn to represent 1,000 men is both twice as tall *and* wide as the man for 500 men. Why is this graph misleading?

As children work with different types of graphs, emphasize the following:

- A graph does not present data as accurately as a table.
- Bar graphs should be constructed with all bars the same width. Spaces of equal width should appear between the bars.
- All graphs should include a title. Many graphs require a key for the symbols used. Vertical and horizontal scales should be labeled.
- Neatness is important.
- The scale should be uniform and appropriate for the values depicted.

In Figure 19.23, the same data is used with three different scales, and the most appropriate scale is easily identified.

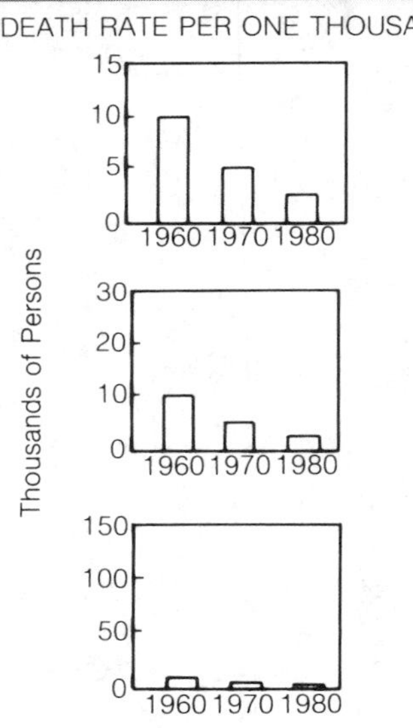

FIGURE 19.23
Data Shown with Three Different Scales

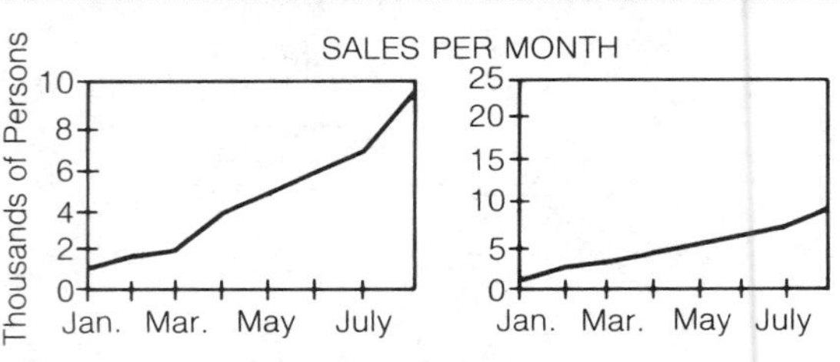

FIGURE 19.24
Graph Used for Emphasis

Each child can make predictions by looking at known values on a graph and finding the value for a unit beyond those recorded, a process called *extrapolation*. A prediction is justified when there is a clear trend. For example, the population of the United States in 1990 might be predicted from Figure 19.25. However, emphasize that extrapolating *can* lead to inaccurate results.

As children gain experience in reading and interpreting graphs, examine those that require *interpolation*, the process of estimating the value for a unit between two dots on a graph.

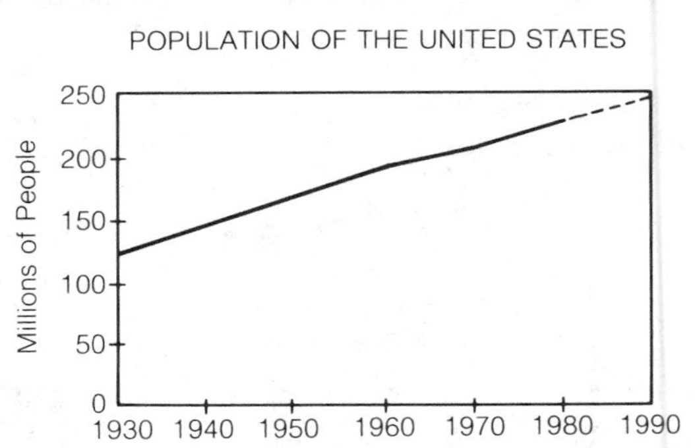

FIGURE 19.25
Graph Used to Extrapolate Data

EXCURSION
Using Graphs for Emphasis

Graphs are often used to emphasize a point or trend. Therefore, the choice of a scale is critical. Which graph in Figure 19.24 most clearly suggests the product increased in popularity during the period depicted?

This important skill will allow children to investigate all of the information represented on a line graph.

In introducing children to circle graphs, emphasize that this type of graph shows how a whole is separated into parts. In Figure 19.26, the circular region represents a 24-hour day. Even children who have not studied angle measurement can answer some questions from this graph.

- The greatest amount of time in Christa's day is spent doing what?
- Christa spends about as much time playing as doing what?
- Christa spends as much time eating and doing homework as what?

FIGURE 19.26
Circle Graph

After children have learned to measure angles and multiply fractions, they can construct circle graphs. Set up a table that shows each fractional part and its number of degrees before beginning construction. Figure 19.27 is a chart used to construct the circle graph illustrated.

Applications of everyday statistics are easy to find. For example, you can collect a variety of data both in and out of school with questionnaires. The limitations of these questionnaires, however, are not always evident to the child. The wording can affect both the response and the ease or difficulty of the reply. Questions X and Y in Figure 19.28 were designed to elicit the same information, but subtle differences affected the results of polling.[4]

As children develop their own questionnaires, they will learn general principles of writing good questions and to spot errors in design. They will also appreciate the limitations of newspaper and television opinion polls.

HOW SCOTT SPENDS HIS ALLOWANCE

	AMOUNT	FRACTION	DEGREES OF CIRCLE
LUNCH	$ 6.00	$\frac{3}{8}$	135°
CLOTHING	4.00	$\frac{1}{4}$	90°
ENTERTAINMENT	2.00	$\frac{1}{8}$	45°
SAVINGS	4.00	$\frac{1}{4}$	90°
TOTAL	$ 16.00	1	360°

FIGURE 19.27
Circle Graph Constructed from a Table

Consolidating and Transferring Activities

The study of statistics and graphs is made more interesting when it is correlated with different subjects in the curriculum. Activities should require each child to critically examine situations, make decisions regarding the best method for collecting data, and then proceed to construct a table or a graph.

Science

Compare the growth patterns of plants under different circumstances by varying the amount of sunlight, water, or fertilizer. Small groups of children can be asked to devise the experiment and define the problem. Some questions that might be asked are:

4. Peter Holmes et al., "Teaching Statistics to Eleven-to-Sixteen-Year-Olds," *Teaching Statistics and Probability,* 1981 Yearbook of the National Council of Teachers of Mathematics (Reston, Va.: National Council of Teachers of Mathematics, 1981), pp. 18–24.

	Yes	No
X. I'd rather be a humble doctor than a successful pop singer.	8	22
Y. I'd rather be a doctor than have to sing for my living.	20	10

	Yes	No	Don't Know
X. Should all pupils be allowed to stay at school until they are 16?	14	8	8
Y. Should all pupils be forced to stay at school till 16, even when they don't want to?	8	20	2

FIGURE 19.28
Data Affected by Subtle Differences in Questions

- Under what circumstances do plants grow the fastest?
- Which of the three ingredients (light, water, or fertilizer) appears to have the greatest influence on plant growth?
- After one month, which plant showed the greatest growth?
- Were the results the same after the second month?
- What is the average growth per week?

As children watch plants grow, they can make periodic measurements, record data, and make predictions about future growth. After the experiment, they can organize and analyze the data collected, and attempt to make generalizations about the effects of the three ingredients on the growth of plants. Graphs can be constructed as a final activity.

Physical Education

Gather data and compare the running, jumping and throwing abilities of children. Or, collect readily available data on favorite football or baseball teams.

Social Studies

Often, community issues—such as a new traffic light or senior citizen recreation center—need to be researched. Class activities that focus on the community help children see a real need to collect and analyze statistical data.

One such project is described by Marks et al.[5] At the request of the school parent organization, a class conducted a traffic survey on number of cars that pass the school. When the teacher described the project to the class, the following questions arose and had to be answered in order to obtain accurate, nonbiased data.

- At what time or times should the survey be made?
- Do we count all vehicles for a single total, or should we classify them in some way and report for each classification?
- Where do we count the vehicles—in front of the main entrance to the school, or in a corner of the school yard at the intersection of two streets?
- If a car stops at the school, should it be counted?

The final report included three parts:

1. A description of how, when, and where the data were collected.
2. A compilation of the raw data, including the different kinds of vehicles (such as cars, trucks, fire engines) for each hour and the total number for each hour.

5. John L. Marks et al., *Teaching Elementary School Mathematics for Understanding*, 4th ed. (New York: McGraw-Hill, 1975), pp. 255–56.

3. Graphs: (a) bar graphs showing the number of each kind of vehicle that passed by each hour; (b) a broken line graph showing the total number of vehicles during each hour; (c) a circle graph showing the percent of each kind of vehicle that passed in front of the school each day.

PLANNING ASSESSMENT

The topic of this chapter, unlike most of the others presented in this book, can be taught at almost any elementary level. Differences in presentation and assessment will depend on the age and maturity of children, and on mastery of prerequisite skills.

Assessment of ability to make tables and draw graphs and to find measures of central tendency is straightforward. However, assessment of each child's understandings is equally important, and can include the following questions.

Gathering and Organizing Data

1. Why do we make frequency-distribution tables?
2. In what way is a frequency-distribution table a more effective way to present data?

Analyzing Data

1. Tell how to find the mean, median, and mode of a set of data.
2. If you know only the number of scores in a set of data and the mean, can you find the total of the scores?
3. Must the mode of a set of data be equal to one of the scores? Must the mean of a set of data be equal to one of the scores? Must the median be equal to one of the scores?
4. Can the mean of a set of data ever be equal to the greatest number in the set? Can the mode of a set of data ever be equal to the greatest number in the set? The smallest number in the set?
5. Give an example in which the mode is a more representative score than the mean.
6. Give an example in which the median is the most effective way to indicate the central tendency of a set of data.
7. Can you find the mean, median, and mode from the data pictured on this graph? What are they?

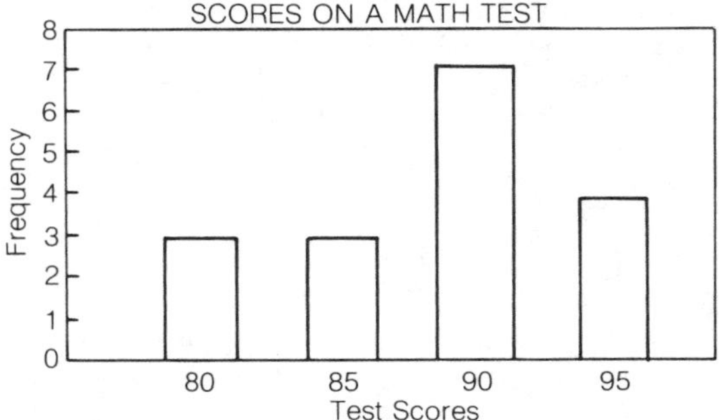

Reporting Data

1. For what purposes would you select a bar graph to present a set of data visually? A line graph? A circle graph?
2. Why is using a graph an effective way to present data? What is the disadvantage for interpreting graph data?
3. A child constructs a circle graph to show how she plans to spend her allowance. If her allowance is increased, explain if this will require a change in the circle graph.
4. Is a picture graph more advantageous than a bar graph?
5. A store collected this data on the size of dresses sold.

	Sold		*Sold*
Size 5	3	Size 11	14
Size 7	12	Size 13	6
Size 9	24	Size 15	2

Which type of graph should be used? What measure of central tendency is the best?

Other Concerns of the Teacher

VI

20

Management and Assessment

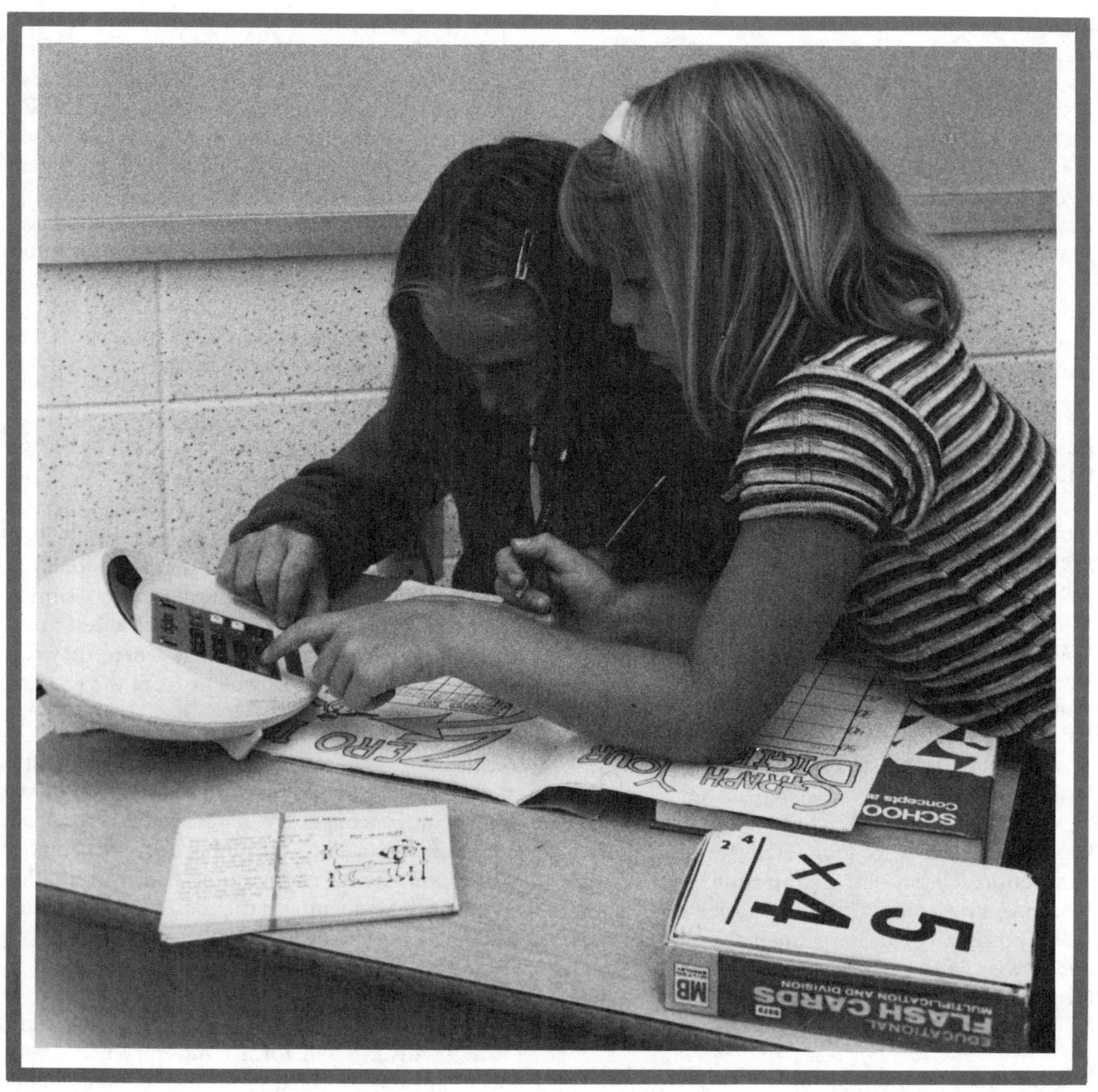

Each child's success and enjoyment in learning mathematics depends not only upon our selection of appropriate mathematical content and carefully sequenced activities, but also upon our ability to manage the program of mathematics instruction within a classroom setting, and to monitor progress toward goals.

The latter two themes have been emphasized in sections entitled *Tips on Managing the Classroom Situation* and *Planning Assessment*. However, suggestions were specific to the content of the chapter. We need to take a broader look at classroom management for mathematics instruction and to consider our overall program for gathering data and evaluating each child.

MANAGING-ORCHESTRATING CLASSROOM ACTIVITIES

Our management of the classroom situation will greatly influence how well we are able to apply the guiding models that have been presented and used. We must recognize that managing a program of mathematics instruction within a classroom setting is extremely complex. Many factors must be considered simultaneously and kept in balance. The task is so similar to that of a composer arranging music for an orchestra that the authors use the term *managing—orchestrating*.

When selecting content and activities, keep in mind factors that relate to personalized instruction, classroom organization, and use of materials and equipment. Emphasize one factor, then another. Try to keep them related in a balanced way so there is that harmony that characterizes a classroom in which each child learns and enjoys mathematics.

Personalizing Instruction

Throughout this text the authors have emphasized strengths and needs of *each* child. This has been done because of a deep conviction that instruction must be personal. Each child has unique needs.

In the past, some programs have emphasized teaching content according to the individual's development. More recently, the idea of providing the least restrictive environment for each child (mainstreaming) has led to an even greater range of knowledge and skill development within classrooms.

Children differ in many ways beyond surface concepts and skills. They differ in aptitude, in socioeconomic and geographic experiences, in motivation for classroom learning, in attitude toward mathematics, and in learning style. All of these differences enter into decisions about the appropriate nature of instruction.

Although an individual program of instruction will need to be planned to reflect strengths and needs, our emphasis should be on personalizing instruction for each child and not on arranging for each child to work alone. Personalized instruction involves planning for individuals, but permits children with similar strengths and needs to work together at least part of the time within group settings where they can learn from each other as well as from the teacher.

Grouping Children for Instruction

Teachers should restructure the total class into smaller groups based upon ability (both mixed and homogeneous), by achievement, for peer teaching, and so on. Grouping for mathematics instruction should normally be based on mathematical achievement and diagnosed needs of each student.

Because a child's strengths and needs in mathematics vary over time, use a flexible approach, regrouping students as appropriate. Occasionally we will form temporary groups for specific purposes. Such groups may include children having difficulty with a specific cluster of objectives, or those with comparable achievement levels for a consolidating game. Needs other than mathematics concepts and skills should also be considered. For example, children with a specific learning style may be placed in more independent laboratory type activities.

Some teachers tend to place children in a few groups and keep them there.[1] If we are flexible in our approach to grouping, we are more likely to actually plan for the needs of each child. Focus on your information about the achievement and needs of each child, then think about what you have learned concerning individual instruction.

In contrast, Baratta-Lorton states that whole-class teaching should be used to avoid reinforcing the concept of nonlearners, and to provide a greater number of people with whom children can share their ideas.[2] A child's self-concept as a learner or nonlearner is important; however, even Baratta-Lorton advocates a form of grouping as he describes differentiated tasks that enable each child to be challenged during a whole-class lesson.

Independent Work

A teacher can only work directly with one small group at a time. Therefore, other children must have work they can do independently, with a partner or with another small group. At other times, children can be involved in such activities and the teacher can circulate among them, asking questions and providing guidance.

Help children use logical thinking skills and focus on attributes of objects and events that are relevant to the concept being developed. Children can work more independently when they are taught strategies. For example, when a teacher stresses the meanings of the four operations, properties of the operations such as commutativity and distributivity, and the many special relationships that can be observed, a child typically learns strategies for figuring out more than just easily recalled basic facts.

Independent work is often a consolidating activity; conversely, when children are truly ready for consolidating activities they have little difficulty working independently. Independent work can also be a transferring activity. Horizontal transfer activities such as the Parts Pages for different sums or the worksheets for multiplications facts (used with different facts) can be done while we work with other children needing a directed activity.

Involving Each Child

If instruction is to be truly personalized, make sure that each child is *actively* involved in the learning process. One procedure that works is to increase *waiting time*. Whenever we ask a group of children a question, we tend to wait only a second before calling on a child to respond. If this time is increased to at least 3 seconds, *more* children are ready with longer and more thoughtful responses.[3] Behavior modification strategies can sometimes also be used to increase involvement, as can working in pairs or peer tutoring. Learning centers are another frequently used method.

We can increase group participation by having each child respond simultaneously. One procedure is to use the Show-Me Cards described in Chapter 6. Each child can arrange digit cards in a Show-Me Card, then place it face down until you ask all children to simultaneously display their cards. Or, assign the same task to each child, but encourage children to make individual responses. For example, children can make graphs with arrows.[4] Figure 20.1 illustrates different responses made by children when ⟶ means + 4 and ➡ means X 2. Although each child in the group is given the same task, each one responds at a different level of sophistication.

Sometimes instructional materials need to be modified to increase active involvement. For example, some children become confused and unable to respond whenever a text or worksheet is crowded with symbols. It may help such children to present each exercise on a separate page

---

1. Anna O. Graeber, Eui-Do Rim, and Nancy J. Unks, *A Survey of Classroom Practices in Mathematics: Reports of First, Third, Fifth, and Seventh Grade Teachers in Delaware, New Jersey, and Pennsylvania* (Philadelphia, Pa.: Research for Better Schools, 1977).
2. Robert Baratta-Lorton, *Mathematics. . . A Way of Thinking* (Menlo Park, Calif.: Addison-Wesley, 1977), p. 286.
3. See Robert B. Sund and Arthur Carin, *Creative Questioning and Sensitive Listening Techniques: A Self-Concept Approach*, 2nd ed. (Columbus: Charles E. Merrill, 1978), pp. 24–26.
4. See Comprehensive School Mathematics Program, *CSMP in Action* (St. Louis, Mo.: CEMREL, Inc., 1978), pp. 25–46. The Comprehensive School Mathematics Program (CSMP) makes extensive use of colored arrows to present relations and function in a precise, nonverbal way. This example is adapted from materials in the CSMP elementary school curriculum and reproduced here with permission of the publisher CEMREL, Inc., St. Louis, Missouri.

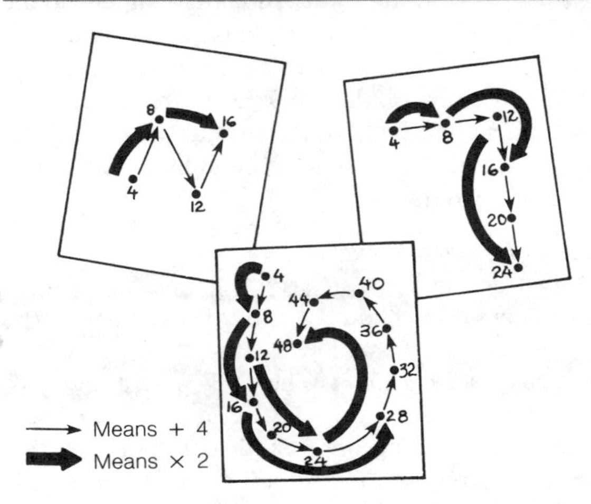

FIGURE 20.1
Arrow Graphs

or card, gradually increasing the number of exercises until regular material can be used successfully. Others are helped if the number of alternative responses is carefully controlled. Begin by having the child select one of the two possible responses. Later, proceed to three, then four, possibly to incomplete responses or clues, and finally to open-ended responses such as a box (variable) in which the child is to write a numeral. Prepare a worksheet for each choice-level, or construct a learning center with a series of stations that gradually introduce choices.

We must also provide prompt feedback concerning each child's responses. Possibilities include answer cards, confirmation by another child, and the use of a calculator. It is usually easy to build immediate feedback into a learning center.

Organizing the Classroom

Some schools have rows of self-contained classrooms; others have very large areas without walls and are called *open space* schools. While the design or physical setting of a school will have some influence on your teaching, more important is your philosophy about who learns and how learning is facilitated. A teacher who supports a personalized learning environment will arrange the classroom so that the instruction often takes place in small groups or with individuals.

Be flexible in organizing the classroom furnishings; desks and tables should be easily rearranged if necessary to help achieve specific instructional goals. Baratta-Lorton describes a basic plan for a self-contained classroom in which tables are available for learning centers or small group work, with instructional materials available to all.

> Rows of student desks are positioned to give everyone a clear view of the overhead projector screen or blackboard. The desks are surrounded by tables, chairs, and bookcases. Tables are usually in short supply in schools, but card tables or doors with legs attached work as well. To eliminate the need for students to carry their desk chairs to the tables, the folding chairs that usually sit unused in school auditoriums may be used. Bricks and boards make usable bookcases.
>
> Bookcases separate the tables from one another and provide ample storage space for tiles, cubes, geoboards, and other materials. They also store the activity-time equipment. . . . Closed cupboards are not desirable, because students should be able to see the stored materials.
>
> If the teacher does not wish students to have general access to a certain material it is best not stored in the classroom. To establish an atmosphere of trust, everything in the classroom should be open and available to all.[5]

For small group activities, have chairs surround a table. When these are teacher-directed, place the child who has difficulty learning directly across from you, as in Figure 20.2. Whenever a group of children work on their own or with another adult, discuss what is expected of them.

5. Baratta-Lorton, *Mathematics,* p. 286.

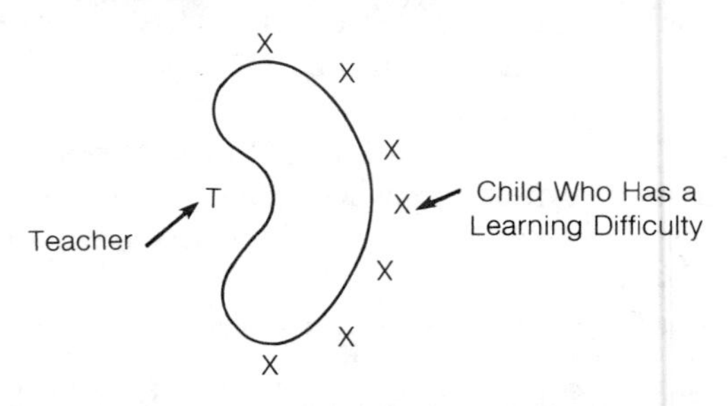

FIGURE 20.2
Placement of Children for Teacher-Directed Activity

Learning Centers

You can have learning centers whether you teach in an open-space setting or in a self-contained classroom. Such centers provide an excellent way to adapt instruction to individual needs and can be used for regular developmental instruction, diagnosis, remedial instruction, or enrichment activities.

A learning center is a corner with one or more desks or tables with interesting tasks for children to do on their own. A center can have a single task or may contain several stations to be approached at random or in a specified order. It also has appropriate materials, including special forms on which each child can record the results.

It is easiest to design learning centers for diagnosing, schematizing, consolidating, and transferring activities. Children usually need more direct teacher guidance during initiating and abstracting activities, especially since vocabulary and symbols are being introduced. After children understand an idea or a procedure, prepare learning centers so that individuals who need special help or are ready for a challenge can independently proceed with activities at their own rate.

You can easily design a learning center by including the following: objectives, directions, task cards, manipulatives, reference material, and a way to record or report what was done.[6] An example of a learning center is depicted in Figure 20.3.

---

6. See C. M. Charles, *Individualizing Instruction,* 2nd ed. (St. Louis, Mo.: C. V. Mosby, 1980), pp. 172–93; John I. Thomas, *Learning Centers: Opening Up the Classroom* (Boston: Holbrook Press, 1975); and Louise F. Waynant and Robert M. Wilson, *Learning Centers . . . A Guide for Effective Use* (Paoli, Pa.: Instructo, 1974).

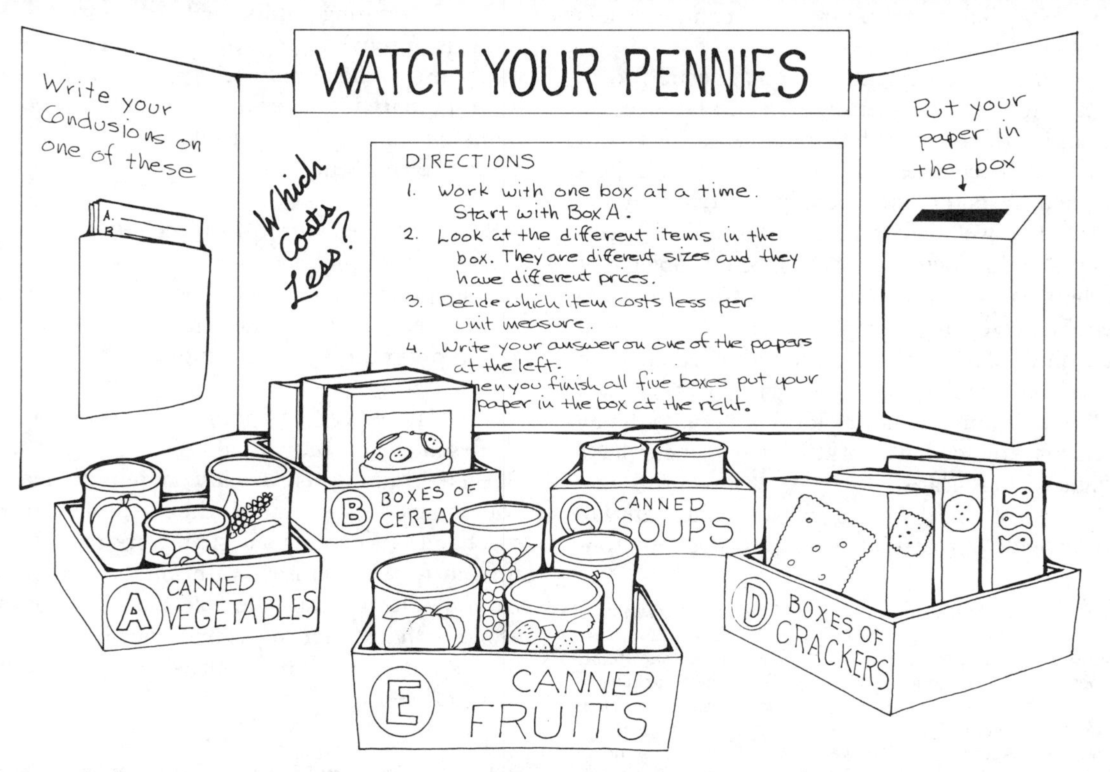

FIGURE 20.3
Learning Center for a Transferring Activity

A Math Lab

Sometimes a school provides a special room—a laboratory-like environment—for math instruction called a *math lab*. Many of the characteristics of a math lab can be seen in the regular classrooms of creative teachers.

Children may come to a math lab as a class or as individuals. When children come as a class, the regular teacher often works with another teacher who is in charge of the lab. Parent volunteers may also be present. If children come as individuals, they typically work with children from other classes who have similar strengths and needs.

A math lab should provide a rich learning environment in which children are usually involved in groups at tables or at one of many learning centers. They work on activities or experiments, at least some of which relate to their geographical situation. For example, there may be tasks that involve maps of the neighborhood around the school or problems faced by local businesses. Greater quantities of classroom materials are available in a math lab; a computer terminal may also be present.

Pay special attention to diagnostic procedures and to record keeping. A math lab needs a system for diagnosing the strengths and needs of children. Special learning centers and teacher-designed tests or interview protocols may be included as well as commerical tests. Data gathering should focus on the specific objectives teachers have adopted for math instruction. Keep records of specific completed activities, noting the child's development with reference to each objective. Such systems are best worked out within the context of the local school, but pay attention to accounts of the way others have organized math labs.[7]

For a child who has difficulty, a math lab can provide an attractive, supportive environment with many activities he can do successfully. For an exceptionally capable child, it can provide an atmosphere that stimulates and challenges with many transferring activities that involve realistic problems. A math lab personalizes instruction for all children. Children are highly motivated without grades on papers, and are often proud of what they have produced when they complete a task.

Individualized Learning Systems

During recent years a number of individualized learning systems have been developed for teaching mathematics. Included are already sequenced materials and a means for recording the progress of each child. Though often costly, a commercially published system contains materials that would take you months to develop.

These systems are not usually developed locally; therefore, they may not relate well to local objectives or to the needs of your children. Therefore, select a learning system with great care.

Evaluating individualized learning systems is a challenging process, partly because many are rather complex. Houser and Heimer have prepared a checklist for evaluation with items clustered under ten headings.[8] Note especially the following three:

- Scope and sequence of objectives is consistent with school philosophy.
- Student interaction with peers and teacher is provided at regular intervals.
- The program is easy to implement.

These headings point to some of the difficulties of using individualized learning systems. Special training is usually required: teachers sometimes feel so much time is spent on the mechanics of making a system work that they are unable to work closely with children. However, careful selection and flexible use of a system—along with keeping the specific needs of children in mind—can help provide personalized instruction.

---

7. See Nuffield Mathematics Project, *I Do and I Understand* (New York: John Wiley, 1967); Joan Y. Schussheim, "An Annotated Math Lab Inventory," *School Science and Mathematics* 80, no. 6 (October 1980): 513–21. Also see articles in the December 1971 issue of *The Arithmetic Teacher;* Thomas R. Post, "A Model for the Construction and Sequencing of Laboratory Activities," *The Arithmetic Teacher* 21, no. 7 (November 1974): 616–22; and Robert E. Reys and Thomas R. Post, *The Mathematics Laboratory: Theory to Practice* (Chicago: Science Research Associates, 1970).

8. Larry L. Houser and Ralph T. Heimer, "A Model for Evaluating Individualized Mathematics Learning Systems," *The Arithmetic Teacher* 26, no. 4 (December 1978): 54–55.

Systems have different orientations. One system, Developing Mathematical Processes (DMP)[9], is a component of Individually Guided Education (IGE) developed by the Wisconsin Research and Development Center for Cognitive Learning at the University of Wisconsin. DMP is an activity-oriented program emphasizing problem solving, and incorporates games and materials for learning centers. Instructional activities can be adapted to individuals and small or large groups. Other systems provide individual assignments based on a carefully structured sequence of tests.[10] Individual worksheets or workbooks are usually involved.

Using Materials and Equipment

Throughout this text the authors have emphasized the need to use varied exemplars in teaching specific content. The following guidelines may help you create instructional materials.

1. *Provide attractive materials and thereby increase attention span.* During initiating, abstracting, and schematizing activities, have children handle objects that are simple geometric shapes with a minimum of irrelevant and potentially distracting attributes. Toys can be used later for application settings.

2. *Make sure you have enough materials for each child or each group.* Inventory them before you start the lesson, and have them near by.

3. *Provide adequate time for each child to manipulate objects before moving to symbols.* Mathematical ideas grow slowly, so let children measure, count, make comparisons and so on at their own rate. Discuss what they are doing.

4. *Relate two-dimensional exemplars to the objects they represent.* Children—especially younger ones—are not always able to relate a specific drawing or photograph to its three-dimensional counterpart. Make sure children can match two-dimensional representations with the real objects.

5. *Provide each child with a number line.* Each child should have a number line covered with clear contact paper on top of his desk. Arrows that are drawn can be erased and the number line used again.

6. *Use the chalkboard.* Pastel shades of colored chalk erase easily and are usually interesting. You may also want to pull a shade or a projection screen over part of the material on the chalkboard, raising it when the material is needed. You can make an unknown by fastening a piece of paper over the answer with masking tape.

7. *When using a game, group children of comparable ability together.* Games make learning mathematics fun for children; but if participants are to be interested and experience a measure of success, they must have comparable concept and skill development.

8. *Make sure children can see demonstrations.* Demonstrations that involve manipulatives should be on a vertical surface such as a flannel or magnetic board.

9. *Use the overhead projector.* Objects can be placed on the stage of the projector and moved about for all to see. If the objects are translucent, such as colored plastic tiles for mosaics, they project in color; otherwise, shadows can be observed. Objects can be used in this way to demonstrate grouping and commutativity, and can illustrate two-dimensional shapes within three-dimensional shapes. Part of the stage can be covered with a paper mask when division is developed from an array. Also, a cut out mask can be used to focus on a 3×3 array of numbers within a transparency of a monthly calendar. Ask children to add the numbers in the array and find the relationship between the sum and the numbers exposed.[11] A special set of Cuisenaire rods

---

9. *Developing Mathematical Processes*, developed by the Analysis of Mathematics Instruction (AMI) Project of the Wisconsin Research and Development Center for Cognitive Learning at the University of Wisconsin (Chicago: Rand McNally, 1974–76).

10. *Individually Prescribed Instruction,* developed by Research for Better Schools, Inc. (Philadelphia, Pa.: Research for Better Schools).

11. Max A. Sobel and Evan M. Maletsky, *Teaching Mathematics: A Sourcebook of Aids, Activities, and Strategies* (Englewood Cliffs, N.J.: Prentice-Hall, 1975), pp. 206–7.

are prepared for an overhead projector. Commercial transparencies are also available, but you can make most of your own transparencies with transparency pens.

10. *Use bulletin boards.* Bulletin boards can be used to teach specific concepts or skills in the classroom, the math lab, or a hallway. Design the display so children can interact with it: pose questions that involve patterns to be discovered or applications to be made. Make immediate feedback available by using a card that a child raises after responding to the questions. On one bulletin board you could provide a space where each child displays his best work in mathematics. Many books and magazines have ideas for creative bulletin boards.[12]

11. *Select audiovisual aids in terms of your objectives.* When you use films and filmstrips, make sure they relate to your instructional objectives. Prepare your children by raising questions; give them something to look for. Films and filmstrips are usually commercially prepared, but you can make slides of the neighborhood, for example, to help children observe shapes. Children can work out problems together while looking at the menu of a local fast-food outlet.

ASSESSING PUPIL PROGRESS

Success in managing or orchestrating classroom activities depends on what we know about each child. Similarly, our ability to select suitable content and activities depends on what we know about the child's strengths and weaknesses. We should collect useful data, then evaluate pupil progress.

Data Gathering and Evaluation

Assessment is sometimes thought of as testing, but it is much more. It includes different forms of data gathering and judgments based on the data collected. Data can be gathered from paper-and-pencil devices, interview protocols, anecdotes, information from parents and other teachers, and existing records. The best judgments made about a child are usually based upon varied and plentiful data.

Assessment takes place in relation to specified objectives regarding knowledge, skills, and attitudes. It also includes evaluation, a decision-making process in which judgments are made as to the quality of data at hand, the relative importance of different topics for an individual, and how adequately the child has met specified objectives. Interpretations of data and judgments are affected by our beliefs concerning the child, mathematics, and learning itself.

Assessment should reveal a child's strengths and weaknesses. It should show how far a child has come in the sequence of steps in developing an algorithm; how well the child performs a particular skill and comprehends the concepts involved. Assessment should also reveal the child's attitude toward mathematics. A thorough assessment will even suggest what kind of instruction is most likely to be effective.

Types of Testing Instruments

Much of the data we collect for each child will be gathered with a test. Some tests are norm-referenced; others are criterion-referenced. Some are published standardized tests, and others are teacher-made tests. The published norms for standardized tests allow us to compare a child with the average performance of the larger group.

Norm-referenced tests are useful in evaluating how much the child has learned. They allow us to compare the performance of one child with others of the same age or grade level. Derived scores, such as grade-equivalent scores, percentile ranks, or stanine scores are used. Norms have been established using a large number of children from different geographic areas and socioeconomic levels.

Criterion-referenced tests are useful in evaluating what the child has learned. Instead of comparing a child's performance to that of others, they allow comparisons against a list of

12. See Seaton E. Smith, Jr., *Bulletin Board Ideas for Elementary and Middle School Mathematics* (Reston, Va.: National Council of Teachers of Mathematics, 1977).

EXCURSION
Do Standardized Tests Measure the Right Things?

The Arithmetic Teacher is the journal of the National Council of Teachers of Mathematics that is of special interest to mathematics teachers in elementary schools. Occasionally, it publishes two articles with seemingly opposing views. Obtain a copy of the April 1973 issue and read the responses by Epstein and Wilson to the question: Do standardized tests measure the wrong things?

predetermined behaviors or competencies in a given skill area. Criterion-referenced tests contain items for each competency on the list. This item-by-item identification with goal behaviors makes these tests especially useful.

These tests are often developed by a state department of education or by a school district to fit a list of specific competencies. Unfortunately, a school or district will sometimes adopt such a list but use a published norm-referenced test to gather data for evaluation, resulting in a poor match between the content sampled on the test and the list of competencies.

Published Tests

There is no tidy scheme for classifying published tests. However, achievement tests and diagnostic tests are particularly applicable to teaching mathematics.

An *achievement test* is used primarily to determine how well children have attained the goals of instruction. Published achievement tests are usually standardized instruments with norms for interpreting test results. Comparable forms are usually available for retesting. There is also a manual to guide administration and interpretation. A text series publisher will often have available achievement tests that focus on the content of the texts, sometimes without norms, however.

Norm-referenced achievement tests are carefully developed, with data available concerning validity and reliability. Because they focus on common classroom objectives at a given level, they may not accurately reflect the objectives selected for a particular school or classroom. The manual for a standardized achievement test may state that it can be used for diagnosis. However, Chase notes:

> Standard achievement tests sometimes claim they can be used to diagnose special deficiencies in a given skill area. However, no standardized test contains enough items on a specific skill—e.g., adding unlike fractions—to produce reliable subscores on each of the component skills.[13]

Norm-referenced achievement tests do make it possible to compare the mathematics performances of each child with a sample of children from across the nation. They may also be used for curriculum assessment. Typically, published mathematics achievement tests are available as parts of comprehensive batteries.

In recent years, increasing interest has led to the publication of a limited number of criterion-referenced tests. Although a few such tests have been commercially published, most have been developed at the state or school district level so children can demonstrate skill with a specified set of behaviors or competencies. Most states include an assessment in mathematics in minimum competency-testing programs.

A limited number of published *diagnostic tests* are available. These tests are typically longer, for they must sample more extensively from specific skills to be reliable. Subtest scores suggest categories of strength and weakness for each child (Level II diagnosis as described in Chapter 3); however, more intensive diagnosis within categories of weakness (Level III diagno-

13. Clinton I. Chase, *Measurement for Educational Evaluation*, 2nd ed. (Reading, Mass.: Addison-Wesley, 1978), p. 299.

sis) is necessary before instruction can be planned for many individuals. This may require use of teacher-made tests and forms of data gathering other than paper-and-pencil tests.

Examples of published diagnostic tests include the KeyMath Diagnostic Arithmetic Test[14] and the Stanford Diagnostic Mathematics Tests.[15] The KeyMath is for elementary grades and is individually administered to children. Because of its attractive format, a child's interest is easy to maintain. The test has fourteen subtests and is useful for screening and for identifying general strengths and weaknesses. The Stanford Diagnostic Mathematics Tests are paper-and-pencil instruments administered to groups; different tests are available for different grade levels.

Teacher-Made Tests

In constructing *any* teacher-made test, begin with a set of objectives. To prepare an achievement test, sample the set of objectives for the period of instruction. Make sure the number of items for each cluster of objectives reflects the stress placed on each. Whenever the curriculum is defined in terms of a specific list of competencies, a criterion-referenced test can be constructed by preparing test items for *each* behavior on the list.

To construct test items, first describe behaviors for each objective. If the curriculum is described with content objectives, prepare behavioral indicators. If the curriculum uses behavioral objectives, determine specific but related behaviors for each objective; designate an exemplar in each behavior. Once behaviors are described, preparation of test items is relatively simple. Each test item is merely a format for eliciting a specific behavior.

Acquisition of a concept or skill is not cut and dried; no one behavior adequately indicates acquisition. More than one behavior—consequently, more than one test item per objective—is preferred. This way we obtain a clearer picture of how well a child understands the concept or can perform the skill.

14. See *KeyMath Diagnostic Arithmetic Test* (Circle Pines, Minn.: American Guidance Service, 1971).

15. See *Stanford Diagnostic Mathematics Tests* (New York: Harcourt, Brace, Jovanovich, 1976).

For a paper-and-pencil test, describe only paper-and-pencil behaviors; the range of behaviors is limited. However, a paper-and-pencil test can incorporate two-dimensional representations. Consider the following objective:

> **7.8**
> Each digit in a multidigit numeral names a number that is the *product* of its face value and the place value for its position.

For this objective, one behavioral indicator is:

> Given a four-digit numeral, the child identifies the number and kind of base 10 blocks suggested by each digit.

It would be easy to have a child pick out blocks in an interview situation. For a group of students, a test item can be designed that uses a two-dimensional representation of base ten blocks, as in Figure 20.4.

Items picturing base 10 blocks can be prepared with the help of rubber stamps purchased from school supply houses. If necessary, directions for each item can be read to the whole group.

Many teacher-made tests are prepared for diagnostic purposes. When given before a unit of instruction, they can show what each child is ready to learn. During a unit of instruction they can be used to find out which concepts and skills have been and need to be learned. Diagnostic tests are also valuable after instruction; a periodic general review is more effective if a diagnostic test can determine which concepts and skills need special attention. A published diagnostic test can sometimes be used, but our own tests best reflect the particular objectives we select.

A diagnostic test is constructed using procedures similar to those already described. Decide upon the objectives, then determine behavioral indicators for one or more objectives. Finally, write test items for each behavior to enable us to observe which tasks can be performed by each child. This way, we can infer which concepts are understood and which skills are mastered. (See the Sample Diagnostic Test on pp. 458-63.)

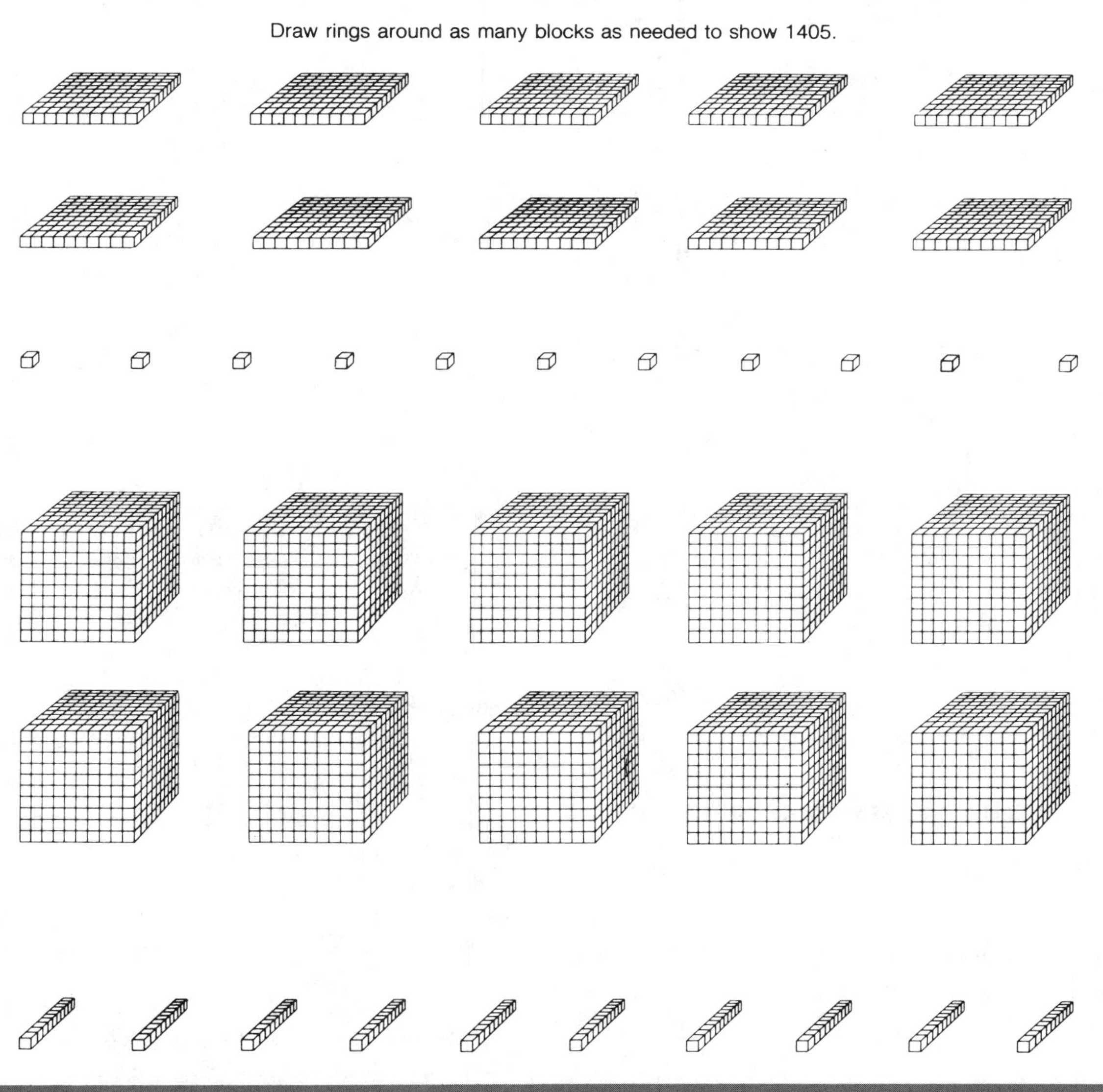

Draw rings around as many blocks as needed to show 1405.

FIGURE 20.4
Sample Test Item

Diagnostic tests that focus on paper-and-pencil computational skills should be based on task analysis procedures. For a skill such as subtraction of whole numbers, this involves identifying the sequence of skill acquisitions that eventually enable a child to subtract *any* whole number from a larger number. Consult the lists of ordered content that appear in chapters 8 and 10 in preparing these tests. For each specific identified skill, prepare three test items. Then assemble the test by ordering the items from simplest to most complex. Figure 20.5 illustrates a teacher-made test prepared by consulting the list of specific skills in chapter 8. Subtraction skills 15–20 were used for rows A–F.

After a child completes a test of computational skill, score the test to identify strengths and weaknesses and look for patterns among the

NAME ____________

A.	482 − 261	748 − 425	637 − 134
B.	391 − 78	273 − 46	482 − 69
C.	429 − 73	817 − 64	534 − 81
D.	872 − 437	354 − 128	965 − 516
E.	528 − 337	759 − 282	436 − 194
F.	636 − 487	354 − 198	726 − 249

FIGURE 20.5
Teacher-Made Test of Selected Subtraction Skills

errors, gaining further clues as to the child's thinking and additional instruction required. *Error Patterns in Computation* illustrates this process for many different algorithms.[16]

Paper-and-pencil devices can also suggest a child's attitudes towards mathematics. Typically, instruments developed for research purposes have been Thurstone or Likert scales appropriate for older students or for a clinic setting. Yet a paper-and-pencil device can be designed for a group of children in the primary grades. Each child is given a paper with rows of faces as illustrated.

16. Robert B. Ashlock, *Error Patterns in Computation,* 3rd ed. (Columbus, Ohio: Charles E. Merrill, 1982).

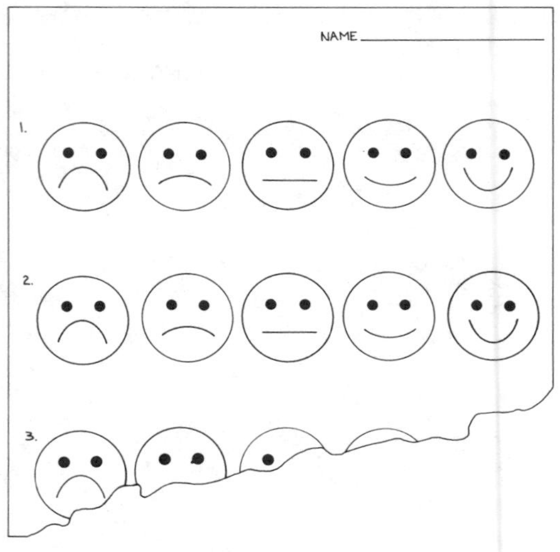

Each child marks the face that shows how he feels about a stimulus that is read. Pose questions similar to the following:

- How do you feel when you do math?
- How do you feel in math class when you tell what the answer is?
- How do you feel when you use base 10 blocks?

Other Data-Gathering Procedures

Most data-gathering discussed are paper-and-pencil procedures generally used for assessing groups of children. Other data-gathering procedures that focus on each individual should be used regularly.

The structured interview, introduced in Chapter 3, follows a planned protocol. Thereby, a more detailed Level III diagnosis can be conducted within a content category. Interview protocols point out things we cannot obtain from paper-and-pencil tests: a child's thinking patterns, the nature of specific misunderstandings, what strategies a child uses for solving problems, and particularly helpful exemplars.

When planning an interview protocol, use the Exemplar-Class Translation Matrix found in Chapter 3. Both stimulus and response can be three-dimensional, two-dimensional, or symbolic. Plan a *variety* of tasks so you can better understand what the child knows and can do.

Keep a day-by-day record of each child's cognitive and affective growth. Make notations on a 3 × 5 card whenever his behavior suggests knowledge or skill in a key area. Notations can also be made whenever the child says or does something that suggests how he feels about mathematics and instruction.

Similarly, make a record of observations on a weekly chart like the one shown in Figure 20.6. Such a chart can be particularly useful when planning instruction or conducting conferences with parents.

A child can be systematically observed through a checklist. Items on the list can be checked with a 2 if the activity is frequently observed, 1 if seldom observed, and 0 if never observed. Examples of items include:

_____ Volunteers in class.
_____ Asks a question if something is not understood.
_____ Appears happy when engaged in math activities.
_____ Works independently.
_____ Applies what has already been learned.
_____ Undertakes optional assignments.
_____ Completes optional assignments undertaken.
_____ Mentions math-related activities undertaken outside of school.

(text continued on page 464)

CHILD	M	TU	W	TH	F
Ann	Did well measuring computing ratio	Excellent diagram of problem solution			
Jim	Not sure how to measure	absent			
Mike	Measured easily, computed ratio with difficulty	Good diagram with help from Julie			
Julie	Led group in solving the problem	Excellent diagram worked independently			
Virginia	Measured could not compute ratio	Made a picture and story for her record.			

FIGURE 20.6
Chart for Recording Observations

EXCURSION
Multiple-Choice Items

Although multiple-choice items are often used when preparing paper-and-pencil tests, such items can be tricky. A child may pick the wrong alternative for any number of reasons.

First, read the following articles, then select a content objective and prepare three different multiple-choice items for the objective. Administer the items to a child, then interview the child to determine why each response was chosen. Will you revise any of your items? How?

- Leroy G. Callahan, "Test-Item Tendencies: Curiosity and Caution," *The Arithmetic Teacher* 25, no. 3 (December 1977): 10–13.
- William D. McKillip, "Teacher-Made Tests: Development and Use," *The Arithmetic Teacher* 27, no. 3 (November 1979): 38–43.

A Sample Diagnostic Test for a Unit on Money[17]

Mrs. Foster taught a unit on money. Content objectives for the unit were as follows:

Content Objectives

A. Identification

1. **a.** This represents a penny.
b. It is worth 1 cent.

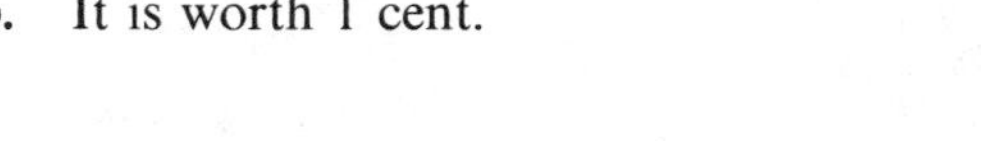

2. **a.** This represents a nickel.
b. It is worth 5 cents.

3. **a.** This represents a dime.
b. It is worth 10 cents.

4. **a.** This represents a quarter.
b. It is worth 25 cents.

5. **a.** This represents a half-dollar.
b. It is worth 50 cents.

B. Symbols

1. The ¢ (cents) is used to indicate cents, and follows the numerals.
2. **a.** The $ (dollars) is used to indicate dollars, and appears before the numeral.
b. The $ (dollars) is said after the amount.

17. Adapted from the unpublished work of four teachers: Carol Kochheiser, Linda Raivel, Linda Stanley, and Nancy Wilson.

C. Relations

1. One nickel and 5 pennies are equivalent in value.
2. One dime; 2 nickels; 1 nickel and 5 pennies; and 10 pennies are equivalent.
3. One quarter; 2 dimes and 1 nickel; .... 25 pennies are equivalent in value.
4. One half-dollar, 2 quarters; .... are equivalent in value.
5. The total value of a set of coins is the sum of their individual values.
6. The value of a coin is not consistently related to its size.

D. Making Change

1. Change is given in the fewest number of coins possible.

For the content objectives, Mrs. Foster prepared the following behavioral indicators. Can you match each behavior with one of the content objectives? Some content objectives are already noted.

Behavioral Indicators

A1a-5a	Given a collection of coins, the child will identify a penny (nickel, dime, quarter, half-dollar).
______	When shown a coin, the child will state the name of the coin.
A1b-5b	When shown a coin, the child will state its value.
B1	The child will write 45¢ when told to write, *forty-five cents.*
______	The child will write $1.45 when told to write, *one dollar and forty-five cents.*
______	The child will say, "One dollar and forty-five cents" when shown a card with $1.45 written on it.
C1-4	Given a coin other than a penny, the child will indicate how many pennies are required to equal the value of that coin.
______	Given less than one dollar in coins, the child will state the total value.
______	Given a value, the child will show it with different sets of coins.
______	Given a dime and a penny (dime and nickel; quarter and nickel), the child will point to the one with the larger value.
D1	Given an item such as a toy, with an indicated cost, and the amount handed to a clerk, the child will choose the fewest possible coins as change.

Test items were devised for the behavioral indicators. The following oral items are to be individually administered, and the paper-and-pencil items presented to a group.

Oral Items

Spread out 4 quarters, 10 dimes, 10 nickels, and 20 pennies on a table.

1. Ask the child to show various coins; such as a nickel.
2. Point to various coins and ask the child to identify them.
3. Ask the child which coin is worth the most, the least.
4. Point to various coins and ask what they are worth.
   Set the coins aside.
5. Show the child a card with *35¢* on it. Ask the child to read the card.
6. Repeat for $1.45 and 46¢.

7. Ask the child to write forty-five cents using the correct signs. Ask him if he knows another way to write the same amount using different signs.
8. Ask the child to write one dollar and thirty-eight cents using the correct signs.

Paper-and-Pencil Items

1. Draw lines from each coin to its amount and to its name.

25¢		Nickel
1¢		Quarter
10¢		Dime
5¢		Penny
50¢		Half Dollar

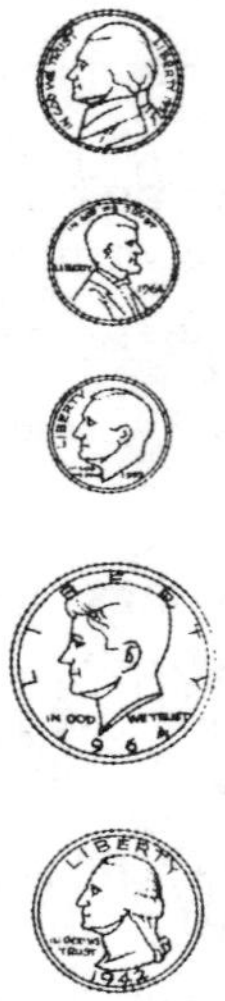

2. The correct way to write one dollar and forty-five cents is
   a. $.145
   b. $145
   c. $1.45

3. The correct way to say $.73 is
   a. seventy-three dollars
   b. seventy-three cents
   c. seventy-three

4. Circle the two sets of coins that show the same amount of money.

5. Circle all the sets that show amounts equal to a dime.

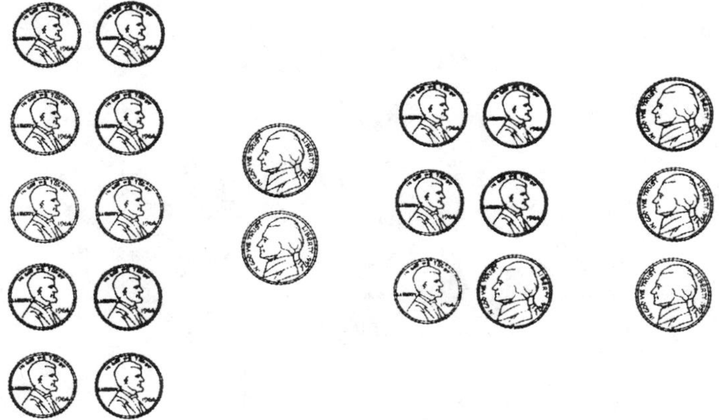

6. Circle the sets that show amounts equal to a quarter.

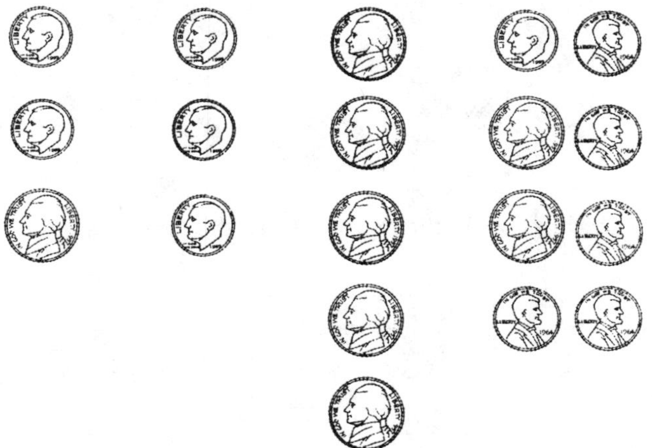

7. Circle the set of coins that is worth 63¢.

8. The total value of all these coins is
a. 39¢
b. 34¢
c. 70¢

9. Circle the set of coins that shows 57¢.

10. In each box, circle the coin that has the larger value.

11. Lisa had a dollar. She bought a book for 35¢. Which set of coins shows her change?

12. Jim had a quarter. He bought a pencil for five cents. How much change will he get? Circle the set of coins which shows the change he is most likely to receive.

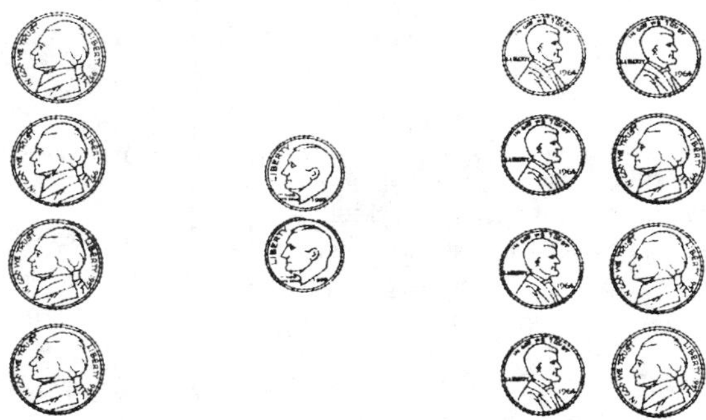

EXCURSION
Measuring Attitudes

Make a device for measuring a child's attitude toward mathematics. Focus on each of the following:

1. Enjoyment of mathematics
2. Security and confidence with mathematics
3. Appreciation of the usefulness and value of mathematics[18]

For ideas about attitude scales, questionnaires, and the use of incomplete sentences, consult one or more of the following:

- Mary Corcoran and E. Glenadine Gibb, "Appraising Attitudes in the Learning of Mathematics," in *Evaluation in Mathematics,* 26th Yearbook (Washington, D.C.: National Council of Teachers of Mathematics, 1961), pp. 105–22.
- William A. Dunlap, "An Attitudinal Device for Primary Children," *The Arithmetic Teacher* 23, no. 1 (January 1976): 29–31.
- Wilber H. Dutton and Martha Perkins Blum, "The Measurement of Attitudes Toward Arithmetic with a Likert-Type Test," *The Elementary School Journal* 68 (February 1968): 259–64.
- Linda A. Michaels and Robert A. Forsyth, "Measuring Attitudes Toward Mathematics? Some Questions to Consider," *The Arithmetic Teacher* 26, no. 4 (December 1978): pp. 22–25.

When you have completed your device for measuring attitudes, administer it to a few children. What did you learn? What suggestions would you make regarding instruction?

18. Linda A. Michaels and Robert A. Forsyth, "Measuring Attitudes Toward Mathematics? Some Questions to Consider," *The Arithmetic Teacher* 26, no. 4 (December 1978): 22–25.

An excellent context for making observations about each child is an everyday situation in which groups are solving problems that require measurement, computation, and record keeping. Consider the following:

PROBLEMS

What is the relationship between a person's height and the distance between the fingertips of his left and right hands when both arms are extended as far apart as possible? Be sure to measure several people before you decide.

Find the height of a tree in the playground. Have one person stand straight, then place a mirror flat on the ground so that person can see the top of the tree in the mirror. How tall is the tree?

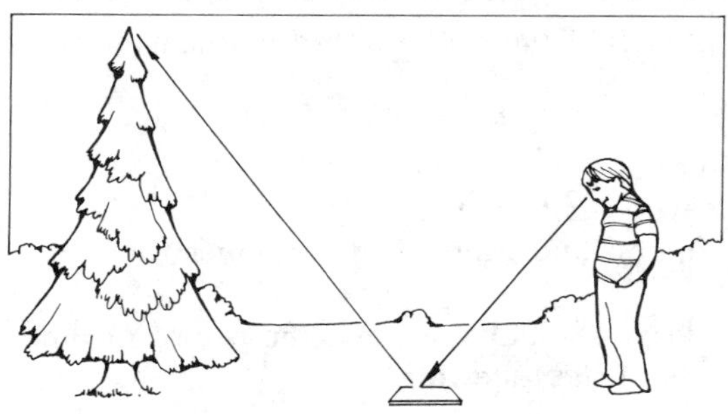

Such tasks typically involve computation of one form or another; different kinds of records can be made. Have students write a paragraph to explain how the task was completed, or prepare a product such as a graph, model, or scale drawing. Careful observation of children's problem-solving methods will help us plan further instruction.

Assessment and Instruction

Do not gather data just once and, on the basis of that data, plan instruction for the weeks ahead. Published or teacher-made paper-and-pencil tests may initially be used to identify areas of instruction for each child. But we must keep on collecting data and revising instructional plans as we learn more about each child.

Figure 20.7 illustrates the long and short-term cycle of collecting data, making professional judgments in view of observed patterns, collecting and planning instruction for each child. After instruction, we collect more data and revise our instruction. Even during a single interaction, we say something (instruct), note the child's response (data-gathering), then revise our comments (instruction). We can help *each* child learn mathematics if we view instruction as a part of a continuous cycle that includes assessment.

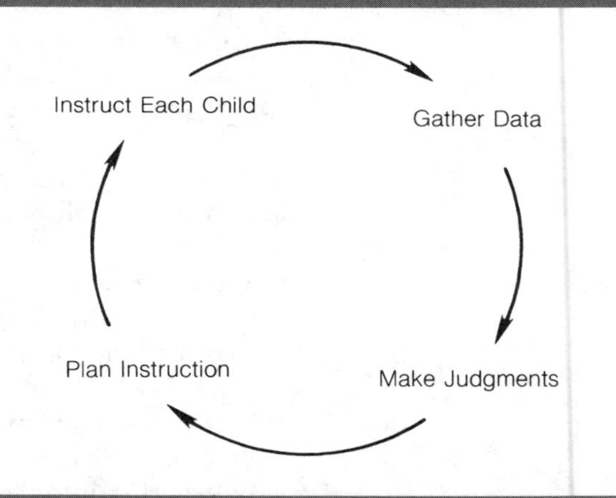

FIGURE 20.7
Cycle of Assessment and Instruction

21

In Your Future

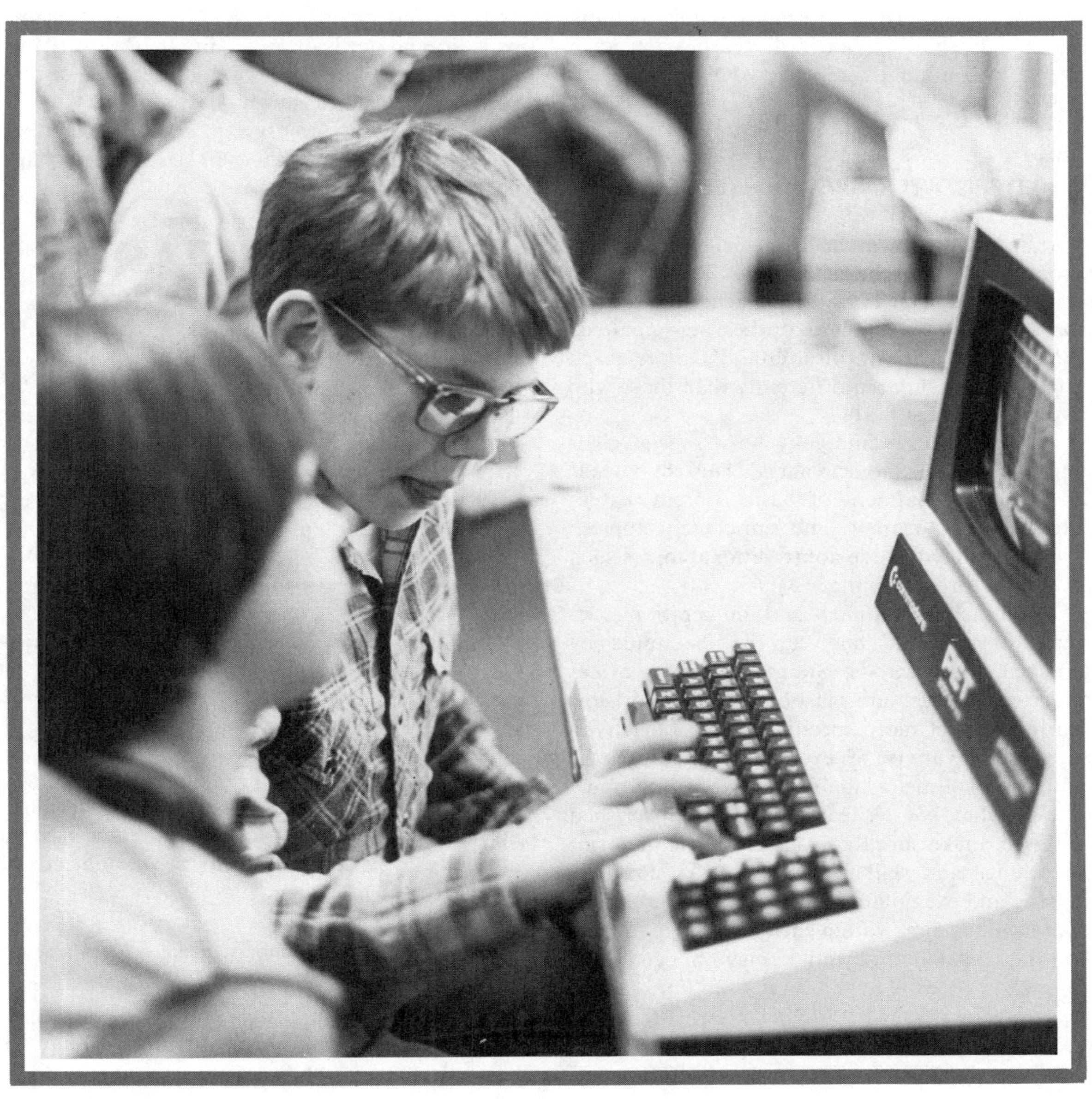

As you complete this text, take a brief look down the road ahead. There are many things to think about—the children you will teach, your own efforts to continually improve your skills.

CHILDREN WITH SPECIAL NEEDS

Certain children will have special needs. What should we do with the gifted child? The child who learns very slowly? How can we adapt instruction for a child who is blind or hearing-impaired? What about learning disabilities? Do we teach impoverished children differently than those who live in affluent suburbs?

The gifted child may be a special challenge as she seems to leap ahead. This child needs an in-depth treatment of basic content, extensions of that content, and enrichment topics.[1] The latter may include nontraditional topics such as BASIC programming.[2]

Teaching number and numeration concepts to a blind child can be facilitated by notched number rods (Stern Blocks, for example), base blocks, and plastic numerals. For slow learners, select more specific content objectives, and use a variety of exemplars. (Because the amount of time available is fixed, you must decide what *not* to teach.) With impoverished children, make sure that each child has the prior knowledge and skill needed for tasks. Problem solving and exemplars need to focus on situations that have meaning within each child's experience; but the selection of objectives may not need to be altered.

1. Joseph Payne, "The Mathematics Curriculum for Talented Students," *The Arithmetic Teacher* 28, no. 6 (February 1981): 18–21.
2. See James H. Wiebe, "BASIC Programming for Gifted Elementary Students," *The Arithmetic Teacher* 28, no. 7 (March 1981): 42–44.

Remember that all children learn and come to understand an idea in basically the same way. Special needs often require that we vary our plans, but children search for meaning during instruction, even those who have difficulty learning. We need to adjust our choice of content objectives and exemplars to fit the situation. This is true for *each child.*

ADDITIONAL WORK WITH CONTENT OBJECTIVES

As you strive to improve your teaching of mathematics, you will need to create your own content objectives. Actually stating the content objective and its behavioral indicators helps you focus more precisely. For example, the following objective is important when teaching a child an algorithm for adding whole numbers.

8.5
When the sum of the numbers in a column (place) is greater than 9, the sum must be regrouped or renamed.

The authors did not state *everything* a child must know in order to regroup. If a child is having difficulty, you may need to create objectives similar to the following:

When adding whole numbers, make a *10* whenever possible; that is, *10* ones, *10* tens, *10* hundreds, etc.

When regrouping in addition, only the number of ones is recorded in the sum, and the 1 ten is placed at the top of the next column to the left.

State what you want each child to understand in your own words, even if you feel your vocabulary is too advanced. Then restate the objective in simpler terms when talking with a child. Be accurate so that children will apply the idea appropriately. The following objective is accurate as stated, but it should not be expressed this way when working with young children:

> 5.26
> It is *not* true that the order of two whole numbers has no effect on the difference between the two numbers.

Another way of stating the same idea when considering it with six- or seven-year-olds is:

> Trading places does not work when you are subtracting one number from another.

As children work with exemplars and come to understand a relationship, they are often able to express it in their own words. They can help us think of ways to express the idea that are accurate and understandable.

MATHEMATICS AS A LANGUAGE

As you study your own teaching and how you can become even more effective, think of mathematics as a *language*. Children *read* mathematical expressions and *write* symbols on paper. Like words, symbols need to be associated with meanings. However, many of the words they write are associated with observable objects or actions. Numerals are associated with mental ideas, such as how many and which one.

The meaning of a symbol or a symbolic expression is often stated with words in different ways. This can be very confusing to young children who do not easily recognize the equivalence of such expressions. For example, Preston notes that ways of expressing the meaning of "5 + 3" include the following:[3]

5 plus 3	5 and 3 together
3 greater than 5	3 more than 5
The total of 5 and 3	Add 3 to 5
5 add 3	5 count on 3
The sum of 5 and 3	Increase 5 by 3
5 and 3	

Mathematical symbols are very efficient. This is why some mathematics educators emphasize the importance of oral language in early instruction.[4] Exercise care in selecting texts and worksheets, and in discussing lessons; children must clearly understand both mathematical words and written symbols.

Many of the symbols used in mathematics express operations and relationships that are complex and cumbersome when stated in ordinary language. It takes time for a child to understand that simple mathematical symbols can express complex ideas. The equals sign, for example, is small but says much. It means *the symbols or symbolic expressions on either side are two different ways of expressing the same idea*, but children often come to think of it as just a way of writing *the answer is*.

Numerals such as 5 are used as simple nouns, as are multidigit numerals. An expression like $2 + 3$ or $270 - 1$ or $2 + (\frac{1}{4} \times 4)$ is used as a compound noun.[5] In the statement $2 + 3 = 5$, a compound and a simple noun mean the same thing. In fact, for every simple noun (or standard number name) there are an infinite number of nonstandard ways to state the same number idea. A child who understands this will sense something of the power of mathematics.

Mathematics is also language in terms of number sentences. Some sentences are open; they include a variable. Others are closed.

Open sentence	$8 + \square = 15$
Closed sentence	$8 + 7 = 15$

3. Mike Preston, "The Language of Early Mathematical Experience," *Mathematics in School* 7, no. 4 (September 1978): 31–32.

4. See Dora H. Skypek, "Teaching Mathematics: Implications from a Theory for Teaching the Language Arts," *The Arithmetic Teacher* 28, no. 7 (March 1981): 13–17.

5. John V. Trivett, "The End of the Three Rs," *Mathematics Teaching*, no. 83 (September 1978): 14–19.

We use synonyms freely as we speak and write, substituting one noun for another. Similarly, we substitute one noun for another in mathematics in using an equivalent name for the same idea. For example, when computing 65 − 28 we usually replace 60 + 5 with 50 + 15, a synonymous expression. Also, in adding unlike fractions we often substitute a fraction in the form *n/n* for the number one before multiplying.

Spoken and written language have many elements and structures that are similar to mathematics. If we help each child observe these similarities, we should increase the effectiveness of our teaching.

CALCULATORS AND MICROCOMPUTERS

Hand-held calculators are gradually becoming accepted in classrooms. Calculators are used for checking answers, for applications that involve large numbers, and for exploring number patterns.

In the future calculators can be used for initiating, abstracting, and schematizing activities to help children understand mathematical concepts and operations. For example, most inexpensive calculators have a counting function; they count by ones when you press [1][+][1][=] [=] [=] . . . Other calculators will count if you press [1][+][1][+][1][+] . . . The counting function can help a child understand addition as he counts one addend and then the other. Calculators also count by twos, fives, tens, etc., and can thus help a child understand multiplication.[6] Inverse operations are easily observed as well.

Microcomputers are also rapidly becoming a part of our lives, and can be found in most secondary schools. They are becoming more and more commonplace in elementary classrooms. Even some preschoolers use these machines.[7] Familiarize yourself with microcomputers, recognizing their limitations and potential for facilitating instruction. For instance, they can provide diagnoses of difficulties and identify prescriptions for instruction. The two-dimensional format is a limitation for diagnosis, but advances in technology may eventually overcome this difficulty. Microcomputers can help manage the complex task of keeping records as we increasingly focus instruction on *each* child. As Martin notes, "Rather than envision them as instructional delivery systems that impress experience, we must see within them the power to give expression to experience." Microcomputers will eventually be able to reflect the learning patterns of *each* child; children will be able to program the machine rather than respond to a program already in the machine.[8]

Select computer activities carefully as you integrate them into the mathematics program. Incorporate computer-related instruction where it solves educational problems more efficiently.

A variety of programs is available. Often software is written by programmers who do not have training and experience in preparing appropriate classroom aids; at other times, software is written by classroom teachers who lack programming skills. Keep these guidelines in mind when you select software.[9]

1. *Theoretical base.* Is the program based on the same theories of learning and instruction as your own instructional program?
2. *Reading level.* Will children who are ready to study the content be able to read the material displayed?
3. *Student interaction.* Are the lessons personalized? Will they draw the child into the learning process?
4. *Use of technology.* Does the program use the available technology, or is it merely an electronic worksheet?
5. *Suitability.* Is the program the most suitable way to teach the content?

---

6. James H. Wiebe, "Using a Calculator to Develop Mathematical Understanding," *The Arithmetic Teacher* 29, no. 3 (November 1981): 36–38.
7. Kathleen Martin, "The Learning Machines," *The Arithmetic Teacher* 29, no. 3 (November 1981): 41, 43.
8. *Ibid.*, p. 43.
9. Jay W. Dean, "What's Holding Up the Show?" *Today's Education: Mathematics/Science Edition* 71, no. 2 (April–May 1982): 23–24.

If you want to gain some experience in programming, courses are available at local colleges and in adult education programs. You should also read *Guidelines for Evaluating Computerized Instructional Materials*, a recent publication of the National Council of Teachers of Mathematics.[10]

MORE HELP IS AVAILABLE

Always be alert for ways to improve your teaching, both when you make mistakes and when lessons go well. Talk to experienced teachers and supervisors; they want to help you become as effective as possible.

Professional associations also stimulate confidence and creativity. The largest national association is the National Council of Teachers of Mathematics (NCTM). They publish a magazine, *The Arithmetic Teacher*, and hold annual and regional meetings.

Two smaller national associations are the School Science and Mathematics Association, which publishes *School Science and Mathematics*, and the Research Council for Diagnostic and Prescriptive Mathematics. The latter organization focuses primarily on helping individuals who have difficulty learning mathematics. Professional organizations are also available at the state and local level.

You can further sharpen your skills by attending workshops and courses; for example, enroll in a class in measurement and evaluation if you have not already had one. Some graduate schools offer special programs that specifically focus on mathematics instruction. These include supervised clinical activities in which teachers work with children who are gifted or have learning problems.

PERSPECTIVE

The years ahead will bring significant changes in the materials we work with. Microcomputers will become more available in the classroom; they may even be tied in with other information-handling technologies, thereby bringing in an unbelievable wealth of resources. As exciting as these possibilities are, we must continue to focus our attention on *children* and their needs.

The authors hope that the guiding models you learned about and applied while reading this text will help you plan for each child, and that you are now better able to articulate your own beliefs about teaching mathematics to children.

10. Instructional Affairs Committee of the National Council of Teachers of Mathematics, *Guidelines for Evaluating Computerized Instructional Materials* (Reston, Va.: The Council, 1981).

APPENDIX A

A Visit to Mrs. Brown's Classroom

Mrs. Brown teaches a second grade class at nearby Forest Hill Elementary School. She has been teaching there since completing her bachelor's degree in elementary education three years ago.

Mrs. Brown's classroom is especially attractive, with bulletin boards that emphasize reading, nutrition, and mathematics. There are also learning centers that appear to involve children in a variety of activities. Numerous manipulative aids are visible as well.

At this time Mrs. Brown is instructing a small group in the corner of the room where desks are arranged in a semicircle.

Each child has a small flannel board constructed within a manila folder, a collection of felt pieces (some green, others yellow), and paper and pencil. Mrs. Brown is seated so that she can be easily seen by all children. A large flannel board on the chalkboard tray beside her shows various combinations of yellow and green felt pieces arranged to show the number 9 in different ways.

Today's lesson is on addition and subtraction facts with 9 as the sum. The model being used is a set of 9 pieces of felt separated into two disjoint subsets. The number for each subset or part is noted and recorded. Each child has been using his flannel board to help generate the different combinations on Mrs. Brown's board. Note the following dialogue between Mrs. Brown and new students.

Mrs. Brown: (pointing to the display on the flannel board) We have a total of nine in each row. Some are yellow and some are green. Begin with the top row and name the addition fact shown with the flannel pieces. Write the number for the part shown with yellow pieces, then write the number for the part shown with green pieces.

(Mrs. Brown writes the numbers on the chalkboard to the left of her flannel board, and the children write them on their papers.)

Mrs. Brown: George is putting a plus sign between the two numerals. You'll want to make a plus sign, too. It means that the numerals on both sides tell about the parts; they are both addends. Be sure to write the sum, the simple name for the total number. How can we show that *9 + 0* and *9* are both names for the same number?

Peggy and *Jill*: Make equals.

Mrs. Brown: Yes, an equals sign will show that *9 + 0* is the same as *9*. What is the addition fact?

Jan: Nine plus zero equals nine.

Mrs. Brown: Good. Look at the next row. Can you write the addition fact?

(The lesson continues. Clearly, these children have done this sort of thing before. They have had similar lessons for lesser sums. Before long Mrs. Brown and the children have written all ten addition facts.)

Mrs. Brown: Very good! Now let's write the subtraction facts that are also shown. We wrote addition facts as *part plus part equals total* or *addend plus addend equals sum.* We write subtraction facts as *total minus part equals part* or *sum minus addend equals addend.* Let's begin with the second row. What is the total?

Children: Nine.

Mrs. Brown: (covering the green piece with her hand) The yellow part has eight pieces. How many are in the green part?

Children: One.

Mrs. Brown: How do you know? Who can tell me?

Mark: Eight and *one* is nine.

Mrs. Brown: (uncovering the green piece) Mark is correct, isn't he? We can write nine (the total) minus the number for the part we know (eight) equals the number for the other part (one). We write $9 - 8 = 1$. (Mrs. Brown writes the subtraction fact on the chalkboard to the right of the second row, and the children also write the fact.)

Mrs. Brown: Now let's write the other subtraction facts with sums of nine. Start with the top row.

Jimmy: Nine minus nine equals zero?

Mrs. Brown: Is Jimmy correct?

(Mrs. Brown and the children continue. Soon the display at the chalkboard looks like Figure A.1. She then begins to question, focusing on the relationship between the addition and subtraction sentences.)

Mrs. Brown: (pointing to the sentences) The addition sentences have numbers for two parts and the whole, and the subtraction sentences also have numbers for two parts and a whole. We learned that the plus sign is always placed between the two numbers for parts. Where do we place the minus sign? Look back at the sentences we have written.

Peggy: Between the total and the part we know.

Mrs. Brown: Hmm. Between the number for the

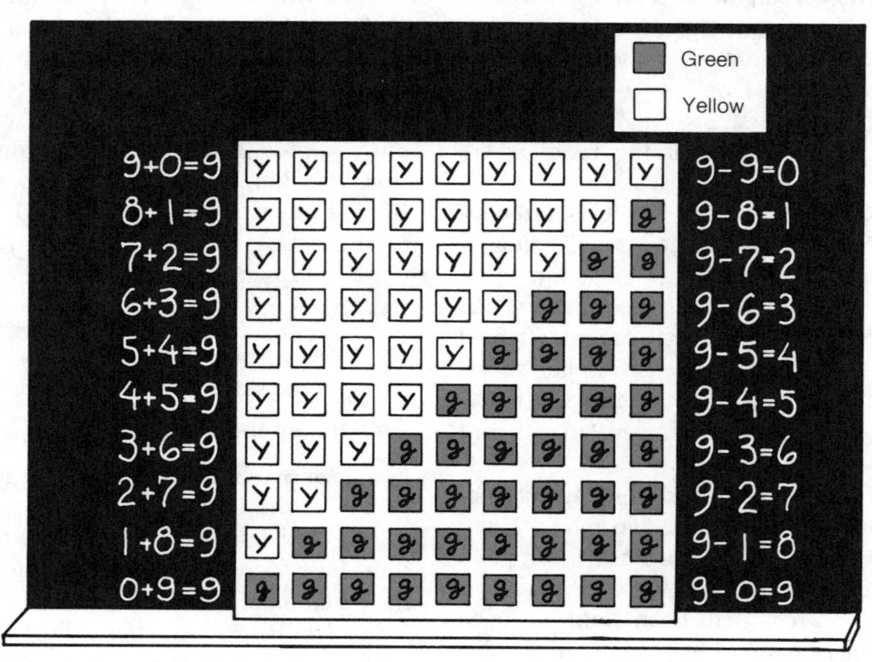

FIGURE A.1
Completed Display

whole set and the number for the part we know. . . . Is that *always* true?

Children: (after a pause) Yes.

Mrs. Brown: Remember this as you write another set of facts—the facts with *ten* as the sum. Use your flannel board to show rows of ten for all the addition and subtraction facts with ten as the sum, then write the facts on your paper. Do you have any questions?

(As each child in the group begins, Mrs. Brown calls a different group of children to another part of the room. Soon she is seated with eight students around a table that has sticks in bundles of 10 and loose sticks. After she writes a two-digit subtraction example on the portable chalkboard near the table, she has each child use the sticks to model the procedure for subtracting.)

Mrs. Brown: Look at our example, *35 − 17 =* □. What number is the total?

Children: Thirty-five.

Mrs. Brown: What does the seventeen tell us?

Bill: Part.

Mary: The number for the part we know.

Mrs. Brown: Do we start with the number for the part and subtract, or do we start with the total and then subtract the number for the part?

Julie: We start with the total.

Mrs. Brown: Is that true?

Children: Yes.

Mrs. Brown: Good! Let's each show the total with bundles of ten and loose sticks. I'll write our example on the chalkboard with the total on top as we do when we subtract to find the number for the other part. (See Figure A.2.)

Mrs. Brown: Look at our example. When we subtract we begin with the digits in the ones place. Is seven the number for part of five things?

Children: No.

Mrs. Brown: So we do not think *five minus seven.* We need to have more units if we are going to show seven units in the part we know. What must we do?

Written on Chalkboard

35 − 17 = □

35 Total
−17 Known Part

Shown with Sticks

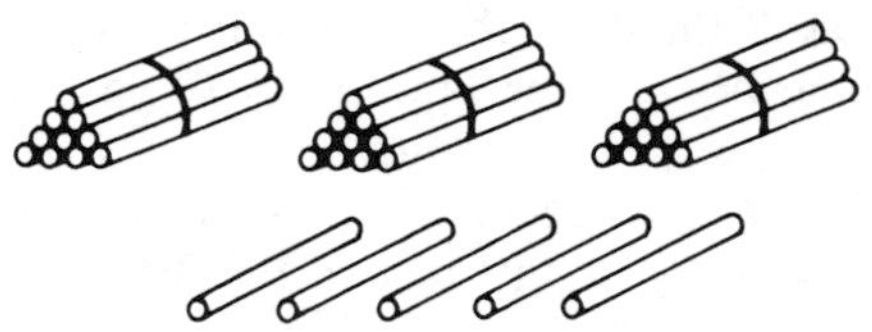

FIGURE A.2
The Problem Displayed in Different Ways

Charley: Seven minus five?

George: Take a bundle of ten and make it into ones. There will be fifteen ones.

Mrs. Brown: Let's all do that. You show it with the sticks, and I'll show it with our record at the chalkboard.

(Each child undoes one bundle of 10 and puts the 10 sticks with the other 5 sticks. Mrs. Brown records this regrouping at the chalkboard as seen in Figure A.3.)

Written on Chalkboard

35 − 17 = □

2 15
~~3~~~~5~~ Total
−17 Known Part

Shown with Sticks

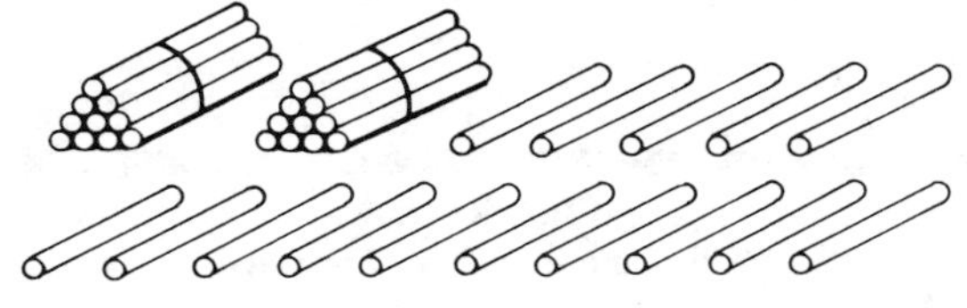

FIGURE A.3
Regrouping Recorded

Mrs. Brown: Is seven a part of 15?

Children: Yes.

Mrs. Brown: Then take seven sticks from the fifteen, and put them in a separate pile. This pile will show the part we *do* know.

Jan: Eight are left for the other part.

Mrs. Brown: Do you all have eight left in the other part? I'll show that with our record on the chalkboard here. (See Figure A.4.)

Mrs. Brown: Now, let's think about the tens column. Is one ten a part of two tens?

Children: Yes.

Mrs. Brown: Then we *can* subtract one ten to find the other part. With your sticks, remove one bundle of ten and put it with the part we know.

George: One ten is left. The answer is eighteen.

Mrs. Brown: Is that true?

Children: Yes.

Mrs. Brown: (as she completes the record at the chalkboard) So, thirty-five minus seventeen equals eighteen. (See Figure A.5.)

Mrs. Brown: How do we check a subtraction problem?

Julie: By adding the numbers for the parts to see if we get the total.

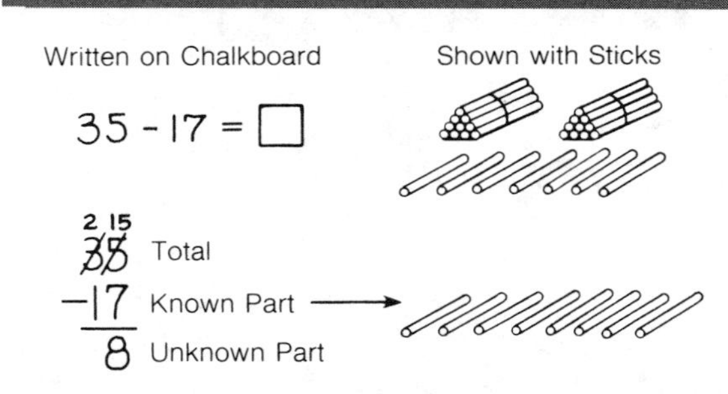

FIGURE A.4
The Unknown Part Recorded

Mrs. Brown: Good. Let's add seventeen and eighteen and see if the total is thirty-five. Use

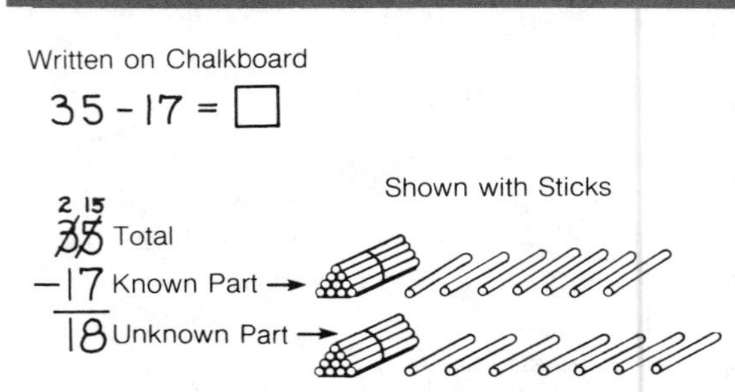

FIGURE A.5
The Problem Completed

your sticks. Combine the two parts: regroup if you can make a ten.

(Soon the children have completed the task and note that there *are* 35 in the total. Mrs. Brown also writes the check as an addition example on the chalkboard.)

Mrs. Brown: Very good! Now I want each of you to try a subtraction example. Use your sticks, then make a record with pencil and paper. Martha, try this one. (Mrs. Brown writes 42 − 17 = □ on Martha's paper.) You had a little trouble with it yesterday. . . .

As we reflect upon our visit, it becomes obvious that Mrs. Brown has developed management procedures that make it possible for her to teach a small group while other children in the class work independently. She plans carefully in order to create this type of learning environment.

The two groups we observed were involved in subtraction activities at different levels. Recall that the first group was generating basic subtraction facts that have 9 as the sum. Although these subtraction facts are usually introduced and often mastered before this point in the second grade, the content was exactly what these particular children were ready to learn. The second group was studying subtraction of two-digit numbers with regrouping, a more advanced topic.

In order for Mrs. Brown to place her children into instructional groups, she made an initial diagnosis. She knows, as do all good teachers, that it is nearly impossible to provide appropriate mathematics instruction for each student when teaching one group of thirty. Her continuous use of diagnosis and flexible grouping reflects her diagnostic-prescriptive philosophy of teaching.

Mathematics involves interrelated ideas. Did you notice how Mrs. Brown emphasized mathematical relationships? She helped children understand that in an *addition* situation the addends are given and the operation of addition tells us the missing sum; whereas, in a *subtraction* situation the sum and one addend are given, and the operation of subtraction tells us the other addend. She chose to develop these related concepts rather than the erroneous idea that subtraction is taking away. Mrs. Brown develops computational skills by emphasizing this relationship between addition and subtraction in her discussion of two-digit subtraction with the second group.

Note that Mrs. Brown had to determine the sequence of content; she had to decide what each group was ready for next. Her concern is obviously with each child, and as a result of this personal evaluation, she notes changes and regroups the children from time to time.

Mrs. Brown is also aware that manipulative aids can help children understand mathematical ideas. She chose a variety of three-dimensional exemplars and used them correctly. For example, she introduced symbolic expressions as a record of observations made while children worked with sticks.

Mrs. Brown's children appeared to be eager to learn. She encouraged them through her careful questioning and well-placed feedback.

APPENDIX B

Selected Content Objectives

3.1
Numerals are symbols or sets of symbols used to name or represent numbers.

3.2
A *digit* is a numeral with only one symbol.

3.3
The number to which a digit is assigned is called the *face value* of the digit.

3.4
Each power of 10 has a fixed position or place assigned to it rather than a symbol.

3.5
Horizontally arranged positions are assigned to numbers with respect to a reference called a *decimal point.* When no decimal point is recorded, as is usually the case for whole numbers, it is assumed to be immediately to the right of the numeral.

3.6
The first position to the left of the decimal point in the Hindu-Arabic system is assigned the number 1.

3.7
The second position to the left of the decimal point in the Hindu-Arabic System is assigned the number 10.

3.8
A multidigit numeral names a number that is the sum of the products of each digit's face and place value.

3.9
The non-negative rational numbers are defined as the set of numbers that may be named in the form a/b, where a represents a whole number and b represents a counting number.

3.10
A fraction is used to express the number for part of a whole or of a set.

3.11
The denominator below the fraction bar names the fractional part under consideration. It tells how many parts of the same size are in the whole (or how many members are in the set).

3.12
The numerator above the fraction bar tells how many of the parts are being considered.

3.13
Cardinal numbers are whole numbers used to indicate how many.

3.14
A cardinal number is the property common to all sets in a class of equivalent sets.

3.15
Sets such as (#, $), (a, b), (*, X) are representative of the class whose property is *twoness.*

3.16
The number property of a finite set may be determined by counting.

4.1
A *set* is any well-defined collection of real or representative objects or events.

4.2
The objects included in the set are called the *elements* or *members* of the set.

4.3
A set may be defined by stating the criterion for membership, for instance, stating the property common to all members.

4.4
The members of the set may or may not have common attributes.

4.5
A set with no members is called the *empty* or *null set.*

4.6
The union of two sets, A and B, is the set C such that C contains all elements belonging to A or to B (or to both).

4.7
If every member of set A can be paired with a distinct member of set B such that each member of set B is also paired with a distinct member of A, sets A and B are in one-to-one correspondence.

4.8
The set that has unmatched elements remaining after the pairing of its elements with the elements of a second set is said to have *more* elements than the second set.

4.9
The set that has a deficiency of elements after the pairing of its elements with the elements of a second set is said to have *fewer* elements than the second set.

4.10
Sets that can be placed in one-to-one correspondence are *equivalent.*

4.11
A relation is a set of ordered pairs.

4.12
Relations that are reflexive, symmetric, and transitive are *equivalence* relations.

4.13
Relations that are nonreflexive, nonsymmetric, and transitive are *order* relations.

4.14
A *cardinal* number is the property common to all sets in a class of equivalent sets.

4.15
Sets such as {1}, {1, 2}, {1, 2, 3}, . . . are called counting sets.

4.16
The last number of a counting set is the *number property* of that set.

4.17
Counting is the process of matching one-to-one the members of a set with the members of one of the counting sets.

4.18
Numerals are symbols or sets of symbols used to name or represent numbers.

4.19
A *digit* is a numeral using only one symbol.

4.20
Each digit is assigned to or names a unique whole number.

4.21
Each digit has a unique shape.

4.22
The number to which a digit is assigned is called the *face value* of the digit, or the *digit value.*

4.23
The set of digits used in the Hindu-Arabic numeration system is {0, 1, 2, 3, 4, 5, 6, 7, 8, 9}.

4.24
The Hindu-Arabic digits are assigned to the set of whole numbers 0 through 9.

5.1
Addition of whole numbers is an operation that associates each pair of whole numbers with a unique whole number.

a. The pair of numbers added are called *addends.*
b. The unique number associated with the addends is called the *sum.*
c. Addition is used to find the sum when its addends are known.

5.2
The union of disjoint sets is a model for addition on whole numbers.

a. One addend indicates the number of elements in one set.
b. The second addend indicates the number of

elements in the other set.

c. The sum is the number of elements in the union of the sets.
d. Addition is an operation used to find the total number of elements in a union of two disjoint sets when the number of elements in each set is known.

5.3
A set may be partitioned into two disjoint subsets, even if one or both of the subsets is empty. The subsets and the original set model the relationships between two addends and their sum.

5.4
All possible pairs of addends associated with a given sum can be found by partitioning a set into all possible pairs consisting of a subset and its complement.

a. The sum indicates the number of elements in the set to be partitioned.
b. The numbers related to a subset and its complement set form a pair of addends.
c. All such pairs of addends are the possible pairs of addends for that given sum.

5.5
Beginning at the zero point, the union of successive segments of a number line to the right is a model for addition on whole numbers.

5.6
The + (read *plus* or *add*) is a symbol that indicates addition.

5.7
An *addition phrase* consists of numerals or variables separated by the plus sign.

$$3 + 4 \qquad 53 + \square \qquad a + b$$

a. The numerals or variables on each side of the plus sign in a phrase indicate addends.
b. An addition phrase names a single number, a sum.

5.8
A single numeral (digit or multidigit) used to name a sum is called the *simplest name, standard name,* or *standard numeral* for the sum.

5.9
An addition *equation,* or *equality sentence,* consists of addition phrases, variables and/or standard numerals, and an equals sign.

$$3 + 4 = 7 \qquad 53 = 17 + 36 \qquad 4 + \square = 13$$

a. An addition equation states that numerals and phrases on each side of the equals sign name the same sum.
b. An equation containing a variable is called an *open equation* or *open sentence.*
c. Addition equations may be written in horizontal or vertical form.

$$3 + 5 = \square \quad \text{or} \quad \begin{array}{r} 3 \\ +\ 5 \\ \hline \square \end{array}$$

d. In the vertical form, the bar under the phrase serves as an equals sign.

5.10
Closure—The sum of any two whole numbers is always a whole number.

5.11
Commutative property—The order or sequence in which two whole numbers are added has no effect on their sum.

5.12
Associative property—The way in which three whole numbers are grouped together in addition has no effect upon their sum.

5.13
Identity element—Zero added to any whole number yields a sum equal to the original number.

5.14
Subtraction of whole numbers is an operation that associates a pair of whole numbers with a unique whole number.

a. One of the pair of numbers is a sum (sometimes called a *minuend*) and the other number is one of the addends (sometimes called a *subtrahend*).
b. The sum's other addend (sometimes called the *remainder* or *difference*) is the third number associated with the sum and one of its addends.
c. Subtraction is an operation used to find an addend (remainder or difference) when the sum (minuend) and its other addend (subtrahend) are known.

5.15
The partition of a universal set into a subset and its complement, or remainder set, is a model for subtraction.
a. The sum (minuend) is the number of elements in the universal set (the starting set).
b. The known addend (subtrahend) is the number of elements in one subset.
c. The other addend (remainder) is the number of elements in the complement, or remainder set.
d. Subtraction is an operation used to find the number of elements in the complement, or remainder set (one addend), when the number of elements in the universal set (the sum) and number of elements in one of its subsets (known addend) are known.

5.16
Comparison by matching one-to-one the elements of two disjoint sets is a model for subtraction of whole numbers.
a. The known sum (minuend) is the number property of one of the disjoint sets (the greater set if not equivalent).
b. The known addend (subtrahend) is the number property of the second disjoint set (the lesser set if not equivalent).
c. The unknown addend (difference) is the number property of the subset of unmatched elements (the *difference subset*) remaining in the first set after the elements of the second set are matched one-to-one with elements of the first set.
d. Subtraction of whole numbers is an operation used to find the number associated with the difference between two disjoint sets when the number of members in each of the disjoint sets is known.

5.17
The partition of a line segment from zero on a number line into two successive segments is a model for subtraction on whole numbers.

5.18
Subtraction is the inverse of addition.
a. Addition is used to find a sum when its addends are known; subtraction is used to find one of the addends when the sum and its other addend are known.
b. The inverse of each model for addition is a model for subtraction.

5.19
Elements in each row of an array of objects can be successively partitioned into a subset and its complement. This forms a model for all the possible pairs of addends for a given sum and their inverses for subtraction.

5.20
The $-$ (read *minus* or *subtract*) is a symbol that indicates subtraction.

5.21
A *subtraction phrase* consists of numerals and variables separated by the $-$ sign.

$$7 - 2 \quad 52 - \square \quad a - b$$

a. The numeral or variable to the left of the minus sign always indicates a sum (minuend).
b. The numeral or variable to the right of the minus sign always indicates one of the addends (subtrahend) of the sum.
c. A subtraction phrase names one number.

5.22
A single numeral (digit or multidigit) used to name an addend (remainder, difference) is called the *simplest name, standard name,* or *standard numeral* for the addend.

5.23
A simple *subtraction equation* or *equality sentence* consists of a subtraction phrase, a standard numeral or variable, and an equals sign.

$$7 - 2 = 5 \quad 8 - 3 = \square \quad 52 - \square = 20$$

a. A subtraction equation states that the numeral and phrase on each side of the equals sign name the same addend (remainder, difference).
b. In a simple subtraction sentence, phrases and standard numerals or variables may be written on either side of the equals sign.

$$7 - 2 = 5 \quad \text{or} \quad 5 = 7 - 2$$
$$\square - 5 = 8 \quad \text{or} \quad 8 = \square - 5$$

c. An equation containing a variable is called an *open equation* or *open sentence.*
d. Subtraction equations may be written in horizontal or vertical form. For example,

$$7 - 2 = \square \quad \text{or} \quad \begin{array}{r} 7 \\ -\underline{2} \\ \square \end{array}$$

e. In the vertical form, the bar under the phrase serves as an equals sign.

5.24
For every simple subtraction sentence involving unequal addends, there are three other sentences that state the same relationship among the sum and its addends; one subtraction sentence and two addition sentences.

$7 - 2 = 5$	$9 - \square = 6$
$7 - 5 = 2$	$9 - 6 = \square$
$2 + 5 = 7$	$\square + 6 = 0$
$5 + 2 = 7$	$6 + \square = 9$

5.25
Closure–It is *not* true that the difference of any two whole numbers is always a whole number.

5.26
Commutativity–It is *not* true that the order of two whole numbers has no effect on the difference between the two numbers.

5.27
Associativity–It is *not* true that the way in which three or more whole numbers are grouped together for subtraction has no effect on the difference.

5.28
Identity element–Zero subtracted from any sum yields an addend equal to the original sum.

5.29
A *basic* (or *primary*) *fact* for addition is an addition equation of the form:

$$a + b = c \quad \text{or} \quad \begin{array}{r} a \\ \underline{+\,b} \\ c \end{array}$$

Addends have only one digit, and the equation does not contain a variable.

5.30
A *basic* (or *primary*) *fact* for subtraction is a subtraction equation of the form:

$$x - y = z \quad \text{or} \quad \begin{array}{r} x \\ \underline{-\,y} \\ z \end{array}$$

Addends have only one digit, and the equation does not contain a variable.

5.31
Compensation in addition–If the number added to one addend is subtracted from the other addend, the sum is unchanged.

5.32
Compensation in subtraction:

a. If a number is added to both the sum and the known addend, the unknown addend is unchanged.
b. If a number is subtracted from both the sum and the known addend, the unknown addend is unchanged.

5.33
When an addend is increased by 1 and the other addend is unchanged, the sum is increased by 1.

5.34
When an addend is decreased by 1 and the other addend is unchanged, the sum is decreased by 1.

5.35
Basic addition facts with sums less than 10.

5.36
Basic addition facts with 10 as the sum.

5.37
Basic addition facts with sums greater than 10.

5.38
Basic subtraction facts with sums less than 10.

5.39
Basic subtraction facts with 10 as the sum.

5.40
Basic subtraction facts with sums greater than 10.

5.41
Addend + addend = $\boxed{\text{sum}}$

5.42
Sum − addend = $\boxed{\text{addend}}$

7.1
A *fixed position* within the numeral is assigned to each power of 10.

7.2
The power of 10 assigned to each fixed position is called the *place value* for that position.

7.3
The fixed positions to which place values are

assigned are arranged horizontally, with values assigned in ascending order from right to left.

7.4
The *base* of a numeration system is a whole number greater than 1 used as a factor to yield place values.

7.5
In our decimal numeration system, each place has a value that is 10 times as great as the place to its right, 100 times as great as the place two positions to its right, and so forth.

7.6
In our decimal numeration system, each place has a value that is 1/10 as great as the place to its left, 1/100 as great as the place two positions to its left, and so forth.

7.7
In a numeral consisting of more than one digit, only one digit is written in each position to which a place value is assigned.

7.8
Each digit in a multidigit numeral names a number that is the *product* of its face value and the place value for its position.

7.9
A multidigit numeral names a number which is the *sum* of the products of each digit's face value and place value.

7.10
A period is any cluster of three adjacent positions such that the first period consists of the positions to which 1, 10, and 100 are assigned, the second period consists of the next three positions to the left, and so forth.

7.11
The names of the periods in our decimal numeration system are, from right to left:

- First period–ones
- Second period–thousands
- Third period–millions
- Fourth period–billions
- Fifth period–trillions

7.12
The names of places within each period in our decimal numeration system are, from right to left: period name, ten period name, and hundred period name.

7.13
The three digits in the ones period are read individually from left to right as the name of the digit (except zero) followed by the place value of the digit (except ones).

7.14
The three digits within periods, other than the ones period, are read as if the three digits were in the ones period, followed by the period name.

7.15
Any whole number named by a *standard multidigit numeral* may be expressed in a nonstandard form by renaming one of the tens (or hundreds, etc.) as 10 ones (or 10 tens, etc.) and adding these to the number of ones (or tens, etc.) already indicated in the standard numeral.

367 = (2 hundreds + 1 hundred) +
(5 tens + 1 ten) + 7 ones
(2 hundreds) + (10 tens + 5 tens) +
(10 ones + 7 ones)
= 2 hundreds + 15 tens + 17 ones

7.16
Any whole number expressed in a nonstandard form may be named by a *standard numeral* by renaming 10 of the ones (or tens, etc.) as 1 ten (or hundred, etc.) and adding this to the number of tens (or hundreds, etc.) expressed in the nonstandard forms.

2 hundreds + 14 tens + 16 ones =
2 hundreds + (10 tens + 4 tens) +
(10 ones + 6 ones) =
(2 hundreds + 1 hundred) +
(4 tens + 1 ten) + 6 ones = 356

7.17
Each place value in our Hindu-Arabic numeration system can be written in exponential notation.

7.18
The "3" in 10^3 is called an *exponent* and indicates how many times 10 is used as a factor. It is written above and to the right of 10.

7.19
We read an expression in exponential form as follows:

a. 10^2 is read as *ten to the second power* or *ten squared.*
b. 10^3 is read as *ten to the third power* or *ten cubed.*
c. 10^4 is read as *ten to the fourth power*; and so on.

8.1
An *algorithm* is a process for computing an unknown number by writing numerals and other mathematical symbols in a fixed sequence of steps.

8.2
Addition is a binary operation that associates a pair of whole numbers with a third whole number called the sum.

8.3
The order in which the addends are associated does not change the sum.

8.4
To add whole numbers, add ones with ones, tens with tens, hundreds with hundreds, etc.

8.5
When the sum of the numbers in a column (place) is greater than 9, the sum must be regrouped or renamed.

8.6
If the number added to one addend is subtracted from the other addend, the sum is unchanged.

8.7
To subtract whole numbers, subtract ones from ones, tens from tens, hundreds from hundreds, etc.

8.8
When the product value of any digit in the subtrahend (known addend) is greater than the product value of the corresponding digit in the minuend (sum), the number named in the minuend must be regrouped (renamed). Subtraction in this case is sometimes called *compound subtraction.*

8.9
If the same number is added to both the minuend and the subtrahend, the difference between the two remains the same.

9.1
Multiplication on whole numbers is an operation that associates a pair of whole numbers with a third unique whole number.
a. The pair of numbers multiplied are called *factors.*
b. The third number associated with the factors is called the *product.*
c. Multiplication is used to find a product when its factors are known.

9.2
The *union of equivalent disjoint sets* is a model for multiplication on whole numbers.
a. One factor indicates the number of equivalent sets.
b. The other factor indicates the number of elements in each of the equivalent sets.
c. The product is the number of elements in the union set.
d. Multiplication is an operation used to find the number of elements in the union set when the number of equivalent sets and the number in each equivalent set are known.

9.3
Repeated addition of equal addends is a model for multiplication on whole numbers.
a. One factor indicates the number of equal addends.
b. The other factor indicates the value of each addend.
c. The product is the sum of the equal addends.

9.4
The union of successive congruent line segments beginning at the zero point on a number line is a model for multiplication.

9.5
The *Cartesian product of two sets* is a model for multiplication of whole numbers.
a. One factor indicates the number of elements in one set.
b. The other factor indicates the number of elements in another set.
c. The product is the *number of ordered pairs* of elements in the Cartesian set.
d. Multiplication is used to find the number of ordered pairs in a Cartesian set when the number of elements in each of two sets is known.

9.6
An *array* is a model for multiplication on whole numbers.

a. One factor indicates the number of rows in the array.
b. The other factor indicates the number of columns in the array.
c. The product is the number of elements in the array.
d. Multiplication is an operation used to find the number of elements in an array when the number of rows and columns are known.

9.7
An *array* is a model that shows the relationships of the models in content objectives 9.2, 9.3, and 9.5.
a. For CO 9.2, all the rows in an array are examples of equivalent disjoint sets.
b. For CO 9.3, the numbers of elements in each row of an array are examples of equal addends.
c. For CO 9.5, each element in an array can represent one of the ordered pairs in the Cartesian set.

9.8
The × is a symbol that indicates multiplication and is read *multiply* or *times*.

9.9
A multiplication phrase consists of numerals or variables separated by the × sign, such as 3 × 4; 25 × □; *a* × *b*.
a. The numerals or variables on each side of the × in a phrase indicate factors.
b. Multiplication phrases name a single number, a product.

9.10
Other ways to write multiplication phrases are:
a. Place a dot midway up and between the numerals or variables naming factors. For example, 3 • 4 means the same as 3 × 4.
b. Enclose each factor name in a parenthesis and juxtapose; for example, (3)(4) means 3 × 4.
c. If one or more factors are indicated by variables, simply juxtapose; for example, 3□ means 3 × □; *ab* means *a* × *b*.

9.11
A single numeral (one digit or multidigit) used to name a product is called the *simplest name, standard name*, or *standard numeral.*

9.12
A multiplication equation consists of multiplication phrases, variables and/or standard numerals separated by an equals sign. For example, 3 × 4 = 12; 24 = 4 × 6; □ = 4 × 5.
a. When a variable such as a □ or a letter is used in an equation, the equation is called an open equation or an open sentence.
b. A multiplication equation states that the numerals and phrases on each side of the equals sign name the same product.
c. Multiplication equations may be written in horizontal or vertical form, such as

$$3 \times 4 = \square \quad \text{or} \quad \begin{array}{r} 4 \\ \times\ 3 \\ \hline \square \end{array}$$

d. In the vertical form, the bar under the phrase serves as an equal sign.

9.13
A multiplication equation may be read in different ways. For example, 3 × 4 = 12 may be read as *three fours equals (is) twelve, three multiply four equals (is) twelve,* or *three times four equals (is) twelve.*

9.14
Closure—The product of any two whole numbers is in every case a whole number.

9.15
Commutativity—The order or sequence in which two whole numbers are multiplied has no effect on their product.

9.16
Associativity—The way in which three whole numbers are grouped together in multiplication has no effect on their product.

9.17
Identity—The product of any whole number and 1 is equal to the given whole number.

9.18
Distributivity—Whole numbers under the operation of multiplication are distributive with respect to addition.

9.19
Zero property—The product of any whole number and zero is zero.

9.20
Division on whole numbers is an operation that associates a pair of whole numbers with a third unique whole number.
a. One number in the pair is a product (sometimes called *dividend*) and the other number is one of the product's factors (called a *divisor*).
b. The third number associated with the product and one of its factors is the product's other factor (called a *quotient*).
c. Division is an operation used to find one factor (quotient) when the product (dividend) and its other factor (divisor) are known.

9.21
Division is not defined when zero is the divisor.

9.22
Division is the inverse of multiplication.
a. Multiplication is used to find a product when its factors are known, whereas division is used to find one of the factors when the product and its other factor are known.
b. The inverse of each model for multiplication is a model for division.

9.23
The partition of a universal set into equivalent disjoint subsets is a model for division on whole numbers.
a. The product (dividend) is the number of elements in the universal set.
b. One factor is the number of equivalent subsets into which the universal set is partitioned.
c. The other factor is the number of elements in each of the equivalent disjoint subsets.
d. Division is an operation used to find the number of elements in each equivalent subset (quotient) when the number of elements in the universal set (dividend) and the number of equivalent subsets formed (the divisor) are known (*partitive model*).
e. Division is an operation used to find the number of equivalent subsets contained in a universal set (quotient) when the number of elements in the universal set (dividend) and number of elements in each equivalent subset are known (*measurement model*).

9.24
Repeated subtraction of equal addends is a model for division on whole numbers.
a. The product (dividend) is the sum of equal addends.
b. One factor (divisor) is the value of each addend.
c. The other factor (quotient) is the number of equal addends subtracted from the product.
d. Division is an operation used to find the number of equal addends subtracted from a product when the value of each addend (divisor) and the product (dividend) are known.

9.25
The partition of a line into successive congruent line segments of a number line is a model for division on whole numbers.

9.26
The partition of the ordered pairs in a Cartesian set into their respective sets of single elements is a model for division on whole numbers.

9.27
An *array* is a model for division on whole numbers.
a. The product (dividend) is the number of elements in an array.
b. One factor (divisor) is the number of rows in the array.
c. The other factor (quotient) is the number of columns in the array.
d. Division is used to find the number of rows (or columns) in an array when the number of elements in the array and the number of columns (or rows) are known.

9.28
An *array* is a model that shows the relationships of the models in objectives 9.23, 9.24, and 9.26.

9.29
The symbol ÷ indicates division and is read *divided by*.

9.30
A division phrase consists of numerals or variables separated by the ÷ sign. For example, $12 \div 3$; $\square \div 5$; $a \div b$.
a. The numeral (or variable) to the left of a ÷ sign names a product (dividend).
b. The numeral or variable to the right of a ÷ sign indicates a factor (divisor).
c. A division phrase names one number.

9.31
The $\overline{)\ \ }$ is also a symbol indicating division.

a. In a division phrase, the numeral for the product (dividend) is written inside the symbol and the numeral for one factor (divisor) is written to the left of the vertical bar. For example, in $3\overline{)12}$ the 3 is the factor and the 12 is the product.
b. The number named by a division phrase using the $\overline{)\ \ }$ is the *indicated factor* or *quotient*.

9.32
A single numeral (one digit or multidigit) used to name the quotient (one factor) is called the *simplest name* or *standard numeral* for the quotient.

9.33
A *division equation* consists of phrases, variables, and/or standard numerals separated by an equals sign. For example, $12 \div 3 = 4$; $6 = 24 \div 4$; $24 \div 6 = \square$; $24 \div \square = 6$.

a. A division equation states that the phrases and standard numerals on each side of the equals sign name the same factor.
b. In a division equation using the $\overline{)\ \ }$ sign, the horizontal bar serves as an equals sign and the numeral for the quotient is written directly over the horizontal bar.

9.34
Closure–Whole numbers are *not* closed under division; that is, a whole number product (dividend) and factor (divisor) will *not always* yield a whole number quotient.

9.35
Commutativity–Whole numbers are *not* commutative under the operation of division.

9.36
Associativity–Whole numbers are *not* associative under the operation of division.

9.37
Identity–A whole number divided by 1 yields a quotient equal to the original number.

9.38
Distributivity–Whole numbers are right distributive for division with respect to addition.

9.39
A basic (or primary) fact for multiplication is a multiplication equation of the form $a \times b = c$ or $\begin{array}{r} b \\ \times\ a \\ \hline c \end{array}$ that does not contain a variable, in which numerals for the factors have only one digit.

9.40
A basic (or primary) fact for division is an equation of the form $x \div y = z$ or $y\overline{)x}$ with z above $\quad (y = 0)$ which does not contain a variable, and in which numerals for the factors have only one digit.

9.41
If one factor is multiplied and the other factor divided by the same number, the product is unchanged.

9.42
If the dividend and divisor are both multiplied or divided by the same nonzero whole number, the quotient remains unchanged.

9.43
Basic multiplication facts with *zero as a factor*.

9.44
Basic multiplication facts with *1 as a factor*.

9.45
Basic multiplication facts for which *skip counting* is a useful strategy.

9.46
Basic multiplication facts for which *repeated addition* is a useful strategy.

9.47
Basic multiplication facts for which *one more* is a useful strategy.

9.48
Basic multiplication facts for which *twice as much* is a useful strategy.

9.49
Basic multiplication facts for which *facts of 5* is a useful strategy.

9.50
Basic multiplication facts with *9 as a factor*.

9.51
Factor × factor = $\boxed{\text{product}}$

9.52
Product ÷ factor = $\boxed{\text{factor}}$

10.1
To multiply a number greater than 10, rename the number as a sum. Then, apply the distributive property; multiply each term.

10.2
When we multiply a number greater than 10, the product is found in parts called *partial products.* The partial products are added to obtain the final product.

10.3
Places to the left of the decimal point are assigned to special products of 10 in a decimal system, and are called *powers of 10.*

10.4
When multiplying by a power of 10 or by a multiple of same, multiply the nonzero numbers and annex as many zeros to the product as zeros in the factor.

10.5
The special products of 10 and any other whole number factor are multiples of 10.

10.6
When multiplying two multiples of 10, multiply the nonzero numbers and annex the number of zeros in both factors.

10.7
The estimated product of two factors has upper and lower bounds.

a. The *upper bound* is the product of the two rounded up factors.
b. The *lower bound* is the product of the two rounded down factors.

10.8
When dividing by the subtractive method, the divisor or a multiple of same is subtracted from the dividend until the remainder of the subtraction is less than the divisor.

10.9
When dividing larger numbers by the subtractive method, the quotient is found in parts called *partial quotients.* These partial quotients are added to obtain the final quotient.

10.10
The whole numbers are right distributive for division with respect to addition.

10.11
When dividing by the distributive method, begin at the left of the dividend. To model the procedure, partition the set indicated by the dividend into the number of equivalent subsets indicated by the divisor.

10.12
For any division statement, the dividend is equal to the product of the divisor and the quotient, plus the remainder.

11.1
A *natural number* is any number in the set {1, 2, 3, 4, 5, 6, 7, . . .}. The set of natural numbers is also called the set of *counting numbers.*

11.2
A *whole number* is any number in the set {0, 1, 2, 3, 4, 5, . . .}.

11.3
The whole numbers that end in 0, 2, 4, 6, or 8 are called *even numbers.* The set of even numbers can be represented as {0, 2, 4, 6, 8, . . .}.

11.4
Any whole number that is not an even number is an *odd number.* The set of odd numbers can be represented as {1, 3, 5, 7, 9, 11, . . .}.

11.5
The set of whole numbers with their opposites are called the set of *integers.* They can be represented as {. . . , $^-3$, $^-2$, $^-1$, 0, 1, 2, 3, 4, . . .}.

11.6
The set of *rational numbers* contains the set of integers and the set of *fractional numbers.*

11.7
A *triangular number* is any number in the set {1, 3, 6, 10, 15, 21, . . .}.

11.8
A *square number*, or *perfect square*, is any number in the set {1, 4, 9, 16, 25, 36, . . .}.

11.9
A *pentagon number* is any number in the set {1, 5, 12, 22, 35, . . .}.

11.10
If two or more numbers are multiplied, each number is a *factor* of the product.

11.11
A *multiple* is the product of two or more numbers.

11.12
A whole number that has exactly two different factors is called a *prime number*.

11.13
A whole number greater than 1 that is not a prime number is called a *composite number*.

11.14
Every composite number can be expressed as a product of prime numbers in a unique way. This is called the *Fundamental Theorem of Arithmetic*.

11.15
Two numbers are *relatively prime* to each other if they have no factors in common except 1.

11.16
The *Greatest Common Factor* (G.C.F.) of two numbers is the greatest of the factors common to both numbers.

11.17
The *Least Common Multiple* (L.C.M.) of two or more numbers is the smallest number evenly divisible by each.

11.18
If the last digit in a whole number is even (0, 2, 4, 6, or 8), the number is divisible by 2.

11.19
If the sum of the digits of a whole number is divisible by 3, the original number is divisible by 3.

11.20
If the last two digits (tens and ones) of a whole number are divisible by 4, the original number is divisible by 4.

11.21
If the last digit of a whole number is either 0 or 5, the number is divisible by 5.

11.22
If the last digit in a whole number is even and the sum of the digits is divisible by 3, the original number is divisible by 6.

11.23
If the sum of the digits of a whole number is divisible by 9, the original number is divisible by 9.

11.24
Reasoning in which a general conclusion is made after considering specific examples is called *inductive reasoning*.

11.25
Reasoning in which a specific conclusion is made from other ideas or assumptions is called *deductive reasoning*.

12.1
The fraction a/b may be interpreted as a of b equal size parts. This is the fraction or partition interpretation.

12.2
A fraction may be used to express the number for part of a whole thing, or part of a set of things.

12.3
A fraction may be used to express the number for more than one whole.

12.4
The *denominator*, indicated by the numeral below the fraction bar, gives the name to the fraction. It tells how many parts of the same size are in the whole.

12.5
The *numerator*, indicated by the number above the fraction bar, tells how many of the particular fractional parts are indicated.

12.6
Partitioning a region or a line segment, or a solid into congruent subregions, line segments, or solids is a model for the fractional part(s) interpretation of a non-negative rational.

12.7
A fraction may be interpreted as an indicated division: $a/b = a \div b$.

12.8
A rational number is determined by dividing a whole number (the numerator) by a counting number (the denominator); therefore, zero cannot be a denominator.

12.9
The numerator indicates a product; the denominator a known factor.

12.10
Partitioning a set of discrete elements into equivalent disjoint subsets is a model for the division interpretation of a non-negative rational number.

12.11
A fraction may be interpreted as a *ratio* or *rate pair*: a to b.

12.12
The size (length, weight, etc.) of two things may be compared by using a ratio.

12.13
The ratio of one number to another may be shown by a fraction.

12.14
The numerator indicates the number in one set. The denominator indicates the number in a second set being compared by the ratio of the first set to the second set.

12.15
Comparing sets (many-to-many correspondence) is a model for the ratio or rate pair interpretation of a non-negative rational.

12.16
Two fractions that name the same non-negative rational are called *equivalent fractions.*

12.17
A set of all equivalent fractions is called an *equivalence class.*

12.18
The set of whole numbers is a subset of the non-negative rationals.

12.19
Two fractions $\frac{a}{b}$ and $\frac{c}{d}$ are equivalent if and only if the product of the numerator of the first and the denominator of the second equals the product of the denominator of the first and the numerator of the second; that is, $\frac{a}{b} = \frac{c}{d}$ if and only if $ad = bc$.

12.20
Two fractions $\frac{a}{b}$ and $\frac{c}{d}$ are equivalent if there is a nonzero number n such that $\frac{a}{b} \times \frac{n}{n} = \frac{c}{d}$.

12.21
As fractional parts get smaller, the denominator of the fraction gets larger.

12.22
Four of the fourths (or two halves, three-thirds, seven-sevenths, and so on) make a whole.

12.23
Two fractional parts identified with the same fraction are the same size only when they are parts of the same whole object, or of some other whole object of identical size. Parts of wholes identified with the same fraction are *not* the same size if they are parts of different sized wholes.

12.24
If two fractions have the same numerators, the fraction with the smaller denominator names the greater number.

12.25
If two fractions have the same denominators, the fraction with the greater numerator names the greater number.

12.26
The number of rational numbers between any two rational numbers is limitless.

12.27
If a and b are non-negative rationals and $a < b$, there is an infinite set of rationals such as $\frac{a + b}{2}$, $\frac{a + b}{3}$, $\frac{a + b}{4}$, . . . between a and b.

12.28
A *rational number* is an idea; the symbol for this number idea is called a *numeral* or a *fraction.*

12.29
The term *fraction* is commonly used to refer to either the symbol or to the rational number.

12.30
A fraction has three parts: a bar, a numeral above it, and another numeral below it.

12.31
A number using both whole number and fraction names a *mixed numeral* or *mixed form.* These expressions are other names for the *sum* of two numbers. For example, $2 = 2\frac{1}{2} + \frac{1}{2}$.

12.32
If the numerator of a fraction is greater than the denominator, the rational number can be renamed as a whole number or in mixed form.

12.33
A fraction for a rational number is in *simplest form* when it shows a numerator and a denominator that have no common factor greater than 1.

12.34
A fraction can be renamed to *higher* or *lower terms* by multiplying both the numerator and denominator by the same number, specifically, $\frac{n}{n}$.

12.35
In a place value numeration system, each digit within a numeral is assigned a place or position.

12.36
Rational numbers can be expressed with *decimals.* Decimals extend the Hindu-Arabic numeration system to places with values less than 1.

12.37
The digits in a decimal are arranged horizontally with respect to a reference point called the decimal point.

12.38
In a decimal, each place has a value 10 times as great as the place to its right.

12.39
The first three places to the right of the decimal point are tenths, hundredths, and thousandths.

12.40
The decimal has an implied numerator and denominator. The name of the place of the last digit to the right in a decimal indicates the denominator. The numeral itself names the numerator.

12.41
When two non-negative decimals have the same denominator, the decimal with the greater numerator is the larger of the two.

12.42
Fractions and mixed numerals may be renamed as *decimals.*

12.43
Fractions and decimals can be renamed as *percents.*

12.44
Every rational number can be expressed either as a terminating or repeating decimal.

12.45
A *terminating decimal* can be written with a finite number of digits, such as, .25 or .69324.

12.46
A *repeating decimal* cannot be written with a finite number of digits. Some digits or series of digits repeat infinitely, such as, $.33\overline{3}$ or $.2727\overline{27}$.

12.47
Numbers written with a large number of digits can be written in a shorter form by using exponential notation.

12.48
Numbers represented in *exponential notation* contain a number called the base, and a superscript (raised numeral) called the exponent.

12.49
An *exponent* tells how many times the base is taken as a factor.

12.50
Numbers expressed in *scientific notation* are written as the product of two factors: one is a number between 1 and 10, and the other is a power of 10 expressed in exponential notation.

13.1
Closure—The sum of any two rational numbers is always a rational number.

13.2
Commutative—The order or sequence in which two rational numbers are added has no effect on their sum.

13.3
Associative—The way in which three rational numbers are grouped together in addition has no effect upon their sum.

13.4
Identity—Zero added or subtracted from any rational number yields a sum or difference equal to the original number.

13.5
Decimals are added and subtracted much in the same manner as whole numbers.

13.6
When decimals are added or subtracted, decimal points are aligned vertically.

13.7
Tenths are added to tenths, with the resulting sum in tenths; hundredths are added to hundredths, with the sum in hundredths; and so on.

13.8
Tenths are subtracted from tenths, with the resulting difference in tenths; hundredths are subtracted from hundredths, with the difference in hundredths; and so on.

13.9
Closure—The product of any two rational numbers is always a rational number.

13.10
Commutative—The order or sequence in which two rational numbers are multiplied has no effect on their product.

13.11
Associative—The way in which three rational numbers are grouped together in multiplication has no effect upon their product.

13.12
Identity—The product or quotient of any rational number and the number 1 is equal to the given number.

13.13
Distributivity—Rational numbers under the operation of multiplication are distributive with respect to addition.

13.14
The product of a whole number and a decimal in tenths is expressed in tenths; the product of two decimals in tenths is expressed in hundredths; the product of a decimal in tenths and a decimal in hundredths is expressed in thousandths; and so on.

13.15
The algorithm for multiplying decimals may be considered in two phases.

a. Ignore the decimal points and compute as though the factors were whole numbers.
b. Locate the decimal point in the product with as many digits to the right of the point as there are to the right of both decimal points in the factors.

13.16
The algorithm for division with decimals may be considered in two phrases.

a. Locate the decimal point in the quotient.
b. Ignore the decimal point and divide as though the dividend and divisor were whole numbers.

13.17
If both dividends and divisor are multiplied by the same nonzero number, the quotient does not change.

13.18
Multiplying a decimal by 10 or a power of 10 has the effect of moving the digits in the numeral the same number of places to the left of the decimal point as there are zeros in the power of 10.

13.19
Dividing a decimal by 10 or a power of 10 has the effect of moving the digits in the numeral the same number of places to the right of the decimal point as there are zeros in the power of 10.

13.20
Every basic percent problem involves one of three situations, determined by the unknown term in the basic relationship.

a. Finding a part or percent of a number. For example,

   25% of 20 is ____ .

b. Finding what part or percent one number is of another. For example,

   ____% of 15 is 5.

c. Finding a number, when a certain part or percent of that number is known. For example,

   20% of ____ is 6.

14.1
To add or subtract rational numbers expressed as fractions, the fractions must have the same denominators; they must be *like* fractions.

14.2
When adding like fractions, add the numerators to get the numerator of the sum. The denominator of the sum is the same as the denominator in each fraction.

14.3
When subtracting like fractions, subtract the numerators to get the numerator of the difference. The denominator of the difference is the same as the denominator in each fraction.

14.4
A multiple of a rational number is a product of that number and a whole number.

14.5
The Least Common Denominator (L.C.D.) for two or more fractions is the Least Common Multiple of their denominators.

14.6
To add (or subtract) when denominators are different:
a. Find the L.C.D.
b. Write equivalent fractions using the L.C.D.
c. Add (or subtract).
d. Simplify the sum (or difference).

14.7
To add with mixed numerals:
a. Find the Least Common Denominator.
b. Write equivalent fractions for the fraction part.
c. Add the fraction part.
d. Add the whole number part.
e. Simplify the sum.

14.8
To subtract a number expressed as a mixed numeral from a whole number:
a. Rename the whole number as a mixed numeral.
b. Subtract the fraction part.
c. Subtract the whole number part.
d. Simplify the difference.

14.9
To subtract one number expressed as a mixed numeral from another:
a. Find the Least Common Denominator.
b. Write equivalent fractions using the L.C.D.
c. If necessary, rename the larger mixed numeral.
d. Subtract the fraction part.
e. Subtract the whole number part.
f. Simplify the difference.

14.10
Repeated addition of equal fraction addends is a model for multiplication of a fraction by a whole number.

14.11
An array is a model for multiplication of a fraction by a fraction or a whole number by a fraction.

14.12
The ratio concept of multiplication is a model for multiplication when one or both of the factors are fractions.

14.13
Inverse—Every positive rational number has a reciprocal or multiplicative inverse. The product of a positive rational number and its multiplicative inverse is 1, the identity for multiplication.

14.14
To multiply with fractions, multiply the numerators to obtain the numerator of the product, and multiply the denominators to find the denominator of the product.

14.15
To simplify a fraction before multiplying, divide both the numerator and the denominator of the fraction by the same nonzero number.

14.16
Whenever a factor is expressed as a mixed numeral or a whole number, rename the factor as a fraction to multiply.

14.17
A fraction *of* a number and a fraction *times* a number have the same result.

14.18
The partition of a universal set into equivalent disjoint subsets is a model for division with fractions.

14.19
Repeated subtraction is a model for division with fractions.

14.20
If a nonzero number is expressed as a fraction, interchange the numerator and denominator to find the reciprocal.

14.21
The quotient of a given number divided by 1 is the given number.

14.22
When the numerator and the denominator of a fraction are multiplied by the same nonzero number, the value of the fraction remains the same.

14.23
To divide one fraction by another, multiply the dividend by the reciprocal of the divisor.

15.1
In geometry, *point, line,* and *plane* are undefined.

15.2
A *line segment* is a portion of a line consisting of two points together with all points in between.

15.3
A *ray* is a half-line that includes its endpoint–a closed half-line.

15.4
The union of two noncolinear rays (not on the same line) having a common endpoint is called an angle.

15.5
A *curve* in a plane is a set of points that can be traced without lifting the pencil from the paper.

15.6
A curve is a *closed curve* if the tracing begins and ends at the same point.

15.7
A curve is a *simple closed curve* if it is a closed curve that does not cross or retrace itself.

15.8
A *circle* is a simple closed curve having a point in its interior that is equidistant from each point on the curve.

15.9
A *polygon* is a simple closed curve consisting of the union of line segments. Each line segment is called a *side* of the polygon.

15.10
Polygons are named as follows: 3 sides–triangle, 4 sides–quadrilateral, 5 sides–pentagon, 6 sides–hexagon, 7 sides–heptagon, 8 sides–octagon, 9 sides–nonagon, 10 sides–decagon.

15.11
A polygon having all sides equal in length is called *equilateral.*

15.12
A polygon having all angles equal is called *equiangular*.

15.13
Polygons that are both equiangular and equilateral are *regular* polygons.

15.14
Any two plane figures that are exact replicas of each other are said to be *congruent*.

15.15
A congruence is a description of how figures can be made to coincide – a correspondence that tells which points (line segments, angles) on one figure correspond to which points (line segments, angles) on another figure.

15.16
Two figures are congruent if they have the same shape and size.

15.17
A *prism* is a polyhedron having N faces. At least two faces must be parallel, and at least N - 2 faces must be parrallelograms.

15.18
A *pyramid* is a polyhedron having all but one surface bounded by triangles that intersect at one point.

15.19
A *regular polyhedron* is a polyhedron with equal surfaces; each surface is congruent to every other surface.

16.1
A *measurable attribute* of an object is a characteristic that can be quantified by comparing it to some standard unit.

16.2
Measuring is the process of comparing an attribute of a physical object to some unit selected to quantify that attribute.

16.3
A *unit* is a fixed quantity, value, or size and need not have physical substance.

16.4
A *standard* is a physical representation of a unit.

16.5
A *measurement* is the result of measuring. As such, a measurement involves both a number and the unit, whereas a *measure* is a number alone.

16.6
Accuracy refers to the discrepancy between the true value and the result obtained by measurement. Accuracy is affected by the measuring instrument and the way the instrument is used.

16.7
Precision refers to the agreement among repeated measures of the same physical quantity.

16.8
A year is the period of time required for the earth to make one complete revolution around the sun.

16.9
A month is approximately the period of time it takes for the moon to make one revolution around the earth.

16.10
A day is the period of time that it takes for the earth to make one complete revolution on its axis.

16.11
AM and PM refer to before noon and after noon, respectively.

16.12
The face of the clock is the area that has the twelve numerals and the hands.

16.13
The short hand of the clock shows the hours.

16.14
The long hand of the clock shows minutes.

16.15
When the minute hand points to the numeral 12, the time is at the hour; when the hour hand points directly to a numeral, the hour of the day is indicated.

16.16
When the minute hand points to the numeral 6, the time is at the half hour.

16.17
As the long hand moves from numeral to numeral, the time lapse is five minutes.

16.18
When there is a second hand, the time lapse for one complete revolution around the clock is one minute.

16.19
The time lapse for the short hand to move from one numeral to the next is one hour.

17.1
Two line segments are defined as being *congruent* if their measures are equal.

17.2
The symbol $\cong$ is used to indicate congruence, e.g., $\overline{AB} \cong \overline{CD}$.

17.3
Two angles are defined as being *congruent* if their measures are equal.

17.4
An angle having a measure of 90° is a *right angle.*

17.5
An angle having a measure less than 90° is an *acute angle.*

17.6
An angle having a measure greater than 90° but less than 180° is an *obtuse angle.*

17.7
An angle having a measure of 180° is a *straight angle* (also named a *straight line*).

17.8
Two angles having the same vertex and a common side are *adjacent angles.*

17.9
Angle measurement is additive; that is, $m(\angle BAC) + m(\angle CAD) = m(\angle BAD)$.

17.10
Two angles are defined as *complementary* if the sum of their measures equals 90°.

17.11
Two angles are defined as *supplementary* if the sum of their measures equals 180°.

17.12
If the angle formed when two lines intersect is equal to 90°, the lines are *perpendicular.*

17.13
Two lines that do not intersect within a plane are *parallel.*

17.14
Two lines that do not intersect within different planes are *skew lines.*

17.15
The sum of the measures of the three interior angles of a triangle is 180°.

17.16
A triangle having a 90° angle (a right angle) is called a *right triangle.*

17.17
A triangle having an obtuse angle is an *obtuse triangle.*

17.18
A triangle in which all three angles are acute is an *acute triangle.*

17.19
A triangle having all three angles equal is an *equiangular triangle.*

17.20
A triangle having all three sides equal is an *equilateral triangle.*

17.21
A triangle having at least two sides congruent is an *isosceles triangle.*

17.22
A triangle in which all sides differ in length is a *scalene triangle.*

17.23
Two triangles are congruent if corresponding sides and angles are congruent. For example, $\triangle ABC$ is congruent to $\triangle DEF$ if $\angle A \cong \angle D$, $\angle B \cong <E$, $\angle C \cong \angle F$ and $\overleftrightarrow{AB} \cong \overleftrightarrow{DE}$, $\overleftrightarrow{AC} \cong \overleftrightarrow{DF}$, $\overleftrightarrow{BC} \cong \overleftrightarrow{EF}$.

17.24
Two triangles are *similar* if corresponding angles are congruent; sides may or may not be congruent.

17.25
A quadrilateral having four right angles *and* four congruent sides is a *square.*

17.26
A quadrilateral having four right angles and opposite sides congruent is a *rectangle.*

17.27
A quadrilateral having opposite sides congruent and parallel is a *parallelogram.*

17.28
A quadrilateral with all sides congruent is a *rhombus.*

17.29
A quadrilateral having two sides parallel is a *trapezoid.*

17.30
A quadrilateral may have no parallel or congruent sides.

17.31
A polygon having all sides and angles congruent is a *regular polygon* (regular pentagon, regular hexagon, etc.).

17.32
The perimeter of a square can be found by using the formula $P = 4s$, where s is the length of a side.

17.33
The perimeter of a rectangle can be found using the formula $P = 2l + 2w$, where l is the length and w is the width.

17.34
Generally, the perimeter of a polygonal figure is found by using the formula $P = s_1 + s_2 + \ldots + s_n$, where s is the length of a side and n is the number of sides.

17.35
The area of a square or a rectangle is found by using the formula $A = lw$, where l is the length and w is the width.

17.36
The area of a rhombus or a parallelogram is found by using the formula $A = bh$, where b is the base and h is the height or altitude.

17.37
The area of a triangle is found by using the formula, $A = \frac{1}{2}bh$, where b is the length of the base and h is the height.

17.38
The area of a trapezoid is found by using the formula $A = \frac{1}{2}h(b_1 + b_2)$, where h is the height and b_1 and b_2 are the bases, respectively.

17.39
The area of a circle is found by using the formula $A = \pi r^2$, where π is a constant (about 3.14) and r is the radius of the circle.

17.40
The volume of a rectangular prism is found by using the formula, $V = lwh$, where l is the length, w is the width and h is the height. This definition can be shortened to $V = Bh$, where B is the area of a base and h is the height.

18.1
The set of integers contains the set of natural numbers, their opposites, and zero.

18.2
Integers may be associated with equidistant points on a line.

18.3
Numbers to the right of zero are called *positive numbers* or *positive integers.* Numbers to the left of zero are called *negative numbers* or *negative integers.*

18.4
A + (read *positive*) placed before a numeral denotes a positive number; and − (read *negative*) placed before a numeral denotes a negative number.

18.5
An integer is greater than a given integer if its position on the number line is to the right of the given integer.

18.6
Addition of integers is a binary operation that associates each pair of integers with a unique integer.

18.7
To add two positive integers, add the absolute values. The sum is positive.

18.8
To add two negative integers, add the absolute values. The sum is negative.

18.9
To add a positive and a negative integer, subtract to find the difference between their absolute values. The sign for the sum of the integers is the same as the sign of the integer with the greater absolute value.

18.10
Subtraction of integers is a binary operation that associates each pair of integers with a unique integer.

18.11
Subtraction is the inverse of addition.

18.12
Subtraction may be thought of as an operation used to find one addend when the sum and the other addend are known.

18.13
To subtract one integer from another, add the opposite of the integer to be subtracted (subtrahend or known addend) to the other integer (minuend or sum).

18.14
Multiplication of integers is a binary operation that associates each pair of integers with a third integer.

18.15
Multiplication is used to find the product when its factors are known.

18.16
To multiply two positive or two negative integers, multiply their absolute values. The product is positive.

18.17
To multiply a positive and a negative integer, multiply their absolute values. The product is negative.

18.18
Division of integers is a binary operation that associates a pair of integers with a third integer.

18.19
Division is used to find a missing factor

(quotient) when the other factor and the product are known.

18.20
To divide two positive or two negative integers, divide their absolute values. The quotient is positive.

18.21
To divide a negative integer by a positive integer, or to divide a positive integer by a negative integer, divide their absolute values. The quotient is negative.

18.22
Additive Inverse–If a is an integer, then there exists a unique integer, $-a$, such that $a + -a = 0$.

18.23
If a, b and c are integers, than $a - b = c$ if and only if $b + c = a$.

18.24
Closure for Subtraction–If a and b are integers, then $a - b$ is an integer.

18.25
If a, b and c are integers, and $b \neq 0$, then $a \div b = c$ if and only if $b \times c = a$.

19.1
The number of items in a set of data is called the *frequency of the data*. An organized summary of the data is called a *frequency distribution*.

19.2
To find the arithmetic *mean* of a set of numbers, divide their sum by the number of numbers in the set.

19.3
To find the *median* of a set of numbers, arrange them in order, smallest to largest, and find the middle number.

19.4
The *mode* for a set of numbers is the number that occurs most frequently.

19.5
Bar graphs compare the size or frequency of two or more quantities.

19.6
Line graphs show change over time and relationships between two or more variables that involve continuity.

19.7
Circle graphs display the relation of parts to the whole and to each other.

Index

*This topic is also found in the glossary of the accompanying Student Handbook.

ABOUT THE AUTHORS

Prior to his work in teacher education, Dr. Robert Ashlock served as an elementary school teacher and as a school principal in Indiana. With degrees from Butler University (B.S., 1957; M.S., 1959) and Indiana University (Ed.D., 1965) he began fifteen years of service as a professor at the University of Maryland, where he established and directed the Arithmetic Center and began a clinic program for children having difficulty learning mathematics. He is presently Professor of Education in the Graduate School of Education of Reformed Theological Seminary in Jackson, Mississippi.

In addition to articles in journals such as *The Arithmetic Teacher* and *School Science and Mathematics*, Dr. Ashlock's writings include *Current Research in Elementary School Mathematics* (1970) and *Error Patterns in Computation* 3rd ed. (1982). The latter reflects his special interest in diagnosis and corrective instruction in elementary mathematics. Dr. Ashlock was featured in two special education television films that focused on teaching math to children who have difficulty learning. From 1979–81 he served as president of the Research Council for Diagnostic and Prescriptive Mathematics.

Dr. Ashlock has served as a consultant, conducted workshops, and addressed school districts and professional associations across the nation. His work has involved him with private and parochial schools as well as public schools.

Dr. Martin L. Johnson is an Associate Professor of Education at the University of Maryland, College Park, Maryland. He received both masters and doctorate degrees in mathematics education from the University of Georgia, Athens. At the University of Maryland, Dr. Johnson teaches both graduate and undergraduate mathematics education courses to elementary, middle, and special education students and teachers. He is currently the Coordinator of the Mathematics Education Faculty at UMCP, Director of the Center for Mathematics Education, and Director of the Arithmetic Clinic.

Dr. Johnson is an active writer and a popular consultant and speaker. He has published in the *Arithmetic Teacher*, *Journal for Research in Mathematics Education*, *School Science and Mathematics*, and numerous publications of the Research Council for Diagnostic and Prescriptive Mathematics. He serves as referee for numerous journals, including *Journal for Research in Mathematics Education, School Science and Mathematics, Investigations in Mathematics Education*, and served on the editorial board of *Psychological and Educational Research*. He is a founding member of RCDPM and has held many offices in that organization, including Vice-President for Research.

Dr. Johnson has taught at elementary, middle school, secondary, college, and university levels. He is actively sought for inservice workshops and teacher education programs by schools and school systems. The many years of teaching experience are reflected in the practical suggestions for teaching in *Guiding Each Child's Learning of Mathematics*.

Dr. Wilson began his teaching career as a second grade classroom teacher in the Onondaga Indian Reservation School near Syracuse, New York. After graduate work at Syracuse University, he joined the faculty and taught mathematics education courses. He worked with Dr. Vincent J. Glennon in developing a clinic for children under-achieving in arithmetic, the first clinic of its kind in the United States at that time.

Later he taught at the University of South Florida. The last years of his career were spent at the University of Maryland. There he worked with his colleagues Dr. Ashlock and Dr. Johnson, as well as graduate and undergraduate students in mathematics education.

As director of the Arithmetic Clinic he developed several models for use in diagnosis and remediation such as the Content Taxonomy and Activity Type Cycle found in this book. He was also coauthor of the Maryland Diagnostic Arithmetic Test with Dr. Barbara Sadowski.

Wilmer L. Jones holds a B.S. degree (1953) from Towson State University, an A.B.A. degree (1960) from Baltimore University, and an M.Ed. (1962) in Administration and Supervision from Loyola College. He received his Ph.D. in Mathematics and Early Childhood Education from the University of Maryland in 1976.

Dr. Jones has taught mathematics at the junior high school level since 1953. He has served as a Mathematics Department Head, and as a Unit Principal, a Mathematics Specialist, and a Mathematics Supervisor at Herring Run Junior High School. Since 1973, he has served as Coordinator of Mathematics, K–12.

Drawing on his long career in teaching mathematics, Dr. Jones has published extensively, and his credits include many materials for inservice application. He has served in professional organizations as president of the Maryland Council of Teachers of Mathematics, and is a member of the National Council of Teachers of Mathematics and of the Maryland Association for Educational Uses of Computers.